Planning for Community-based Disaster Resilience Worldwide

We are witnessing an ever-increasing level and intensity of disasters from Ecuador to Ethiopia and beyond, devastating millions of ordinary lives and causing long-term misery for vulnerable populations.

Bringing together 26 case studies from six continents, this volume provides a unique resource that discusses, in considerable depth, the multifaceted matrix of natural and human-made disasters. It examines their bearing on the loss of human and productive capital; the conduct of national policies and the setting of national development priorities; and on the nature of international aid and bilateral assistance strategies and programs of donor countries. In order to ensure the efficacy and appropriateness of their support for disaster survivors, international agencies, humanitarian and disaster relief organizations, scholars, non-governmental organizations, and members of the global emergency management community need to have insight into best practices and lessons learned from various disasters across national and cultural boundaries.

The evidence obtained from the numerous case studies in this volume serves to build a worldwide community that is better informed about the cultural and traditional contexts of such disasters and better enabled to prepare for, respond to, and finally rebuild sustainable communities after disasters in different environments. The main themes of the case studies include:

- the need for community planning and emergency management to unite in order to achieve the mutual aim of creating a sustainable disaster-resilient community, coupled with the necessity to enact and implement appropriate laws, policies, and development regulations for disaster risk reduction;
- the need to develop a clear set of urban planning and urban design principles for improving the built environment's capacities for disaster risk management through the integration of disaster risk reduction education into the curricula of colleges and universities;
- the need to engage the *whole community* to build inclusive governance structures as prerequisites for addressing climate change vulnerability and fostering resilience and sustainability.

Furthermore, the case studies explore the need to link the existence and value of scientific knowledge accumulated in various countries with decision-making in disaster risk management; and the relevance and transferability from one cultural context to another of the lessons learned in building institutional frameworks for *whole community* partnerships.

Adenrele Awotona, Professor of Urban Planning and Community Studies, is the founder and Director of the Center for Rebuilding Sustainable Communities after Disasters, and a former Dean of the College of Public and Community Service at the University of Massachusetts, Boston, USA. He was previously a Director of Studies for the British Council International Seminars ("Reconstruction after disasters") in the United Kingdom, where he has also served at the University of Newcastle upon Tyne as a Director of Graduate Studies in architecture and urban design. Furthermore, he was an Educator/Coordinator of Seminars (on community architecture) at the annual American Institute of Architects National Conventions for several years.

Planning for Community-based Disaster Resilience Worldwide

Learning from Case Studies in Six Continents

Edited by Adenrele Awotona

Routledge
Taylor & Francis Group

LONDON AND NEW YORK

First published 2017
by Routledge
2 Park Square, Milton Park, Abingdon, Oxon OX14 4RN

and by Routledge
711 Third Avenue, New York, NY 10017

First issued in paperback 2018

Routledge is an imprint of the Taylor & Francis Group, an informa business

British Library Cataloguing in Publication Data
A catalogue record for this book is available from the British Library

Library of Congress Cataloging in Publication Data
Names: Awotona, Adenrele A., editor.
Title: Planning for community-based disaster resilience worldwide :
    learning from case studies in six continents / Adenrele Awotona, editor.
Description: First Edition. | New York : Routledge, 2016. | Includes
    bibliographical references and index.
Identifiers: LCCN 2016021349 | ISBN 9781472468154 (hardback : alk.
    paper) | ISBN 9781315600734 (ebook)
Subjects: LCSH: Disaster relief—Developing countries—Case studies. |
    Community development—Developing countries—Case studies. |
    Sustainable development—Developing countries—Case studies.
Classification: LCC HV555.D44 P53 2016 | DDC 363.34/7091724—dc23
LC record available at https://lccn.loc.gov/2016021349

ISBN 13: 978-1-138-60158-1 (pbk)
ISBN 13: 978-1-4724-6815-4 (hbk)

Typeset in Sabon
by Apex CoVantage, LLC

Contents

Figures

Tables

Contributors

Adenrele Awotona, PhD (Cantab.) is a professor at, as well as the founder and director of the Center for Rebuilding Sustainable Communities after Disasters, University of Massachusetts Boston, US.

Iankaa Aguse, BSc (BSU), is a tutor at Trinity School, Dagba-Adagi, Kwande Local Government Area, Benue State, Nigeria.

Fatima M. Alfa is a doctoral candidate, Graduate School of Management and Technology, University of Maryland University College, Adelphi, MD, US.

Elyse Baker, formerly an early career public health researcher in the Jack Brockhoff Child Health and Wellbeing Program at the University of Melbourne, Australia, is now a primary care Consultant – Population Health Planning in Western Victoria.

Benjamin Bechtel is a research associate in University of Hamburg, Center for Earth System Research and Sustainability, Institute of Geography, Hamburg, Germany.

Nitin Bharadwaj, MBBS, MD, DNB, is Senior Resident, Department of Hospital Administration, Sanjay Gandhi Post Graduate Institute of Medical Sciences, Lucknow, India.

Karen Block is a postdoctoral researcher in the Jack Brockhoff Child Health and Wellbeing Program in the Centre for Health Equity at the University of Melbourne, Australia.

Richard Bryant is Scientia Professor and ARC Laureate Fellow at the School of Psychology, University of New South Wales, Australia.

Sweta Byahut, PhD, is Assistant Professor in the Graduate Program in Community Planning at Auburn University, Auburn, AL, US.

Hem Chandra, MBBS, MHA, MBA, PhD, DLitt, FAIMA, FIMSA, FAHA, is Professor and Head, Department of Hospital Administration, Sanjay Gandhi Post Graduate Institute of Medical Sciences, Lucknow, India.

James J. Cochran, PhD, is Professor of Applied Statistics in the Department of Information Systems, Statistics and Management Science, University of Alabama, US.

Victoria Cornell, PhD, is Research Associate, Centre for Housing, Urban and Regional Planning, School of Social Sciences, University of Adelaide, Australia.

Karen Da Costa, PhD (Graduate Institute, Geneva), is Research Associate in International Law, Faculty of Laws, University College London, UK.

Thomas C. Cross completed a BS in Civil Engineering at Merrimack College in North Andover, MA, US.

David Forbes is Professor in the Department of Psychiatry, University of Melbourne, and Director of Phoenix Australia: Centre for Posttraumatic Mental Health, Australia.

Hugh Colin Gallagher is a postdoctoral researcher in the social networks laboratory (MelNet) at Melbourne School of Psychological Sciences, University of Melbourne, Australia.

Lisa Gibbs is Associate Professor and Acting Director of the Jack Brockhoff Child Health and Wellbeing Program in the Centre for Health Equity at the University of Melbourne, Australia.

Lou Harms is Professor and Deputy Head of the Department of Social Work, University of Melbourne, Australia.

Seth H. Holmes, Assistant Professor at University of Hartford, CT, US, holds a Bachelor of Architecture (Roger Williams University) and a Master of Design Studies (Harvard University).

Mahmood Hosseini, PhD (Azad), is Associate Professor at the Structural Engineering Research Center, International Institute of Earthquake Engineering and Seismology, Tehran, Iran.

Greg Ireton is Senior Consultant for the Centre for Disaster Management and Public Safety, University of Melbourne, and has an honorary position in the Jack Brockhoff Child Health and Wellbeing Program at the University of Melbourne, Australia.

Yasamin O. Izadkhah, PhD (Cranfield), is an Associate Professor at the Risk Management Research Center, International Institute of Earthquake Engineering and Seismology, Tehran, Iran.

James Kaklamanos, PhD, is Assistant Professor of Civil Engineering at Merrimack College in North Andover, MA, US.

Giedrius Kaveckis, MSc, is a PhD candidate and research associate at the University of Hamburg, Center for Earth System Research and Sustainability, Institute of Geography, Hamburg, Germany.

Connie Kellett is a social worker who works with trauma and disaster. She is currently completing a PhD in the Department of Social Work at the University of Melbourne, Australia.

Jorge León, PhD, Melbourne School of Design, Faculty of Architecture, Building and Planning, University of Melbourne, Victoria, Australia.

'Maseka Lesaoana, PhD (University of Southampton), is Associate Professor at the Department of Statistics and Operations Research, University of Limpopo.

John "Jack" Lindsay is Associate Professor at the Department of Applied Disaster and Emergency Studies, Brandon University, Canada.

Dean Lusher is a social network analyst at the Center for Transformative Innovation, Swinburne University of Technology, Australia.

Colin MacDougall is Professor of Public Health at Flinders University; he is also an executive member of the Southgate Institute of Health, Society, and Equity, and an Honorary Principal Fellow, Jack Brockhoff Child Health and Wellbeing Program, University of Melbourne, Australia.

Daniel Maposa, MSc (NUST, Zimbabwe), Lecturer, Department of Statistics and Operations Research, University of Limpopo, South Africa.

Alan March, Associate Professor at the Faculty of Architecture, Building, and Planning at the University of Melbourne, has practiced in the public and private sectors in Australia and the UK.

Leela Masih, BSc MSc, is Nursing Superintendent, Department of Hospital Administration, Sanjay Gandhi Post Graduate Institute of Medical Sciences, Lucknow, India.

Kenichi Nakagami, PhD (Engineering), is a specially appointed professor, College of Policy Science, Ritsumeikan University, Japan.

Abiodun Olukayode Olotuah (BSc, MSc, PhD) is Professor of Architecture at the Federal University of Technology, Akure, Nigeria.

Jürgen Ossenbrügge is a professor at the University of Hamburg, Center for Earth System Research and Sustainability, Institute of Geography, Hamburg, Germany.

Ankita Pandey, MTech, is a senior research fellow, Indian Institute of Toxicology Research, Lucknow, India.

Patricia E. Perkins (PhD Economics, University of Toronto) is a professor at the Faculty of Environmental Studies, York University, Toronto, Canada.

Thomas Pohl is an academic advisor and research associate at the University of Hamburg, Center for Earth System Research and Sustainability, Institute of Geography, Hamburg, Germany.

Paulina Pospieszna, PhD (University of Alabama), is Assistant Professor, Department of Political Science and Journalism, Adam Mickiewicz University, Poznan, Poland.

Xuepeng Qian, PhD (Engineering), is Associate Professor, College of Asia Pacific Studies, Ritsumeikan Asia Pacific University, Japan.

Matías Ezequiel Barberis Rami, PhD, is a researcher for the Center of Urban, Territorial and Environmental Research, University of Ferrara, Italy.

John F. Richardson is the National Coordinator of Emergency Preparedness for the Australian Red Cross. He is also Honorary Research Fellow in the Jack Brockhoff Child Health and Wellbeing Program at the University of Melbourne, Australia.

Sarika Sharma, Master in Hospital Administration, Department of Hospital Administration, Sanjay Gandhi Post Graduate Institute of Medical Sciences, Lucknow, India.

Yutaka Sho is Assistant Professor, Syracuse University School of Architecture, and a partner and co-founder of the GA Collaborative, US.

Hans Skotte, Dr Ing (PhD), M. Arch, is a professor in the Department of Urban Design and Planning, Faculty of Architecture and Fine Art, Norwegian University of Science and Technology, Trondheim, Norway.

Vikki Sinnott is a senior policy advisor in the Victorian Department of Health and Human Services, Australia.

Marita Smith completed her Master of Social Work (Research) at the University of Melbourne, with additional supervision assistance from Australian Red Cross.

Darwin H. Stapleton, Executive Director Emeritus, Rockefeller Archive Center, was Professor of History at the University of Massachusetts, Boston (2010–2014).

Melissa A. Surettte (Savilonis), DLP, MSEM, CEM, is a graduate of Northeastern University's Doctor of Law and Policy Program, US.

Abraham Adeniyi Taiwo (BSc, MSc, PhD) is Associate Professor of Architecture at the Federal University of Technology, Akure, Nigeria.

Fariha Tariq, Assistant Professor at the University of Management and Technology, Lahore, has a Bachelor's degree in City and Regional Planning, a Master's degree in Architecture (UET, Lahore), and a PhD (NC State University, US).

Supriya Trivedi, BHMS, is a DHA Student, Department of Hospital Administration, Sanjay Gandhi Post Graduate Institute of Medical Sciences, Lucknow, India.

Bernard Tarza Tyubee, PhD (Nigeria), is a senior lecturer, Department of Geography, Faculty of Environmental Sciences, Benue State University, Makurdi, Nigeria.

Derin N. Ural, PhD, is Professor and Founding Director, Center of Excellence for Disaster Management, MEF University, Istanbul, Turkey.

Zanzan Uji is Professor of Architecture at the University of Jos in Plateau State, Nigeria, and a practicing lawyer.

Yu Wang earned his PhD in 2015 at the Department of Urban Design and Planning at the Norwegian University of Science and Technology.

Elizabeth Waters (dec. September 2015) was the Jack Brockhoff Chair of Child Public Health, Director of the Brockhoff Child Health and Wellbeing Program, University of Melbourne, and leader of the Cochrane Collaboration Public Health Group, Australia.

Weisheng Zhou, PhD (Engineering), is a professor at the College of Policy Science, Ritsumeikan University, Japan.

Preface

We are witnessing an ever-increasing level and intensity of disasters, from Chile to Mozambique and beyond, devastating millions of ordinary lives and causing long-term misery for vulnerable populations. In 2013, Asia was the continent most often hit by natural disasters (40.7 percent), followed by the Americas (22.2 percent), Europe (18.3 percent), Africa (15.7 percent), and Oceania (3.1 percent). In the same year, Asia accounted for 90.1 percent of global disaster victims, followed by Africa (5.1 percent).[1] Additionally, population trends and climate change are increasing the world's vulnerability, and all stakeholders, especially the public and private sectors, must find ways to reduce risks from all hazards, particularly natural hazards. On March 7, 2015, the United Nations Office for Disaster Risk Reduction (UNISDR) reported that the vast majority of disasters globally are climate-related, which now account for over 80 percent of all disaster events and contribute enormously to economic losses as well as population dislocations generated by disaster events.[2]

Indeed, representatives from many nations of the world recently met in Sendai City, Miyagi Prefecture, Japan, from March 14 to 18, 2015, at the *Third UN World Conference on Disaster Risk Reduction* to examine their respective disaster risk reduction strategies, with an emphasis on how to "build the resilience of nations and communities to disasters."[3] The thousands of participants included representatives of governments, parliaments, civil society, the International Red Cross and Red Crescent Movement, non-governmental organizations, national platforms for disaster risk reduction, focal points for the Hyogo Framework for Action, local governments, scientific institutions, and the private sector, as well as organizations of the United Nations system and intergovernmental organizations. According to the UNISDR, the conference, amongst other activities, finished the assessment and review of the implementation of the Hyogo Framework for Action; deliberated on the knowledge gained through the regional and national strategies/institutions, plans for disaster risk reduction, and their recommendations, as well as relevant regional agreements within the implementation of the Hyogo Framework of Action; adopted a post-2015 framework for disaster risk reduction; identified modalities of cooperation based on commitments to implement a post-2015 framework for disaster risk reduction; and determined modalities for a periodic review of the implementation of a post-2015 framework for disaster risk reduction.

Accordingly, the purpose of this book, *Planning for Community-Based Disaster Resilience Worldwide: Learning from Case Studies in Six Continents,* is to provide a unique resource in which a wide array of case study material from six continents discusses and surveys, in considerable depth, the multifaceted matrix of natural and human-made disasters and, of critical significance, their bearing on the loss of human and productive capital, the conduct of national policies, the setting of national development priorities, and the nature of international aid (through multilateral lending and technical cooperation) and bilateral assistance strategies and programs of donor countries. In order to ensure the efficacy and appropriateness of their support for disaster survivors, international agencies, humanitarian and disaster relief organizations, scholars, non-governmental organizations, and members of the global emergency management community need to have insight into best practices and into lessons learned from various disasters across national and cultural boundaries, which is provided by the large number of authors in this book from around the world. Similarly, the evidence obtained from the numerous case studies in *Planning for Community-Based Disaster Resilience Worldwide* serves to build a worldwide community that is both better informed about the cultural and traditional contexts of such disasters and better enabled to prepare for, respond to, and finally rebuild sustainable communities after disasters in different environments.

The distinct contributions of this volume are its emphasis on the built environment (architecture, landscape, urban design, community planning, land-use modeling, and engineering, amongst others) whilst being multidisciplinary in approach; the fresh perspectives that the authors from around the globe bring to bear on the subject matter through their in-depth empirical work; and its value as a resource (reference book) for academics and those researching and looking for original source material on the different aspects of disaster studies in a wide variety of cultural, economic, social, and geographical contexts.

In addition to these contributions, the book's unique benefits to the audience include the provision of insights into how local communities globally develop long-term resilience to disasters and manage catastrophes in Africa, the Americas, Asia, Australia, New Zealand, and Europe; and its provision, through its numerous innovative interdisciplinary case studies, of needed scientific data regarding the causes and consequences of disasters, as well as strategies and long-term plans that have been implemented world-wide to prepare for, mitigate, respond to, and reconstruct communities after disasters and build resilience to their occurrence.

An additional value of the book derives from the fact that, although the different contributing authors address the phases of the disaster management spectrum from vastly different conceptual and operational perspectives, they all converge around the issue of the applicability and relevance of lessons learned in other cultural and environmental contexts intercontinentally.

The structure of the edited book

This volume consists of an introduction, five sections (each with its own brief section introduction), and a conclusion. In totality, they present different approaches to disaster risk reduction and all the phases of disaster management in various contexts.

The components of the book are:

- Introduction
- Part 1: Africa
- Part 2: The Americas
- Part 3: Asia

- Part 4: Australia
- Part 5: Europe and Multi-Continental Studies
- Conclusion

Adenrele Awotona

Notes

1 D. Guha-Sapir, P. Hoyois, and R. Below (2014), *Annual Disaster Statistical Review 2013: The Numbers and Trends* (Brussels, Belgium: Centre for Research on the Epidemiology of Disasters).
2 "Ahead of Global Risk Reduction Conference, UN Review Finds Vast Majority of Disasters Climate-Related" (2015, March 7), *Australian News.Net* [Website], available at http://www.australiannews.net/index.php/sid/230848467.
3 Ibid.

Bibliography

"Ahead of Global Risk Reduction Conference, UN Review Finds Vast Majority of Disasters Climate-Related." (2015, March 7). *Australian News.Net* [Web site]. Available at http://www.australiannews.net/index.php/sid/230848467.
Guha-Sapir, D., P. Hoyois, and R. Below. (2014). *Annual Disaster Statistical Review 2013: The Numbers and Trends*. Brussels, Belgium: Centre for Research on the Epidemiology of Disasters (CRED). Available at http://www.cred.be/sites/default/files/ADSR_2013.pdf.

Introduction

Aspects of community-based disaster management and disaster resilience

Adenrele Awotona

Community-based disaster management (CBDM) is a bottom-up approach under-pinned by the goal of disaster risk reduction. The central objective is to lessen people's susceptibilities and reinforce their ability to deal with vulnerabilities. Thus, the basis of policies, programs, and projects aimed at vulnerability decrease is a systematic and comprehensive appraisal of a community's exposure to hazards and an investigation of their defenseless conditions and skillfulness to cope with them. Although the community prepares for, mitigates, responds to, and undertakes recovery measures after disasters as needed, minimizing disaster risks is of prime importance.

Community members, especially the vulnerable populations who are the crucial actors in a community, actively participate in all the phases of disaster management, from planning and policy formation to implementation.[1]

The key features of CBDM are as follows:

- Disaster risk reduction is at the heart of CBDM, which is community-focused and in which the local community itself is the key resource that plays a central role throughout the process. The reduction of vulnerabilities and their root causes are achieved by developing and strengthening the capabilities, assets, and survival strategies of the community. These ensure that the community is the principal beneficiary of disaster risk reduction.
- CBDM links together poverty reduction, the development process, and disaster management, leading to the improvement of the quality of both the natural environment and the life of the local community as well as community empowerment politically, socially, and economically.
- CBDM uses a variety of multi-sectoral and multidisciplinary approaches to bring together a multitude of stakeholders for disaster risk reduction, thereby expanding the community's resource base. The local community works with a variety of sectors (public, private, and non-governmental) at all levels (from local to international) to address the complexity of developmental and disaster vulnerability issues.
- CBDR mobilizes the poor and the most vulnerable members of a community to actively participate in the decision-making process.[2]

Disaster risk management

Understanding *risk* (which comprises four elements: hazard, exposure, vulnerability, and consequence) and *risk management* (which includes risk reduction strategies) is the basis of building disaster resilience.[3] In the US, the key actors responsible for managing disaster risk include the government (at all levels), members of the construction industry, the private sector, homeowners, emergency managers, and researchers. Table I.1 illustrates some of their duties and challenges.

According to the US National Institute of Building Sciences (NIBS), every dollar spent on the execution of risk-informed mitigation measures results in averting four

Table I.1 Responsibilities, challenges, and opportunities of key interacting parties in risk management (in the US)[4]

Interested Party	Responsibility	Challenges	Opportunities
Federal government	Provides, and in some cases operates, protection structures for communities; supports NFIP; provides disaster assistance	No comprehensive or coordinated approach to disaster risk management	Stemming the growth in outlay of post-disaster recovery funds
State and local governments	Ensure public health and safety in use of land, zoning, land-use planning, enforcement of building codes, development of risk management strategies	Reluctance to limit development; difficulty in controlling land use on privately owned land	Reaping benefits of multiple ecosystem services by investing and strengthening natural defenses
Homeowners and businesses in hazard-prone areas	Take action to reduce vulnerability and increase resilience of property	May be unaware of or underestimate the hazards that they face	Creating demand for disaster-resistant or retrofitted structures that have increased value
Emergency managers	Oversee emergency preparedness, response, recovery, and mitigation activities	More focused on immediate disaster response than risk management	Reorientation of training and roles to balance focus toward prevention and overall disaster resilience
Construction and real estate	Incorporate resilience into designs; inform clients of risk	Actions may increase cost and reduce likelihood of sales	New opportunities in niche market
Banks and financial institutions	Require hazard insurance	No incentives to require insurance	Reduce overall risk in their portfolios
Private insurers and reinsurers	Offer hazard insurance at actuarial rates; identify risks	Limits may be placed on rate structures	Greatly expanded and risk-reduced market by offering incentives such as premium reductions for retrofit measures
Capital markets	Catastrophe bonds and other alternative risk transfer instruments	Availability limited due to globalized financial markets	Large resource base and new investment opportunities that could be directed in an anticipatory way
Insurance rating agencies	Identify stability of insurers	May negatively impact insurer position	Transparency to enable informed decisions on the part of consumers
Researchers	Collect, analyze, and communicate data, forecasts, and models about risk, hazards, and disasters	Insufficient or dispersed datasets; understanding how to share scientific information with broad audiences	Increased forecasting capability and improved data-based models of physical processes leading to disasters

dollars' worth of disaster losses, on average.[5] As disaster mitigation refers to methods of lessening the impact of disasters on the built environment, the components of risk assessment include: sabotage and terrorist attacks; natural phenomena such as hurricanes and floods; and any force that is likely to destroy, inflict casualties, and cause loss of function in the built environment.[6]

There are two types of mitigation measures: active (which focus on the hazard) and passive (which focus on the potential damage).[7] Active measures may be structural or non-structural. Passive measures, such as hazard mapping in land-use planning, are used to reduce the potential loss from a disaster by, for example, modifying the spatial and temporal character of either the damage produced by floods and debris flows or the related defenselessness. Disaster risk management strategies and measures for disaster risk reduction and mitigation may thus be structural (construction-related), nonstructural (non-construction-related), or a combination of the two, depending on the specific social, economic and environmental contexts, and are contingent on three main factors: hazards identified, the location and construction type of a proposed building or facility, and the specific performance requirements for the building.[8]

NIBS recommends a number of measures that should be taken into consideration when integrating disaster reduction measures into building design for earthquakes, hurricanes, typhoons, tornadoes, flooding, rainfall and wind-driven rains, differential settlement (subsidence), landslides and mudslides, wildfires, tsunamis, and areas of refuge.[9] It also provides a list of relevant codes and standards when designing for each of these natural phenomena in the US.[10] It should, however, be noted that

> compliance with regulations in building design is not sufficient to guarantee that a facility will perform adequately when impacted by the forces for which it was designed. Indeed, individual evaluation of the costs and benefits of specific hazard mitigation alternatives can lead to effective strategies that will exceed the minimum requirements. Additionally, special mitigation requirements may be imposed on projects in response to locale-specific hazards.[11]

Structural mitigation measures include the following:

- Hazard-resistant construction and design (e.g., earthquake-proof structures, engineering systems, and vibration-control systems). Examples of structural mitigation measures include building material and technique selections for all building construction types (*Masonry* – Unreinforced Masonry, Reinforced Masonry; *Reinforced Concrete [RC]* – RC Moment Resisting Frame (MRF), RC MRF with Unreinforced Masonry Infill, RC Shear Wall, Steel and RC Composite Frame, Precast MRF; *Steel* – Steel Frame, Steel MRF with Unreinforced Masonry, Steel-Braced Frame, Light Metal Frame), building code compliance, and site selection.
- Building designs that are fireproof both to prevent fires from occurring and to prevent them from spreading (e.g., fireproof structures and use of unburnable materials).
- Hazard-conscious ("Smart") building.
- Disaster-resistant construction and retrofitting existing building stock.[12]
- Stilted houses to reduce flood damage.

- Check dams, slit dams, retention basins, seawalls, and/or levees.
- Stone walls and heavy roof construction to prevent damage from strong winds (caused by typhoons, cyclones, or hurricanes) and salt corrosion.

As noted by NIBS President Henry L. Green, it is the responsibility of the building industry to guarantee that the design, construction, operation, and maintenance of facilities safeguard the health, safety, and welfare of members of the community.[13]

Non-structural mitigation comprises the following:

- Land use (living away from disaster-prone areas; the application of building regulations on active faults to avoid earthquake damage and on coastal areas to reduce damage from tsunamis and hurricanes; etc.).[14]
- Strategies that focus on risks arising from damage to non-load-bearing building components, including architectural elements such as partitions, decorative ornamentation, and cladding; mechanical, electrical, and plumbing components such as HVAC, life safety, and utility systems.[15] Mitigation actions (prescriptive, engineered, or non-engineered) include securing these elements to the structure in order to minimize damage and functional disruption.[16]
- Building standards and codes, tax incentives/disincentives, zoning ordinances, preventive healthcare programs, and public education to reduce risk.[17]

Other non-structural measures include:[18]

- Natural defenses (such as wetlands and swamps for storing excess waters from riverine flooding and decreasing downstream effects; coastal sand dunes for protecting structures built behind them and reducing coastal attrition, amongst others);
- Risk mapping; zoning ordinances (to prohibit building or rebuilding in hazard-prone locations);
- Hazard and vulnerability disclosure (to inform potential home buyers that the property they are buying is located in the pathway of a potential hazard);
- Hazard forecasting and early warning systems;
- Insurance (to allow financial risk to be transferred from a single entity to a pooled group of risks through a contract. For example, people residing in flood-prone areas are required to purchase flood insurance as a condition for obtaining a mortgage); and
- Catastrophe bonds (used by insurers, reinsurers, and governments to deal with a catastrophic loss).

David R. Godschalk has also classified mitigation measures as "voluntary" and "non-voluntary."[19] Voluntary mitigation programs

rely upon individuals, organizations and communities to recognize the dangers posed by hazards and to reduce their exposure to the risk. Tax incentives, information concerning hazards and how to avoid them, and information on safe building practices, for example, only work if individuals, organizations and communities decide that the risk of certain behaviors (such as building in wildfire areas) outweighs the benefits. Studies of floodplain management generally find that people will not limit development on the floodplains without strict regulations and the threat of punishment, e.g., withdrawal of eligibility for low-cost flood insurance or eligibility for disaster assistance.[20]

On the other hand, non-voluntary or mandatory mitigation programs use threats of punishment to enhance compliance with established standards.

However, a combination of mitigation tools (such as comprehensive and hazard-specific planning, technical assistance, mitigation grant programs, structural measures, and hazard insurance) are necessary to deal with existing older buildings that were constructed before modern building codes and standards were established.[21]

It is important to note that buildings and infrastructure should be designed to be resilient when subjected to both natural and man-made disasters; that is, they should be able to absorb and speedily recover from a disruptive event. Continuity of operations is at the core of resilience.[22]

Disaster risk reduction and mitigation in the US National Mitigation Framework

Disaster mitigation has always existed at every level in the US, "from the family that creates a sheltering plan in case of a tornado, to corporate emergency plans for opening manufacturing plants to the community, to local codes and zoning that systemically address risks in a community's buildings."[23] What is new, however, is that the National Mitigation Framework elevates building widespread resilience throughout communities to a national priority, with all members – individuals, businesses, non-profit organizations and local, state, tribal, territorial and Federal governments – working together in a coordinated manner to address how the nation manages risk through mitigation capabilities. This Framework also defines mitigation roles and strategies for increasing risk awareness and leveraging mitigation products, services, and assets throughout the whole community.

Building and land-use codes constitute an important non-structural tool that can be used in hazard mitigation. Consequently, and as a contribution to strengthening the provisions contained in the National Mitigation Framework, the Association of State Floodplain Managers (ASFPM) presented its review of existing codes to the US House Transportation and Infrastructure Committee Subcommittee on Economic Development, Public Buildings, and Emergency Management on July 24, 2012.[24] Its first observation was the current voluntary nature and wide variability of building code adoption. The second observation was that state adoption did not necessarily equal local adoption or enforcement of codes.

Table I.2 lists risk management tools, actors, time frames, benefits, and potential adverse impacts.

Table I.2 Illustrative risk management tools, actors, time frames, benefits, and potential adverse impacts[25]

Illustrative risk management tools	Relevant individuals, groups and organizations	Time frame	Potential benefits	Potential adverse impacts
Structural (construction-related)				
Levees, dams, and floodways	USACE, USGS, FEMA; state, county, and local governments; researchers; private sector	One to two years evaluation and decision; three to 50 years construction	Flood risk reduction	Belief that levee will fully protect against all floods
Disaster-resistant construction and retrofitting of existing building stock	Federal government, local officials, researchers, private sector (risk management firms and engineering firms), professional organizations, individuals	On the order of weeks to years, depending on measures employed and size of structure	Mitigation against extreme weather events, other natural hazards such as earthquakes and wildfires, floods, and hurricanes	Cost of the measures, belief that energy dissipation systems will fully protect against all hazards
Hazard-conscious ("smart") building	Engineering and construction firms, individual businesses and homeowners, federal and state governments, professional organizations	Similar to or slightly longer time to build than other new homes or businesses	Mitigation against a variety of natural hazards; reduce losses	Cost may be higher than with non-hazard-conscious buildings
Securing building components and contents	Building owners, tenants	At the time the building is occupied	Reduce earthquake (or other) damage for low investment	None
Well-enforced building codes	Local officials working in conjunction with USGS, NOAA, USFS, FEMA, NIST; engineering firms and professional safety and engineering organizations; and businesses and homeowners	Code development: months to several years to review and revise existing codes relative to existing hazard risk Code enforcement: continuous	Home owners and business owners adopt mitigation measures; seals of approval on homes to show that property meets mitigation standards	Inability of some residents to afford compliance and lack of safety net
Nonstructural (non-construction-related)				
Natural defenses	Communities, regions, states, federal government	One to four years; evaluation and decision; from three years to many lifetimes	Protect structures built behind them by reducing impacts of disasters (wind, water, fire)	May prevent building new structures on protected areas; requires long-term perspective

Illustrative risk management tools	Relevant individuals, groups and organizations	Time frame	Potential benefits	Potential adverse impacts
Risk mapping	FEMA, USACE, NOAA, NASA, USFS, USGS in conjunction with state and local authorities; engineering firms	Weeks to several years, depending upon quality and availability of data and map area covered	Communication of the hazard risk to the community	Overreliance on accuracy of maps
Zoning ordinances	Local and state governments	Immediate	Prohibits building or rebuilding in hazard-prone locations	May prevent lucrative construction of homes or businesses in specific areas
Hazard and vulnerability disclosure	Private sector; federal, state, and local governments	Immediate if adopted freely by the private sector; several years or more if new legislation is required to implement	Allows buyers to identify potential hazards or construction known to be vulnerable to such hazards before the purchase of a home or business; increases the value of disaster-resistant buildings	May hinder sales or lower property values in areas where hazards are revealed or for vulnerable construction types
Economic and tax incentives	Federal, state, and local governments	May be quickly adopted and implemented if political will, competing demands for resources, and public acceptance align; realization of returns on investment may be months to years	Subsidies, grants, fines, or tax rebates can provide incentives to homeowners and businesses to install hazard-mitigation measures	Negative incentives (fines, penalties) may not be acceptable to residents or businesses; positive incentives (subsidies, grants, rebates) incur immediate costs to the government with delayed return on investment

(Continued)

Table I.2 (Continued)

Illustrative risk management tools	Relevant individuals, groups and organizations	Time frame	Potential benefits	Potential adverse impacts
Hazard forecasting and warning systems	NOAA, USGS, USACE, NASA, USFS, state agencies, and the private sector	Constant data collection and monitoring	Allows forecasts of potential events and their impacts to be made; when communicated in a timely way, warning systems can save lives	Complex disasters and natural systems, increasing populations and potential longer-term impacts require increased data precision and better forecasting models
Insurance	FEMA, state insurance commissioners, private insurance industry, banks	Policies currently are issued on an annual basis but some consideration is being given to multiyear insurance tied to the property	Risk-based pricing that communicates level of risk to people in hazard-prone areas; vouchers for lower-income owners	Continued public financial assistance to those who do not buy insurance
Catastrophe bonds	Insurers, banks, investors	Typically one to three years	Risk is transferred to a broad investor base in the case of a catastrophic event; allows access to large fund amounts fairly quickly	Investors lose invested funds if a catastrophic event occurs; insurers pay bond amount with interest if the event does not occur

Note: FEMA = Federal Emergency Management Agency, NASA = National Aeronautics and Space Administration, NIST = National Institute of Standards and Technology, NOAA = National Oceanic and Atmospheric Administration, USACE = US Army Corps of Engineers, USFS = US Forest Service, and USGS = US Geological Survey.

Building a resilient future through disaster risk management

All disasters are local, and it is at the local community level that disaster risk management and resilience are most needed. Table I.3 illustrates the recommended features of a local resilience plan.

In its own work, the Global Facility for Disaster Reduction and Recovery (GFDRR) uses a five-action approach to increase the resilience of people to natural disasters. They are:

- Determination and comprehension of disaster risks by facilitating vulnerable people's improved access to "information about disaster and climate risks and greater capacity to create, manage and use this information."

Table I.3 Suggested elements of a local resilience plan[26]

Program element	Attributes
Community organization	Reflects community structure and leadership
Standards and codes	Represents current and needed building and development codes, standards, and zoning ordinances, where compliance and enforcement are emphasized
Performance metrics and resilience rating system	Represents assessment status and needs for essential progress in building resilience and desired performance of critical services and infrastructure following disruption
Education and communication	Represents critical education, outreach, and communication plans and practices for resilience to reach all community members
Local capacity	Designed to establish baselines and close essential capacity gaps in the community
Resource management	Integrates resources such as human and financial capital, mutual aid agreements, asset management strategies, essential relationships within interdependent communities and agencies

- Avoidance of generating new risks and lessening of prevailing risks in the community by better protecting vulnerable populations "through improved planning, better building practices and increased investments in vulnerability reduction."
- Improvement of the warning system and management of disasters at all levels (national, local, and community) "through the provision of more accurate and timely early warning, and through civil protection agencies capable of mobilizing a quick response in the event of a disaster."
- Increasing the fiscal resilience of both the public and private sectors by improving "the financial resilience of communities to the impact of natural disasters, with improved post-disaster financial response capacity and stronger domestic catastrophe insurance markets."
- Ensuring speedier, more resilient recovery by improving "the quality and timeliness of recovery and (post-disaster) reconstruction."[27]

Similarly, the US Department of Commerce's National Institute of Standards and Technology has developed a *Disaster Resilience Framework* to enhance community resilience by providing "a starting point for stakeholders to advance from current practice to resilience-based approaches that can be adapted by communities of varying size and complexity."[28] The four pillars of the Framework are: societal needs; performance goals for buildings and infrastructure lifelines, including their return to functionality; emergency communication systems and plans; and economic factors.

As a document, the Framework consists of nine sections: the Community (social aspects of resilience); Community Disaster Resilience for the Built Environment; Interdependencies and Cascading Effects; the Building Sector; the Transportation Sector; the Energy Sector; the Communication and Information Sector; the Water and Wastewater Sector; and Economic Considerations for Community Resilience.[29]

The US Environmental Protection Agency (EPA) and the Federal Emergency Management Agency (FEMA) have also developed Smart Growth Approaches for Disaster-Resilient Communities.[30] In their 2012 response to Tropical Storm Irene, which devastated infrastructure, communities, and lives in Vermont in 2011, the two agencies incorporated smart growth principles into state policies, local development regulations, and Hazard Mitigation Plans to increase community flood resilience. These were "measures taken to reduce the vulnerability of communities to damages from flooding and to support long-term recovery after an extreme flood," such as protecting vulnerable undeveloped lands, siting development in safer locations, and designing development so it is less likely to be damaged in a flood.[31] The agencies recommend that, in order to enhance their flood resistance, communities should:

- Update and integrate their community or comprehensive land-use plans with Hazard Mitigation Plans, ensuring that the comprehensive plan identifies future growth areas in safer locations and that hazard mitigation activities are consistent with the comprehensive plan priorities. Uncoordinated plans might unintentionally result in conflicting actions.
- Ensure that their policies, regulations, and budgets are aligned with the flood resilience goals outlined in their community development plans and Hazard Mitigation Plans.
- Appropriately modify their existing policies, regulations, and budgets or create new ones that help achieve the flood resilience goals delineated in their plans.

Furthermore, the agencies recommend precise local land-use policy options to improve flood resilience for four different geographic areas in a community:

- River Corridors: Conserve land and discourage development in particularly vulnerable areas along river corridors such as flood plains and wetlands.
- Vulnerable Settlements: Where development already exists in vulnerable areas, protect people, buildings, and facilities to reduce future flooding risk.
- Safer Areas: Plan for and encourage new development in areas that are less vulnerable to future floods.
- The Whole Watershed: Implement enhanced storm water management techniques to slow, spread, and infiltrate floodwater.[32]

In its *Oregon Resilience Plan*, the Oregon Seismic Safety Policy Advisory Commission in the US defined its seismic resilience goal as follows:

> Oregon citizens will not only be protected from life-threatening physical harm, but because of risk-reduction measures and pre-disaster planning, communities will recover more quickly and with less continuing vulnerability following a Cascadia subduction zone earthquake and tsunami.[33]

In order to ascertain the "steps needed to eliminate the gap separating current performance from resilient performance and to initiate . . . policy changes so that the inevitable natural disaster of a Cascadia earthquake and tsunami will not deliver a catastrophic blow to Oregon's economy and communities,"[34] the Commission developed a detailed assessment of the likely physical effects of a great (magnitude 9.0) Cascadia subduction zone earthquake and tsunami on different sectors of the community, and assessed the workplace integrity, workforce mobility, and building systems performance needed to allow Oregon's businesses to remain in operation following a Cascadia earthquake and tsunami and to drive a self-sustaining economic recovery (amongst other factors).

On the basis of the finding that Oregon would be far from resilient if faced with the impacts of a great Cascadia earthquake and tsunami, the Commission advised that methodical efforts be made to evaluate Oregon's buildings, lifelines, and social systems. A sustained program to make Oregon resilient should also be developed to replace, retrofit, and redesign existing facilities.

Measuring community resilience

An understanding of and ability to measure community resilience are critical to determining the most effective interventions and allow development experts to prioritize aid and invest in projects that build adaptive capacities where they are most needed. They also enable a better understanding of why one community is more resilient than others. The more resilient a community is, the fewer resources will be needed for disaster response and recovery.

There is, however, no standard definition of "resilience." The US Agency for International Development (USAID) defines it as "the ability of people, households, communities, countries and systems to mitigate, adapt to and recover from shocks and stresses in a manner that reduces chronic vulnerability and facilitates inclusive growth."[35] Jordan and Javernick-Will define it as the "ability to withstand disaster impacts as well as to cope with those impacts and recover quickly."[36] According to Paton and Johnston, it can be thought of "as a function of *inherent resilience*, the ability to withstand impacts without extensive losses, and *adaptive resilience*, the ability to adapt and access resources to cope with a disaster and recover."[37] Thus, as community resilience acts to counter vulnerability, the four key adaptive capacities that are essential for a resilient community are: economic development, social capital, information and communication, and community competence.[38]

Communities need to increase their resilience in order to:

- Save lives and the money needed to respond to a disaster by taking action before an event occurs, and build stronger, safer, and more secure communities.
- Help in understanding current levels of exposure and potential impacts from adverse events, thereby helping a community take responsibility for its own disaster risk.
- Allow for identification of the community's capacity to cope with adverse effects and where improvements are needed.
- Nurture a culture of self-sufficiency, helping-behavior, and betterment.
- Encourage collaboration among community members.[39]

In the US, there are a number of models and metrics that have been developed for measuring progress toward resilience. These include the Community Assessment of Resilience Tool, the Community Resilience System, the Resilience Capacity Index, and the Community Disaster Resilience Index.[40] Table I.4 illustrates some of the international resilience metrics and indicators. Although many tools have been

Table I.4 Selected summary of international metrics and indicators for vulnerability, risk, and resilience[41]

United Nations Development Program (UNDP) Disaster Risk Index (DRI)	The DRI, introduced in 2004, measures the *average risk of death* per country in three types of disasters (earthquakes, tropical cyclones and floods). It is a measure of vulnerability to a specific hazard that also accounts for the role of sociotechnical-humanistic and environmental issues that could be correlated with death and may point toward causal processes of disaster risks. The key steps in determining the DRI for *a specific hazard* include calculation of physical exposure in terms of number of people exposed to a hazard event in a given year; calculation of relative vulnerability in terms of number of people killed to number of people exposed; and calculation of vulnerability indicators using 26 variables. Based on the value of the DRI, and for a given specific hazard, countries are ranked according to their degree of physical exposure, relative vulnerability and degree of risk.
Inter-American Development Bank Disaster Deficit Index (DDI)	The DDI, introduced in 2005, is an indicator of a country's economic vulnerability to disaster. It is limited to Latin America and the Caribbean. DDI is a measure of the likely economic loss related to a disaster in a given time period and for the economic coping capacity of the country.
Inter-Agency Standing Committee (IASC) In-Country Team Self-Assessment Tool for Natural Disaster Response Preparedness	Established in 1994, the IASC was created to be the primary mechanism for interagency coordination of humanitarian assistance at the international level. It is composed of representatives of all 14 leading UN agencies, non-UN humanitarian agencies and three consortia of nongovernmental organizations. The In-Country Team Self-Assessment Tool for Natural Disaster Response Preparedness consists of a support chart and a checklist of issues and questions to self-assess the level of international standards. It also provides resources to address key concerns and priority areas for disaster preparedness and response. See http://humanitarianinfo.org/iasc/
United Nations University Institute for Environment and Human Security, World Risk Index	The World Risk Index, introduced in 2011, indicates the probability that a country or region will be affected by an extreme natural event (earthquakes, storms, floods, droughts, and sea-level rise). It also focuses on (i) the vulnerability of the population (levels of poverty, education, food security, infrastructure, economic framework) to natural hazards; (ii) its capacity to cope with severe and immediate disasters as a function of governance, disaster preparedness, early warning systems, medical services, and social and economic security; and (iii) its adaptive precautionary measures against anticipated future natural disasters. The World Risk Index is also combined with local and project risk indexes.

developed both in the US and internationally, communities rarely use them because they are too "complex, too computationally intensive, or too simple and do not provide the right information." Communities actually need a resilience measurement tool that can:

- Assess and help prioritize needs and goals;
- Establish baselines for monitoring progress and recognizing success;

- Evaluate costs (investments) and benefits (results); and
- Assess the effects of different policies and approaches.[42]

In order to be effective, a chosen model should: be open and transparent; be aligned with the community's goals and visions; be simple and well-documented (evidence-based); be replicable; be able to address multiple hazards; be representative of a community's geographical extent, physical characteristics, and diversity; and be adaptable and scalable to different community sizes, compositions, and changing circumstances.[43]

Indeed, the National Research Council has identified four critical dimensions of a consistent system of resilience indicators or measures:

1   Vulnerable Populations – elements that capture the different needs of individuals and groups, based on their socio-economic and demographic circumstances, amongst others.
2   The ability of critical and environmental infrastructure (such as water and sewage, transportation, power, communications) to recover from disaster events.
3   Social factors that enhance or limit a community's ability to recover, including components such as social capital, education, language, governance, financial structures, culture, and workforce.
4   The ability of built infrastructure to withstand the impacts of disasters.[44]

However, there is currently no consistent basis for measuring resilience that includes all of these dimensions in the US. This makes it difficult for communities to observe advances or fluctuations in their resilience closely.[45]

Notes

1   A. Awotona, ed. (2012), *Rebuilding Sustainable Communities with Vulnerable Populations after the Cameras Have Gone: A Worldwide Study* (Newcastle upon Tyne, UK: Cambridge Scholars Publishing); A. Awotona (2016), "Design for Disaster Preparation and Mitigation," in *The Routledge Companion for Architecture Design and Practice: Established and Emerging Trends*, M. Kanaani and D. Kopec, eds., pp. 339–60 (New York, NY: Routledge).
2   Adapted from S. Yodmani (2001), *Disaster Risk Management and Vulnerability Reduction: Protecting the Poor*, paper presented at the Asia and Pacific Forum on Poverty, organized by the Asian Development Bank – The "Social Protect[iton] Workshop 6: Protecting Communities – Social Funds and Disaster Management," February 5–9, 2001, Manila, the Philippines, pp. 9–10.
3   The National Research Council (2012), *Disaster Resilience, A National Imperative* (Washington, DC: The National Academies Press), pp. 26–27.
4   Reprinted with permission from *Disaster Resilience, A National Imperative* (2012), pp. 39–40), by the National Academy of Sciences, courtesy of the National Academies Press, Washington, DC.
5   Whole Building Design Guide Secure/Safe Committee (2013a), *Natural Hazards and Security*, National Institute of Building Sciences (NIBS), retrieved July 24, 2014, from http://www.wbdg.org/design/resist_hazards.php
6   Ibid.
7   F. Zollinger (1985), "Debris Detention Basins in the European Alps," in *Proceedings of the International Symposium on Erosion, Debris Flow and Disaster Prevention*, Tsukuba, Japan, pp. 433–38 (Tokyo, Japan: Erosion Control Engineering Society); J. Hübl, A. Strauss, M. Holub, and J. Suda (2005), "Structural Mitigation Measures," in *Proceedings of the Third Probabilistic Workshop: Technical Systems + Natural Hazards*, University of Applied

Sciences and Natural Resources, November 24–25, 2005, Wien, Austria, retrieved March 13, 2015, from http://www.baunat.boku.ac.at/fileadmin/data/H03000/H87000/H87500/files/arbeitsgruppen/schutzbauwerke/bemessung/2005_02_Huebl_etal.pdf

8 WBDG (2013a).

9 Ibid.

10 Ibid.; **Earthquakes:** Public Law 95–124, *Earthquake Hazards Reduction Act* (1977); Executive Order 12699, *Seismic Safety of Federal and Federally Assisted or Regulated New Building Construction* (January 5, 1990); United States Code Title 42, Chapter 86, *Earthquake Hazards Reduction*; Public Law 101–614, *National Earthquake Hazards Reduction Program Reauthorization Act* (1990); Executive Order 12941, *Seismic Safety of Existing Federally Owned or Leased Buildings* (December 1, 1994); Public Law 103–374, *Earthquake Hazards Reduction Act of 1977, Authorization and Amendment* (1994); Public Law 106–503, *Earthquake Hazards Reduction Authorization Act* (2000); UFC 3–301–01 *Structural Engineering* (2013); UFC 3–310–04 *Seismic Design for Buildings* (2013).

   Hurricanes, Typhoons and Tornadoes: Public Law 108–146, *Tornado Shelters Act* (2003).
   Floods: Public Law 93–234, *National Flood Insurance Act* (1968).
   Tsunamis: Public Law 109–424, *Tsunami Warning and Education Act* (2006).
   Rainfall and Wind-Driven Rain: National Institute of Standards and Technology, *NISTIR 4821 Envelope Design Guidelines for Federal Office Buildings: Thermal Integrity and Airtightness* (1993); DOE-STD-1021–93 *Natural Phenomena Hazards Performance Categorization Guidelines for Structures, Systems and Components* (1993); DOE-STD-1022–94 *Natural Phenomena Hazards Characterization Criteria* (1994); DOE-STD-1023–95 *Natural Phenomena Hazards Assessment Criteria* (1995); DOE-STD-1020–2002 *Natural Phenomena Hazards Design and Evaluation Criteria for Department of Energy Facilities* (2002); UFC 3–440–05N *Tropical Engineering* (2004).

11 Ibid.

12 Hübl et al. (2005); the National Research Council (2012), pp. 45–47; O. Murao (2008), "Case Study of Architecture and Urban Design on the Disaster Life Cycle in Japan," paper presented at the 14th World Conference on Earthquake Engineering, October 12–17, 2008, Beijing, China, retrieved January 3, 2015, from http://www.iitk.ac.in/nicee/wcee/article/14_S08–032.PDF; WBDG Secure/Safe Committee (2013b), *Overview*. NIBS WBDG Website, retrieved July 26, 2014, from http://www.wbdg.org/design/secure_safe.php

13 On Tuesday, May 13, 2014, leaders of the US design and construction industry, along with building owners and operators, for the first time "agreed to promote resilience in planning, building materials, design, construction and operational techniques as the solution to making the nation's aging infrastructure more safe and secure." This was contained in a signed joint statement by the CEOs of almost two dozen leading design and construction industry associations – with a combined membership of more than 750,000 professionals, who are responsible for generating almost $1 trillion in GDP – to mark the occasion of "Building Safety Month." Among other things, they also committed the industry to educating itself through continuous learning; to advocating effective land- use policies; to responding to disasters alongside first responders; and to planning for future events, with strategies for fast recovery. The list of organizations that signed the joint statement included:

American Council of Engineering Companies
American Institute of Architects
American Planning Association
American Society of Civil Engineers
American Society of Interior Designers
American Society of Landscape Architects
American Society of Plumbing Engineers
American Society of Heating, Refrigerating, and Air-Conditioning Engineers (ASHRAE)
Associated Builders and Contractors
Associated General Contractors of America
Building Owners and Managers Association
International Code Council

International Facility Management Association
International Interior Design Association
Lean Construction Institute
National Association of Home Builders
National Institute of Building Sciences
National Society of Professional Engineers
Royal Institute of Chartered Surveyors
Urban Land Institute
US Green Building Council

Source: "CEOs Announce Major Commitment to Promote Resilient Planning and Building Materials" (2014, May 13), *NIBS Website*, available at http://www.nibs.org/news/172768/CEOs-Announce-Major-Commitment-to-Promote-Resilient-Planning-and-Building-Materials.htm

14 Murao (2008).
15 WBDG Secure/Safe Committee (2013b).
16 Ibid.
17 C. Berginnis (2012, July 24), *A Review of Building Codes and Mitigation Efforts to Help Minimize the Costs Associated with Natural Disasters*. Testimony before the House Transportation and Infrastructure Committee Subcommittee on Economic Development, Public Buildings and Emergency Management, Association of State Floodplain Managers, Washington, DC.
18 The National Research Council (2012), pp. 47–52.
19 D.R. Godschalk (2007), "Mitigation," in *Emergency Management: Principles and Practice for Local Government*, 2nd edn., W.L. Waugh, Jr. and K. Tierney, eds., pp. 89–112 (Washington, DC: International City/County Management Association).
20 Ibid.
21 Berginnis (2012).
22 WBDG Secure/Safe Committee (2013b). Some of the relevant codes and standards that design, construction, and operation of buildings and infrastructure must abide by in order to prepare for and provide a unified response to disasters and emergencies in the US include the following:

- *ASIS GDL BC 01–2005* Business Continuity Guideline – A Practical Approach for Emergency Preparedness, Crisis Management and Disaster Recovery.
- *ASIS GDL CSO 04–2008* ASIS Chief Security Officer Guideline.
- *ASIS SPC.1–2009* Organizational Resilience: Security Preparedness and Continuity Management Systems – Requirements with Guidance for Use.
- *ASIS/BSI BCM.01–2010* Business Continuity Management Systems: Requirements with Guidance for Use.
- *NFPA 1600 Standard on Disaster/Emergency Management and Business Continuity Programs*, 2010 edition.
- *NFPA 72 National Fire Alarm and Signaling Code*, 2013 edition.

Some of the relevant publications are:

- *The Buildings and Infrastructure Protection Series (BIPS)* by the Department of Homeland Security:
- BIPS 01 Aging Infrastructure: Issues, Research and Technology;
- BIPS 02 Integrated Rapid Visual Screening of Mass Transit Stations;
- BIPS 03 Integrated Rapid Visual Screening of Tunnels;
- BIPS 04 Integrated Rapid Visual Screening of Buildings;
- BIPS 05 Preventing Structures from Collapsing;
- BIPS 06 / FEMA 426 Reference Manual to Mitigate Potential Terrorist Attacks Against Buildings;
- BIPS 07 / FEMA 428 Primer to Design Safe School Projects in Case of Terrorist Attacks and School Shootings;

- BIPS 08 Field Guide for Building Stabilization and Shoring Techniques;
- BIPS 09 Blast Load Effects in Urban Canyons: A New York City Study (FOUO); and
- BIPS 10 High Performance Based Design for the Building Enclosure.
- *Facilities Standards for the Public Buildings Service, P100* by the General Services Administration (GSA).
- *FEMA 386 Series, Mitigation Planning How-To Guide Series.*
- *FEMA 386–2 Understanding Your Risks: Identifying Hazards and Estimating Losses.*
- *FEMA 452 Risk Assessment – A How-To Guide to Mitigate Potential Terrorist Attacks Against Buildings.*
- *International Building Code*, International Code Council, 2012.
- *The National Strategy for "The Physical Protection of Critical Infrastructure and Key Assets,"* The White House, February 2003.
- National Institute of Standards and Technology (NIST) Publications.
- B.D. McDowell, A.C. Lemer, Committee on Risk Appraisal in the Development of Facilities Design Criteria, National Research Council, eds. (1991), *Uses of Risk Analysis to Achieve Balanced Safety in Building Design and Operations* (Washington, DC: National Academy Press).

Source: WBDG Secure/Safe Committee (2013b), "Overview".

23 Presidential Policy Directive (PPD) 8: *National Preparedness* was released in March 2011 with the goal of strengthening the security and resilience of the US through systematic preparation for the threats that pose the greatest risk to the security of the nation. PPD-8 defines five mission areas – Prevention, Protection, Mitigation, Response, and Recovery – and mandates the development of a series of policy and planning documents to explain and guide the nation's approach to ensuring and enhancing national preparedness. US Department of Homeland Security (2013), *National Mitigation Framework, May 2013* (Washington, DC: FEMA).
24 Berginnis (2012).
25 Table taken from the National Academy of Sciences (2012), *Disaster Resilience, A National Imperative* (Washington, DC: National Academies Press), pp. 54–57. Reprinted with permission.
26 Ibid., pp. 126–27. Reprinted with permission.
27 Global Facility for Disaster Reduction and Recovery (2014, July), *Managing Disaster Risks for a Resilient Future: A Work Plan for the Global Facility for Disaster Reduction and Recovery 2015–2017*, retrieved February 20, 2015, from https://www.gfdrr.org/sites/gfdrr/files/publication/Updated%20GFDRR%20Work%20Plan%202015–17%2010.24.14_0.pdf, pp. 13–36.
28 National Institute of Standards and Technology (2014), *Disaster Resilience Framework Document*, retrieved February 20, 2015, from http://www.nist.gov/el/building_materials/resilience/framework.cfm
29 Ibid.
30 US Environmental Protection Agency (2014), *Planning for Flood Recovery and Long-Term Resilience in Vermont: Smart Growth Approaches for Disaster-Resilient Communities*, EPA 231-R-14–003 July, retrieved February 21, 2015, from www.epa.gov/smartgrowth
31 Ibid.
32 Ibid., p. 3.
33 Oregon Seismic Safety Policy Advisory Commission (2013, February), *The Oregon Resilience Plan: Reducing Risk and Improving Recovery for the Next Cascadia Earthquake and Tsunami*, Report to the 77th Legislative Assembly, Salem, Oregon, retrieved February 21, 2015, from http://www.oregon.gov/OMD/OEM/osspac/docs/Oregon_Resilience_Plan_draft_Executive_Summary.pdf
34 Ibid.
35 US Agency for International Development, *Resilience* [Web page], retrieved February 28, 2015, from http://www.usaid.gov/resilience
36 E. Jordan and A. Javernick-Will (2012), "Measuring Community Resilience and Recovery: A Content Analysis of Indicators," *Construction Research Congress*, ASCE,

retrieved February 28, 2015, from http://rebar.ecn.purdue.edu/crc2012/papers/pdfs/-63. pdf, p. 2192.
37 D. Paton and D. Johnston (2006), *Disaster Resilience: An Integrated Approach* (Springfield, IL: Charles C. Thomas Publishing).
38 F.H. Norris (2008), *Capacities that Promote Community Resilience*. National Consortium for the Study of Terrorism and Responses to Terrorism, University of Maryland, retrieved February 28, 2015, from https://www.orau.gov/dhssummit/presentations/plenary/March11/norris_fran.pdf
39 The National Research Council (2015), *Developing a Framework for Measuring Community Resilience: Summary of a Workshop*, D.A. Brose, Rapporteur, Committee on Measures of Community Resilience: From Lessons Learned to Lessons Applied, Resilient America Roundtable, Policy and Global Affairs Division. (Washington, DC: The National Academies Press), p. 6.
40 Ibid., pp. 102–03.
41 Table taken from the National Academy of Sciences (2012), *Disaster Resilience, A National Imperative* (Washington, DC: National Academies Press), pp. 109–10. Reprinted with permission.
42 Ibid., p. 7.
43 The National Research Council (2012), p. 92.
44 Ibid.
45 The National Research Council (2015).

Bibliography

Awotona, A., ed. (2012). *Rebuilding Sustainable Communities with Vulnerable Populations after the Cameras Have Gone: A Worldwide Study*. Newcastle upon Tyne, UK: Cambridge Scholars Publishing.
———. (2016). "Design for Disaster Preparation and Mitigation." In *The Routledge Companion for Architecture Design and Practice: Established and Emerging Trends*. M. Kanaani and D. Kopec, eds. pp. 339–60. New York, NY: Routledge.
Berginnis, C. (2012, July 24). *A Review of Building Codes and Mitigation Efforts to Help Minimize the Costs Associated with Natural Disasters*. Testimony before the House Transportation and Infrastructure Committee Subcommittee on Economic Development, Public Buildings and Emergency Management, Association of State Floodplain Managers, Washington, DC.
Global Facility for Disaster Reduction and Recovery (GFDRR). (2014, July). *Managing Disaster Risks for a Resilient Future: A Work Plan for the Global Facility for Disaster Reduction and Recovery 2015–2017*. Retrieved February 20, 2015, from https://www.gfdrr.org/sites/gfdrr/files/publication/Updated%20GFDRR%20Work%20Plan%202015–17%2010.24.14_0.pdf
Godschalk, D.R. (2007). "Mitigation." In *Emergency Management: Principles and Practice for Local Government*. 2nd edn. W.L. Waugh, Jr. and K. Tierney, eds. pp. 89–112. Washington, DC: International City/County Management Association.
Hübl, J., A. Strauss, M. Holub, and J. Suda. (2005). "Structural Mitigation Measures." In *Proceedings of the Third Probabilistic Workshop: Technical Systems + Natural Hazards*. University of Applied Sciences and Natural Resources. November 24–25, 2005, Wien, Austria. Retrieved March 13, 2015, from http://www.baunat.boku.ac.at/fileadmin/data/H03000/H87000/H87500/files/arbeitsgruppen/schutzbauwerke/bemessung/2005_02_Huebl_etal.pdf
International Bank for Reconstruction and Development/World Bank. (2010). *Natural Hazards – UnNatural Disasters: The Economics of Effective Prevention*. Washington, DC: World Bank.
Jordan, E., and A. Javernick-Will. (2012). "Measuring Community Resilience and Recovery: A Content Analysis of Indicators." *Construction Research Congress*, ASCE, pp. 2190–2199. Retrieved February 28, 2015, from http://rebar.ecn.purdue.edu/crc2012/papers/pdfs/-63.pdf

Murao, O. (2008). "Case Study of Architecture and Urban Design on the Disaster Life Cycle in Japan." Paper presented at the 14th World Conference on Earthquake Engineering, October 12–17, 2008, Beijing, China. Retrieved January 3, 2015, from http://www.iitk.ac.in/nicee/wcee/article/14_S08–032.PDF

National Institute of Standards and Technology (NIST). (2014). *Disaster Resilience Framework Document*. Retrieved February 20, 2015, from http://www.nist.gov/el/building_materials/resilience/framework.cfm

National Research Council, the. (2012). *Disaster Resilience, A National Imperative*. Washington, DC: The National Academies Press.

———. (2015). *Developing a Framework for Measuring Community Resilience: Summary of a Workshop*. D.A. Brose, Rapporteur, Committee on Measures of Community Resilience: From Lessons Learned to Lessons Applied, Resilient America Roundtable, Policy and Global Affairs Division. Washington, DC: The National Academies Press.

Neal, D.M. (1997). "Reconsidering the Phases of Disaster." *International Journal of Mass Emergencies and Disasters* 15(2), pp. 239–64.

Norris, F.H. (2008). *Capacities that Promote Community Resilience*. National Consortium for the Study of Terrorism and Responses to Terrorism, University of Maryland. Retrieved February 28, 2015, from https://www.orau.gov/dhssummit/presentations/plenary/March11/norris_fran.pdf

Oregon Seismic Safety Policy Advisory Commission. (2013, February). *The Oregon Resilience Plan: Reducing Risk and Improving Recovery for the Next Cascadia Earthquake and Tsunami*. Report to the 77th Legislative Assembly, Salem, Oregon. Retrieved February 21, 2015, from http://www.oregon.gov/OMD/OEM/osspac/docs/Oregon_Resilience_Plan_draft_Executive_Summary.pdf

Paton, D., and D. Johnston. (2006). *Disaster Resilience: An Integrated Approach*. Springfield, IL: Charles C. Thomas Publishing.

US Agency for International Development (USAID). *Resilience* [Web page]. Retrieved February 28, 2015, from http://www.usaid.gov/resilience

US Department of Homeland Security. (2013). *National Mitigation Framework, May 2013*. Washington, DC: FEMA.

US Environmental Protection Agency (EPA). (2014). *Planning for Flood Recovery and Long-Term Resilience in Vermont: Smart Growth Approaches for Disaster-Resilient Communities*. EPA 231-R-14-003 July. Retrieved February 21, 2015, from www.epa.gov/smartgrowth

Whole Building Design Guide (WBDG) Secure/Safe Committee. (2013a). *Natural Hazards and Security*. National Institute of Building Sciences (NIBS). Retrieved July 24, 2014, from http://www.wbdg.org/design/resist_hazards.php

———. (2013b). *Overview*. NIBS WBDG Website. Retrieved July 26, 2014, from http://www.wbdg.org/design/secure_safe.php

World Bank. (n.d.). *Building Resilient Communities: Risk Management and Response to Natural Disasters through Social Funds and Community-Driven Development Operations*. Retrieved February 20, 2015, from http://siteresources.worldbank.org/INTSF/Resources/Building_Resilient_Communities_Complete.pdf

Yodmani, S. (2001). *Disaster Risk Management and Vulnerability Reduction: Protecting the Poor*. Paper presented at the Asia and Pacific Forum on Poverty, Organized by the Asian Development Bank – The "Social Protect[tion] Workshop 6: Protecting Communities – Social Funds and Disaster Management." February 5–9, 2001, Manila, The Philippines.

Zollinger, F. (1985). "Debris Detention Basins in the European Alps." In *Proceedings of the International Symposium on Erosion, Debris Flow and Disaster Prevention, Tsukuba, Japan*, pp. 433–38. Tokyo, Japan: Erosion Control Engineering Society.

Part 1
Africa

Introduction

Adenrele Awotona

Britain's Overseas Development Institute's report, *The Geography of Poverty, Disasters and Climate Extremes in 2030*, declares that poverty and disasters are closely connected and that terminating extreme poverty is contingent upon the governments of the poorest countries coming to terms with their increased risk of natural disasters. This is because there is a very close overlap between the countries that are expected to still have very high levels of poverty in 2030 and those most unprotected from natural vulnerabilities.[1]

The African Union and the Regional Economic Communities have committed themselves to the goals of poverty alleviation and disaster risk reduction in their core mandates. Nonetheless, over 60 percent of Africans continue to live in abject poverty, in areas that are most at risk for disaster-induced poverty and on a continent that recorded 147 disasters in 2011 and 2012, causing economic losses of US$1.3 billion.[2] In 2012 alone, more than 34 million people were affected by drought and extreme temperatures, which compounded other vulnerabilities and hazards such as storms and disease transmission.[3]

Africans are thus adversely affected by environmental disasters and are very vulnerable to the negative impacts of global climate change, such as declining harvests and suitable land for pasture, food insecurity, land degradation, spread of disease, and decreasing water supplies, and are without an effective central capacity to manage these problems. They therefore suffer from disproportionate numbers of deaths, transpositions, and destruction of infrastructure. With very few assets, a weak social safety net to help them cope with multiple and interdependent forms of vulnerability and catastrophes (hurricanes, earthquakes, landslides, droughts, flooding, and biological hazards), and no insurance to cover the loss of their possessions, they are forced to live in locations that are subject to numerous risk factors. These places include unplanned urban areas, substandard informal housing and settlements on the periphery of cities, inaccessible rural areas with little or no effective early warning programs, and locations near infrastructure that is easily damaged when natural disasters strike.

Africa currently has the highest rate of urbanization in the world, with 40 percent of the population living in cities or urban areas; this will increase to 50 percent by 2050.[4] However, according to the UNISDR, development strategies are not keeping pace with this exponential physical and demographic development, and most governments have yet to undertake massive national and trans-boundary comprehensive multi-hazard risk assessments to inform their disaster risk programs and to implement effective programs to reduce the underlying risk factors of disasters, in spite of the increasing acknowledgment of the connection between poverty and vulnerability to natural disasters.[5] There is also a lack of integration of risk analysis and disaster risk reduction measures into national and local development programs. Governments' inability in this regard is a result of inadequate resources and lack of data on vulnerability; lack of political will; fiscal limitations; inadequate technical and operational capacity; absence of a multi-sectoral capacity-development plan for strategic government institutions; ineffective institutional, legislative, and policy frameworks; non-prioritization of disaster risk reduction at both the national and local levels; insufficient or inactive involvement of multi-sectoral and multi-stakeholder participation in national platforms; and an inability to engage with relevant non-governmental actors such as civil society organizations, the private sector, and communities at risk to implement local initiatives.

Consequently, in order to prepare for and mitigate the impacts of natural disasters, reduce risks, diminish poverty, protect economic growth, save livelihoods as well as lives, and decrease the effects of climatological hazards due to climate change, there is an urgent need to improve urban planning, encourage afforestation and water conservation, apply stringent building standards, reinforce social support programs, and develop long-term initiatives to combat climate change. It is also essential to integrate disaster risk reduction into the educational curricula of schools, colleges, and universities. Currently, very few university degree programs exist that have a focus on disaster risk science and sustainable development.[6]

The chapters in this section focus on some of these issues, especially flood preparedness and response, the assessment of the vulnerability of the rural poor to climate hazards, and capacity-building for sustainable post-disaster reconstruction through public education.

In Chapter 1, Abiodun Olukayode Olotuah and Abraham Adeniyi Taiwo examine how the architectural curricula in Nigerian universities can produce graduates with a strong grounding in sustainable architecture (e.g., environmentally conscious design techniques; affordable, eco-friendly, and sustainable housing designs and construction; and healthier and smarter buildings which are flood-resilient). Similarly, Chapter 2, by Zanzan Akaka Uji, seeks to introduce vulnerable rural communities to technologically appropriate design and construction. The aim is to help create sustainable communities as an effective response to the human-made disasters that result from the incessant crises and conflicts in Nigeria. The ultimate goal is to incorporate these ideas into university curricula and programs in the construction industry and disciplines in order to ensure proper training in building for sustainability in human-made, disaster-prone environments. In Chapter 3, Bernard Tarza Tyubee and Iankaa Aguse focus on disaster preparedness, response and recovery, environmental protection for vulnerable groups in vulnerable places, and climate change adaptation strategies. They investigate major climate hazards, assess the level of vulnerability of rural households to these climate hazards, and define strategies households can use to adapt to climate hazards. In Chapter 4, Daniel Maposa, James J. Cochran, and 'Maseka Lesaoana focus on the lower Limpopo River basin of Mozambique, which is characterized by alternating extreme floods and severe droughts: these extreme natural hazards are the major causes of destruction of crops and human lives. Floods and droughts account for about 90 percent of all people affected by all natural disasters, and the purpose of this chapter is to report the results of a flood frequency analysis using flood heights (water levels) data series where the gauging instruments either may not be functioning due to lack of servicing or may simply be unavailable due to budget constraints. This study reveals that in countries where there is a scarcity of quality rainfall and river discharges data series records, flood heights can be used to make important decisions in flood management and risk reduction.

Notes

1 Voice of America (2013), *Report: Natural Disasters Worsen Poverty*, retrieved March 14, 2015, from http://www.voanews.com/content/natural-disasters-poverty-17oct13/1770717.html
2 United Nations Office for Disaster Risk Reduction (UNISDR) (2013, June 6), *Disaster Risk Reduction in Africa – Status Report on the Implementation of Africa Regional Strategy and Hyogo Framework for Action*, Executive Summary, retrieved March 14, 2015, from http://reliefweb.int/report/world/disaster-risk-reduction-africa-status-report-implementation-africa-regional-strategy
3 Ibid.
4 Ibid.
5 Ibid.
6 Ibid.

Bibliography

Environment and Poverty Times. (n.d.). *Natural Disasters: "At the Whim of Nature" – Between Drought and Flood*, No. 1, p. 12. Retrieved March 14, 2015, from http://www.grida.no/files/publications/environment times/poverty%20No1%20-%20page12.pdf
United Nations Office for Disaster Risk Reduction (UNISDR). (2013, June 6). *Disaster Risk Reduction in Africa – Status Report on the Implementation of Africa Regional Strategy and Hyogo Framework for Action*. Executive Summary. Retrieved March 14, 2015, from http://reliefweb.int/report/world/disaster-risk-reduction-africa-status-report-implementation-africa-regional-strategy
Voice of America. (2013). *Report: Natural Disasters Worsen Poverty*. Retrieved March 14, 2015, from http://www.voanews.com/content/natural-disasters-poverty-17oct13/1770717.html

1 Architectural curricula and the sustainable reconstruction of flood-devastated housing in Nigeria

Abiodun Olukayode Olotuah
and Abraham Adeniyi Taiwo

Introduction

A flood occurs when a body of water overflows land that is not normally submerged.[1] Increases in the water levels of rivers, lakes, swamps, and oceans usually result in flooding of land that is normally dry. The flow of a river usually is confined to a well-defined channel. River floods occur when the supply of water to the normal river channel exceeds the hydraulic carrying capacity of the channel.[2] Heavy rainfall can cause the river to overflow into its floodplain, which is a wide flat area adjacent to the channel. Floodplains are the natural floodways of rivers, used to carry away flood-waters. Human activities often reduce the natural carrying capacity of river channels because floodplains naturally invite human occupancy. Structures such as bridges, highways, railroad embankments, and buildings considerably reduce the flood-carrying capacity of river channels. The construction of communities on floodplains further reduces the flood-carrying capacity of river channels. This is encouraged by the fact that rivers overrun their floodplains only at rare intervals. Such communities grow into large settlements, which are subject to huge losses when floods do occur. The rapid growth of cities has contributed to such developments on floodplains. The aftermath of flooding has often been destruction of buildings and the displacement of large numbers of people; the entire floodplain then requires re-planning and the buildings must be reconstructed to make them habitable again. The skills required by architects to accomplish these tasks are acquired during their training in architecture schools. This chapter examines the curricula of architectural programs in Nigerian universities to assess the extent of instruction in flood-resilient housing and reconstruction after flooding as important components of disaster management and disaster risk reduction. It appraises the need for dynamic curricula improvement in architectural education to reflect the societal needs and technological development of Nigeria.

Flooding incidents in Nigeria

The impacts of flooding have risen to threatening proportions, resulting in the loss of lives and property in Nigeria.[3] Houses, school buildings, markets, and bridges often collapse due to flooding. Farmlands are swept away while schools and market places are submerged. Flooding incidents occur perennially in almost all states in Nigeria with varying degrees of impact. Severe damage to buildings has been reported, often accompanied by the loss of lives.

Poor urban conditions in Nigeria are often a major root cause of flooding incident. The development control mechanisms in place are inadequate and incompetently enforced. These lead to settlements being constructed with little consideration for storm water drainage. As adequate housing is often out of the economic reach of the urban poor, they reside in informal settlements located on marginal low-lying lands, river banks, and floodplains that are vulnerable to flooding.

The incidence of rapid urbanization taking place in Nigeria has given rise to poor urban services and infrastructure, poor housing conditions, and a generally degraded environment. Rural-urban drift is mainly responsible for the high rate of urbanization. Rural areas generally lack vital infrastructure (water, electricity, roads, and telecommunication), as well as social services and facilities (such as education, health, entertainment, and marketing). The absence of these amenities makes migration by rural dwellers to urban centers (which have a higher concentration of these amenities) inevitable.[4]

Severe housing poverty characterizes the living conditions of Nigerian urban dwellers.[5] Rural-urban drift has created demand for, and put pressure on, the limited houses available, resulting in overcrowding in the existing stock and the sprouting of makeshift shacks lacking minimum structural and normative quality. People live in squalid, substandard, and poor housing in deplorable conditions in insanitary residential environments. The urban centers are characterized by a high density of buildings, air pollution, surface water, noise, and excessive solid waste.[6]

The resultant effect on poorly planned urban centers in Nigeria is that large areas are occupied by buildings and paved surfaces, giving rise to peak storm discharge. Drains to take rain runoff are inadequate in most urban centers, and those that do exist are poorly designed. With inadequate natural and artificial channels, the water from surface runoff is discharged into streams, resulting in overflow.

Flood-resilient housing concepts

Housing design concepts and construction technology have been developed to meet the challenges of flooding. These have arisen in response to severe cases of flooding involving the loss of lives and property. Below are four examples of these concepts.

Floating homes and amphibious housing

Floating homes are designed in such a way that they can float on the water in the event of flooding. They are designed with a foam base covered with concrete, making them light enough to float while the concrete base provides stability. Floating houses are fixed to a suspension mechanism to prevent horizontal motion. Amphibious houses are fixed to thick steel posts or anchored masts to hold the houses in position. In the event of flooding, an amphibious house will rise with the water up to a height of 5 m; it will return to its original placement on land when the water recedes, coming to rest on a concrete foundation. Amphibious houses could be designed to harness tidal energy, allowing the water to rise safely and using an alert system to warn occupants of rising water levels. Energy harnessed as the water rises could be used as an additional energy supply to the houses and for desalination. Such harnessed energy is useful during flooding, when water and power supplies are cut off.[7]

Dikes

Dikes are embankments for holding back flooding from a river, sea, or ocean. Dutch engineers designed an ingenious network of dams, sluices, and barriers after the North Sea floods devastated the southwest of the Netherlands in 1953, killing hundreds of people.[8] The dikes were a new kind of barrier against the sea. It was a system that took into consideration the fact that dikes require regular maintenance, constant monitoring, and an understanding of the behavior of storms, rivers, and oceans. The dikes were multifunctional, with retail shops and offices designed on top of them.

Going with the flow

In the concept of controlled flooding, water is allowed in rather than stopped[9] – working with nature rather than against it. Interlinked infrastructure, comprising lowered dikes and spillways, allows water to flow with little hindrance. A good example is the Overdiepse Polder in Netherlands. By lowering the dike along the northern edge of the 3 km² of the Overdiepse Polder, the Bergse Maas Canal is able to spill in, lowering the water level in the canal by 30 cm and sparing the residents of Deb Bosch in the event of a flood.[10]

Pole houses

A pole house is an example of integrated engineering in which a house's skeletal structure carries its entire load. The skeletal structure is designed to be independent of the walls, making all of the interior and exterior walls completely non-load-bearing. Large-diameter wood poles are bolted on reinforced concrete footings that rise through the entire structure to carry the suspended floor slabs, beams, and the roof structure. It applies some principles of post-and-beam timber frame construction, with large cantilevered sections. It is designed to be economical to build and flood-resistant. Pole houses survive in the event of conventional floods.[11] Because timber components are prone to water damage, manufactured or engineered timber components are used. If natural timber is to be used, waterproof glue bonding is applied to make it moisture-resistant.

Sustainable housing reconstruction

Sustainable housing reconstruction requires a bottom-up participatory approach. Architects at the local or grassroots level should be involved, particularly in providing local housing prototypes, which would inculcate the use of appropriate construction technology and materials. The building design should be based on core housing principles, leading to incremental or gradual housing projects. The architects should be involved in managing the reconstruction process.

In sustainable re-housing, people at the grassroots level must be given the opportunity to participate. Without reference to the perceptions and capabilities of local people, housing programs often fail.[12] This is because local communities are in the best position to identify their own needs and order their priorities. Attitudes towards space and its use and organization are all linked to cultural traditions that are often best understood by the local people themselves. Local communities have valuable experience and a special understanding of their environment, their local building resources, and how to make the best use of them.

The goal of the sustainable housing reconstruction process is to decrease the environmental costs incurred as a result of inadequate construction systems and solutions, and to improve users' comfort. The process involves environmentally friendly and community-based practices. It reduces the reconstructed houses' negative impact on the environment through the use of sustainable building materials and environmental designs. The reduction of carbon dioxide emissions in housing reconstruction is a critical issue in environmental management, in view of climate change implications. Taking key sustainable factors into consideration ensures the satisfaction, wellbeing, and productivity of the users. Sustainability in housing reconstruction is thus a crucial aspect of the architectural curriculum, necessary to enhance the quality of the built environment.

Architecture curricula in Nigerian schools of architecture

Architecture curricula in Nigerian schools of architecture are largely fashioned after the British and American models of architectural education; however, the original programs have undergone tremendous changes to reflect national needs and aspirations. The aim has been to ensure that the programs are relevant to the nation's aspirations and that they can meet technological developments at any point in time. To assess the societal relevance of a course, curriculum evaluations are carried out from time to time.[13]

The Nigerian National Universities Commission stipulates the minimum standards that guide schools of architecture in Nigerian universities in their curriculum design. Courses are categorized into seven course modules, namely: Architectural Design, Arts and Drawing, Historical and Theoretical Studies, Building Systems Technology, Humanities and Social Studies, Environmental Control Systems, and Physical Sciences. Schools of architecture have over one hundred course titles to choose from in formulating their programs.

Architectural education is provided in over 20 federal, state, and private universities in Nigeria. These programs are all very similar, as courses are drawn from the seven modules stipulated by the National Universities Commission (NUC). The NUC provides a benchmark called the Benchmark Minimum Academic Standard (BMAS), which stipulates the minimum standards to which all architecture schools must conform. Schools may add to the courses in the BMAS in line with the philosophy, vision, mission, and specialties of their universities.

At the Federal University of Technology, Akure (a public university) there are 256 credit units taken for the Bachelor and Master degrees (B.Tech. and M.Tech.). Only two credit units are taken in Housing, in a course entitled *Housing Seminar*. The course is taught in the second semester of the final year of the Bachelor's degree (semester 10). The course examines the phenomenon of squatter settlements in developing nations, urban population growth, and the demand for shelter. It also examines the incidence of rapid urbanization and the poverty of rural communities in developing nations.

Courses in construction are part of the Building Systems Technology module, and construction courses are taken as Building Components and Methods I in the second year, Building Components and Methods II in the third year, Building Components and Methods III in the fourth year, and Building Components and Methods IV in the fifth year of the first degree (B.Tech.). Elements and components of construction are also taught as part of Building Materials I in the second year and Building Materials II in the third year of the first degree (B.Tech.).

At the Department of Architecture Covenant University, Ota (a private university in Nigeria), students must successfully complete a minimum of 256 credit units (175 units for the first four years of the BSc and 81 units for the last two years of the MSc) to graduate. Three courses have two credit units each. One of the courses, *Environmental Sciences*, is taught at the undergraduate level. It is a theory course taught in the second semester (known as the Omega semester) of the second year of the Bachelor's degree. It explains the philosophy of environmental sciences and how environmental sciences deal with planning, design, construction, and management of man-made and natural environments.

The other two courses relevant to housing studies and sustainability are taught at the graduate level (the first year of the MSc). These are ARC 818 *Housing Studies* and ARC 863 *Housing and Urban Renewal*. *Housing Studies* is a theory course taught in the first semester (known as the Alpha semester). Topics taught include housing in third-world countries, housing policies and practices in Africa, South America and South East Asia, housing construction waste materials and recycling, garbage, energy implications of design, and solar energy utilization. *Housing and Urban Renewal* is also a theory course and is taught as part of the MSc (first year) during the second semester. Topics taught include general concepts of housing, architectural and planning considerations in both public and private housing, and policies and problems connected with the supply and financing of housing.

Building Component Courses (ARC 316, ARC 326, ARC 416, and ARC 426) are taught to make students conversant with the techniques of construction. Emphasis is placed on fundamental principles of construction, which make the scope broad-based and encompass more innovative approaches for any new materials being developed.[14]

The program of architectural education at the Imo State University, Owerri (a state university) incorporates housing into Human Settlements Studies. The course is taken as *Human Settlement I* in the second semester of the second year of the Bachelor degree; *Human Settlement II* in the first semester of the third year; and *Human Settlements III* in the second semester of the same third year of the first degree. The Human Settlements course constitutes three course units out of a total of 176 units taken in the first degree of architecture. At the Master's degree level, Housing Studies are taught in a course taken in the second semester of the first year entitled *Human Spatial Organization* (one unit course). The total number of course units required to earn a Master's degree is 73.[15]

Design studio at Federal University of Technology, Akure, and Covenant University

For the first degree, the design studio is taught in four academic sessions (second year to fifth year) at the Federal University of Technology, Akure, Nigeria. It is a course that runs during both semesters of the academic session. Course mentors are appointed to teach and guide the students in the course. Appropriate methods of teaching are used to mentor the students in the various classes.

Housing studies are included in the architectural design curriculum. Design of residential buildings is done as part of the design studio in the second year (200L), third year (300L), and fifth year (500L). In 200L, the course is ARC 201B-Architectural Design II, which is taken in the second semester of the second year (semester 4). In 300L, taken in the first semester of the third year (semester 5), the course is ARC 301A – Architectural Design III, in which the minor project is a housing project. The design studio for 500L is ARC 501 – Advanced Design studio, in which the major project is a typical housing scheme for a fairly large population. The course is taken in the first and second semesters of the fifth year (semesters 9 and 10).[16]

At Covenant University, the curriculum is run similarly to that of the Federal University of Technology, Akure; the courses, however, run on a semester basis. Mentors are appointed by the department, and they guide and teach the students using appropriate methodologies.

Recommended improvements in curricula

The curriculum is designed so that students can achieve, as far as possible, certain educational ends or objectives. Stone and Nielsen[17] assert that a curriculum is an integrated collection of courses and activities developed in response to social needs and related to the needs of the culture that supports the school. As Tanner and Tanner[18] affirm, a curriculum comprises planned and guided learning experiences and intended learning outcomes, formulated through the systematic reconstruction of knowledge and experiences, to support the learner's continuous growth in personal and social competence.

Curricula design and execution are concerned with the substance and method of education. These, however, cannot be pursued in isolation from the purpose, organization, and administration of education. Architecture curricula should thus be informed by national educational goals and aspirations that accept tradition as the starting point of creative independent thinking. The curriculum should be adapted to local conditions and emphasize the special character of the natural and social environment.[19]

The curriculum of architectural education in Nigeria, despite being tailored to meet local needs, has been a subject of debate. Its adequacy in dealing with emerging urban problems and its ability to adapt to the changing socio-economic situation in the country have been critically assessed.[20] The capability of architectural education to adjust to new demands on the profession has been identified as an inherent problem. Rapid developments in science and technology have posed daring challenges for architecture in the country. Continuous evaluation of curricula is necessary to ensure that they remain relevant to their set objectives while meeting current technological developments.[21] Madike[22] affirms that periodic evaluation of learning should be carried out to determine whether gaps exist between specific instructional and expressive objectives, on the one hand, and expected learning outcomes, on the other.

In the light of the challenge of flooding incidences in Nigeria and the imperative of sustainable reconstruction thereafter, new courses designed to address emerging issues should be introduced into architectural curricula. This is because at the Federal University of Technology, Akure, sustainable construction is absent from the synopsis of the courses, and sustainability is not taught in any course. Flood resilience is absent from housing design and construction, while reconstruction after disaster does not feature in any form in the course curriculum. Similarly, at Covenant University, sustainable construction is also absent from the synopsis of the courses and sustainability is taught peripherally in only one course at the Master's degree level. Flood resilience is absent from housing design and construction. This is a general reflection of what is featured in Nigeria's architecture curricula.

New courses designed to meet these pressing needs could form a part of the housing modules in Architectural Design Studios to emphasize flood-resilient design and the reconstruction of houses. Building Components and Methods courses should also include flood-resilient technology and materials. Urban renewal courses should include floodplain management to emphasize floodplain land policy, flood hazards, risk assessment, and related courses.

Courses on sustainable development should be included in curricula to enhance knowledge of ecological and environment-friendly architectural development. Some of the courses could be *Low-Carbon and Renewable Energy*, *Green Architecture*, *Green Building Design Studio*, *Green Building Materials, and Construction Methods and Practical Applications for Green Roofs and Green Walls*.

The curricula should emphasize research for the generation and dissemination of new knowledge. Sanoff[23] asserts that architecture should be based on the knowledge of people's needs, and "without research, scholarship and a vigorous knowledge base, the profession cannot take stands on significant health, economic, social, political or ethical issues."[24]

Disaster management, disaster risk reduction, and decision-making: Policy implications of study

Disaster management and disaster risk reduction

Disaster management refers to measures aimed at mitigating the occurrence of a disaster event and for preventing such an occurrence from having harmful effects on communities. Disaster management encompasses a body of policies, regulations, and operational activities for dealing with disasters,[25] including the classical management functions of planning, organizing, staffing, leading, and controlling.

The modern paradigm for disaster management is disaster risk reduction, which encourages a wider and deeper understanding of why disasters happen, accompanied by more integrated and holistic approaches to reduce their impact on society. It seeks to build resilience and reduces vulnerability, and therefore offers capacities to support adaptation with regard to coping with extreme events such as droughts, floods, and storms, as well as addressing longer-term issues such as ecosystem degradation that increase vulnerability to these events. Disaster risk reduction is an all-encompassing term for disaster preparedness, mitigation, and management. It is also part of sustainable development. For development activities to be sustainable, they must also reduce disaster risk.

Disaster risk and the adverse impacts of natural hazards can be reduced by monitoring, systematically analyzing, and managing the causes of disasters, including avoiding hazards, reducing social and economic vulnerability, and improving preparedness for response to adverse hazard events.[26] Disaster risk reduction is a systematic approach to identifying, assessing, and reducing the risks of disaster. It aims to reduce socio-economic vulnerabilities to disaster as well as dealing with the environmental and other hazards that trigger them. The United Nations Office for

Disaster Risk Reduction (UNISDR) defines disaster risk reduction as "the conceptual framework of elements considered with the possibilities to minimize vulnerabilities and disaster risks throughout a society, to avoid (prevention) or to limit (mitigation and preparedness) the adverse impacts of hazards, within the broad context of sustainable development."[27] The aim of disaster risk reduction is thus to reduce the damage caused by natural hazards like earthquakes, floods, droughts, and cyclones, through an ethic of prevention. Risk arising from climatic hazards can be addressed by preventative measures such as avoiding settlement in floodplains and constructing strong buildings; monitoring, early warning, and response measures to manage extreme events; and risk transfer, including insurance, to cope with unavoidable impacts.

The UNISDR avers that disaster risk reduction and its impact on sustainable development require sound foundations in scientific, social, and economic knowledge and understanding. This includes developing relevant scientific and technical capacity-building, especially in developing countries such as Nigeria. The impacts of natural hazards on society can be significantly reduced through the application of sound, evidence-based investment in disaster risk reduction.[28] Architecture and construction engineering have contributed a significant body of knowledge in the development of flood-resilient housing and buildings and ingenious networks of embankments, dams, sluices, and barriers. These have been developed in response to the challenges of severe flooding incidences in Europe, particularly the Netherlands. This research has shown the daunting inadequacy of the curricula of Nigerian schools of architecture in this respect, which limits the ability of architects in flooding disaster preparedness, mitigation, and recovery. There is thus a nexus between decision-making in disaster risk reduction and the research outcomes.

Decision-making

The national policy on disaster management

In 1999 the Federal Government of Nigeria formulated a national policy on disaster management, which spelled out the responsibilities of the federal, state, and local governments in disaster management. The goal of government in formulating the national policy is to integrate disaster management into the national development process in order to facilitate a quick and coordinated response to such situations as may be required, so as to save as many lives as possible when disasters occur. The policy is designed to establish and strengthen disaster management institutions, partnerships, networking, and mainstreaming disaster risk reduction in the development process so as to strengthen the resilience of vulnerable groups and their ability to cope with disasters. The policy also highlights the coordination of disaster risk reduction initiatives within a unified policy framework in a proactive manner at all levels of government.[29]

The National Emergency Management Agency (NEMA)

The National Emergency Management Agency (NEMA) was established in 1999. Its establishment was one major step towards a holistic approach in addressing issues relating to disasters and emergencies in the country. This was geared towards mainstreaming disaster risk reduction into sustainable development in the country. This is in line with the Hyogo Framework for Action 2005–2015, which stressed that efforts to reduce disaster risk must be systematically integrated into policies, plans, and programs for sustainable development and poverty reduction.[30]

The Federal Government of Nigeria recognizes the fact that disaster management is multidisciplinary and that successful response can be achieved only through an integrated and coordinated approach. Disaster management responsibilities therefore involve the collaborative efforts of all stakeholders.

NEMA has recognized the need for knowledge generation, dissemination, and application in disaster management and risk reduction. One major step taken by NEMA is the mainstreaming of disaster risk reduction into university education for the purpose of capacity-building, public education, and awareness. The agency collaborated and signed a memorandum of understanding (MOU) with six universities (one in each of the six geopolitical zones of the country) on November 12, 2009. The universities are:

1   University of Maiduguri, Maiduguri (Northeast zone);
2   Ahmadu Bello University, Zaria (Northwest zone);
3   Federal University of Technology, Minna (North-central zone);
4   University of Ibadan (Southwest zone);
5   University of Port Harcourt (South-south zone); and
6   University of Nigeria, Nsukka (Southeast zone).

Architecture schools

The deficiencies in the architecture curricula for training architects in disaster management are a strong pointer to an increased acknowledgment of the need for improvement in the school programs. A review of curricula would be expedient for the schools to meet the roles they are expected to play in manpower development in disaster mitigation and recovery, particularly in the design and reconstruction of sustainable housing. Disaster recovery is the process by which a disaster-stricken area returns to its pre-disaster state. It involves the reconstruction of buildings, the restoration of essential services and infrastructure damaged by the disaster, and the rehabilitation of stricken people or displaced persons.

New degree and postgraduate diploma programs focusing particularly on disaster mitigation and recovery with an emphasis on architectural components would suffice in addressing the needs of architects. The National Universities Commission (NUC) stipulates the minimum academic standards and regulates the establishment of degree and postgraduate diploma programs in Nigerian universities. The NUC would have to acknowledge the worth and significance of the desired changes to give the necessary approval. The benchmark of academic standard (BMAS) would have to be improved upon and enlarged to incorporate more courses at the undergraduate level. The development of new programs and the strengthening of existing ones should be done with a cue drawn from the Centers for Disaster Risk Management and Development Studies, with recourse to NEMA. Important decisions need be taken at the appropriate national levels to properly integrate disaster risk reduction into architecture schools' programs.

Professional bodies

The Nigerian Institute of Architects (NIA) and the Architects Registration Council of Nigeria (ARCON) are the professional bodies that regulate the practice of architecture and the standard of architectural education in Nigeria. Both bodies organize programs for the Continued Professional Development (CPD) of architects in the country. These include annual seminars, conferences, workshops, exhibitions, and colloquia that are beneficial for career development of architects and for the contribution they make to society. State chapters of the Nigerian Institute of Architects hold similar programs and play host to the national body at quarterly dinners and other functions. The professional bodies are responsible for ensuring that architectural education meets national needs and is socially relevant. They also ensure that there is the right nexus between education and practice. In this regard, the professional bodies closely monitor the standards of architectural education in Nigeria through visitations, accreditation, and participation in external examinations. They also ensure that schools of architecture play important roles in the activities of the professional bodies.

As leading members of the built environment disciplines, architects ought to play important roles as disaster responders. The professional bodies should recognize that architects must also be able to mitigate disasters through their professional training and practice. In this regard the development problems, particularly in urban settings, that the professional bodies have consistently looked at in their programs would have to be extended to disaster management and risk reduction, in concert with NEMA. Architectural education programs need to be broadened to accomplish this change and professional bodies will have to superintend the development in view of the glaring deficiencies unearthed in this study.

The NIA and ARCON need to coordinate education and training that help architects make effective contributions to communities preparing for, responding to, and rebuilding after disaster. Architects need to utilize their skills in disaster response environments and better serve as leaders in their community. Architects require the tools, training, and leadership skills to make meaningful contributions to disaster mitigation and recovery. They need to collaborate with other professionals such as engineers and planners, because inefficient land-use regulations and urban planning are some of the root causes of flooding disasters in Nigeria, while engineers play an important role in the design of flood-resilient housing.

Conclusion

Incidences of flooding are a perennial problem in Nigeria and must be tackled head-long by relevant government agencies, research institutes, and universities. These are occasioned by the socio-economic circumstances pervading the Nigerian nation, ranging from the rapid rate of urbanization and urban poverty to inadequate development control mechanisms. Tackling the devastating effects of flooding on housing necessitates sustainable reconstruction, which is an important aspect of architectural curricula. The appraisal of the curricula used in Nigerian schools of architecture shows a deficiency in this regard. Flood-resilient housing concepts should be introduced into the architectural design studio, and new courses that address the reconstruction of flood-devastated housing and emerging trends of developing healthier and smarter buildings need to be introduced into the architectural curricula of Nigerian schools of architecture. Disaster management and disaster risk reduction can be achieved through architectural education, with instructive decisions taken at the national policy level, while the National Universities Commission and the professional bodies have important roles to play.

Notes

1 R.C. Ward (1978), *Floods: A Geographical Perspective* (London, UK: Macmillan).
2 G. Fadairo and A.O. Olotuah (2005), "Flooding and the Urban Environment: An Empirical Investigation," *Journal of Applied Sciences* 8(1), pp. 4511–18.
3 G. Fadairo (2008), "Impact of Flooding on Urban Housing: A Focus on Ala River in Akure, Nigeria," Unpublished Ph.D. thesis, Department of Architecture, Federal University of Technology, Akure, Nigeria.
4 A.O. Olotuah (2005), "Sustainable Urban Housing Provision in Nigeria: A Critical Assessment of Development Options," in *Proceedings of the Africa Union of Architects Congress: African Urbanization*, pp. 64–74. May 23–28, Abuja, Nigeria.
5 I.P.C. Asiodu (2001), "The Place of Housing in Nigerian Economic and Social Development," *Journal of the Nigerian Institute of Quantity Surveyors* 35(2), pp. 2–7.
6 M.O. Filani (1987), "Accessibility and Urban Poverty in Nigeria," in *The Urban Poor in Nigeria*, P.K. Makinwa and O.A. Ojo, eds., pp. 128–38 (Ibadan, Nigeria: Evans Brothers); A.L. Mabogunje, J.E Hardoy, and P.R. Mistra (1978), *Shelter Provision in Developing Countries* (Surrey, UK: The Gresham Press); D.O. Olanrewaju (2001), "Urban Infrastructure: A Critique of Urban Renewal Process in Ijora Badia," *Habitat International* 20, pp. 517–30; A.O. Olotuah (1997), "The House: Accessibility and Development – A Critical Evaluation of the Nigerian Situation," in *The House in Nigeria*, Proceedings of the National Symposium, Obafemi Awolowo University, Ile-Ife, Nigeria, July 23–24, B. Amole, ed., pp. 312–17 (Ile-Ife, Nigeria: Obafemi Awolowo University); A.O. Olotuah (2002), "An Appraisal of the Impact of Urban Services on Housing in Akure Metropolis," *Journal of Science, Engineering and Technology* 9(4), pp. 4570–82.
7 C. Zevenbergen (2007), *Adapting to Change: Towards Flood-Resilient Cities*, Inaugural Address, UNESCO-IHE Institute for Water Education, Delft, the Netherlands, December 14, 2007.
8 B.H. Khan (2012), "Web-Based Instruction 9WBI: What Is It and Why Is It?" in *Web-Based Instruction*, B.H. Khan, ed., pp. 5–18 (Englewood Cliffs, NJ: Educational Technology Publications).
9 H. Aglan, R. Wendt, and S. Livengood (2012), *Field Testing of Energy-Efficient Flood-Damage-Resistant Envelop System* (Oak Ridge, TN: Oak Ridge National Laboratory).
10 Ibid.
11 H. Doole (2009), *Flood Risk in an Increasingly Globalized World*, available at http://www.rms.com/publications/No Flood Risk.pdf
12 A.O. Olotuah and A.O. Aiyetan (2006), "Sustainable Low-Cost Housing Provision in Nigeria: A Bottom-up, Participatory Approach," in *Proceedings of 22nd Annual ARCOM Conference, 4–6 September, Birmingham, UK*, D. Boyd, ed., pp. 633–39 (Birmingham, UK: Association of Researchers in Construction Management).
13 A.O. Olotuah and A.O. Ajenifujah (2009), "Architectural Education and Housing Provision in Nigeria," *CEBE Transactions* 6(1), pp. 86–102.
14 A.A. Taiwo, A.O. Olotuah, A.B. Adeboye, Y.M.D. Adedeji, G. Fadairo, and D.A. Ayeni (2013), "Inculcating Housing Sustainability into Architectural Education in Nigeria," paper presented at the Higher Education Authority STEM Conference, April 17–18, Birmingham, UK.
15 Imo State University (2001), *University Prospectus and Calendar* (Owerri, Nigeria: Imo State University).
16 Olotuah and Ajenifujah (2009).
17 D.R. Stone and E.C. Nielsen (1982), *Educational Psychology – The Development of Teaching Skill* (New York, NY: Harper and Row).

18 D. Tanner and L.N. Tanner (1975), *Curriculum Development: Theory to Practice* (New York, NY: Harcourt Brace).
19 Y.M.D. Adedeji, A.A. Taiwo, G. Fadairo, and A.O. Olotuah (2012), "Architectural Education and Sustainable Human Habitat," in *Sustainability Today*, C.A. Brebbia, ed., pp. 89–99 (Southampton, UK: Wessex Institute of Technology Press).
20 A.O. Olotuah (2006), "At the Crossroads of Architectural Education in Nigeria," *CEBE Transactions* 3(2), pp. 80–88, doi:10.11120/tran.2006.03020080
21 A.Z. Uji (2001), "Beyond the Critiques of the Curriculum of Architectural Education in Nigeria," in *Architects and Architecture in Nigeria: A Tribute to Prof. E.A. Adeyemi*, U.O. Nkwogu, ed., pp. 109–22 (Akure, Nigeria: Association of Architectural Educators in Nigeria).
22 F.U. Madike (1983), "Curriculum Orientation and Planning Strategies: An Analytical Survey," in *Professional Education: A Book of Readings*, Nduka Okoh, ed., pp. 149–56 (Benin, Nigeria: Ethiope Publishing).
23 H. Sanoff (2003), *Three Decades of Design and Community* (Raleigh, NC: North Carolina State University).
24 A.M. Salama (2006), "A Lifestyle Theories Approach for Affordable Housing Research in Saudi Arabia," *Emirates Journal for Engineering Research* 11(1), pp. 67–76.
25 S.G. Mohammed (2012, July 27), "National Policy on Disaster Management," *Environmental Health Watch* [Blog], available at http://tsaftarmualli.blogspot.com/2012/07/national-policy-on-disaster-management.html
26 The Informal Taskforce on Climate Change of the Inter-Agency Standing Committee and the International Strategy for Disaster Reduction (2008), *Disaster Risk Reduction Strategies and Risk Management Practices: Critical Elements for Adaptation to Climate Change*, submission to the UNFCCC Ad hoc Working Group on Long-Term Cooperative Action.
27 United Nations Office for Disaster Risk Reduction (2004), *Living with Risk: A Global Review of Disaster Reduction Initiatives* (New York, NY: UNISDR).
28 United Nations Office for Disaster Risk Reduction (2014), "Evidence-Based Investments in DRR," available at http://www.unisdr.org/partners/academia-research
29 Mohammed (2012).
30 The University of Maiduguri (2014), *The Centre for Disaster Risk Management and Development Studies* [Website], available at www.unimaid.edu.ng/root/Disaster_Management.html

Bibliography

Adedeji, Y.M.D., A.A. Taiwo, G. Fadairo, and A.O. Olotuah. (2012). "Architectural Education and Sustainable Human Habitat." In *Sustainability Today*. C.A. Brebbia, ed. pp. 89–99. Southampton, UK: Wessex Institute of Technology Press.
Aglan, H., R. Wendt, and S. Livengood. (2012). *Field Testing of Energy-Efficient Flood-Damage-Resistant Envelop System*. Oak Ridge, TN: Oak Ridge National Laboratory. Available at http://www.cseweb.greennet.org.uk/pdfs/017/017060.pdf
Asiodu, I.P.C. (2001). "The Place of Housing in Nigerian Economic and Social Development." *Journal of the Nigerian Institute of Quantity Surveyors* 35(2), pp. 2–7.
Doole, H. (2009). *Flood Risk in an Increasingly Globalized World*. Retrieved July 12, 2012, from http://www.rms.com/publications/No Flood Risk.pdf
Fadairo, G. (2008). "Impact of Flooding on Urban Housing: A Focus on Ala River in Akure, Nigeria." Unpublished Ph.D. thesis, Department of Architecture, Federal University of Technology, Akure, Nigeria.
Fadairo, G., and A.O. Olotuah. (2005). "Flooding and the Urban Environment: An Empirical Investigation." *Journal of Applied Sciences* 8(1), pp. 4511–18.
Filani, M.O. (1987). "Accessibility and Urban Poverty in Nigeria." In *The Urban Poor in Nigeria*. P.K. Makinwa and O.A. Ojo, eds. pp. 128–38. Ibadan, Nigeria: Evans Brothers.

Imo State University. (2001). *University Prospectus and Calendar*. Owerri, Nigeria: Imo State University.

Informal Taskforce on Climate Change of the Inter-Agency Standing Committee (IASC), the, and the International Strategy for Disaster Reduction (ISDR). (2008). *Disaster Risk Reduction Strategies and Risk Management Practices: Critical Elements for Adaptation to Climate Change*. Submission to the UNFCCC Adhoc Working Group on Long-Term Cooperative Action. Retrieved October 31, 2014, from http://preventionweb.net/go/7602

Khan, B.H. (2012). "Web-Based Instruction 9WBI: What Is It and Why Is It?" In *Web-Based Instruction*. B.H. Khan, ed. pp. 5–18. Englewood Cliffs, NJ: Educational Technology Publications.

Mabogunje, A.L., J.E. Hardoy, and P.R. Mistra. (1978). *Shelter Provision in Developing Countries*. Surrey, UK: The Gresham Press.

Madike, F.U. (1983). "Curriculum Orientation and Planning Strategies: An Analytical Survey." In *Professional Education: A Book of Readings*. pp. 149–56. Benin, Nigeria: Ethiope Publishing.

Mohammed, S.G. (2012, July 27). "National Policy on Disaster Management." *Environmental Health Watch* [Blog]. Available at http://tsaftarmuhalli.blogspot.com/2012/07/national-policy-on-disaster-management.html

Olanrewaju, D.O. (2001). "Urban Infrastructure: A Critique of Urban Renewal Process in Ijora Badia." *Habitat International* 20, pp. 517–30.

Olotuah, A.O. (1997). "The House: Accessibility and Development – A Critical Evaluation of the Nigerian Situation." In *The House in Nigeria, Proceedings of the National Symposium, Obafemi Awolowo University, Ile-Ife, Nigeria, July 23–24*. B. Amole, ed. pp. 312–17. Ile-Ife, Nigeria: Obafemi Awolowo University.

———. (2002). "An Appraisal of the Impact of Urban Services on Housing in Akure Metropolis." *Journal of Science, Engineering and Technology* 9(4), pp. 4570–82.

———. (2005). "Sustainable Urban Housing Provision in Nigeria: A Critical Assessment of Development Options." In *Proceedings of the Africa Union of Architects Congress: African Urbanization*, pp. 64–74. May 23–28, Abuja, Nigeria.

———. (2006). "At the Crossroads of Architectural Education in Nigeria." *CEBE Transactions* 3(2), pp. 80–88, doi:10.11120/tran.2006.03020080

———, and A.O. Aiyetan. (2006). "Sustainable Low-Cost Housing Provision in Nigeria: A Bottom-up, Participatory Approach." In *Proceedings of 22nd Annual ARCOM Conference, 4–6 September, Birmingham, UK*. D. Boyd, ed. pp. 633–9. Birmingham, UK: Association of Researchers in Construction Management.

———, and A.O. Ajenifujah. (2009). "Architectural Education and Housing Provision in Nigeria." *CEBE Transactions* 6(1), pp. 86–102, doi:10.11120/tran.2009.06010086

Salama, A.M. (2006). "A Lifestyle Theories Approach for Affordable Housing Research in Saudi Arabia." *Emirates Journal for Engineering Research* 11(1), pp. 67–76.

Sanoff, H. (2003). *Three Decades of Design and Community*. Raleigh, NC: North Carolina State University.

Stone, D.R., and E.C. Nielsen. (1982). *Educational Psychology – The Development of Teaching Skill*. New York, NY: Harper and Row.

Taiwo, A.A., A.O. Olotuah, A.B. Adeboye, Y.M.D. Adedeji, G. Fadairo, and D.A. Ayeni. (2013). "Inculcating Housing Sustainability into Architectural Education in Nigeria." Paper presented at the Higher Education Authority STEM Conference, April 17–18, Birmingham, UK.

Tanner, D., and L.N. Tanner. (1975). *Curriculum Development: Theory to Practice*. New York, NY: Harcourt Brace.

Uji, A.Z. (2001). "Beyond the Critiques of the Curriculum of Architectural Education in Nigeria." In *Architects and Architecture in Nigeria: A Tribute to Prof. E. A. Adeyemi*. U.O. Nkwogu, ed. pp. 109–22. Akure, Nigeria: Association of Architectural Educators in Nigeria.

United Nations Office for Disaster Risk Reduction (UNISDR). (2004). *Living with Risk: A Global Review of Disaster Reduction Initiatives.* New York, NY: UNISDR. Retrieved November 15, 2015, from http://www.unisdr.org/we/inform/publications/657

———. (2014). *Evidence-Based Investments in DRR.* Retrieved October 31, 2014, from http://www.unisdr.org/partners/academia-research

University of Maiduguri, the. (2014). *The Centre for Disaster Risk Management and Development Studies* [Website]. Retrieved October 31, 2014, from www.unimaid.edu.ng/root/Disaster_Management.html

Ward, R.C. (1978). *Floods: A Geographical Perspective.* London, UK: Macmillan.

Zevenbergen, C. (2007). *Adapting to Change: Towards Flood-Resilient Cities.* Inaugural Address, UNESCO-IHE Institute for Water Education, Delft, the Netherlands, December 14, 2007.

2 Crises and conflict-induced disasters in Nigeria

The challenges of curricular development on rebuilding devastated communities for sustainability

Zanzan Akaka Uji

Introduction

Elsewhere in much of the developed world, when the word "disaster" is mentioned, it usually conjures images of natural calamities such as earthquakes, hurricanes, typhoons, tsunamis, seismic earth movements, etc. Barring some subtle non-specific warnings from weather forecasts, these occurrences are usually sudden, unannounced, and unexpected, accompanied by varying degrees of unpredictable and devastating effects on both human beings and the land on which the victims live.

In several areas of Africa, especially Nigeria, some of the most well-known natural disasters are desertification, deforestation, erosion, and (recently) flooding. Except for the last one, the emergence of the others is subtle, surreptitious, gradual, and even usually anticipated, due to their being partly linked to human activities (which thus makes many of them both natural and man-made).

However, here in Nigeria, there are disasters that are purely the result of man-made activities and actions, whose effects are no less devastating and calamitous (involving the loss of hundreds of lives and a large amount of property, as well as the displacement of vast swathes of communities) than the natural disasters mentioned. Such activities and actions include:

- disputes over ownership of land, land resources, and boundary adjustments, and clashes between nomads and sedentary farmers over grazing and the destruction of farmers' crops by nomads' cattle;[1]
- political disagreements, usually arising from election malpractice and/or bad governance;
- communal, civil, and ethnic strife arising from disagreements over traditional ownership of or authority and rights over certain resources, city roads, markets, or ancestral lands;
- religious conflicts arising from alleged blasphemy of deities or perceived official favoritism of one religion over others by the government;
- reprisal attacks by one community against another in mob action arising from (e.g.) mass hysteria, resulting in the lynching of persons suspected of acts ranging from petty theft to bizarre accusations such as witchcraft, etc.

Unfortunately, the government's own idea of conflict management is still usually restricted solely to the restoration of law and order so as to allow the people to go about their normal duties without hindrance. The people are left on their own to rebuild (or, if they choose, to relocate from) their destroyed homes, perhaps with supplements of small grants that are insufficient to make a significant difference to the people affected. It is therefore our aim in this paper to go beyond mere conflict resolution and to suggest strategies for rebuilding devastated communities more formidably and more sustainably so that they can serve as self-regenerating bulwarks against any further possible future conflict-induced devastation.

Our ultimate goal, though, is that ideas for rebuilding communities after disasters, using sustainable principles and best practices on disaster mitigation, be introduced in the curricula used to train young people in institutions of higher learning, in relevant disciplines in the built environment, as well as some others in the social sciences, whose concern also involves a quest for a more sustainable living environment.

Communities' vulnerability to physical attacks in times of crisis

There are some obvious factors that predispose communities so easily to physical attacks whenever disputes degenerate into conflict. These factors include (though are not necessarily limited to): location of the communities, settlement patterns, lack of good access into (or out of) the communities, the materials used for the construction of dwellings, the quality of the physical environment, and quality of life of the people themselves. We shall briefly take a look at these factors.

The location of communities and settlements

For most of our rural communities, settlements are usually located in vulnerable areas such as remote isolated hinterlands, hillsides, swampy areas, or deep within forests, rendering the homes invisible and inaccessible by vehicular traffic, or around valleys that make the homes vulnerable to any form of attack. To worsen matters, instead of settling in larger clusters of home layouts with the characteristics of townships, the communities consist of small, nucleated pockets of homesteads, making each isolated

Figure 2.1 Isolated locations and settlement patterns as major factors of vulnerability.[2]

Figure 2.2 Typical village setting of scattered homesteads.[3]

homestead an easier target for invading enemies. In times of crisis, they suffer enormous property losses: homes, farms, and human lives. They usually return surreptitiously after the crisis and rebuild their homes the same way, thus remaining vulnerable to future rounds of crisis (which are bound to erupt again, sooner or later).

Materials used for building the residences

Most of the houses occupied by residents within the areas that are most prone to conflicts are made of unprocessed local materials – mud, reeds, and thatching of grass and leaves. Most of these materials are transient and highly vulnerable to fire hazards. Thus, in times of crisis, there is hardly any hope of any of them (especially the roofs made from those materials) effectively resisting or withstanding destruction.

Physical, social, and economic conditions of and within the living environment

Most residents are local subsistence farmers, using primary means of farming with hardly any opportunities to graduate to mechanized systems of secondary/tertiary farming. There are no good roads for the conveyance of their farm produce to the markets, no pipe-borne water, no electricity, no health facilities, and no schools or any other social services. This makes the quality of life generally very low.

The drudgery and drabness of rural life, as encapsulated here, tends to push the youth out into the cities in search of a better life, leaving mostly the elderly and children who are, naturally, not strong enough to withstand attackers with any effective force.

Even in urban areas, however, most of those communities that come under attack in times of crisis are the settlements occupied by the low-income earners and the very

Figure 2.3 Local transient building materials.[4]

Figure 2.4 A typical city slum settlement, with congestion and difficult access to homes as major factors of vulnerability.[5]

dregs of society; the low-density, high-income settlements and the so-called Government Reservation Areas (GRAs) may not even be aware of any conflicts until the crisis is over. The settlements that are attacked are usually informal settlements or squatter neighborhoods whose size, haphazard nature, and uncontrolled (planless) growth make them highly vulnerable to attack and destruction. The settlements are generally so congested and so crammed together in an apparently haphazard fashion dictated by the uneven terrain that vehicular or even pedestrian access to them is hardly ever available, making them that much more vulnerable to destruction by fire or any other mode of attack.

With this scenario of vulnerable villages and cities coming under attack in times of crisis, it is easy to see why the destruction of such communities, whenever a crisis occurs, is almost always a *fait accompli*.

The aftermath of crises and strategies for rebuilding devastated communities sustainably

In order to even hope to achieve true sustainability, the strategies for rebuilding such devastated communities must take the social, economic, and physical factors of the communities into consideration.

Social factors

People become psychologically attached to the place and the environment where they have lived for a long time. Friendships have been made, alliances formed, relationships fostered, etc., and it is only natural that, even after the place has been devastated by war or some other disaster, returning to it would be the first option for most people, even if that means moving back into the path of danger. This perhaps forms one of the major challenges in the quest to rebuild communities sustainably after a disaster.

However, where an alternative place for relocation can be readily found, with all the advantages that the former place lacked, including easier accessibility to the outside world, security, the possibility of easier access to infrastructure, amenities, and, above all, if the new place has the promise of reducing the people's vulnerability to attacks and destruction, the wisest thing to do would be for the people to embrace relocation as the first option of reprieve. Thus, by accepting relocation and reconfiguring homesteads into larger clusters with better networks of surfaced access roads, services, and security, much of their vulnerability to incessant attacks would be considerably reduced, if not entirely eliminated.

Physical factors

The first strategy for rebuilding communities after disaster strongly advocates that, where relocation is impossible for practical reasons, a new and more practical manner of settlement may be introduced to replace the scattered homesteads spread out randomly around and within the inaccessible hinterlands. For such communities, there will be the need to come together and rebuild a close-knit set of settlements with modern principles of planning. This way, it will even be easier to provide access roads as well as other amenities and infrastructure. A close-knit settlement will itself also constitute a good form of security and defense for the community against future predatory attacks.

In the urban, high-density, or low-income areas that may also come under attack in times of crisis, farmlands do not exist. Relocation is, therefore, rarely an option. Thus, ideally, strategies for rebuilding such communities after disaster will entail the following:

- design for deconstruction;
- use of industrialized building systems techniques;
- pulsation of architectural structures; and
- employment of intelligent buildings.

Design for deconstruction

According to Adebayo and Iweka,[6] the *design for deconstruction* is to design buildings to accommodate and facilitate future changes and the possibility of eventual dismantling (in part or wholly) that will permit the recovery of systems, components, and materials while losing as little of them as possible. To be most effective, (such) deconstruction will ensure that any material or component recovered afterward is in good enough condition to ensure that its economic value can be maximized and environmental impacts can be humanized through subsequent recovery and reuse (or repair, re-manufacture, and recycling, where necessary).[7] To be able to achieve this, such buildings must be designed with the following characteristics as a guide:

- designed joints must be simple and with a limited number of different material types and component sizes, and must be easily accessible;

- chemical joints must be avoided or minimized – bolted, screwed, and nailed connections should instead be the norm;
- mechanical, electrical, and plumbing (MEP) systems should be separated and also made easily accessible;
- the form of the structure should be as simple as possible, with many building systems and materials that are similar and laid out in regular repetitive patterns; and
- materials should be easily separable into reusable components – this is why mechanical fasteners are preferable to adhesives.

When fighting breaks out following some conflicts, the people are usually the prime targets, before any attention ever shifts to their property (or homes). Thus, where the people are able to repel the attacks effectively, their properties may suffer little damage. It is only when they cannot effectively defend themselves that they are killed, and then their properties are destroyed as "icing on the cake." However, by the time the attackers shift their attention to the properties of their victims, they are themselves in a hurry to move on, as quickly as they are able to, so as not to risk any reversals of fortune from unforeseen quarters. Consequently, there is not much time to completely demolish buildings. The most common mode of destruction is torching and burning. This generally affects mainly the roofs of buildings, although parts of the walls still suffer serious damage. However, though partially destroyed, large portions of walls of buildings in such devastated communities remain standing, and, if made of components that can be disassembled, could be potentially salvageable. This means that, with good planning and foresight in the design of buildings in crisis-prone communities, a lot less would be lost after a crisis and so rebuilding the communities would be made that much easier.[8]

It is proposed here that to achieve easier deconstruction, as shown by Pulaski et al.,[9] wall components may be made of industrialized building materials, with a variety of possibilities, as shall be shown presently.

Industrialized building systems and rebuilding for sustainability

Industrialized building systems (IBS) is a principle of design and construction of buildings where building elements are mostly prepared in special industrial enterprises or factories using machines and plants and carried to the construction sites where these elements are then assembled either wholly by machines or partly by manual labor.[10] One of the most effective approaches to quick production of buildings, it ensures reduction of the procurement period, manual labor, and the costs of building production, while increasing production quantity and achieving stability, strength, thermal and acoustic efficiency, fire-resistance, and resistance to climatic and environmental elements with simple methods of connecting, jointing, and assembly, with interfaces intentionally made easy for the possibility of future disassembly.[11]

What underscores the efficacy of these positive characteristics of industrialized building systems is the fact that they employ the common factors of planning and production, such as a unified modular system of measurements, unification, typification, standardization, and nominalization,[12] all of which are aimed at reducing the number of differences in component sizes, thereby achieving commonality and uniformity in types, standards, etc., in order to enhance and facilitate mass production and ease of assembly.

The materials used include: concrete (pre- and post-tensioned) prefabricated components, metals, timber, and composites made of many materials, ranging from aluminum, fiberglass, particle board, polystyrene, and galvanized wire mesh for the production of all components of buildings ready for on-site assembly.[13] The following are a few possibilities that may be considered as specific industrialized systems most suitable for this purpose.

Interlocking bricks and wall panels

Interlocking bricks can be produced in the same way that they are usually made for laying over a horizontal floor or other flat surface area. The interlocking bricks can be made of concrete mix, such as are made for pavement tiles, or using compressed earth, as is the norm for compressed earth bricks. These can then be used in an interlocking pattern to erect vertical building walls without the need for much application of concrete mortar for binding, thus providing greater opportunity for future deconstruction, should the need arise, for disassembly, with the prospect of faster retrieval of the wall components for reuse (see Figure 2.5).

Polystyrene wall panel system

Polystyrene wall panel systems (introduced in Nigeria by *Cubic Homes* Nigeria Ltd., around the year 2000) are meant to replace the traditional walling system made of sandcrete blocks, masonry, or bricks (sun-dried adobe or burnt bricks). It is a unique type of walling system that employs expanded galvanized steel mesh fastened by electro-welds on both sides of the panel, bonded together by transverse welds to give the panel a reinforced composite effect that covers every square meter of the panel. The meshed polystyrene panel is then sprayed with a thin film of concrete, so that it achieves its full structural capability. The panels are then erected using a very simple technique (requiring very little labor and a fraction of the time it would take to build a wall of the same size with the traditional sandcrete blocks or bricks). The erection is achieved by simply standing the panels end-to-end and fastening them to each other and to the starter bars in the floor using binding wire. They are then made plumb with a spirit level and temporary bracings are fixed before concrete mortar is sprayed on the required surfaces. Compared to a typical heavy prefabricated panel, a polystyrene panel is extremely lightweight, weighing between 4 and 10 kg/m². The panel is also versatile, strong, fire-resistant, insulated, easy to assemble, and inexpensive.

Figure 2.5 Models of interlocking wall panels, as proposed by the author.[14]

For ease of construction and disassembly, the polystyrene wall panel system can be designed with serrated edges that can interlock, as shown in the example of the interlocking blocks/brick systems illustrated in Figure 2.5 above.

Sand-filled plastic water bottle walls

An ingenious and imaginative local walling system has been developed by local builders at an outpost station called Yelwa, on the outskirts of Kaduna in northern Nigeria, where discarded empty plastic bottles that once contained table water have been filled and compacted with sand to form the main walling component, in the same way that bricks are normally used. A firm concrete foundation is first laid to ensure that the structure is firm and stable, then the sand is thoroughly sieved to remove small stones to ensure that maximum compacting capacity is achieved. The sand-filled bottles are then placed side-by-side, one on top of two below, and bound together with mud mortar. Each house (one bedroom, living room, kitchen, toilet, and bathroom) utilizes an estimated 7,800 plastic bottles. The characteristics of the sand-filled bottle "bricks" are as follows:

- it costs only about one-third of what a similar house made of normal sandcrete blocks would cost;
- compacted sand inside a plastic bottle is nearly 20 times stronger than bricks and can be used to build buildings of even up to two stories;
- It is ideal for the hot Nigerian climate because the sand insulates the room from the sun's heat, helping to keep the room temperature low; and
- because of the compacted sand, the walls are also bullet-proof, which will be a boost to the security situation in any crisis-ridden areas of the country.

With such systems as described here – depending on the fixing methods of components used for the various industrialized building systems – it is deconstruction, rather than destruction, of the building components that will occur in the event of disasters.

The pulsation of architectural structures

If buildings can be designed for possible deconstruction or disassembly, with the use of industrialized building systems, and where it is possible to reuse the disassembled components and materials, then it stands to reason that the distinct "sections" (for example, the walls) of such buildings constructed using such disassemblable components can themselves be made re-adjustable. In other words, portions of the buildings can be shifted and rearranged, possibly to create smaller or larger spaces within the same building if circumstances so demand. Buildings that possess characteristics of such adaptability in response to changing needs are said to be capable of pulsation in the face of functional load oscillations. Such buildings are said to be capable of providing opportunities for the reversible transformation of their spaces, volumes, and structures in response to oscillations of functional loads. This is what is referred to as the pulsation of architectural structures.[15]

The concept of pulsation demands that when there are oscillations of functional loads in time and space, the building structures may be physically made to respond by being reconfigured to accommodate the impact of such load oscillations. Oscillations, Diogu stresses, are necessary for pulsation to occur in structures.[16] Oscillation of functional loads implies the existence of minimum and maximum functional loads at peak and lowest periods of use respectively, and these require corresponding space parameter changes, hence the concept of "optimization." Optimization requires the determination of space parameters that will ensure successful performance of functions at both peak and lowest periods without "fallow" intervals and wastages of part or all of the architectural space.

The residential environment in Nigeria, for instance, apart from its main function of domestic shelter, may sometimes have to contend with functional load changes. Reception of guests during certain family and social ceremonies (births, deaths, social meetings, marriages, etc.) are some of the functional loads that may not have been fully taken into consideration at the design stage of the residential environment. Accommodating them therefore requires reasoned functional reconfigurations with transformational implications, requiring the residential environment to adjust to these additional functions. These can be reversed and the environment restored to its former state, upon withdrawal or reduction of the functional loads. The micro-residential system at the family level may also pulsate in response to additional loads placed on it. For instance, within the Nigerian context (and perhaps in most of Africa), visitors and extended family may arrive without prior notice to stay for an unspecified number of days. Tradition demands that such visitors have to be accommodated. When this happens, a functional load oscillation has occurred, constituting a demand for pulsation of the structure. Towards that end, therefore, the living room may serve as a bedroom and additional bed spaces may be created out of existing bedrooms for this period. When the "heat" is over, the space arrangements revert to their original uses.

Communal buildings that have the most dynamic and the highest incidence of functional oscillations are commercial, transport, audio-visual exhibition, and recreational building complexes. The design of such communal complexes challenges the creative ability of the designer to take into account all possibilities of functional load oscillations and the corresponding responses of possible future reversible transformation of the structures. For such buildings to have reasonable prospects of sustainability, the designer must take great care to take into account all the variables that may affect their functional oscillations and reversible transformations. (Please note that the modules of interlocking wall panels depicted in Figure 2.5 are also meant to support the pulsation of architectural structures as described here.)

The links between the concepts discussed and sustainability

As the aim in this paper is not only to work towards the rebuilding of devastated communities but to attain sustainability in the same regard, it is important to reiterate the importance of the concepts advanced in this paper for that purpose – Design for Deconstruction (DfD), Industrialized Building Systems (IBS), and Pulsation of Architectural Structures.

Although the term "sustainability" generically covers an energy- and ecologically conscious approach to the design of the built environment, our concern here is the sustainability of the physical structures of rebuilt devastated post-disaster communities. In particular, the focus is on the building materials used and how these support the realization of the concepts of DfD, IBS, and the pulsation of architectural structures in the face of functional load oscillations in order to build successful, sustainable post-disaster communities.

Some of the most common examples of sustainable materials (as described by Wikipedia)[17] include recycled denim, blown-in fiberglass insulation, sustainably harvested wood, grass, linoleum, wool, concrete (high- and ultra-high-performance roman self-heating concrete panels made from paper flakes), baked earth, clay, vermiculite, linen, sisal, sea-grass, cork, expanded clay grains, coconut, wood fiber plates, calcium, sandstone, and locally obtained stone, rock, and bamboo (which is one of the strongest and fastest-growing woody plants). Indeed, in areas where there is an abundance of timber and/or bamboo as well as unskilled labor (this is actually true of most rural African areas), the availability of simple machines and just a few skilled technicians allows the production of pre-cut and prefabricated small and linear timber/bamboo components such as posts, beams, plates, panels, etc., to be made by artisans at workshops or in small factories. These prefabricated elements, which may be made in modular panels for floors, walls, ceilings, or roof trusses, can be easily and cheaply transported to the desired site. They are light enough to be handled by two to four workers, and can be quickly assembled with simple machinery such as pulleys or non-mechanized cranes.[18]

Curricular development for rebuilding communities for sustainability after disaster

Teaching and training in all curricula of architectural education, almost everywhere, tends to hold on to the notion that buildings are meant to be permanent or at least to outlive the longest-living human – or worse, to have a lifespan beyond the normal course of living that may be contemplated by man. Because of this, curricula usually concentrate on ensuring that the mental element of permanence forms the fulcrum of students' understanding of structural stability of buildings. Because of this perception, reactions of befuddlement arise when, against all expectations, a building fails and collapses due to structural failure or aging (when its lifespan may have been reached), demolition to give way to some other development, or, in fact, disaster (natural or man-made), as is being considered here. On account of this, a subject such as DfD, as valuable as it is for sustainability, may appear rather surreal and far-fetched.

The curriculum envisioned here, therefore, while not necessarily undermining the necessity for structural stability, must still bring the principle of DfD into the mainstream of architectural training and all the advantages the profession stands to gain from a clearer understanding of its locus in the sustainability of the built environment.

By the same token, the value of IBS in architectural training has remained stymied due to the fact that, all along, the mere mention of pre-fabrication conjures in the mind images of heavy equipment and buildings larger than the human scale. For that reason, it is not very popular, and hence has remained useful mainly for building heavy industrial complexes such as factories, airports/hangars, shipping complexes, stadia, etc. Only through innovative approaches such as those envisaged here can a new way of thinking and a new image begin to be created of IBS on a human scale to make it an acceptable subject of discourse in curricula in institutions dealing with the building construction industry in Nigeria. The emphasis will be on innovative approaches to a clear understanding and utilization of associated concepts of typification, unification, standardization, modularity, and the modular coordination of units. With respect to how poor communities in danger of repeated attacks can benefit from such curricula, the emphasis will be on possibilities of teaching local skills to those people with the least skills to manufacture and assemble the building components. It will also focus on the development and production of appropriate machinery and equipment compatible with the production of building elements and components. This will underscore the essence of sustainability in the buildings that will potentially be produced from these components.

As for the pulsation of architectural structures, the closest studies have come to this concept is a mere exhortation to bear in mind the flexibility and adaptability of space and spatial organization at the design stage; much of this is again subsumed under the concern for aesthetics or the constraints of site conditions and other environmental elements. These constraints, however, play a role only to the extent that the consciousness of pulsation is not built into the evolution process of the entire planning scheme. The concern for the curriculum in question, therefore, will be how to evolve the most suitable schemes to implement the idea of pulsation, balancing it with the necessity of achieving privacy, flexibility, good circulation, etc., as may be deemed most appropriate for any given circumstance.

Summary and conclusion

This paper focused on strategies for carrying out the rebuilding process in communities that have been devastated by man-made disasters arising from conflicts and various forms of crises, eliminating the communities' vulnerability in the environment, improving the quality of the social, economic, and physical environment and, hence, the quality of life of the communities' residents. The strategies also focused on improving the prospects of the community's future generations through innovative strategies for rebuilding their homes to ensure that very little is lost during crises and a lot more is reclaimed from the rubble of the destroyed homes afterwards. Such innovative strategies comprise the application of concepts such as the *Design for Deconstruction*, the ability of structures to achieve *pulsation in response to functional load oscillations*, and innovative forms of *industrialized building systems*. The links between all these concepts were identified as the major hinge upon which school curricula could be developed to assure the local communities of eventual transmission of the knowledge to them through appropriate communal training in the various new techniques and thus ensure the attainment of sustainability in the built environment.

Sustainable redevelopment using these strategies, it was shown, would thus allow traumatized people to begin taking control of their future, ensuring that control is available to their future generations without any unnecessary anxiety regarding the prospects of other crises that may yet cause further future disasters.

Notes

1 Z.F. Zirra and U. Garba (2006), "Socio-Economic Dimension of Conflicts in the Benue Valley: An Overview of the Farmers-Nomads Conflict in Adamawa Central, Adamawa State of Nigeria," in *Conflicts in the Benue Valley*, T.T. Gyuse and A. Oga, eds., p. 147 (Makurdi, Nigeria: Benue State University Press).

2 Google Earth Image, GIS Laboratory, Department of Geography, University of Jos, Nigeria April, 2014.

3 Image produced by the author, December 2013. Reprinted by permission.

4 Ibid.
5 Left-hand, center images: Images produced by the author, March, 2014. Reprinted by permission. Right-hand image: Google Earth Map, GIS Laboratory, Department of Geography, University of Jos, Nigeria, March 2014.
6 K.A. Adebayo and A.C.O. Iweka (2009), "Sustainable Design and Construction of Buildings: The Concept of Design for Deconstruction," *Architects Registration Council of Nigeria Colloquium Proceedings*, April 25–29, Abuja, Nigeria, p. 149.
7 See C. Morgan and F. Stevenson (2005), *Design and Detailing for Deconstruction – SEDA Design Guides for Scotland No. 1* (Edinburgh, UK: Scottish Ecological Design Association), p. 4; M. Pulaski, C. Hewitt, M. Horman, and B. Guy (2004), "Design for Deconstruction," *Modern Steel Construction*, June, available at http://msc.aisc.org/modernsteel/archives/2004/june/, p. 2.
8 See A. Papakyriakou and L. Hopkinson (2012, September 18), "The Potential of Integrating Design for Deconstruction as a Waste Minimization Strategy into the Profession of the Architect," paper presented at the Second Network for Comfort and Energy Use in Buildings: People and Buildings, London, UK.
9 Pulaski et al. (2004), p. 7.
10 J.O. Diogu (2005), "Techno-Professional Base for Adopting Industrialized Building Systems in Nigeria: The Architectural Design Approach and Principles," in *Proceedings of the International Conference in Engineering*, p. 572. Lagos, Nigeria: ICE.
11 See also S.A.S. Zakaria, G. Brewer, and T. Gajendran (2012), "Contextual Factors in the Decision-Making of Industrialized Building System Technology," *World Academy of Science and Technology* 6(7), pp. 34–42; Pulaski et al. (2004), p. 48; K.A.M. Kamar, Z.A. Hamid, M.N.A. Azman, and M.S.S. Ahamad (2011), "Industrialized Building System (IBS): Revisiting Issues of Definition and Classification," *International Journal of Emerging Science* 1(2), p. 125.
12 T.P. Olson (2010), *Design for Deconstruction and Modularity in a Sustainable Built Environment* (Pullman, WA: Washington State University), p. 5.
13 H. Schreckenbach and J.G.K. Abankwa (1982), *Construction Technology for a Tropical Developing Country* (Eschborn, Germany: Deutsche Gesellschaft for Technische Zusammenarbeit GmbH).
14 Images produced by the author, March 2013. Reprinted by permission.
15 Diogu (2005), p. 31.
16 Ibid.
17 "Sustainable design," Wikipedia, retrieved December 20, 2015, from https://en.wikipedia.org/wiki/Sustainable_design
18 Schreckenbach and Abankwa (1982), p. 167.

Bibliography

Adamu, A. (2010, January 29). "Jos, Jang, and Genocide (I)." *Daily Trust Newspapers*. Available at http://allafrica.com/stories/201001290165.html
———. (2010, February 5). "Jos, Jang, and Genocide (II)." *Daily Trust Newspapers*. Available at http://allafrica.com/stories/201002050310.html
Adebayo, K.A. and A.C.O. Iweka. (2009). "Sustainable Design and Construction of Buildings: The Concept of Design for Deconstruction." *Architects Registration Council of Nigeria (ARCON) Colloquium Proceedings*, pp. 148–68. April 25–29, Abuja, Nigeria.
Agzaku, C.B. (2006). "The Kwalla-Tiv Ethnic Conflict in Qua-an-Pan Local Government Area of Plateau State." In *Conflicts in the Benue Valley*. T.T. Gyuse and O. Ajene, eds., pp. 81–82. Makurdi, Nigeria: Benue State University.
Ahire, P.T., (Ed). (1993). *The Tiv in Contemporary Nigeria*. Zaria, Nigeria: The Writers Organization.
Avav, T. (1993). "The Tiv and Their Neighbors." In *The Tiv in Contemporary Nigeria*. P.T. Ahire, ed. pp. 34–41. Zaria, Nigeria: The Writers Organization.
Ayua, I.A. (2006). "The Historic and Legal Roots of Conflicts in the Benue Valley." In *Conflicts in the Benue Valley*. T.T. Gyuse and A. Oga, eds. pp. 23–33. Makurdi, Nigeria: Benue State University.

Becker, W.S. and R.F. Stauffer. (1994). *Rebuilding for the Future: A Guide to Sustainable Redevelopment for Disaster-Affected Communities*. Washington, DC: US Department of Energy.

Cubic Homes. (2008). *The Future of Building Technology: Versatile Building System*. Abuja, Nigeria: Unending Experiences.

Diogu, J.O. (2005). "Techno-Professional Base for Adopting Industrialized Building Systems in Nigeria: The Architectural Design Approach and Principles." In *Proceedings of the International Conference in Engineering*, pp. 567–77. Lagos, Nigeria: ICE.

———. (2006). "Pulsating and Intelligent Buildings – The Space-Time Changes of the Architectural Environment." *Environmental Research and Development Journal* 1(1), pp. 31–38.

Dwyer, D.J. (1979). *People and Housing in Third World Cities: Perspectives on the Problem of Spontaneous Settlements*. London: Longman.

Ekeh, C. (1999). *Nigeria: Aguleri-Umuleri Conflict – The Theater of Traditional War*. Retrieved February 19, 2016, from www.connflictprevention.net

Gyuse, T.T. and A. Oga, (Eds). (2006). *Conflicts in the Benue Valley*. Makurdi, Nigeria: Benue State University Press.

Haruna, M. (2010). "Jang, the Media and the Genocide in Jos." *The Nation* 4(1), p. 286.

Institution of Engineering and Technology (IET). (2012). *Intelligent Buildings: Understanding and Managing the Security Risks*. IET Sector Insights, Version 2. Available at www.theiet.org/sectors

Iyo, J. (1993). "The Tiv Warrior Tradition: Historical Perspective." In *The Tiv in Contemporary Nigeria*. P.T. Ahire, ed., pp. 93–100. Zaria, Nigeria: The Writers' Organization.

Jibo, M. (2009). *Chieftaincy and Politics: The Tor Tiv in the Politics and Administration of Tivland*. Frankfurt, Germany: Peter Lang GmbH.

Kamar, K.A.M., Z.A. Hamid, M.N.A. Azman, and M.S.S. Ahamad. (2011). "Industrialized Building System (IBS): Revisiting Issues of Definition and Classification." *International Journal of Emerging Science* 1(2), pp. 120–32.

Ler, E.L. (2006). "Intelligent Building Automation System." A Dissertation Submitted to Faculty of Engineering and Surveying, University of Southern Queensland towards the Degree of Bachelor of Mechanical Engineering.

Lloyd, P. (1979). *Slums of Hope? Shanty Towns of the Third World*. Harmondsworth, UK: Penguin.

Morgan, C., and F. Stevenson. (2005). *Design and Detailing for Deconstruction – SEDA Design Guides for Scotland No. 1*. Edinburgh, UK: Scottish Ecological Design Association.

"Natural Disaster." *Wikipedia* [Website]. Retrieved November 17, 2015, from en.wikipedia.org/wiki/natural_disaster

Olson, T.P. (2010). *Design for Deconstruction and Modularity in a Sustainable Built Environment*. Pullman, WA: Washington State University.

Oluyemi-Kusa, D., F.O. Iheme, P.O. Opara, L. Obafemi, and J. Ochogwu, (Eds). (2008). *Strategic Conflict Assessment of Nigeria: Consolidated and Zonal Reports*. Abuja, Nigeria: Institute for Peace and Conflict Resolution.

Papakyriakou, A., and L. Hopkinson. (2012, September 18). "The Potential of Integrating Design for Deconstruction as a Waste Minimization Strategy into the Profession of the Architect." Paper presented at the Second Network for Comfort and Energy Use in Buildings: People and Buildings, London, UK. Available at http://www.nceub.org.uk/mc2012/mc12_papers.html

Piette, M.A., J. Granderson, M. Wetter, and S. Kiliccote. (2012). "Intelligent Building Energy Information and Control Systems for Low Energy Operations and Optimal Demand Response." In *IEEE Design and Test of Computers*. E.O. Lawrence and Berkeley National Laboratories, eds., pp. 1–11. Berkeley, CA: Ernest Orlando Lawrence Berkeley National Laboratory.

Powerwall Systems Ltd. (2002). *Technical Information Manual: Specialist Façade Systems*. Vancouver, Canada: Author.

Pulaski, M., C. Hewitt, M. Horman, and B. Guy. (2004). "Design for Deconstruction." *Modern Steel Construction*, June. Retrieved November 17, 2015, from http://msc.aisc.org/modernsteel/archives/2004/june/

Schreckenbach, H., and J.G.K. Abankwa. (1982). *Construction Technology for a Tropical Developing Country*. Eschborn, Germany: Deutsche Gesellschaft for Technische Zusammenarbeit (GTZ) GmbH.

Uji, Z.A. (2010). *Building Collapse in Nigeria and the Collapse of National Values*. Jos, Nigeria: Ichejum Publications.

Uji, Z.A., and M.M. Okonkwo. (2007). *Housing the Urban Poor in Nigeria: User Involvement in the Production Process*. Enugu, Nigeria: EDPCA Publications.

Umaru, I. (2006). "Exploring the Economic Underpinnings of the Toto Ethnic Conflicts in Nasarawa State of Nigeria: A Primer." In *Conflicts in the Benue Valley*. T.T. Gyuse and A. Oga, eds. pp. 35–46. Makurdi, Nigeria: Benue University Press.

Zakaria, S.A.S., G. Brewer, and T. Gajendran. (2012). "Contextual Factors in the Decision-Making of Industrialized Building System Technology." *World Academy of Science and Technology* 6(7), pp. 34–42.

Zirra, Z.F., and U. Garba. (2006). "Socio-Economic Dimension of Conflicts in the Benue Valley: An Overview of the Farmers-Nomads Conflict in Adamawa Central, Adamawa State of Nigeria." In *Conflicts in the Benue Valley*. T.T. Gyuse and O. Ajene, eds., pp. 27–39. Makurdi, Nigeria: Benue State University Press.

3 The vulnerability of rural households to climate hazards in the mountainous area of Kwande local government area, Benue state, Nigeria

Bernard Tarza Tyubee and Iankaa Aguse

Introduction

Mountain systems account for roughly 20 percent of the terrestrial surface area of the globe. They are found on all continents and are source regions for over 50 percent of the globe's rivers.[1] Mountains possess climatic, physical, ecological, and human characteristics that make them unique and sensitive ecosystems where occurrence of extreme events and natural catastrophes are exacerbated.[2] Mountains are high-risk areas, and in recent decades, they have become increasingly disaster-prone with a disproportionally high number of natural disasters.[3] Mountains in many parts of the world are susceptible to the impacts of a rapidly changing climate and provide interesting locations for the early detection and study of the signals of climate change and its impacts on hydrological, ecological, and social systems.[4] Mountain regions are the natural habitat of a diversity of plant and animal species, and provide ecological services to millions of households worldwide.[5]

Nigeria, though a low-lying country, has highlands and mountains along the eastern boundary with the Republic of Cameroon. The border, which extends over 2,000 km, stretches from the Saharan Desert in the north to the Atlantic Ocean in the south, and runs through five states: Borno, Adamawa, Taraba, Benue, and Cross Rivers. There are several mountains on the Nigerian side of the border, including Mandara, Atalantka, Gotel, Wanga, and Sonkwala, ranging in height from 1,200 to 2,000 m. The mountain region has been the home of many indigenous groups or "tribes" for several centuries. Recently, the mountain region has attracted major economic investments in tourism, agriculture, and hydropower generation. The interaction between these economic investments in the region and the increase in human activities can result in environmental deterioration, leading to an increase in hazards and the rapid loss of habitat and species diversity.[6]

Vulnerability, an important component in the disaster risk equation, is fundamental in disaster risk management and climate change adaptation. Without vulnerable people and property, disaster would pose no threat, even in the presence of hazardous events; if there are no hazardous events, then there can be no disaster even where there is a vulnerable population.[7] Vulnerability is crucial in environmental decision-making and policy-making, and a better understanding of the vulnerability of people and systems is key to the design of disaster preparedness and management plans, both of which are vital components of an adaptation strategy.[8]

Location and the physical and socio-economic characteristics of the study area

Benue State has the shortest boundary among the five states in Nigeria that share a boundary with the Republic of Cameroon. It covers an area of 34,000 km² and comprises 23 administrative divisions known as Local Government Areas (LGAs), including Kwande, located between latitudes 6° 30ᶦ and 7° 03ᶦ N and longitudes 9° and 9° 40ᶦ E (Figure 3.1). Kwande covers a total area of 2,300 km²; its total population has risen in recent decades from 180,327 people (1991 census) to 248,642 people (2006 census). Kwande shares boundaries with Katsina Ala LGA, Benue State, in the north, Ushongo LGA, Benue State, in the northwest, Taraba and Cross River States in the east and south, and the Republic of Cameroon in the southeast. It is characterized by highlands and hilly terrain, especially along its boundaries with Cross River State and Republic of Cameroon, where it reaches over 1000 m above mean sea level in many places. The dominant rock is basement complex, and lithosols is the main soil type.[9]

The montane vegetation comprises *Kyaya senegalensis* (mahogany) around the deeper lithosol at the foot of the mountains and along the river courses, and *Daniella oliveri* (soft timber) and *Anacardium occidentale* (wild cashew) on the middle and upper slopes. The area lies on the windward side of the rain-bearing southwest trade wind, and as a result experiences orographic rainfall from April to November. The mountains support the livelihoods of thousands of households of the Tiv tribe of Benue State. Arable crops such as *Dioscorea rotumdata* (yam), *Manihot esculenta* (cassava), *Sorghum bicolor* (Guinea corn), *Ipomoea batatas* (sweet potato), *Oryza sativa* (rice), *Musa sapientum* (banana), *Citrullus lunatus*(melon), *Vigna subterranean* (Bambara nut), and *Zea mays* (maize) are grown. Common tree crops include *Irvingia gabonensis* (bush mango), *Cola nitida* (kola nut), *Dacryodes edulis* (bush butter), *Magnifera indica* (mango), and *Elaeis quineensis* (oil palm). More recently, *Theobroma cacao* (cocoa) has been introduced in the area to supplement household income. Arable farming is characterized by small size, use of simple hand tools such as hoes and cutlasses, low capitalization, and low yield per hectare, and it often involves removal of the vegetation cover.[10] Mixed cropping is commonly practiced, whereby households cultivate more than one crop on the same piece of land close to their settlements. To exploit the environment, most of the Tiv households in the region live together as a family unit and share cooking utensils, usually in groups of two to five mud thatched houses with a rest house called *ate* in the middle.

On the night of October 17, 2010, a thunderous landslide occurred in eight locations in the mountainous area of Kwande LGA, Benue State, Nigeria. The accompanying sound was heard several kilometers away, causing panic and anxiety in the region. The thunderous sound was erroneously reported in the media the following day (October 18, 2010) as an "earthquake" or "volcanic eruption," which further heightened the panic and anxiety in the region. Several thousand tons of debris, comprising soil, rocks, and mud, were carried and deposited downslope. Several arable farms and tree crops were destroyed, and one person was declared missing during the disaster. This event was significant because it showed the region's level of exposure to impacts of climatic hazards and it underscored the need to assess the level of vulnerability to climate hazards of the households living in the area. Such assessment is essential to ensure disaster risk reduction and climate change adaptation.

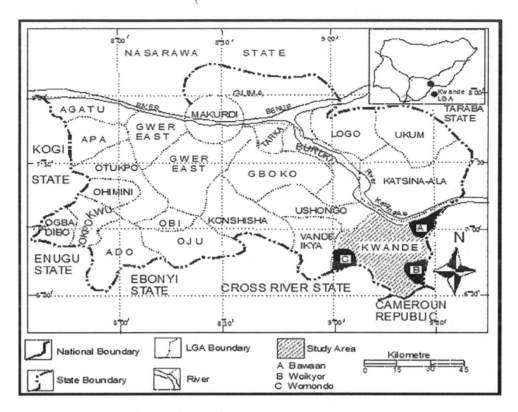

Figure 3.1 Location of the study area.[11]

This study is fundamental to disaster preparedness, response, and recovery as well as to environmental protection, as it focuses on vulnerable groups in vulnerable places or "hot spots." It is also important in prioritizing climate change adaptation strategies. The major objectives of the study are: (1) to investigate major climate hazards; (2) to assess the level of households' vulnerability to climate hazards; and (3) to determine strategies households can use to adapt to climate hazards.

Major climate hazards

In this study, climate hazards include both climate-related and climate-induced events that threaten life, property, and the ecosystem, which supports the livelihoods of households (and also varies among households). These include landslides, flooding, rockfalls, erosion, droughts, and storms. Among these hazards, only droughts and storms were defined for the understanding of the local people. A drought was defined as any year with a shorter duration of rainfall (less than seven months of rainfall) accompanied by a reduction in crop yields.[12] A storm was defined as strong winds (expressed in terms of when a whole tree is in motion, or a wind capable of removing tree branches or uprooting whole trees) lasting for at least 30 minutes, with or without rainfall.

Data on the major climate hazards were collected using focus group discussions (FGDs). FGDs were conducted with all household heads that were at least 50 years of age and had lived in the area for at least 30 years (as of 2010). The households used in the study were those living in the highlands or hilly parts of Kwande LGA, defined in the study as areas with a minimum height of 630 m above sea level. These highlands and hilly parts comprised Bawaan (A), Woikyor (B), and Womondo (C) areas, referred to as "zones" (Figure 3.1). A total of 30 household heads (10 for each zone) were selected for the FGDs. The major climate hazards were classified based on frequency of occurrence: yearly; every two to three years; every three to five years; and more than five years. The cost of damage (the mean cost in Nigerian Naira (NGN) was classified as: no cost, 1–100,000, 101,000–200,000, 201,000–3000,000; 301,000–400,000, and greater than 401,000. Each zone was to organize separate meetings with all the household heads and to record all responses for the zone in the checklist provided. A meeting was then organized between members of each Focus Group with the field workers, during which all responses were collated.

The results showed that the most frequent hazard events occurring yearly in Bawaan and Womondo are erosion, storms, and flooding; in Woikyor, erosion and storms are more common, whereas drought occurred less frequently in all three zones. The occurrence of landslides, however, varied from every three to five years in each of the three zones (Table 3.1). Moreover, in terms of cost of damage, the least expensive hazard is erosion, with zero cost, while the most expensive hazard is landslides, with a mean cost of damage of > 401,000 NGN per event in all three zones (Table 3.2).

Table 3.1 Major climate hazards according to frequency of occurrence

Climate hazards	*Frequency of occurrence (years)*			
	Yearly	*2–3*	*4–5*	*>5*
Bawaan				
Landslide		√		
Flooding		√		
Rock-fall		√		
Erosion	√			
Storms	√			
Drought				√
Woikyor				
Landslide			√	
Flooding	√			
Rock-fall				
Erosion	√			
Storms	√			
Drought				√
Womondo				
Landslide		√		
Flooding	√			
Rock-fall			√	
Erosion	√			
Storms		√		
Drought				√

Table 3.2 Major climate hazards based on the cost of damage

Climate hazard	Aggregate mean cost per event (NGN)					
	No cost	1–100,000	101,000–200,000	201,000–300,000	301,000–400,000	> 401,000
Bawaan						
Landslide						√
Flooding					√	
Rock-fall			√			
Erosion	√					
Storms			√			
Drought			√			
Woikyor						
Landslide						√
Flooding				√		
Rock-fall				√		
Erosion				√		
Storms		√				
Drought	√			√		
Womondo						
Landslide					√	
Flooding				√		
Rock-fall			√			
Erosion			√			
Storms			√			
Drought			√			

The frequency of erosion and flooding in the study area is directly related to the nature of the slope, agricultural activities, and deforestation. Observation suggests that in relation to the slope, the cultivation of crops on the upper slopes may encourage erosion in the region. A further explanation of the yearly occurrence of erosion is related to deforestation, which has removed the vegetation that protected against slope erosion. The high cost of damage due to landslides in Kwande LGA suggests that landslides are a major threat to households' livelihoods. The damage to property and to the ecosystem due to landslides may have a longer and a multiplier effect on the survival and wellbeing of the people. Apart from the economic loss, the households' resilience to future climate and non-climate risks and hazards is reduced, thus increasing their exposure and vulnerability to more hazard events. Efforts at disaster risk reduction in the mountain area of Kwande LGA should therefore be aimed at reducing the exposure of the vulnerable households to landslides.

The vulnerability of households to climate hazards

In the study, vulnerability is defined, from the perspective of climate change adaptation, as the degree to which a system is susceptible to – and unable to cope with – adverse effects of climate change, including climate variability and extremes.[13] Vulnerability is a function of exposure (E), sensitivity (S), and adaptive capacity (AC), and is a product of the interaction of socio-economic and environmental factors.

Nine vulnerability factors that showed significant variation among the households in the study area were selected for the study: slope and distance to river (E), household size, household income, occupation of household head, sources of household water supply (S), healthcare centers, roads, and schools (AD). Data on these nine vulnerability factors were collected from households in Kwande LGA. During a reconnaissance survey, a total of 101 households were identified in the three zones, comprising 28 households in Bawaan, 38 households in Woikyor, and 35 households in Womondo respectively. A table of random numbers was used to select 50 households for the study comprising 14 households in Bawaan, 19 households in Woikyor, and 17 households in Womondo, using the surnames of the household heads.

The following process was adopted in the vulnerability assessment. The nine vulnerability factors were first mapped on a 1–7 vulnerability scale ranging from 1 (very resilient), 2 (resilient), 3 (at risk), 4 (vulnerable), 5 (significantly vulnerable), 6 (very vulnerable), to 7 (extremely vulnerable). In order to obtain an approximate linearity of response for each factor, different response classes were then defined corresponding to different scoring of the factors' raw values on the 1–7 scale.[14] The design of response classes of the 1–7 vulnerability scale was guided by Kaly et al. and Tyubee et al., and was checked for quality control during the reconnaissance survey (see Table 3.6 in the Appendix).[15] None of the nine factors were weighted because they were presumed to have an equal contribution to households' vulnerability in the study. Data were analyzed using descriptive statistics and the aggregation method.

The result of the vulnerability of households to climate hazards is presented in Tables 3.3 and 3.4 and Figure 3.2. The major vulnerability factors in the study area are: household income, occupation, source of domestic water supply, healthcare amenities, roads and schools (with vulnerability scores above 6 being very vulnerable), while household size is the least vulnerable factor, having a vulnerability score of 4.2 (vulnerable). For the group factors, the vulnerability scores varied from 6.2 (adaptive capacity factors), 5.6 (sensitivity factors), to 4.9 (exposure factors) respectively (Table 3.3).

Table 3.3 Vulnerability scores of the major factors

Vulnerability factors	Mean scores
Exposure Factors	
Slope	5.0
Proximity to river	4.8
Mean	**4.9**
Sensitivity Factors	
Household size	4.2
Household income	6.1
Occupation of household head	6.1
Major source of domestic water supply	6.0
Mean	**5.6**
Adaptive capacity factors	
Healthcare center	6.3
Roads	6.2
Schools	6.2
Mean	**6.2**

This suggests that socio-economic factors make a higher contribution to households' vulnerability compared to the physical factors (Table 3.3). This result is not strange, particularly for households whose livelihoods depend mostly on the ecological services of the mountains. Their sustenance depends completely on the availability of vital resources such as forest products, water, and soil, rather than on the absence or presence of risks and hazards. The high vulnerability scores of the S and AD factors are related to the remoteness and inaccessibility of the area, which limits the provision of and access to infrastructure and services. In the absence of healthcare centers and access roads, disaster management in the study area will be greatly impeded. During the landslide of October 17, 2010, for instance, the affected communities complained of delays in relief and other support materials reaching them.

Variations in the level of vulnerability among the households (Figure 3.2) indicated that only four households (8 percent) had the highest vulnerability scores (> 6); 45 households (90 percent) had vulnerability scores between 5.0–5.9, and only one household (2 percent) had a vulnerability score of less than 5.0. Generally, 11 households (22 percent) and 39 households (78 percent) had vulnerability scores below and above the mean vulnerability score of 5.3 (Figure 3.2). This implies that households in all three zones are significantly exposed and vulnerable to climate hazards. The regional variation in households' vulnerability also showed a high vulnerability score (> 5.0) in Woikyor and Womondo, relative to Bawaan (< 5.0) (Table 3.4). This implies that the level of households' exposure to hazard events is higher in Woikyor and Womondo than in Bawaan. Bawaan's lower vulnerability score is related to accessibility by the Trunk C road and the presence of schools and healthcare amenities in the nearby town of Kashimbila, on the Katsina LGA and Taraba State side.

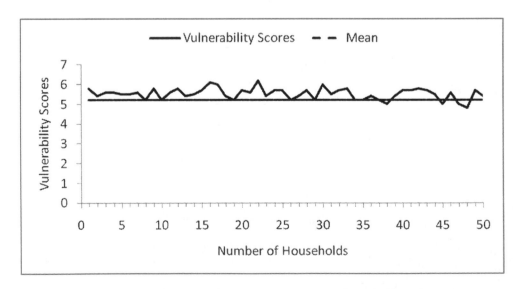

Figure 3.2 Variations in the level of vulnerability among the 50 households in the study area.

Table 3.4 Zonal variations in the levels of
vulnerability among households

Zone	Mean scores
Bawaan	4.8
Woikyor	5.6
Womondo	5.4
Regional mean	5.3

Strategies used by households to adapt to climate hazards

Data on the strategies households used to adapt to climate hazards were collected using questionnaires. Each of the 50 household heads was asked to choose one preferred strategy of adapting to climate hazards. The results showed that a total of 22 households (44 percent) identified changing their agricultural system as a major adaptation strategy, 14 households (28 percent) agreed to relocate downslope, 10 households (20 percent) indicated a change in their major source of livelihood, while four households (8 percent) opted for relocation upslope (see Table 3.5).

Table 3.5 Strategies used by households to adapt to climate hazards

Adaptation strategy	Frequency	Percentage
Relocation downslope	14	28
Relocation upslope	4	8
Change of major livelihood source	10	20
Change in agricultural system	22	44
Total	50	100

The willingness of the majority of the household heads to change their agricultural systems and to relocate downslope is important in disaster risk reduction and the general protection of the mountain ecosystems. A coordinated action plan can be developed by all the stakeholders whereby the people can be resettled from high-risk areas and the abandoned agricultural and settlement lands can be reforested. Sustainable agricultural practices and ecological conservation should also be encouraged in the region. This will reduce degradation and enhance the protective role of vegetation against climate and non-climate hazards. Observations have revealed that a majority of the households showed interest in terrace farming and other sustainable agricultural practices. However, the success of these broader adaptation strategies may depend on several factors. The cost associated with land acquisition and resettlements in Benue State is prohibitively high, and ecological conservation may require legislative backing, the process of which may be unduly delayed due to political interests. Other challenges, such as the disruption of livelihood sources and social networks of the households, will also play a major role in adapting to climate hazards in the mountain area of Kwande LGA.

Conclusion

The broad objectives of this study are to investigate the major climate hazards, to assess the level of vulnerability of households to climate hazards, and to determine the adaptation strategies of households to climate hazards in the mountain areas of Kwande LGA, Benue State, Nigeria. Flooding and erosion are the most frequent hazards, occurring yearly due to relief and agriculture practices. Landslides are the most hazardous event, based on cost of damage, with a mean cost of damage per event exceeding NGN 401,000. Households in the study area are significantly vulnerable (mean V = 5.3) to climate hazards, with socio-economic factors identified as the major drivers of households' vulnerability. The majority of households prefer change in the agricultural system and relocation down-slope as adaptation strategies to cope with the occurrence and impacts of the climate hazards.

The result of the study suggests that the vulnerability of households in the mountain region is driven by the availability of livelihood sources. The lack of healthcare centers and access roads in the region limits disaster response and recovery. Broader adaptation pathways in the region include resettlement, ecological conservation, and sustainable agricultural practices. However, their success depends on how the challenges associated with land acquisition, resettlement, environmental legislation, and disruption of livelihood sources and the social networks of households are resolved.

Appendix

Table 3.6 Linear scoring among the vulnerability classes

Vulnerability factors	Vulnerability classes						
	1	2	3	4	5	6	7
Exposure factors							
Slope (in degrees)	< 3	4–7	8–11	12–15	16–19	20–23	> 23
Proximity to river (m)	> 300	300–250	249–201	200–150	149–100	99–50	< 50
Sensitivity factors							
Household size	< 2	2–3	4–5	6–7	8–9	10–11	> 11
Household income (per year/1,000 naira)	>600	501–600	401–500	301–400	201–300	101–200	<100
Occupation (main)	Civil/public service	–	–	Trading/Shop owner	Carving/Carpentry	Arable/Tree crop farming	Fuel Wood gathering
Source of domestic water supply (main)	Pipe-borne Water	Bore hole	Hand-dug well	–	–	Aquifer	River
Adaptive capacity							
Healthcare centers	Teaching/general hospital	Local hospital	Clinic with maternity	Clinic	Pharmacy	Medical store/dispensary	None
Roads	Truck A	Truck B	Truck C	Motorable (tarred) road	Motorable (untarred) road	Major footpath	Minor footpath
Schools	Tertiary	Secondary (Govt.)	Secondary (community)	Secondary (private)	Primary (Govt.)	Primary (community)	None

Notes

1 Intergovernmental Panel on Climate Change (1995), "Impacts of Climate Change on Mountain Regions," in *The IPCC Second Assessment Report on Scientific-Technical Analysis of Impacts, Adaptation, and Mitigation of Climate Change*. M. Beniston and D.G. Fox, eds., pp. 199–213 (Cambridge, UK: CUP); M. Beniston (2003), "Climate Change in Mountain Regions: A Review of Possible Impacts," *Climate Change* 59, pp. 5–31.

2 IPCC (1995).

3 C. Marty (2009), "Natural Hazards and Risk in Mountains: The Potential Impacts of Climate Change," in *Mountains and Climate Change – From Understanding to Action*. T. Kohler and D. Maselli, eds., pp. 31–39 (Bern, Switzerland: Geographica Bernensia, with support from the Swiss Agency for Development and Cooperation).

4 Beniston (2003).

5 IPCC (1995).

6 Beniston (2003).

7 B. Wisner, P. Blaikie, T. Cannon, and I. Davis (2004), *At Risk: Natural Hazards, People's Vulnerability and Disasters*, 2nd edn. (London, UK: Routledge).

8 F. Villa and H. McLeod (2002), "Environmental Vulnerability Indicators for Environmental Planning and Decision-Making: Guidelines and Applications," *Environmental Management* 29(3), pp. 335–48; A. de Sherbinin, A. Schiller, and A. Pulsipher (2001), "The Vulnerability of Global Cities to Climate Change Hazards," *Environment and Urbanization* 19(1), pp. 39–64.

9 J.L. Nyagba (1995a), "The Geography of Benue State," in *Benue State, the Land of Great Potential: A Compendium*, D.I. Denga, ed., pp. 84–96 (Calabar, Nigeria: Rapid Educational).

10 J.L. Nyagba (1995b), "Soils and Agriculture in Benue State," in *Benue State, the Land of Great Potential: A Compendium*, D.I. Denga, ed., pp. 98–110 (Calabar, Nigeria: Rapid Educational); D.A. Surma (1995), "Agriculture in Benue State," in *Benue State, the Land of Great Potential: A Compendium*, D.I. Denga, ed., pp. 139–49 (Calabar, Nigeria: Rapid Educational).

11 Map produced by the Cartography Unit, Department of Geography, Benue State University, Makurdi. Reprinted with permission.

12 B.T. Tyubee (2006), "An Analysis of Food Crop Yields and Climate Relations in Benue State, Nigeria," *Journal of Nigerian Meteorological Society* 6(1), pp. 13–22.

13 IPCC (2007), "Summary for Policymakers," in *Climate Change 2007: Impacts, Adaptation and Vulnerability. Contribution of Working Group II to the Fourth Assessment Report of the Intergovernmental Panel on Climate Change*, M.L. Parry, O.F. Canziani, J.P. Palutikof, P.J. van der Linden, and C.E. Hanson, eds., pp. 7–22 (Cambridge, UK: CUP).

14 Villa and McLeod (2002).

15 U. Kaly, L. Briguglio, H. McLeod, S. Schmall, C. Pratt, and R. Pal. (1999), "Environmental Vulnerability Index (EVI) to Summarize Natural Environmental Vulnerability Profiles," SOPAC Technical Report No. 275, South Pacific Applied Geosciences Commission, Fiji; B.T. Tyubee, T. T. Gyuse, C.P.K. Basalirwa, and J.G.M. Majaliwa (2010). "An Assessment of the Level of Vulnerability to Climate Risks in a Developing and Unplanned Tropical City using the Environmental Vulnerability Index Model," in *Proceedings of the International Disaster and Risk Conference on Risks, Disasters, Crisis and Global Change: From Threats to Sustainable Opportunities*, pp. 759–62 (Davos, Switzerland).

Bibliography

Beniston, M. (2003). "Climate Change in Mountain Regions: A Review of Possible Impacts." *Climate Change* 59, pp. 5–31.

De Sherbinin, A., A. Schiller, and A. Pulsipher. (2001). "The Vulnerability of Global Cities to Climate Change Hazards." *Environment and Urbanization* 19(1), pp. 39–64.

Intergovernmental Panel on Climate Change. (1995). "Impacts of Climate Change on Mountain Regions." In *The IPCC Second Assessment Report on Scientific-Technical Analysis of*

Impacts, Adaptation, and Mitigation of Climate Change. M. Beniston and D.G. Fox, eds., pp. 199–213. Cambridge, UK: CUP.

———. (2007). "Summary for Policymakers." In *Climate Change 2007: Impacts, Adaptation and Vulnerability. Contribution of Working Group II to the Fourth Assessment Report of the Intergovernmental Panel on Climate Change*. M.L. Parry, O.F. Canziani, J.P. Palutikof, P.J. van der Linden, and C.E. Hanson, eds., pp. 7–22. Cambridge, UK: CUP.

Kaly, U., et al. (1999). "Environmental Vulnerability Index (EVI) to Summarize Natural Environmental Vulnerability Profiles." SOPAC Technical Report No. 275. South Pacific Applied Geosciences Commission, Fiji.

Marty, C. (2009). "Natural Hazards and Risk in Mountains: The Potential Impacts of Climate Change." In *Mountains and Climate Change – From Understanding to Action*. T. Kohler and D. Maselli, eds. pp. 31–39. Bern, Switzerland: Geographica Bernensia, with support from the Swiss Agency for Development and Cooperation (SDC).

Nyagba, J.L. (1995a). "The Geography of Benue State." In *Benue State, the Land of Great Potential: A Compendium*. D.I. Denga, ed. pp. 84–96. Calabar, Nigeria: Rapid Educational.

———. (1995b). "Soils and Agriculture in Benue State." In *Benue State, the Land of Great Potential: A Compendium*. D.I. Denga, ed. pp. 98–110. Calabar, Nigeria: Rapid Educational.

Surma, D.A. (1995). "Agriculture in Benue State." In *Benue State, the Land of Great Potential: A Compendium*. D.I. Denga, ed. pp. 139–49. Calabar, Nigeria: Rapid Educational.

Tyubee, B.T. (2006). "An Analysis of Food Crop Yields and Climate Relations in Benue State, Nigeria." *Journal of Nigerian Meteorological Society* 6(1), pp. 13–22.

———, et al. (2010). "An Assessment of the Level of Vulnerability to Climate Risks in a Developing and Unplanned Tropical City using the Environmental Vulnerability Index Model." In *Proceedings of the International Disaster and Risk Conference on Risks, Disasters, Crisis and Global Change: From Threats to Sustainable Opportunities*, pp. 759–62. Davos, Switzerland.

Villa, F., and H. McLeod. (2002). "Environmental Vulnerability Indicators for Environmental Planning and Decision-Making: Guidelines and Applications." *Environmental Management* 29(3), pp. 335–48.

Wisner, B., P. Blaikie, T. Cannon, and I. Davis. (2004). *At Risk: Natural Hazards, People's Vulnerability and Disasters*, 2nd edn. London, UK: Routledge.

4 A comparative analysis of annual maxima time series models along the lower Limpopo River basin of Mozambique

Daniel Maposa,[1] James J. Cochran, and 'Maseka Lesaoana

In this chapter, we discuss a comparative analysis of six annual maximum (AM) flood heights time series models at two sites: Chokwe and Sicacate in the lower Limpopo River basin of Mozambique. The six AM time series models considered were the annual daily maximum (AM1), annual two-day maximum (AM2), annual five-day maximum (AM5), annual seven-day maximum (AM7), annual ten-day maximum (AM10), and annual 30-day maximum (AM30). A generalized extreme value (GEV) distribution was fitted to each of the six AM time series models. The goodness-of-fit of the GEV distribution at each site was assessed using Anderson-Darling (A-D) and Kolmogorov-Smirnov (K-S) statistics. An analysis of variance (ANOVA) was performed to check for significant differences between the six AM models at each site with reference to skewness, coefficient of variation (CV), excess kurtosis, A-D statistics, and K-S statistics. The results revealed no evidence of significant differences at the 5 percent level of significance among the six AM time series models in terms of skewness, CV, excess kurtosis, and A-D and K-S statistics. A correlation analysis was also performed to check for significant correlations among the time series models: the results revealed high correlations among all six time series models at both sites. These findings suggest that any one of the models at each site can be used in place of the other five annual maximum time series models in flood frequency analyses of the lower Limpopo River. Without losing generality, the annual daily maximum flood heights time series model is used for further analysis to obtain flood frequency curves mainly because of its simplicity and relative ease of use.

Introduction

The recent International Disaster and Risk Conference (IDRC) held in Davos, Switzerland, August 24–28, 2014, emphasized the need for collaborative efforts in disaster risk reduction and building of resilient communities.[2] In his welcome speech, the IDRC Davos 2014 Chairman, Dr. Walter J. Ammann, pointed out that the scope, intensity, and complexity of risks of natural disasters such as floods, earthquakes, and forest fires have been rising in recent years.[3] In addition to this view, the Director-General of UNESCO, Irina Bokova, stated in Arya et al. that:

> Every year, more than 200 million people are affected by natural hazards, and the risks are increasing – especially in developing countries, where a single major disaster can set back healthy economic growth for years. As a result, approximately one trillion dollars have been lost in the last decade alone.[4]

According to Smakhtin, floods and droughts are the major causes of destruction with regard to crop damage and loss of human lives.[5] Floods and droughts account for about 90 percent of all people who are affected by all natural disasters.[6] The lower Limpopo River basin of Mozambique is characterized by extreme natural hazards, alternating between extreme floods and severe droughts.[7] Details of the hydrology of the lower Limpopo River basin are presented in Maposa et al.[8]

The purpose of this chapter is to perform a comparative analysis of the annual maximum sums of flood heights in the lower Limpopo River basin at each of the two sites, the Chokwe and Sicacate hydrometric stations. The annual maximums considered in this study are the daily (one-day), two-day, five-day, seven-day, ten-day, and 30-day annual maximums at each site. The assessment of the six annual maximum time series models was performed with the goal of investigating whether there are significant differences in characteristics among the six time series models at each of the two sites.

Recent advances in block maxima were derived in Ferreira and de Haan, and consistency of maximum likelihood estimators based on block maxima was proved in Dombry.[9] McMahon et al. studied the annual stream flow characteristics of a set of 1,221 global rivers distributed worldwide, including the Zambezi River in Zimbabwe, Southern Africa.[10] The annual flow features examined by McMahon et al. are the mean, variability, skewness, distribution type (gamma or log-normal), flow percentiles, and dependence.[11] The findings by McMahon et al. highlighted differences in annual stream flow characteristics between Australia-Southern Africa (ASA) and the rest of the world (ROW).[12] The approach used by McMahon et al. was quite different from the approach used in this chapter in a number of ways.[13] Firstly, the majority of the features examined in this chapter are different from those in McMahon et al.,[14] except for the skewness and coefficient of variation. Secondly, while in this chapter we compare the features among annual maximum time series models of the same river, McMahon et al.[15] compared the features among different rivers. Baratti et al. performed flood frequency distribution analyses at seasonal and annual scales.[16] While the approach used by Baratti et al.[17] can be useful and applicable in other regions, it may not be relevant in Southern Africa, particularly the Limpopo River, where there are absolutely no floods during the dry season, and all annual maximums belong to the rainy season.

In a study more similar to this chapter, Machiwal and Madan performed a comparative evaluation of 29 statistical tests used to detect hydrological time series characteristics by using them to analyze 46 years of annual rainfall and 47 years of one-day, two-day, three-day, four-day, five-day, and six-day maximum rainfalls at Kharagpur in India.[18] The tests revealed homogeneity in the seven rainfall series, and the time series plots revealed no evidence of trends for any of the seven rainfall series. Machiwal and Madan emphasized the evaluation of statistical tests, but a concluding warning was given against using too many statistical tests for the same objective in time series analyses because this increases the probability of committing a Type-I error – that is, incorrectly rejecting the null hypothesis when it is true.[19] Machiwal and Madan recommended using at least two statistical tests (but not too many) when making decisions about rejecting the null hypothesis.[20] Instead of using rainfall data, our study uses hydrometric data to make inferences about the distribution of extreme flood heights in the lower Limpopo River basin of Mozambique.

The rest of the chapter is arranged as follows: Section 2 presents the materials and methods used in the study, Section 3 presents the results and discussion of the findings,

Section 4 summarizes the value this chapter adds to disaster management and disaster risk reduction, and finally Section 5 presents the concluding remarks.

Materials and methods

In this section, we present the sequential steps used to obtain the block maxima data and briefly discuss the probability framework of the block maxima, descriptive measures of variability, and the goodness-of-fit tests for the GEV distribution.

Study sites and data

The data used in this study were obtained from the Mozambique National Directorate of Water (NAM), the authority responsible for water management in Mozambique under the Ministry of Public Works. The data are hydrometric (in meters) daily flood heights recorded at the sites Chokwe (1951–2010) and Sicacate (1952–2010), hydrometric stations for the lower Limpopo River of Mozambique.[21]

Moving sums and block maxima

The raw data in this study were recorded as daily flood heights for Chokwe and Sicacate hydrometric stations. Because our aim was to compare several annual maximum time series models, sequential steps were taken to obtain the two-day, five-day, seven-day, ten-day, and 30-day moving sums. Further sequential steps were taken to obtain the annual maximum flood heights series for each of the moving sums, including the daily flood heights series. Finally, the following annual maximum flood heights time series models were generated: AM1, AM2, AM5, AM7, AM10, and AM30.

The approach used to determine the annual maximum time series models is known as block maxima. The block maxima (or at-site) approach in flood frequency analysis is preferred to the peaks-over-threshold approach when the data records have sufficiently large sample sizes and the quality of the data is adequate.[22] The data used in this study had sufficiently large annual maxima records, extending to 60 years for the Chokwe hydrometric station and 59 years for the Sicacate. Naturally, in hydrology, when a sample size is sufficiently large, observations are grouped (blocked) by years.[23]

The six annual maximum time series models at both sites were assessed based on selected descriptive statistics measures, namely skewness, excess kurtosis, coefficient of variation (CV), and the goodness-of-fit of the GEV distribution, particularly the Anderson-Darling (A-D) and Kolmogorov-Smirnov (K-S) statistics. The main purpose of this assessment was to investigate whether there were significant differences between the six annual maxima time series models with respect to skewness, excess kurtosis, CV, and A-D and K-S statistics.

Time series plots, probability density plots, and boxplots were used in the visual assessment of the models. The GEV distribution was fitted to all the models and the A-D and K-S statistics were recorded. A one-way Analysis of Variance (ANOVA) was performed on the models at both sites using the six annual maxima time series models as treatments (Pretorius, 2007). A correlation analysis of the six annual maxima time series models was also performed at each site to check for significant correlations among the models.

Overview of the theoretical models

The detailed probability framework of the block maxima was derived in Dombry, Ferreira and de Haan, and Maposa et al.[24] The theoretical set-up of the block maxima was as follows. We considered independent and identically distributed (i.i.d.) random variables $(X_i)_{i\geq 1}$ with the common distribution function $F \in D(G_{\xi_0})$ and corresponding normalization sequences of constants $(a_m > 0)$ and (b_m), such that

$$\lim_{m\to\infty} F^m(a_m x + b_m) = F_\xi(x), \; x \in \Re. \tag{1}$$

The distribution function F in Eqn. [1] satisfies the extreme value condition with index ξ or, equivalently, F belongs to the domain of attraction of G_ξ.[25] According to Coles and Dombry[26] we divided the sequence of i.i.d. random variables $(X_i)_{i\geq 1}$ into blocks of length $m \geq 1$ and defined the kth block maximum by

$$M_{k,m} = \max(X_{(k-1)m+1}, ..., X_{km}), \; k \geq 1. \tag{2}$$

For a fixed $m \geq 1$, the variables $(M_{k,m})_{k\geq 1}$ are i.i.d. with the distribution function F^m and

$$P\left(\frac{M_{k,m} - b_m}{a_m}\right) \to G_{\xi_0}(x), \; \text{as } m \to +\infty \tag{3}[27]$$

The extreme value distribution with index ξ is given in Coles, Dombry, and Maposa et al.[28] as

$$G_\xi(x) = \exp\left(-(1 + \xi x)^{-1/\xi}\right), \xi \in \Re, 1 + \xi x > 0. \tag{4}$$

The closest approximation of the distribution of $M_{k,m}$ or extreme value distribution of F is the GEV distribution with parameters a_m, b_m and ξ_0, commonly estimated by the maximum likelihood method.[29] The GEV cumulative distribution function, G, is given in Eqn. [5] as:

$$G_{(\mu,\sigma,\xi)}(x) = \begin{cases} \exp\left(-\left(1 + \xi\dfrac{x-\mu}{\sigma}\right)^{-1/\xi}\right), 1 + \xi\dfrac{x-\mu}{\sigma} > 0, \xi \neq 0, \\ \exp\left(-\exp\left(-\dfrac{x-\mu}{\sigma}\right)\right), x \in \Re, \xi = 0, \end{cases} \tag{5}$$

where μ, σ and ξ in Eqn. [5] are the location, scale, and shape parameters, respectively, as in Maposa et al.[30]

   The goodness-of-fit of the GEV distribution in this chapter was assessed by the A-D and K-S goodness-of-fit tests, which indicate whether or not it is reasonable to assume that a random sample data comes from a specified distribution (in this case GEV) based on the following null and alternative hypotheses:[31]

   H_0: Sample data come from the specified distribution.
   H_1: Sample data do not come from the specified distribution.

Rejecting H_0 would imply that the specified distribution is not a good fit to the sample data. The A-D test is sensitive to the tails of the distribution, while the K-S test is sensitive to the center of the distribution.[32]

The Pearson's skewness and kurtosis coefficients are given respectively in Eqn. [6] as:

$$\gamma_1 = \frac{\sum_{i=1}^{n}(x-\mu)^3}{n\sigma^3} \text{ and } \gamma_2 = \frac{\sum_{i=1}^{n}(x-\mu)^4}{n\sigma^4} \tag{6}$$

Ricci argues that when the data are standardized, the distribution curves depend mainly on skewness and kurtosis measures.[33] Excess kurtosis refers to kurtosis above or below the normal value, which is taken to be 3.[34]

The coefficient of variation, commonly known as CV in statistics and probability theory, is a measure of relative variability or dispersion of data points relative to the mean in a data series. It is calculated using the formula in Eqn. [7]:

$$CV = \left(\frac{s}{\bar{x}}\right) \times 100\% \tag{7}$$

The CV is a very useful statistic in making comparisons of the extent of variation from one data series to another, even if the means of the series are very different, such as the case in this study.

Detailed information on one-way ANOVA and correlation coefficient was given in Pretorius.[35] The one-way ANOVA serves as an extension of the independent samples t-test when more than two groups (annual maximum time series models) are compared. The null hypothesis indicated that the annual maximum time series models do not differ with regard to their means, while the alternative hypothesis indicated that the annual maximum time series models differ with regard to their means assessed on skewness, excess kurtosis, CV, and A-D and K-S statistics as in Eqn. [8]:

$$H_0: \mu_1 = \mu_2 = ... = \mu_6 \text{ versus } H_1: \mu_1 \neq \mu_2 \neq ... \neq \mu_6. \tag{8}$$

The correlation coefficient between two quantitative variables, or Pearson's product-moment correlation, which is denoted by r, is given in Pretorius as in Eqn. [9]:[36]

$$r = \frac{n\left(\sum xy\right)-\left(\sum x\right)\left(\sum y\right)}{\sqrt{\left[n\left(\sum x^2\right)-\left(\sum x\right)^2\right]\left[n\left(\sum y^2\right)-\left(\sum y\right)^2\right]}}, \text{ where } -1 \leq r \leq 1 \tag{9}$$

A value of r close to 1 indicates a very strong positive correlation, while a value of r close to-1 indicates a very strong negative correlation, and a value of r closer to 0 indicates lack of correlation between the two variables.[37]

Results and discussion

In this section, we present, interpret, and discuss the results of our analysis. The R programming language was used to analyze the results presented in this chapter.[38]

Time series plots of the data

Figures 4.1–4.3 present the time series plots, probability density plots, and boxplots, respectively, for the Chokwe hydrometric station for the period 1951–2010. The annual maximum time series plots in Figure 4.1 exhibit similar variability with respect to trend, cyclic, and random variations, with a few exceptions to AM30 towards the end of the series. However, these minor differences between the six annual maximum time series plots appear to be marginal. All the annual maximum time series models show that the peak flood height occurred in the year 2000.

The probability density plots in Figure 4.2 and the boxplots in Figure 4.3 exhibit positive skewness, in general, for all six models. The probability density plots in Figure 4.2 for models AM1, AM2, AM7, AM10, and AM30 tend to exhibit bi-modality, while the AM5 probability density plot indicates a unimodal distribution. All the boxplots in Figure 4.3 exhibit one outlier, except for AM30, which shows two outliers.

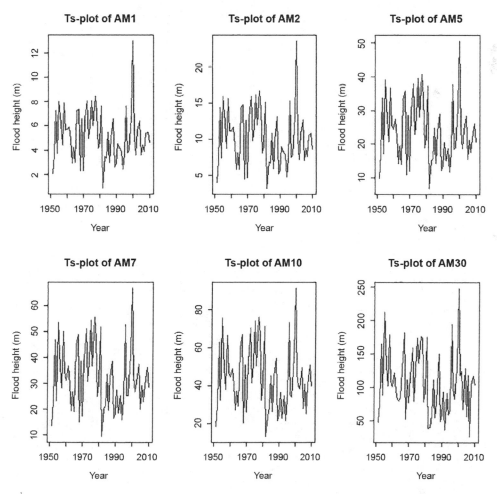

Figure 4.1 Time series plots of the AM1, AM2, AM5, AM7, AM10, and AM30 flood heights at Chokwe hydrometric station, 1951–2010 (respectively, from left to right).

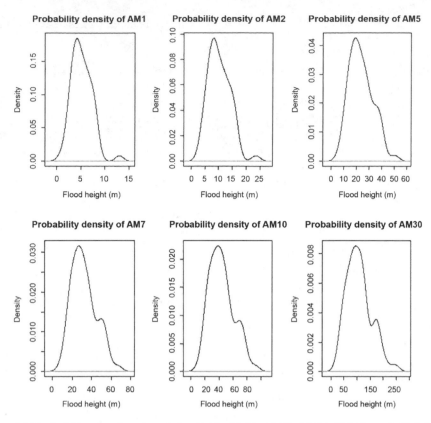

Figure 4.2 Probability density plot of the AM1, AM2, AM5, AM7, AM10, and AM30 flood heights at Chokwe hydrometric station, 1951–2010 (respectively, from left to right).

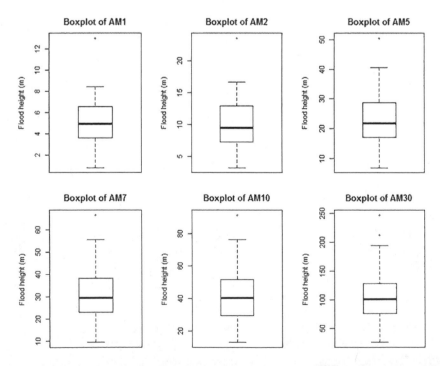

Figure 4.3 Boxplots of the AM1, AM2, AM5, AM7, AM10, and AM30 flood heights at Chokwe hydrometric station, 1951–2010 (respectively, from left to right).

Figures 4.4– 4.6 present the time series plots, probability density plots, and box-plots, respectively, for the Sicacate hydrometric station for the period 1952–2010. The time series plots in Figure 4.4 exhibit very similar variability with regard to trend, cyclic, and random variations. All the annual maximum time series models show that the peak annual maximum flood height occurred in the year 2000 at the site. This is consistent with the results at Chokwe, which is upstream along the same river.

The probability density plots in Figure 4.5 exhibit unimodality and left-skewness for all the models except for AM30, which tended to be symmetric or show slight positive skewness. These differences appear to be marginal. The boxplots in Figure 4.6 indicate negative skewness for all the models except AM30, which appears symmetric, coinciding with the results of the probability density plots.

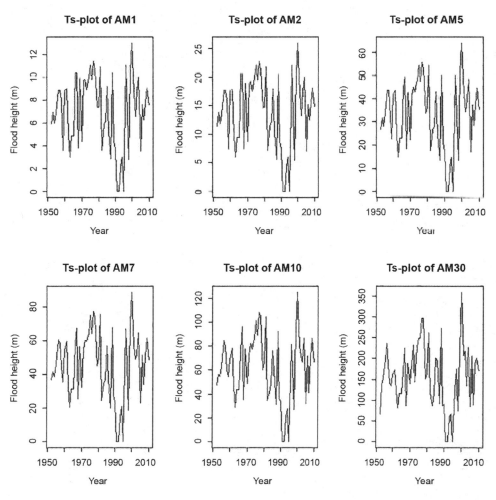

Figure 4.4 Time series plot of the AM1, AM2, AM5, AM7, AM10, and AM30 flood heights at Sicacate hydrometric station, 1952–2010 (respectively, from left to right).

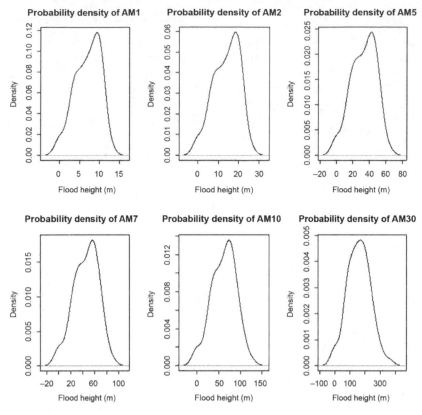

Figure 4.5 Probability density of the AM1, AM2, AM5, AM7, AM10, and AM30 flood heights at Sicacate hydrometric station, 1952–2010 (respectively, left to right).

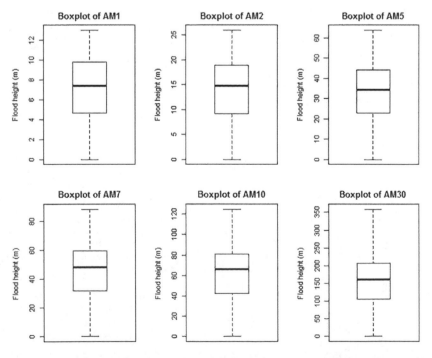

Figure 4.6 Boxplot of the AM1, AM2, AM5, AM7, AM10, and AM30 flood heights at Sicacate hydrometric station, 1952–2010 (respectively, left to right).

One-way ANOVA results

Tables 4.1 and 4.2 present the results of the descriptive statistics for each model and the one-way ANOVA p-values for Chokwe and Sicacate, respectively. The results in Table 4.1 show a one-way ANOVA p-value of 0.3836 (> 0.05), which indicates a lack of sufficient evidence to support the existence of significant differences among the six annual maximum time series models at a 5 percent level of significance. In other words, the six annual maximum time series models at Chokwe hydrometric station do not differ significantly with respect to skewness, excess kurtosis, CV, and A-D and K-S statistics.

Similarly, the results in Table 4.2 show a one-way ANOVA p-value of 0.9581 (>0.05), which suggests a lack of sufficient evidence at 5 percent level of significance to support the existence of significant differences among the six annual maximum time series models at Sicacate hydrometric station. The findings at Sicacate, further downstream the lower Limpopo River, are consistent with those at Chokwe in that the six annual maximum time series models at each site exhibit characteristics that are not significantly (statistically) different with regard to the dispersion measures of skewness, excess kurtosis, coefficient of variation, and the goodness-of-fit of the GEV distribution with respect to Anderson-Darling and Kolmogorov-Smirnov statistics.

Table 4.1 Summary descriptive statistics and one-way ANOVA for Chokwe, 1951–2010

Statistic	AM1	AM2	AM5	AM7	AM10	AM30
A-D	0.0500	0.0500	0.0500	0.0600	0.0600	0.0600
K-S	0.3938	0.2300	0.1800	0.2100	0.2200	0.1800
CV	0.2400	0.3890	0.3870	0.3920	0.3970	0.4450
Skewness	0.4063	0.6880	0.5830	0.5970	0.6030	0.6790
Excess K	1.9940	0.7580	−0.0680	−0.1960	−0.1950	0.2130

Key: p-value = 0.3836 (> 0.05) for the one-way ANOVA of six annual maximum time series models based on the five descriptive measures. A-D and K-S represent Anderson-Darling and Kolmogorov-Smirnov statistics (respectively), based on the fitted GEV distribution. CV represents coefficient of variation, and Excess K represents excess kurtosis.

Table 4.2 Summary descriptive statistics and one-way ANOVA for Sicacate, 1952–2010

Statistic	AM1	AM2	AM5	AM7	AM10	AM30
A-D	0.0913	0.0937	0.0850	0.0574	0.0475	0.0552
K-S	0.3938	0.3697	0.2898	0.2440	0.2162	0.2202
CV	0.4504	0.4508	0.4491	0.4461	0.4459	0.4737
Skewness	−0.4374	−0.4271	−0.3653	−0.3251	−0.2548	0.0820
Excess K	−0.5832	−0.5700	−0.4640	−0.3364	−0.1958	0.0670

Key: p-value = 0.9581 (> 0.05) for the one-way ANOVA of six annual maximum time series models based on the five descriptive measures. A-D and K-S represent Anderson-Darling and Kolmogorov-Smirnov statistics (respectively), based on the fitted GEV distribution. CV represents coefficient of variation, and Excess K represents excess kurtosis.

Pearson's correlation coefficient results

Tables 4.3 and 4.4 present results for the correlation matrices for Chokwe and Sicacate hydrometric stations, respectively. The p-values (< 0.001) at both Chokwe and Sicacate indicate that the correlations between all the models (variables) at both sites are highly significant. The results in both Tables 4.3 and 4.4 reveal strong positive correlations ($r > 0.87$) among the annual maximum time series models at both sites.[39] The strong correlations among the annual maximum time series models suggest that one annual maximum time series model at each site can be used to represent (or in place of) the rest of the annual maximum time series models in flood frequency analyses. In these circumstances, it is common to use the annual daily maximum flood height because it is convenient and easy to obtain.

Flood frequency curves of the annual daily maximum (AM1) models

Figures 4.7 and 4.8 present results for the cumulative distribution function of the GEV distribution at the Chokwe and Sicacate hydrometric stations, respectively. The cumulative distribution function gives the non-exceedance probabilities and their

Table 4.3 Correlation matrix of the annual maximum flood height time series models at Chokwe hydrometric station, 1951–2010

Annual time series models	AM1	AM2	AM5	AM7	AM10	AM30
AM1	1					
AM2	0.9967*	1				
AM5	0.9831*	0.9906*	1			
AM7	0.9732*	0.9821*	0.9980*	1	1	
AM10	0.9617*	0.9713*	0.9910*	0.9959*		
AM30	0.8958*	0.9058*	0.9288*	0.9401*	0.9584*	1

* means all the correlations are highly significant (p-value < 0.001) and are greater than a correlation coefficient of 0.89.

Table 4.4 Correlation matrix of the annual maximum flood height time series models at Sicacate hydrometric station, 1952–2010

Annual time series models	AM1	AM2	AM5	AM7	AM10	AM30
AM1	1					
AM2	0.9997*	1				
AM5	0.9925*	0.9942*	1			
AM7	0.9838*	0.9862*	0.9971*	1	1	
AM10	0.9641*	0.9672*	0.9845*	0.9938*		
AM30	0.8727*	0.8771*	0.9051*	0.9221*	0.9521*	1

* means all the correlations are highly significant (p-value < 0.001) and are greater than a correlation coefficient of 0.87.

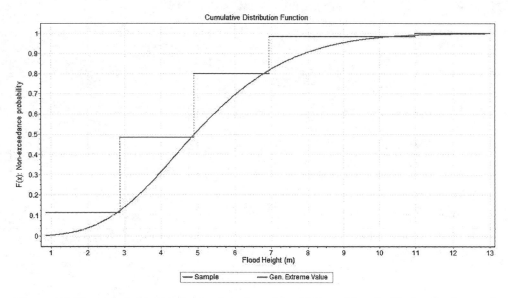

Figure 4.7 The cumulative distribution function of the AM1 model at the Chokwe hydrometric station, 1951–2010.

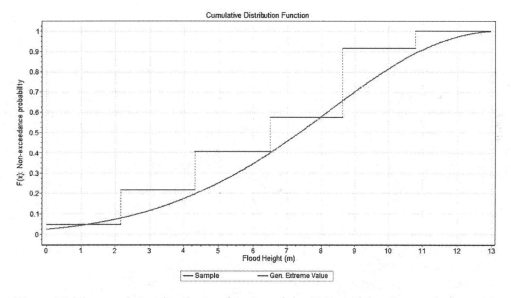

Figure 4.8 The cumulative distribution function of the AM1 model at Sicacate hydrometric station, 1952–2010.

corresponding flood heights (return levels) at each site, based on the maximum like-lihood estimation method. The non-exceedance probability is the probability that a given flood height will not be exceeded in a given return period. Given the non-exceedance probabilities, we can derive the exceedance probabilities by subtracting

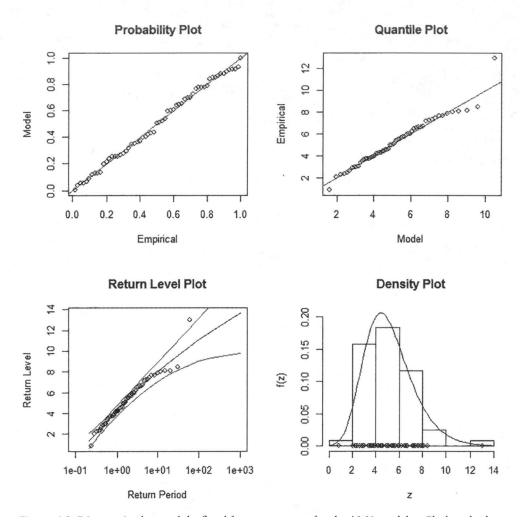

Figure 4.9 Diagnostic plots and the flood frequency curve for the AM1 model at Chokwe hydrometric station, 1951–2010.

non-exceedance probabilities from one, and the reciprocal of exceedance probabilities gives the return periods.[40]

Figures 4.9 and 4.10 present results for the diagnostic plots and flood frequency curves of the AM1 model at the Chokwe and Sicacate hydrometric stations, respectively. In the diagnostics plots, for instance, the probability plots at both sites indicate that the GEV is a good fit to the AM1 model at both sites. The flood frequency curves (return level plots) give return levels (flood heights) and their corresponding return periods. The flood frequency curves at both sites reveal that the 13-meter annual daily maximum flood height which occurred in the year 2000 at both Chokwe and Sicacate in the lower Limpopo River fell outside the 95 percent confidence limits. This indicates that the 13-meter annual daily maximum flood height, which occurred as a result of cyclone Eline and cyclone Gloria in 2000, was a relatively rare event.[41]

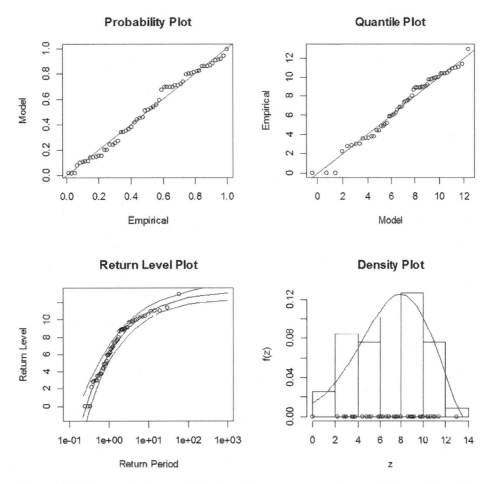

Figure 4.10 Diagnostic plots and the flood frequency curve for the AM1 model at Sicacate hydrometric station, 1952–2010.

Added value for disaster management and disaster risk reduction

The work in this chapter contributes to the decision-making process in disaster management and disaster risk reduction through addressing the gaps and challenges in reducing underlying risk factors. This work can also prepare vulnerable communities for effective response.[42]

While most studies of flood frequency analyses use rainfall (precipitation) and river discharges (flood volumes) data series, this study uses flood heights (water levels) data series. Quality data are not readily available for most countries, particularly for Southern Africa, where the gauging instruments may either be not functioning due to lack of service or simply unavailable due to budget constraints. This study has revealed that in countries where quality rainfall and river discharges data series records are scarce, flood heights can be used to make important decisions in disaster management and disaster risk reduction. Maposa et al. showed that the findings deduced from flood frequency analyses of flood heights data series are closely comparable to those deduced using rainfall or river discharges data series.[43]

Knowledge of the characteristics of annual maximum time series moving sums is crucial in flood frequency analysis. Accurate identification of suitable annual maxima time series moving sums to use in flood frequency analysis helps improve the accuracy of the forecasts of these rare events, such as the return levels and their corresponding return periods. The reader is referred to D. Maposa, J. J. Cochran, and M. Lesaoana (2014), "Investigating the Goodness-of-Fit of Ten Candidate Distributions and Estimating High Quantiles of Extreme Floods in the Lower Limpopo River Basin, Mozambique" and D. Maposa, J. J. Cochran, M. Lesaoana, and C. Sigauke (2014), "Estimating High Quantiles of Extreme Floods in the Lower Limpopo River of Mozambique Using [a] Model-Based Bayesian Approach" for further reading on the return periods of the lower Limpopo River.[44] Mozambique is a developing, flood-prone, and economically challenged country situated in Southern Africa, and therefore natural disaster-related studies in this country will indeed help in substantial reduction of flood events-related losses in human lives and damage to infrastructure such as bridges and other mega constructions in Mozambique.[45]

Improved knowledge of the estimates (forecasts) of the return periods of rare events such as extremely high water levels will help make the vulnerable communities better prepared and therefore respond effectively to the flood-related disasters.

In this section, we have demonstrated the worth and significance of forecasting, which can be applied everywhere in the world using the data available at a particular site. In summary, community preparedness and effective response to natural disasters such as floods will help reduce the associated risks and mitigate the deleterious impacts of these disasters on humans and property. Consequently, this will contribute to a substantial reduction in the amount of aid money spent on post-disaster relief operations.

Conclusion

In this chapter, we have considered the block maxima approach to the extreme value theory, also known as the at-site approach. The study was conducted at two sites, the Chokwe and Sicacate hydrometric stations in the lower Limpopo River basin of Mozambique in Southern Africa. Six annual maximum time series models were considered and compared with respect to skewness, excess kurtosis, coefficient of variation, and GEV goodness-of-fit as measured by Anderson-Darling and Kolmogorov-Smirnov test statistics. The six annual maximum time series models considered at each site were AM1, AM2, AM5, AM7, AM10, and AM30 – that is, annual daily, two-day, five-day, seven-day, ten-day, and 30-day maximum flood heights, respectively.

The findings in this study revealed no sufficient evidence of significant differences among the six annual maximum time series models at both sites with respect to skewness, excess kurtosis, coefficient of variation, and goodness-of-fit of the GEV distribution assessed by Anderson-Darling and Kolmogorov-Smirnov statistics. On the other hand, the study revealed overwhelming evidence of very strong positive correlations among the six annual maximum time series models at both sites.

The study also revealed notable differences between the two sites in overall skewness of the annual maximum time series models, with Chokwe (upstream) dominated by positive skewness and Sicacate (downstream) dominated by negative skewness.

Based on the findings in this study, it can be concluded that using the annual daily maximum flood heights model or any of the other five annual maximum moving sum models in flood frequency analysis has no significant effect on the forecasts of extreme return levels and their corresponding return periods. Without loss of generality, we have used the annual daily maximum flood heights model in flood frequency analysis to plot the flood frequency curves mainly due to its simplicity and the relative ease of obtaining it.

Authors' contribution

D. Maposa conducted the research and did the chapter write-up as part of his PhD thesis. J.J. Cochran and M. Lesaoana, as D. Maposa's PhD supervisors, suggested the comparative analysis of time series annual maxima moving sums, and the two co-authors also reviewed the draft manuscript to ensure that publisher requirements and publishable standards were met.

Acknowledgments

We thank the Mozambique National Directorate of Water (NAM) and Mr. Isac Filimone of NAM, in particular, who provided us with all the necessary data used in this study. We are also indebted to the United Nations Office for the Coordination of Humanitarian Affairs-Southern Africa (OCHA) for providing us with weekly update reports of floods in Southern Africa, particularly for the lower Limpopo River basin of Mozambique.

Notes

1 Please address any correspondence to Daniel Maposa (danmaposa@gmail.com).
2 M. Stal, S. Good, and W. Ammann, eds. (2014), "International Disaster and Risk Conference Davos 2014 Short Abstract Collection: Integrative Risk Management – The Role of Science, Technology, and Practice," in *IDRC Davos 2014 Programme and Short Abstracts*, pp. 81–251. Davos, Switzerland: IDRC.
3 World Meteorological Organization (2013), *Global Climate 2001–2010: A Decade of Climate Extremes – Summary Report*, WMO no. 119 (Geneva, Switzerland: Author); Stal et al. (2014).
4 I. Bokova, cited in A.S. Arya, T. Boen, and Y. Ishiyama (2014), *Guidelines for Earthquake-Resistant Non-Engineered Construction* (Paris, France: UNESCO), p. 6.
5 V. Smakhtin (2014), "Managing Floods and Droughts through Innovative Water Storage Solutions," in *IDRC Davos 2014 Programme and Short Abstracts*, M. Stal, S. Good, and W. Ammann, eds., p. 183 (Davos, Switzerland: IDRC).
6 WMO (2013); Munich Re, (2013, July 9), "Floods Dominate Natural Catastrophe Statistics in First Half of 2013," [Press Release] (Munich, Germany: Author); Smakhtin (2014).
7 Maposa, D., J.J. Cochran, M. Lesaoana, and C. Sigauke (2014b), "Estimating High Quantiles of Extreme Floods in the Lower Limpopo River of Mozambique Using [a] Model-Based Bayesian Approach," *Natural Hazards and Earth System Sciences Discussions* 2, pp. 5401–25.
8 Ibid.

  9 A. Ferreira and L. de Haan (2015), "On the Block Maxima Method in Extreme Value Theory: PWM Estimators," *The Annals of Statistics* 43(1), pp. 276–98; C. Dombry (2013), "Maximum Likelihood Estimators for the Extreme Value Index Based on the Block Maxima Method," *ArXiv.org* [Website], available at http://arxiv.org/pdf/1301.5611.pdf
 10 T.A. McMahon, R.M. Vogel, M.C. Peel, and G.G.S. Pegram (2007), "Global Streamflows – Part 1: Characteristics of Annual Streamflows," *Journal of Hydrology* 347, pp. 243–59.
 11 Ibid.
 12 Ibid.
 13 Ibid.
 14 Ibid.
 15 Ibid.
 16 E. Baratti, A. Montanari, A. Castellarin, J.L. Salinas., A. Viglione, and A. Bezzi (2012), "Estimating the Flood Frequency Distribution at Seasonal and Annual Time Scales," *Hydrology and Earth System Sciences* 16, pp. 4651–60.
 17 Ibid.
 18 D. Machiwal and K.J. Madan (2008), "Evaluation Comparative de Tests Statistiques Pour l'Analyse de Series Temporelles: Application á des Series Temporelles Hydrologiques," *Hydrological Sciences Journal* 53(2), pp. 353–66.
 19 Ibid.
 20 Ibid.
 21 D. Maposa, J.J. Cochran, and M. Lesaoana, (2014a), "Investigating the Goodness-of-Fit of Ten Candidate Distributions and Estimating High Quantiles of Extreme Floods in the Lower Limpopo River Basin, Mozambique," *Journal of Statistics and Management Systems* 17(3), pp. 265–83; Maposa et al. (2014b).
 22 Dombry (2013); Ferreira and de Haan (2015); Maposa et al. (2014a).
 23 Ferreira and de Haan (2015); Maposa et al. (2014b).
 24 Dombry (2013); Ferreira and de Haan (2015); Maposa et al. (2014b).
 25 R.A. Fisher and L.H.C. Tippett (1928), "Limiting Forms of Frequency Distribution of the Largest or Smallest Member of a Sample," *Cambridge Philosophical Society* 24, pp. 180–90; S. Coles (2001), *An Introduction to Statistical Modelling of Extreme Values* (London, UK: Springer); Dombry (2013).
 26 Coles (2001); Dombry (2013).
 27 Coles (2001); Dombry (2013).
 28 Coles (2001); Dombry (2013); Maposa et al. (2014b).
 29 Dombry (2013); Maposa et al. (2014b).
 30 Ibid.
 31 V. Ricci (2005), *Fitting Distributions with R*, available at http://www.researchgate.net/publication/228791072_Fitting_Distributions_with_R; Maposa et al. (2014a).
 32 Ibid.
 33 Ricci (2005).
 34 L.T. DeCarlo (1997), "On the Meaning and Use of Kurtosis," *Psychological Methods* 2(3), pp. 292–307.
 35 T.B. Pretorius (2007), *Inferential Data Analysis: Hypothesis Testing and Decision-Making* (Wandsbeck, South Africa: Reach).
 36 Ibid.
 37 Ibid.
 38 R Core Team (2013), *R: A Language and Environment for Statistical Computing* (Vienna, Austria: R Foundation for Statistical Computing).
 39 Pretorius (2007).
 40 Maposa et al. (2014a).
 41 Maposa et al. (2014a); Maposa et al. (2014b).
 42 World Conference on Disaster Reduction (2005), *Hyogo Framework for Action 2005–2015: Building the Resilience of Nations and Communities to Disasters*. January 18–22, 2005, Kobe, Hyogo, Japan; Maposa et al. (2014b).
 43 Maposa et al. (2014a)
 44 Ibid.; Maposa et al. (2014b).
 45 WMO (2013).

Bibliography

Arya, A.S., T. Boen, and Y. Ishiyama. (2014). *Guidelines for Earthquake-Resistant Non-Engineered Construction*. Paris, France: UNESCO.

Baratti, E., A. Montanari, A. Castellarin, J.L. Salinas, A. Viglione, and A. Bezzi. (2012). "Estimating the Flood Frequency Distribution at Seasonal and Annual Time Scales." *Hydrology and Earth System Sciences* 16, pp. 4651–60, doi:10.5194/hess-16-4651-2012

Coles, S. (2001). *An Introduction to Statistical Modelling of Extreme Values*. London, UK: Springer.

DeCarlo, L.T. (1997). "On the Meaning and Use of Kurtosis." *Psychological Methods* 2(3), pp. 292–307.

Dombry, C. (2013). "Maximum Likelihood Estimators for the Extreme Value Index Based on the Block Maxima Method." *ArXiv.org* [Website]. Retrieved March 10, 2015, from http://arxiv.org/pdf/1301.5611.pdf

Ferreira, A., and L. de Haan. (2015). "On the Block Maxima Method in Extreme Value Theory: PWM Estimators." *The Annals of Statistics* 43(1), pp. 276–98, doi:10.1214/14-AOS1280

Fisher, R.A., and L.H.C. Tippett. (1928). "Limiting Forms of Frequency Distribution of the Largest or Smallest Member of a Sample." *Cambridge Philosophical Society* 24, pp. 180–90.

Machiwal, D., and K.J. Madan. (2008). "Evaluation Comparative de Tests Statistiques Pour l'Analyse de Series Temporelles: Application á des Series Temporelles Hydrologiques." ["Comparative Evaluation of Statistical Tests for Time Series Analysis: Application to Hydrological Time Series."] *Hydrological Sciences Journal* 53(2), pp. 353–66, doi:10.1623/hysj.53.2.353

Maposa, D., J.J. Cochran, and M. Lesaoana. (2014a). "Investigating the Goodness-of-Fit of Ten Candidate Distributions and Estimating High Quantiles of Extreme Floods in the Lower Limpopo River Basin, Mozambique." *Journal of Statistics and Management Systems* 17(3), pp. 265–83, doi:10.1080/09720510.2014.927602

Maposa, D., J.J. Cochran, M. Lesaoana, and C. Sigauke. (2014b). "Estimating High Quantiles of Extreme Floods in the Lower Limpopo River of Mozambique Using [a] Model-Based Bayesian Approach." *Natural Hazards and Earth System Sciences Discussions* 2, pp. 5401–25, doi:10.5194/nhessd-2-5401-2014

McMahon, T.A., R.M. Vogel, M.C. Peel, and G.G.S. Pegram. (2007). "Global Streamflows – Part 1: Characteristics of Annual Streamflows." *Journal of Hydrology* 347, pp. 243–59.

Munich Re. (2013, July 9). "Floods Dominate Natural Catastrophe Statistics in First Half of 2013." [Press Release]. Munich, Germany: Author.

Pretorius, T.B. (2007). *Inferential Data Analysis: Hypothesis Testing and Decision-Making*. Wandsbeck, South Africa: Reach.

R Core Team. (2013). *R: A Language and Environment for Statistical Computing*. Vienna, Austria: R Foundation for Statistical Computing. Retrieved March 10, 2015, from http://www.R-project.org/

Ricci, V. (2005). *Fitting Distributions with R*. Retrieved November 18, 2015, from http://www.researchgate.net/publication/228791072_Fitting_Distributions_with_R

Smakhtin, V. (2014). "Managing Floods and Droughts through Innovative Water Storage Solutions." In *International Disaster and Risk Conference Davos 2014 Programme and Short Abstracts*. M. Stal, S. Good, and W. Ammann, eds., p. 183. Davos, Switzerland: International Disaster and Risk Conference. Retrieved March 10, 2015, from https://idrc.info/archive/idrc-davos-2014/outcomes/conference-proceedings/

Stal, M., S. Good, and W. Ammann, eds. (2014). "IDRC Davos 2014 Short Abstract Collection: Integrative Risk Management – The Role of Science, Technology, and Practice." In *International Disaster and Risk Conference Davos 2014 Programme and Short Abstracts*, pp. 81–251. Davos, Switzerland: IDRC. Retrieved March 10, 2015, from https://idrc.info/archive/idrc-davos-2014/outcomes/conference-proceedings/

World Conference on Disaster Reduction. (2005). *Hyogo Framework for Action 2005–2015: Building the Resilience of Nations and Communities to Disasters.* January 18–22, 2005, Kobe, Hyogo, Japan. Retrieved March 10, 2015, from http://www.unisdr.org/wcdr

World Meteorological Organization. (2013). *Global Climate 2001–2010: A Decade of Climate Extremes – Summary Report.* WMO no. 119. Geneva, Switzerland: Author.

Part 2
The Americas

Introduction

Adenrele Awotona

In order to prepare for, mitigate, and respond effectively to disasters, it is crucial to identify and delineate the demographics of the community. There are "communities of place, interest, belief and circumstance which can exist both geographically and virtually."[1] Disaster operations must take account of the multifarious needs of all these segments of the population. They must also enable emergency management officials to provide better for the needs of those with all types of disabilities (physical, medical, sensory, or cognitive) as well as for the needs of other community members who call for assistance during a disaster (including those without access to transportation, with limited proficiency in English – spoken, written, or read – those with regular medical treatment requirements, as well as those in need of assistance with daily life functions).[2] Reiter has estimated that between 30 and 45 percent of the total population of the US have access or functional needs during disasters.[3]

Thus, the complex nature of the community, the shortcomings in responding to those with certain needs in recent disasters in the US, and President Barack Obama's "Presidential Policy Directive 8 (PPD-8): National Preparedness,"[4] on March 30, 2011 were the main factors responsible for the development of the "Whole-Community" approach to emergency management by the US Department of Homeland Security's Federal Emergency Management Agency (FEMA). According to FEMA,

> Whole Community is a means by which residents, emergency management practitioners, organizational and community leaders and government officials can collectively understand and assess the needs of their respective communities and determine the best ways to organize and strengthen their assets, capacities and interests. By doing so, a more effective path to societal security and resilience is built. In a sense, Whole Community is a philosophical approach on how to think about conducting emergency management.[5]

Furthermore, a Whole Community approach should "engage the full capacity of the private and nonprofit sectors, including businesses, faith-based and disability organizations and the general public, in conjunction with the participation of local, tribal, state, territorial and Federal governmental partners."[6]

However, although the first principle that represents the foundation for establishing a Whole Community approach to emergency management stresses the importance of understanding and meeting the actual needs of the whole community,[7] Melissa Surette (Savilonis) notes in Chapter 5 that prisoners seem to be a forgotten, vulnerable subset of the population; they are often underrepresented, poor, or are members of marginalized groups when it comes to emergency management. Prisons are not prepared to respond to and recover from disasters, leaving prison populations – and the public – at risk. Surette proposes that prisoners require protection during disasters, as they do not have the capability or freedom to make independent decisions to protect themselves. In fact, she continues, pets have received far superior treatment and care than prisoners during disasters. With the majority of prisons across the country facing limited funding, staffing shortages, and a lack of resources, she observes that emergency management planning has fallen to the side. Without some form of federal oversight or guidance, correctional facilities across the country will remain unprepared to respond to and recover from disasters, failing in their duty to protect prisoners as well as the public. Consequently, this chapter focuses on why this is an issue, why this problem affects Americans, and what must be done to ensure that prisons are prepared. It also highlights examples from case studies, presents plausible scenarios, and provides interim recommendations that facilities can implement until federal action is taken.

Similarly, Chapter 6, by Seth H. Holmes, reports the results of a comparative analysis of pro-active disaster mitigation measures by another segment of the community: residents in apartment buildings in Boston, MA, and Washington, DC. Natural disasters such as hurricanes, earthquakes, and severe winter storms often produce power outages in urban and rural areas[8] that can last for days or weeks, similar to those that accompanied Hurricanes Sandy (2012) and Irene (2011) in the US northeast. During power outages, typical HVAC systems in buildings cannot function due to system fan and pump electrical requirements; consequently, these buildings cannot provide adequate indoor thermal comfort and risk losing their indoor livable conditions due to overheating or overcooling depending on the season. This chapter presents a building simulation method that evaluates prolonged blackouts and the associated indoor livable conditions for mid-rise residential building owners to compare various pro-active disaster mitigation measures related to passive heating and cooling (shading, envelope, natural ventilation, passive solar heating, etc.). Internal conditions are predicted and compared for generic mid-rise apartment buildings in Boston and Washington, DC, using a three-day blackout period for summer, equinox, and winter conditions. Holmes concludes that the results allow building owners (1) to understand potential indoor conditions under three-day blackout conditions and (2) to determine the associated risk ranges present when choosing between passive building design solutions aimed at increasing blackout livability.

Engagement and empowerment of all parts of the community, the second principle of the Whole Community approach, calls for public education that will better position stakeholders to be active participants in an emergency management team which is able to plan for and meet the actual needs of a community and strengthen the local capacity to deal with the consequences of all threats and hazards. This requires the integration of disaster studies into the curricula of colleges and universities. James Kaklamanos and Thomas C. Cross explore this theme in Chapter 7 through a case study that focuses on Hurricane Katrina and the Oso landslide, and offer suggestions as to how these disasters can be effectively incorporated into undergraduate civil engineering curricula to illustrate important engineering concepts.

Still on the case study of Hurricane Katrina, in Chapter 8 Fatima M. Alfa explores the social implications of disaster waste management on displaced citizens during the recovery phase from a natural disaster such as a hurricane. It also attempts to identify strategies for effective management of a disaster, as well as waste and citizen displacement, drawn from lessons learned from Hurricane Katrina. In this study, Alfa analyzes the community-wide impact that cut across a wide spectrum of infrastructures and services. She notes that there is a direct link between disaster waste management and displacement, reflected by the slowing of the cleanup effort by the dawdling return of residents and commerce. Consequently, her study shows that there is a cause-and-effect correlation between displacement and disaster waste management by providing multiple scenarios of delayed recovery efforts impacted by property owner absence.

The US Agency for International Development's Office of US Foreign Disaster Assistance has listed floods, tropical storms, earthquakes, landslides, forest fires, volcanic eruptions, and tsunamis as being amongst the most frequent natural hazards in Latin America and the Caribbean.[9] It expresses the view that climate change is expected to "alter [the] frequency, intensity and duration of climate extremes" while non-climatic stressors, such as "population growth, rapid and unplanned urbanization, environmental degradation, natural resource depletion, poverty, and limited opportunities for economic development, may exacerbate current vulnerability and potentially increase exposure to natural hazards."[10]

In Chapter 9, Alan March and Jorge León explore urban planning as a tool for human settlement disaster management in general and as a framework for tsunami evacuation safety in particular, through a case study of two Chilean cities. They demonstrate a method to establish and assess a series of urban design strategies to improve two key response activities: evacuation and safe assembly. Their method identifies critical evacuation areas according to a GIS-based analysis of hazardous conditions and vulnerability; generates a set of urban design interventions for improvement of evacuation and assembly; and assesses these interventions by developing an agent-based computer model which allows the evaluation of different evacuation scenarios according to established criteria. The method is applied to two tsunami-prone Chilean communities, Iquique and Talcahuano (the latter was severely affected by a tsunami in 2010), establishing a comparison between existing and modified scenarios. The preliminary results of March and Leon's study demonstrate two categories of feasible design interventions: modification of the urban networks' configuration (e.g., creation/extension of streets, construction of vertical evacuation facilities) and enhancement of evacuation qualities of routes and open shelter areas (to increase safety, accessibility, and mobility). Their analysis, which was developed in agent-based models, shows that these interventions can increase critical areas' expected evacuation rates by up to 30 percent.

As noted by the United Nations Development Program,[11] Brazil experienced more than 30,000 hazards from 1993 to 2015, an average of 1,363 each year. Of these incidents, 6,771 were floods, resulting in the deaths of 1,066 Brazilians. In 2014 alone, more than 77,000 Brazilians were impacted by a state of emergency that was declared in 120 towns that were badly affected by flooding. Hence, of all the natural disasters in Brazil, flooding poses the greatest danger to the communities. Ederson Franco, Deputy Chief of Civil Defense in the southern Brazilian state of Rio Grande do Sul, has been quoted as saying that "to prevent and to mitigate the risks of disasters together with the population is the main lesson we have learned in the past years."[12] In Chapter 10, Karen Da Costa and Paulina Pospieszna consider to what extent recently adopted laws and policies in Brazil facilitate the lessening of disaster risk in the light of the 2005 Hyogo Framework for Action and the role it assigned to laws and policies in tackling disaster risk reduction. Focusing on social housing initiatives aiming to address the long-standing housing deficit of the lower-income sector of the population, they reveal that Brazil has indeed adopted relevant legislation, policies, and programs with a view to reducing the risk of disasters. However, these measures have yet to be effectively implemented.

Notes

1 Federal Emergency Management Agency (2011), *A Whole Community Approach to Emergency Management: Principles, Themes, and Pathways for Action*, FDOC 104–008–1 / December (Washington, DC: Author).
2 A federal court recently decided that New York City did not do enough to protect the 889,219 individuals with disabilities in the city (11 percent of the city population) during 2012's Superstorm Sandy (comprising 183,651 individuals with a serious hearing difficulty, 210,903 with serious vision difficulties, and 535,840 individuals who have difficulty walking or climbing stairs). The court found that the city specifically violated the Americans with Disabilities Act (ADA) by not adequately protecting the vulnerable disabled population during that disaster – the first such ruling in the US. The court found that the city failed to provide reasonable accommodation to protect these citizens during and after Sandy to ensure that the blind, deaf, and physically disabled were able to get access to post-disaster services such as emergency shelters and transportation, which were available to the able-bodied of the nation's largest city.

> This is how the court described the situation facing the disabled during Superstorm Sandy:
> The question in this lawsuit . . . is whether in planning for, and responding to, emergencies and disasters, the City has adequately addressed the needs of people with disabilities – a segment of the population for which emergency planning is even more challenging . . . These Plaintiffs contend that the City's emergency preparedness program fails to accommodate their needs by, among other things, inadequately planning for the evacuation of people with disabilities, from multi-story buildings and generally; failing to provide a shelter system that is accessible within the meaning of the ADA; ignoring the unique needs of people with disabilities in the event of a power outage; failing to communicate adequately with people with special needs during an emergency; and failing to account for the needs of people with disabilities in recovery operations following a disaster.

The court found that the city violated the ADA when it failed to ensure that the disabled had access to these basic city services.

> Most significantly, the City's plans are inadequate to ensure that people with disabilities are able to evacuate before or during an emergency; they fail to provide sufficiently accessible shelters; and they do not sufficiently inform people with disabilities of the availability and location of accessible emergency services.
> Source: L. Milford (2014, January 23), "Court Finds NYC Disabled Not Adequately Protected After Sandy; Disaster Planning Must Include Vulnerable Populations," *The Huffington Post* [Website], available at http://www. huffingtonpost.com/lewis-milford/court-finds-nyc-disabled-_b_4255402.html

3 N. Reiter (2014, October 7), "Overcoming the Challenges of Protecting Vulnerable Populations During a Disaster," *Emergency Management* [Website], available at http://www.emergencymgmt. com/disaster/Overcoming-Challenges-Protecting-Vulnerable-Populations.html
4 According to FEMA,

> Recognizing that preparedness is a shared responsibility, Presidential Policy Directive / PPD-8: National Preparedness was signed by the President on March 30, 2011. At its core, PPD-8 requires the involvement of everyone – not just the government – in a systematic effort to keep the nation safe from harm and resilient when struck by hazards, such as natural disasters, acts of terrorism and pandemics. This policy directive calls on federal departments and agencies to work with the whole community to develop a

national preparedness goal and a series of frameworks and plans related to reaching the goal. PPD-8 is organized around six elements.

- The *National Preparedness Goal* states the ends we wish to achieve.
- The *National Preparedness System* describes the means to achieve the goal.
- *National Planning Frameworks* and Federal Interagency Operational Plans explain the delivery and how we use what we build.
- An annual *National Preparedness Report* documents the progress made toward achieving the goal.
- An ongoing national effort to build and sustain preparedness helps us maintain momentum.

In addition, a number of new guidance documents will help the general public, businesses and nonprofit organizations and all levels of government make the most of their preparedness activities.

> Source: FEMA (n.d.), *Learn About Presidential Policy Directive-8*, retrieved April 2, 2015, from https://www.fema.gov/learn-about-presidential-policy-directive-8

5 FEMA (2011).
6 Ibid.
7 Ibid.
8 Milford (2014) also reports on a legal class action on behalf of the entire disabled population in the city in which the court found that the City of New York violated the ADA when it failed to ensure that the disabled had access to basic city services. He states that among the

> many issues at trial was the problem of power outages. The lack of reliable electricity alone prevents the disabled from getting the protection of public services in a severe storm – from emergency shelters without power to stalled elevators in public housing to a lack of charging stations to power up wheelchairs and ventilators. On that issue, the court's finding could be an important legal precedent that might lead cities to provide more reliable power to serve vulnerable populations. And if they don't, they might well face lawsuits like this successful one in New York.

The court noted that "when electric power goes out, many disabled are severely affected":

> People with disabilities often depend on access to electricity. For example, some people depend on electricity to power life-sustaining equipment, such as ventilators. And people with mobility disabilities often rely on power wheelchairs or scooters that need to be recharged. In addition, some shelters are only accessible if the elevator is working, and thus if the shelter has power. For many people with disabilities, then, their ability to stay in a shelter depends upon the availability of electricity at the shelter . . . Another essential element that ensures people with certain disabilities are included in general population shelters is the ability to access power (when necessary via generators) for: charging power wheelchairs, scooters and other essential devices, and refrigerating certain medications.
>
> [The Court also noted the] guidance from the Federal Emergency Management Agency ("FEMA") stating that emergency plans "should include strategies to provide power for services that require a back-up power system in an emergency or disaster."

The court went on:

> The City's shelter plans do not include strategies to provide back-up power generators at shelters or to otherwise ensure that electricity will be available at shelters for those who depend on it. (Nothing in the City's written emergency plans "addresses the issue of providing power for people who use medical devices powered by electricity"). . . . During Hurricane Sandy, however, most evacuation centers lacked generators and some, therefore, at times, lacked power.

> Source: Milford (2014).

9 US Agency for International Development's Office of U.S. Foreign Disaster Assistance (2014), *Latin American and Caribbean Disaster Risk Reduction Plan, 2012–2014* (Washington, DC: Author).
10 Ibid.
11 United Nations Development Program (2015, February 26), "Brazil Strengthens the Skills of First Responders, UNDP Headquarters," *PreventionWeb* [Website], available at http://www.preventionweb.net/english/professional/news/v.php?id=42704
12 Ibid.

Bibliography

Federal Emergency Management Agency (FEMA). (2011). *A Whole Community Approach to Emergency Management: Principles, Themes, and Pathways for Action.* FDOC 104–008–1 / December. Washington, DC: Author.

Milford, L. (2014, January 23). "Court Finds NYC Disabled Not Adequately Protected after Sandy; Disaster Planning Must Include Vulnerable Populations." *The Huffington Post* [Website]. Available at http://www.huffingtonpost.com/lewis-milford/court-finds-nyc-disabled-_b_4255402.html

Reiter, N. (2014, October 7). "Overcoming the Challenges of Protecting Vulnerable Populations During a Disaster." *Emergency Management* [Website]. Available at http://www.emergencymgmt.com/disaster/Overcoming-Challenges-Protecting-Vulnerable-Populations.html

United Nations Development Program. (2015, February 26). "Brazil Strengthens the Skills of First Responders, UNDP Headquarters." *PreventionWeb* [Website]. Available at http://www.preventionweb.net/english/professional/news/v.php?id=42704

US Agency for International Development's Office of U.S. Foreign Disaster Assistance (USAID/OFDA). (2014). *Latin American and Caribbean Disaster Risk Reduction Plan, 2012–2014,* Washington, DC: Author.

5 Prisons and disasters in the USA

Understanding why prisons are unprepared to respond to and recover from disasters

Melissa A. Surette (Savilonis)

Introduction

Prisons[1] are not prepared to respond to and recover from disasters. Disasters affecting prisons can include, but are not limited to, disease outbreaks, earthquakes, floods, hurricanes, prison riots, power outages, wild fires, and exposure to hazardous materials. These hazards leave prison populations and the public at risk, as disasters affecting prisons can result in serious and unintended consequences.

There is a broad spectrum of concerns to be considered when responding to and recovering from disasters at prisons. These concerns are especially significant because many prisons throughout the nation house thousands of prisoners, which can make the response and recovery process much more challenging.

The number of individuals housed in correctional facilities throughout the nation is significant. In 2011, there were 1,500,000 prisoners held in over 3,300 Federal, State, and County jails throughout the United States.[2] The National Sheriffs' Association reports that in the United States, a jail is evacuated every six to seven weeks.[3] With over 3,300 correctional facilities geographically dispersed throughout the United States, some located in areas with a high probability of a natural disaster occurring, assuming that a facility will not be affected by a disaster is a costly assumption to make.

The implications of this issue are quite significant. Prisons are legally responsible for the welfare of prisoners and are required to uphold statutory and case laws that protect prisoners' rights. During disasters, prisons need to take action to protect prisoners from preventable harm, so that their rights are not violated. However, this is not being done. The reason for this lack of preparedness is that the Federal government has no policy or guidance in place that prisons can refer to when planning for emergencies.[4] Without some form of Federal oversight or enforcement, correctional facilities across the country will remain unprepared to respond to and recover from disasters, thus failing in their duty to protect prisoners as well as the public.

The recommendations by emergency management and law enforcement officials on how best to address this topic vary, as many recommendations have been formed on the basis of poor assumptions and are contrary to current statutory and case laws. One Federal official interviewed as part of this research felt that it should be written into law that prisoners are left in their cells during a disaster, regardless of the outcome.[5] Despite the varying attitudes of officials, the government is still responsible for reducing and mitigating hazards that threaten society, which includes protecting and preparing for those who are incarcerated. Securing our nation's prisons during disasters is a mission-critical task that must be addressed by the Federal government.

There have been a number of disasters where correctional facilities have been severely impacted, such as the Galveston County Correctional Facility during Hurricane Ike; the Ohio State Penitentiary, which suffered from a large fire; the Orleans Parish Prison during Hurricane Katrina; the Pennsylvania State Correctional Institute, which suffered from a serious prison riot; and the Vermont State Hospital during Hurricane Irene. These facilities suffered damage and encountered a wide range of problems in managing the prison population.

Several investigations and studies concluded that the facilities affected by these disasters were not prepared. Many had no emergency plans in place for responding to and recovering from a disaster; nor were prison officials properly trained to manage the incidents. For many facilities, the failure to plan created a disaster in itself: correctional facilities suffered staffing shortages, as staff were unable to (or chose not to) report to work. Prisons encountered legal issues when evacuating prisoners, especially across State borders, as records were destroyed and States were unable to determine how the sentencing would be upheld under other State laws and authorities, and notification to Consulate General Offices as required by Federal law failed to occur. Prisoners did not know why they were being held for prolonged periods of time when they had not even been charged with a crime. Others were forgotten about in the State jails they were evacuated to, because there was no system in place to track where prisoners were transported to, and the receiving jails did not obtain proper documentation (e.g., legal documentation, personal identification, family information, medical records) to go along with the prisoners they received. These issues resulted in some prisoners serving more time than their sentence required, resulting in several lawsuits filed by prisoners for false imprisonment.

Prisoners also suffered physical, emotional, and mental injuries. There were reports of sexual assaults occurring during prison evacuations. Others failed to receive adequate medical care. Prisoners escaped from their cells; some were never accounted for. The challenges these facilities encountered demonstrate what can go wrong in the absence of a Federal policy specific to securing prison populations during disasters.

There are several reasons for the inadequate response to disasters within prisons in the United States. The reasons for the inadequate response have been continuously repeated and not corrected. These include a lack of communication among the various agencies that play a role in protective action decision-making and emergency response; prison staff not being properly trained in disaster response and recovery; facilities not having emergency plans or continuity of operations plans in place, or plans that were easily accessible or recently updated; facilities not exercising (i.e., practicing) for a disaster; and a shortfall in resources like generators and pumps. While some prison staff should be held directly responsible for the actions taken during these disasters, prison systems, as a whole, have not been incorporated into emergency planning efforts, and therefore the failure rests with the government.

Hurricane Katrina serves as an excellent model for what can go wrong in the absence of planning at a correctional facility. Following the disaster, the Orleans Parish Prison evacuated over 6,500 prisoners. The emergency plans in place at the prison were ambiguous, inadequate, and impractical, and the decisions made by prison officials only exacerbated the problem.

As a result, several issues were encountered when responding to and recovering from the disaster affecting Orleans Parish Prison. Prisoners were abandoned in their cells by prison guards, some prisoners being left up to their necks in sewage-tainted water.

Those who were not able to escape were eventually evacuated days after the storm passed, only to be left on highway overpasses, many unrestrained and unsegregated. Prisoners normally segregated within the confines of the prison were housed together, where prisoner-on-prisoner violence erupted. Prior to the evacuation of the prison, over 600 prisoners managed to escape from Orleans Parish, 260 of whom were sexual offenders. Some of these have never been accounted for.[6] Furthermore, evacuees from Louisiana who were registered as sexual offenders or had warrants out for their arrest were evacuated to other States along with members of the public. For example, the receiving officials at an Air National Guard Base outside of Louisiana had no prior knowledge of evacuees' records. The Air Base received individuals with warrants out for their arrests as well as registered sexual offenders.[7] The failure to properly communicate the evacuee information created significant challenges, as the Air Base was unprepared to house individuals with criminal records separately from the public. Prisoners' records were also lost during the hurricane (e.g., legal documents, medical records, and prisoner identification were stored in basements which were flooded). Without records, many prisoners served more time than their sentences required.[8]

Prisoners are a vulnerable subset of our population,[9] often underrepresented, poor or members of a marginalized group.[10] Prisoners require protection during disasters, as they do not have the capability or freedom to make independent decisions to protect themselves. Prisoners, even though under the care and custody of the government, are still members of society. However, they seem to be a forgotten subset of our population when it comes to emergency management. Despite the fact that 95 percent of prisoners currently incarcerated will return to society,[11] our nation continues to treat prisoners as less than human. Pets have received far superior treatment and care than prisoners. During Hurricane Ike, prisoners were left in their cells at the County jail and suffered greatly, but the local animal shelters were evacuated far ahead of the storm. This inadequate treatment occurs because many officials do not understand the rights of prisoners, the statutory requirements that have been enacted to protect their rights, or the resources that have been developed to prepare for and recover from disasters.

Legal rights of prisoners

Case law and statutes have been enacted to protect the rights of individuals, and this includes those who are incarcerated. Prisoners do not retain all of their rights, although they do retain some, including the right to medical care, the right to freedom from racial discrimination, and the right to due process and access to the courts. Prisoners also retain the right to go before the courts (certain requirements must be first met) and the right to seek damages (monetary or an injunction), the right to equal protection, and the right to be free from cruel and unusual punishment. If prisons have not pre-identified circumstances where rights could be limited during a disaster, as well as methods for addressing these circumstances, prisoners' rights may unnecessarily be violated.

Constitutional and statutory requirements

Prisoners are granted rights under the Constitution. The courts have determined that certain prison conditions, as well as actions taken by prison staff, will lead to violations of these rights. Constitutional Amendments that ensure the protection of prisoners' rights include the Eighth and Fourteenth Amendments. The Eighth Amendment, as applied to prisoners, states that cruel and unusual punishment shall not be inflicted. Under this amendment, prisoners are required to be supplied with basic necessities, many of which serve as the foundation for human survival under Maslow's Hierarchy of Needs: these include food, shelter, medicine, medical care, security, and clean housing.[12] The Fourteenth Amendment as applied to prisoners prohibits correctional facilities and the government from sentencing prisoners without following due process. It also prohibits the deprivation of prisoners' rights even under extreme circumstances. Furthermore, it requires equal protection of the law and procedural fairness. Disasters are likely to produce a scenario where prisoners are deprived of their rights granted under the Eighth and Fourteenth Amendments.

The phrase "cruel and unusual punishment" is a significant statement that applies to this research. The term cruel and unusual was not defined at the time the Eighth Amendment was passed. However, the Supreme Court defined the term in 1972, in *Furman v. Georgia*. Justice Brennan, writing for the court, described cruel and unusual punishment as "degrading, inflicted in arbitrary fashion, clearly rejected throughout society, and patently unnecessary."[13] Prisoners are not to be subject to cruel and unusual punishment, and several cases demonstrate that Courts will – and have – ruled in favor of prisoners who have filed suits for violation of this right. Under the Eighth Amendment, prisoners have the right to be free from inhumane conditions, and during disasters correctional facilities must take action to ensure that minimum standards of living are upheld.

> The Constitution does not require a comfortable prison, but it requires a safe and humane prison. Prison officials must provide humane conditions of confinement; prison officials must ensure that inmates receive adequate food, clothing, shelter, and medical care, and must take reasonable measures to guarantee the safety of the inmates.[14]

Prisoners are required to have certain basic necessities, but when a disaster occurs at a correctional facility, those are necessities prisoners are likely to go without.[15] Despite the constitutional requirements that are in place, our society treats prison populations differently from other vulnerable populations during disasters, often depriving them of these rights. Prisoners are human beings and members of our society. They are mothers, fathers, sisters, brothers, children, and grandparents. While some of those who have been incarcerated have committed violent crimes, many have not.[16] Some have not even been convicted; of those who have, their convictions are often for drug offenses or public-order crimes, not violent crimes. In 2011, approximately 48 percent of the Federal prison population was incarcerated for drug offenses, while 35 percent were incarcerated for public-order crimes.[17] However, prisoners are often lumped together as a group of people who have committed horrific crimes, and as a result are seen as less than human. During Hurricane Ike, Galveston County evacuated the public but did not consider those housed in the jail: "the county decided to evacuate the island to alleviate the suffering of the people, but it didn't consider those in the jail people."[18] The actions taken during this and other disasters demonstrate the disregard society has for prisoners. However, society and the government have a shared responsibility to protect this vulnerable subset of our population.

Officials' opinions on how best to address this issue vary. Prisoners need protection, as they are still members of society. During discussions held with various emergency management, law enforcement, and corrections officials as part of this study, several suggested that cruel and unusual punishment should be allowed during disasters in an effort to manage prison populations. Some even suggested that the term "emergency" waives institutional standards – standards that are meant to protect prisoners and their rights.[19] However, these recommendations are patently contrary to current case and statutory laws. As Justice Stewart stated in *Furman v. Georgia*, "the Eighth and Fourteenth Amendments cannot tolerate the infliction of a sentence of death under legal systems that permit this unique penalty to be so wantonly and so freakishly imposed."[20] Despite desires otherwise, facilities cannot waive certain institutional standards based upon their own interpretation of the term "emergency." Using an emergency to allow cruel and unusual punishments would be to allow the penalty to be wantonly and freakishly imposed. Moreover, as Sharona Hoffman states in *Preparing for Disaster: Protecting the Most Vulnerable in Emergencies*, "[m]any complicated ethical decisions could in fact be avoided with appropriate response preparations."[21] Allowing officials to entertain these freakish recommendations as part of the dialogue on addressing this problem would be a major violation of civil rights, and is a major failure of our government. There are always alternatives, and certainly more ethical decisions that could be made regarding response preparations for prisons.

There are many cases where the courts have set rules regarding the minimal rights of prisoners. These rules require prisons to be responsible for actions taken and for staff to be aware of circumstances that could arise where rights are violated. Courts have required facilities to uphold minimum standards of living at all times, even during scenarios that are likely to unfold during disasters.[22] There are several cases before the courts that have been filed by prisoners for violation of their rights during disasters. Some of these cases have been unsuccessful or have been dismissed due to the rigorous requirements prisoners face when filing claims, including having to demonstrate physical injury or prove deliberate indifference, and the need to meet individual institutional standards.[23]

Federal, State, and local governments have a duty and obligation to protect prisoners by established rules. However, these policies, acts, and directives do not necessarily address emergency preparedness. Prison populations require specific guidance because they are considered a special case while they are under the custody and care of the government. Title 18 U.S.C. 921 §4002, charges the Bureau of Prisons with ensuring that Federal prisoners, under the care of another entity,[24] are provided with suitable living conditions, care, safekeeping, subsistence, and protection. The law, while specific to Federal prisoners, does require State and other non-Federal facilities to meet these standards.

The Federal government provides prisons with Federal tax dollars to supplement expenses incurred at State and County facilities (as well as private non-profit facilities) for housing Federal detainees. As part of receiving grants or subsides, these facilities are expected to adhere to Federal standards. However, a recent report published by the Office of the Inspector General found that many inspections of non-Federal facilities have revealed that these facilities do not meet the standards required for housing Federal detainees or criminals. This shortcoming is a failure of the individual facilities inspected, but more so of the Federal government for not ensuring standards are upheld and enforced. These facilities are supposed to provide "safe, secure, and humane conditions."[25] A report published by the Office of Inspector General, the *Audit of the U.S. Department of Justice's Oversight of Non-Federal Detention Facility Inspections*, stated that risk-based assessments were not incorporated into the inspections. Furthermore, the inadequacies identified during the inspections had no resolution; the facilities did not have to take any sort of corrective action. The United States Marshal's Office argued that the deficiencies were never corrected because the federal government "cannot tell state and local governments" to take corrective action.[26] The report stated that the Department of Justice responded to this finding by stating these inadequacies put the safety and security of Federal prisoners and detainees at risk. Under this law, facilities that contract with the Federal government must meet the same standards as Federal facilities.

Gaps in the law

Prisons have long been assumed to be self-sufficient and unlikely to suffer damage during disasters. As a result, prisons have not been included in emergency management efforts at the Federal, State, or County level, despite reports demonstrating the need for inclusion.[27] The responsibility for planning is placed on each individual facility. However, the Federal government offers no incentives for prisons to adopt and implement emergency management practices. There are Federal laws and policies specific to emergency management that can be applied to prisons, but many of these are ambiguous, arbitrary, broad, and do not clearly define the populations they protect or how the laws are to be implemented.

There are a few documents that prisons can reference when planning for a disaster affecting a prison.[28] However, many facilities are not even aware these documents exist, as they have not been made easily available and are buried deeply within the Internet. These documents are also not required or enforceable, and many are outdated. Despite these efforts, there is still no uniform Federal policy or set of standards specific to protecting prisons as whole communities.

Lack of integration

Prisons have not been included in emergency preparedness activities. This lack of integration during a time when the Federal government has placed emergency management at the forefront of government priorities is a major failure. Prisons have long been assumed to be self-sufficient and unlikely to suffer damage during disasters.[29] Furthermore, there are few prisoners' rights advocates within the Federal government. As Pamela R. Metzger stated in *Doing Katrina Time*, "[n]o one ever got elected by voting for more money for criminal defense."[30] A Federal official advocating additional protective measures for prisoners during disasters would be likely to receive a negative response and a lack of support both publicly and internally within the government.

The Bureau of Prisons (BOP), which falls under the Department of Justice, is responsible only for Federal prisons; it does not regulate the State, County, or local facilities that make up the majority of jails across the country. As the BOP does not have much authority outside the Federal prison system,[31] it is very difficult to integrate these other facilities in emergency planning efforts.

There is also a misconception that correctional facilities will be handled by another agency. An official interviewed as part of this research said, "nobody cares here.

Inmates are another agency's problem."[32] Which agency *is* responsible for addressing this problem has not been identified. No one at the Federal level seems to be able to identify whose responsibility it is to integrate prisons into the planning process. No agency wants to take responsibility for this problem, and no agency recognizes it as an issue facing society.

Federal policy that dissuades planning

Current Federal policies actually deter prisons from implementing emergency management practices. The most disconcerting policy that deters prisons is the Prison Litigation Reform Act (PLRA), passed by Congress in 1996 to address how prisoners file civil suits in federal court.[33] The Act, which was inserted into an appropriations bill, was passed and signed by President Clinton in an effort to reduce the number of frivolous lawsuits filed by prisoners. Since the PLRA has been in effect, the number of cases has dropped to 11 per 1,000 prisoners.[34] The significant decrease in the number of cases filed does *not* mean that the PLRA has been effective; in fact, it has created such a rigorous process for prisoners filing claims that many legitimate claims are never heard – claims that have a striking similarity to ones that could or have occurred during disasters.

The PLRA requires prisoners to take certain steps before a suit can be filed and sets certain criteria and standards that must be met for a suit to be considered lawful. The PLRA requires facilities to establish individual administrative grievance processes that prisoners must adhere to before filing suits in federal court. This exhaustion process is so demanding that even cases most capable of succeeding will not get resolved by the individual facilities or heard by the courts. Under the PLRA, a prisoner who suffered a physical, emotional, or mental health injury during a disaster requires judicial review:

> Thus by cutting off judicial review based on an inmate's failure to comply with his prison's own internal, administrative rules – regardless of the merits of the claim – the PLRA exhaustion requirement undermines external accountability. Still more perversely, it actually undermines internal accountability, as well, by encouraging prisons to come up with high procedural hurdles, and to refuse to consider the merits of serious grievances, in order to best preserve a defense of non-exhaustion.[35]

The administrative process is not standardized across the nation, which means that prisoners are subject to the standards set by each facility or jurisdiction. The period for which a prisoner files a complaint varies: some facilities have a deadline of only a few days from which the act occurred. If prisoners are displaced, as seen during Hurricane Katrina, they may not be able to meet the deadline for filing a complaint. The Courts have held that, regardless of the claim, prisoners *must* meet the internal deadlines for the facilities.[36]

As Schlanger said before the Subcommittee on Crime, Terrorism, and Homeland Security in 2007: "Can anyone reasonably expect a governmental agency to resist this kind of incentive to avoid merits consideration of grievances?"[37] If correctional facilities do not establish emergency management procedures and have established a rigorous exhaustion process, they will not be held accountable by the Federal government for their actions or inactions during disasters, because the complaints prisoners file are very unlikely to make it to the Courts. If the complaints do make it past the exhaustion process, prisoners will still have a number of obstacles to overcome under the PLRA. Prisoners filing suits in Federal court must meet all of the requirements set by the PLRA, including the physical injury requirement.

The physical injury provision presents a huge hurdle for prisoners filing claims following disasters. First and foremost, the courts have not determined what constitutes a physical injury violation. As Giovanna Shay clearly demonstrates in her article *Preserving the Rule of Law in America's Jails and Prisons: The Case for Amending the Prison Litigation Reform Act*, the courts have held even "serious physical symptoms insufficient to allow the award of damages because of the PLRA's physical injury provision."[38] Even sexual assault may not be seen as a physical injury violation by the courts.[39] It remains unclear how the courts will determine what constitutes a physical injury violation during a disaster, because the term "physical injury" has yet to be defined.

Secondly, prisoners cannot file a claim for damages for a mental or emotional injury that occurred during a disaster unless they can demonstrate that they also suffered a physical injury. During disasters, emotional and mental health injuries are common, so much so that the Department of Health and Human Services, as well as the American Red Cross, have resources and teams they deploy to assist individuals mentally or emotionally affected by a disaster.[40] Several reports show that emotional and mental health injury is common during disasters. Disasters can cause individuals to go into shock, leaving them in "an overemotional state that often includes high levels of anxiety, guilt or depression" and "a loss of faith."[41] According to the ACLU, the courts are split over the argument that a mental or emotional health injury in the absence of a physical injury is a constitutional rights violation.[42]

Other obstacles

Prisoners face several other obstacles when bringing a claim against an institution. These hurdles include social obstacles (e.g., lack of funding required to file a claim, representation), proving deliberate indifference, Federal statutes, and State laws. Prisoners do not always have the monetary resources to file suit, and organizations that represent prisoners are very selective when it comes to litigation, as they are often small non-profit organizations whose resources are also limited. Additionally, those prisoners who do not need to meet the requirements of PLRA (those filing in state courts, those who have since been released, those who have not been convicted) face many of the same obstacles under PLRA. Prisoners must show that there was deliberate indifference by the officials against whom the lawsuit is filed (i.e., the official willingly and knowingly caused harm to a prisoner). Prisoners must overcome 42 U.S.C Section 1983, also known as the Qualified Immunity Defense Law, which protects officials from being held liable for the violation of prisoners' constitutional rights. In other words, a prisoner cannot sue an official unless there is deliberate indifference, which can be quite difficult to prove.

Conclusion

Prisons are not only susceptible to disasters, but unprepared to respond to and recover from them. This problem is a Federal issue that requires Federal attention. Every institution is at risk if the Federal government does not address this issue. Correctional facilities, like other governmental entities, are working on restricted budgets and have little or no funding available to voluntarily implement emergency management practices. With the majority of prisons across the country facing the same challenges (limited funding, staffing shortages, lack of resources) as other facilities that provide care (e.g., hospitals, schools, nursing homes), emergency management has and will continue to fall to the side.

Emergency management practices are needed within prison systems. Prisons need some form of guidance or policy reform from the Federal government to plan for, respond to, and recover from disasters. The Federal government may want to consider issuing policies or guidance on protecting and securing prison populations during disasters and providing appropriated grant funding to facilities as a method for programmatic implementation and compliance. If our nation does not plan to manage prison populations during disasters, prisoners' rights will be violated and public safety jeopardized.

In order to resolve this issue, further research is required. The Federal government needs to better understand why this problem exists and what the most effective and practical methods are for addressing it. Identifying the needs and challenges facing prison systems during disasters would allow the Federal government to develop a policy that can be easily adopted and implemented throughout the nation, provided that prisons are receptive to Federal policy on this issue.

Notes

1 For the purpose of this paper, a prison is defined as a facility that houses a person who is under the care and custody of the Federal, State, or County government, and includes jails, correctional facilities, and state hospitals. In some cases, a facility may be contracted out to a private entity that has been granted the same authority as the government.
2 E.A. Carson and W.J. Sabol (2012, December), *Prisoners in 2011*, NCJ 239808 (Washington, DC: Bureau of Justice Statistics, US Department of Justice).
3 Personal interview with an anonymous member of the National Sheriffs' Association, 2008.
4 The National Institute for Corrections has issued checklists and other general guidance documents; however, these are not easily made available to prisons across the nation (many do not know they even exist), and are not required or enforceable.
5 To quote this official directly, he said the government "should just leave them to die." Personal interview with an anonymous Federal official, September 29–November 7, 2012.
6 P. Scharf (2006, October 5), "Responding to Hurricane Katrina: Correctional Response: What If?" [PPT Presentation], presentation given to the New Jersey Chapter of the American Correctional Association, Plenary Session.
7 Personal interview with an anonymous Federal official, September 29–November 7, 2012.
8 See, for example, *Waganfeald v. Gusman*, 674 F .3d 475 (5th Cir. 2012).
9 The term "vulnerable populations" is interchangeable with the term "special populations," which FEMA defines as "*institutionalized persons, the elderly and disabled or those who speak languages other than English*" [emphasis mine]. For the purpose of this study, and to add emphasis, I have included prisoners as a subset of "institutionalized persons." FEMA (1996), *Guide for All-Hazard Emergency Operations Planning*, State and Local Guide 101 (Washington, DC: Author).
10 Personal interview with Giovanna Shay, January 8, 2013.
11 American Correctional Association (2002), *Government and Public Affairs: Returning to Society* (Alexandria, VA: Author).
12 A. Maslow (1943), "A Theory of Human Motivation," *Psychological Review* 50(4), pp. 370–96.
13 Justice Brennan, writing for the court, described cruel and unusual punishment as "degrading, inflicted in arbitrary fashion, clearly rejected throughout society, and patently unnecessary." *Furman v. Georgia*, 408 U.S. 238 (1972).

14 *Blackmon v. Warden Kukua et al.*, No. 11–40316 (5th Cir. 2012).
15 *Prisoners' Constitutional Rights in Disasters: The Failure of Orleans Parish Prison to Provide Protection during Katrina.* (n.d.). San Diego, CA: California Western School of Law.
16 Carson and Sabol (2011).
17 Ibid.
18 "The county decided to evacuate the island to alleviate the suffering of the people, but it didn't consider those in the jail people." Texas Civil Rights Project (2009), *Shelter from the Storm? Galveston County's Refusal to Evacuate Detainees and Inmates at Its Jail during Hurricane Ike* (Austin, TX: Author).
19 See *Waganfeald v. Gusman* (2012).
20 "The Eighth and Fourteenth Amendments cannot tolerate the infliction of a sentence of death under legal systems that permit this unique penalty to be so wantonly and so freakishly imposed." *Furman v. Georgia* (1972).
21 Sharona Hoffman states that "[m]any complicated ethical decisions could in fact be avoided with appropriate response preparations." S. Hoffmann (2009), *Preparing for Disaster: Protecting the Most Vulnerable in Emergencies* (Davis, CA: University of California, Davis).
22 See *Gates v. Cook*, 376 F .3d 323 (5th Cir., 2004); *Brown v. Plata*, No. 09–1233 (2011); *Hutto v. Finney*, 437 U.S. 678. (1978).
23 See, for example, *Frye v. Orleans Parish Prison*, No. 06–5964 (2007); *Kennedy v. Gusman*, No. 10–4198 (2007); *Lloyd v. Gusman*, No. 06–4288 (2007); *Allan v. Gusman*, No. 06–9035 (2007).
24 The federal government contracts out care of federal prisoners and detainees to state, county, and private facilities, under Title 18 U.S.C. 921 §4002.
25 US Department of Justice (2013). *Audit of the U.S. Department of Justice's Oversight of Non-Federal Detention Facility Inspections* (Washington, DC: Author), p. i.
26 Ibid., p. ii.
27 With the exception of some facilities throughout the nation, most prisons do not have comprehensive emergency management plans in place. (Texas Civil Rights Project, 2009); American Civil Liberties Union (2006), *Abandoned and Abused* (New York, NY: Author); American Civil Liberties Union (2007), *Broken Promises* (New York, NY: Author); *Prisoners' Constitutional Rights in Disasters* (n.d.).
28 The Department of Justice has published a document to guide juvenile facilities in preparing for, responding to, and recovering from emergencies; the Department of Health and Human Services published a checklist on influenza planning for correctional facilities; and the National Institute of Corrections has issued general guidance documents that prisons can refer to.
29 F. Navizet and J.C. Gaillard (n.d.), "A Case for Including Prisons and Prisoners in Disaster Risk Reduction" [Slideshow] (Auckland: University of Auckland).
30 P.R. Metzger (2007), "Doing Katrina Time," *Tulane Law Review* 81(4), pp. 1175–218.
31 The exception to this rule is when the Department of Justice contracts with private, state, county, or local facilities to provide care and housing to federal detainees and criminals – these facilities are then required to meet the same standards as Federal facilities. (US DOJ, 2013).
32 Personal interview with an anonymous federal official, September 29–November 7, 2012.
33 The Prison Litigation Reform Act (PLRA) (1996)
34 M. Schlanger and G. Shay (2007), *Preserving the Rule of Law in America's Prisons: The Case for Amending the Prison Litigation Reform Act* (Washington, DC: American Constitution Society).
35 Ibid., p. 9.
36 Ibid.
37 Ibid.
38 Ibid., p. 6.
39 D. Golden (2006, June 7). "The Prison Litigation Reform Act – A Proposal for Closing the Loophole for Rapists," Advance: The Journal of the ACS Issue Groups. (Issue Brief). pp. 95–105.
40 "The American Psychological Association defines trauma as 'an emotional response to a terrible event like an accident, rape or natural disaster.'" S. Babbel (2010), "The Trauma

that Arises from Natural Disasters," *Psychology Today* [Online edition], available at http://
www.psychologytoday.com/blog/somatic-psychology/201004/the-trauma-arises-natural-
disasters, para. 1.
41 Ibid., paras 3, 5.
42 ACLU (2008), *Know Your Rights: PLRA*. New York, NY: Author.

Bibliography

American Civil Liberties Union. (2006). *Abandoned and Abused*. New York, NY: Author.
———. (2007). *Broken Promises*. New York, NY: Author.
———. (2008). *Know Your Rights: PLRA*. New York, NY: Author.
American Correctional Association. (2002). *Government and Public Affairs: Returning to Society*. Alexandria, VA: Author.
American Psychological Association. (2013). *Recovering Emotionally from Disaster*. Retrieved November 18, 2015, from http://www.apa.org/helpcenter/recovering-disasters.aspx
"April 12 – This Day in History: Prisoners Left to Burn in Ohio Fire." (n.d.). The History Channel Website. Retrieved November 19, 2015, from http://www.history.com/this-day-in-history/prisoners-left-to-burn-in-ohio-fire
Associated Press. (2006, August 10). "Report Outlines Prison Horror in Katrina's Wake." *NBC News Website*. Available at http://www.msnbc.msn.com/id/14290778/ns/us_news-katrina_the_long_road_back
Babbel, S. (2010, April 21). "The Trauma that Arises from Natural Disasters." *Psychology Today* [Online edition]. Available at http://www.psychologytoday.com/blog/somatic-psychology/201004/the-trauma-arises-natural-disasters
Benson, C., and J. Twigg. (2007). *Tools for Maintaining Disaster Risk Reduction*. Geneva, Switzerland: Proventium Consortium.
Blackmon v. Warden Kukua et al., No. 11–40316 (5th Cir. 2012).
Bureau of Justice Statistics. (1999, April 1). *Correctional Populations in the United States, 1996*. Part of the Correctional Populations in the United States Series, NCJ 170013. Available at http://www.bjs.gov/index.cfm?ty=pbdetail&iid=743
Carson, E.A., and W.J. Sabol. (2012, December). *Prisoners in 2011*. NCJ 239808. Washington, DC: Bureau of Justice Statistics, US Department of Justice. http://www.bjs.gov/content/pub/pdf/p11.pdf
Cavallo, E., and I. Noy. (2009, December). "The Economics of Natural Disasters: A Survey." IDB Working Paper No. 35. Washington, DC: Inter-American Development Bank.
de Courcy Hinds, M. (1989, October 28). "Rioters Destroy Nearly Half the Buildings in a Pennsylvania Prison." *The New York Times* [Online edition, Reprint]. Available at http://www.nytimes.com/1989/10/28/us/rioters-destroy-nearly-half-the-buildings-in-a-pennsylvania-prison.html
Eberhardt, J.L., P.G. Davies, V.J. Purdie-Vaughns, and S.L. Johnson. (2006). "Looking Deathworthy." *Psychological Science* 17(5), p. 4.
Federal Emergency Management Agency. (n.d.). "Disaster Declarations for 2011." *FEMA Website*. Retrieved November 19, 2015, from http://www.fema.gov/disasters/grid/year/2011?field_disaster_type_term_tid_1=All
———. (n.d.). *Frequently Asked Questions: Threat and Hazard Identification and Risk Assessment (FY2012)*. Retrieved November 19, 2015, from http://www.fema.gov/media-library-data/20130726–1842–25045–7400/nic_faqs_thira_final.txt
———. (1996). *Guide for All-Hazard Emergency Operations Planning*. State and Local Guide 101. Washington, DC: Author.
———. (2013). "Massachusetts – Tropical Storm Irene." *FEMA-4028-DR*. Retrieved November 19, 2015, from http://www.fema.gov/pdf/news/pda/4028.pdf
———. (2013). *Radiological Emergency Preparedness Program Manual*. Washington, DC: Department of Homeland Security.

———. (2015, January 31). "About the Agency." *FEMA Website*. Available at www.fema.gov/about

Federal Prisoners in State Institutions (Public Law 95–624), Supplement 5, Title 18 – Crimes and Criminal Procedure. (2012, January 3). Retrieved November 19, 2015, from https://www.gpo.gov/fdsys/granule/USCODE-2011-title18/USCODE-2011-title18-partIII-chap301-sec4002

Fernandez, M. (2012, June 26). "Two Lawsuits Challenge the Lack of Air-Conditioning in Texas Prisons." *The New York Times* [Online edition]. Available at http://www.nytimes.com/2012/06/27/us/two-lawsuits-challenge-the-lack-of-air-conditioning-in-texas-prisons.html?pagewanted=all&_r=0

Ganderton, P.T. (2004). *Benefit-Cost Analysis of Disaster Mitigation: A Review*. Albuquerque, NM: Department of Economics, University of New Mexico.

Golden, D. (2006, June). "The Prison Litigation Reform Act – A Proposal for Closing the Loophole for Rapists." *American Constitution Society for Law and Policy*. Retrieved November 20, 2015, from http://www.acslaw.org/publications/issue-briefs/the-prison-litigation-reform-act-%E2%80%93-a-proposal-for-closing-the-loophole-for

Goodman, A. (2005, September 27). "Left to Die in New Orleans Prison." *Alternet*. Available at http://www.alternet.org/katrina/26073/

Gram, D. (2012, January 31). "Vermont Struggles to Rebuild Mental Health System after Hurricane Irene." *The Huffington Post*. Available at http://www.huffingtonpost.com/2012/01/31/vermont-mental-health-hurricane-irene_n_1244913.html

Hampton v. Holmesburg Prison Officials. 546 F .2d 1077 (1976).

Harper v. Showers. 174 F .3d 716 (1999).

Hoffman, S. (2009). *Preparing for Disaster: Protecting the Most Vulnerable in Emergencies*. Davis, CA: University of California, Davis.

"Hutto v. Finney." (2015, October 31). The Oyez Project at IIT Chicago-Kent College of Law. Available at http://today.oyez.org/cases/1970–1979/1977/1977_76_1660

Kingdon, J.W. (2003). *Agendas, Alternatives and Public Policies*. New York, NY: Addison-Wesley Educational.

Lehmann, V. (1994). "Prisoners' Right of Access to the Courts: Law Libraries in U.S. Prisons, Madison, Wisconsin." Paper presented at the 60th International Federation of Library Associations General Conference, August 21–27, 1994, Havana, Cuba. Available at http://www.seorf.ohiou.edu/~xx132/iflacorr.htm

Lowe, M. (2012, June 27). "Texas Heat Horror: Inmates Dying from Heat in Texas Prisons and Jail Cells – No AC and No Water, No Fans . . . Unless They Buy Them." *DallasJustice.com* [Website]. Available at http://www.dallasjustice.com/texas-heat-horror-inmates-dying-from-heat-in-texas-prisons-and-jail-cells-no-ac-and-no-water-no-fans-unless-they-buy-them/

Mann, P. (2005). "Hurricane Relief Aid." *The Advocate: Louisiana Association of Criminal Defense Lawyers* 2(4), pp. 3–6.

Maslow, A. (1943). "A Theory of Human Motivation." *Psychological Review* 50(4), pp. 370–96. Available at http://psychclassics.yorku.ca/Maslow/motivation.htm

Metzger, P.R. (2007). "Doing Katrina Time." *Tulane Law Review* 81(4), pp. 1175–218.

Morgan v. Gusman, No. 08–30388 (2009).

National Institute of Corrections, US Department of Justice. (1996). *A Guide to Preparing for and Responding to Prison Emergencies*. Washington, DC: Author.

Navizet, F., and J.C. Gaillard. (n.d.). *A Case for Including Prisons and Prisoners in Disaster Risk Reduction*. [Slideshow]. Auckland: University of Auckland.

Okuyama, Y. (2003). *Economics of Natural Disasters: A Critical Review*. Morgantown, WV: West Virginia University.

Personal Interview with Margo Schlanger, January 11, 2013.Personal Interview with Scott Medlock, January 9, 2013.Pets Evacuation and Transportation Standards Act. (2006).

Post Katrina Emergency Management Reform Act. (2006).

Prisoners' Constitutional Rights in Disasters: The Failure of Orleans Parish Prison to Provide Protection during Katrina. (n.d.). San Diego, CA: California Western School of Law.

PrisonOfficer.org. (2008). *Iowa Floods* [Online discussion forum]. Available at http://forums.prisonofficer.org/general-corrections/3105-iowa-flooding-jail-evacuated.html

Robert T. Stafford Disaster Relief and Emergency Assistance Act (Public Law 93–288), November 23, 1988. https://www.fema.gov/robert-t-stafford-disaster-relief-and-emergency-assistance-act-public-law-93–288-amended

Rundle, E. (2009, September 25). "When Disaster Strikes, Inmates Can Move to the Front Lines of Community Response." *Emergency Management* [Website]. Available at http://www.emergencymgmt.com/disaster/Prisoners-Community-Response.html

Scharf, P. (2006, October 5). "Responding to Hurricane Katrina: Correctional Response: What If?" [PPT Presentation]. Presentation given to the New Jersey Chapter of the American Correctional Association, Plenary Session.

Schlanger, M. (2006). *Civil Rights Injunctions Over Time: A Case Study of Jail and Prison Court Orders*. St. Louis, MI: Washington University.

———, and G. Shay. (2007). *Preserving the Rule of Law in America's Prisons: The Case for Amending the Prison Litigation Reform Act*. Washington, DC: American Constitution Society.

Scott, E., and A. Howitt. (2006, June 19). "Hurricane Katrina (B): Responding to an 'Ultra-Catastrophe' in New Orleans." *Harvard Case Study Analysis Solutions No. 1844.0*. Retrieved November 19, 2015, from http://www.case-study-solutions.com/hurricane-katrina-b-responding-to-an-ultra-catastrophe-in-new-orleans-8725

Sewall, M.P. (2010). "Pushing Execution over the Constitutional Line: Forcible Medication of Condemned Prisoners and the Eight and Fourteenth Amendments." *Boston College Law Review* 51(4), pp. 1278–1322.

Texas Civil Rights Project. (2009). *Shelter from the Storm? Galveston County's Refusal to Evacuate Detainees and Inmates at Its Jail during Hurricane Ike*. Austin, TX: Author.

Twersky-Bumgardner, S. (2010). "Teaching Emergency Management: How to Incorporate Special Populations within a Public Health Framework." [Slideshow]. Presentation given at the FEMA 13th Annual Emergency Management Higher Education Conference, June 7–10.

United Nations. (2009). *Handbook on Prisoners with Special Needs*. New York, NY: Author.

US Department of Health and Human Services. (2007, September 4). *Correctional Facilities Pandemic Influenza Planning Checklist*. Washington, DC: Author. www.flu.gov/planning . . . / business/correctionchecklist.pdf

US Department of Homeland Security. (2008). *The National Response Framework*. Washington, DC: Author. Retrieved November 18, 2015, from http://www.fema.gov/national-response-framework

———. (2011). Presidential Policy Directive/PPD 8: *National Preparedness*. Retrieved November 18, 2015, from http://www.dhs.gov/presidential-policy-directive-8-national-preparedness

———. (2013, June 1). *Guide for Developing High-Quality Emergency Operations Plans for Houses of Worship*. Washington, DC: Author. Available at http://www.fema.gov/media-library/assets/documents/33007US Department of Justice. (2013). *Audit of the U.S. Department of Justice's Oversight of Non-Federal Detention Facility Inspections*. Washington, DC: Author.

———, Office of Justice Programs, Bureau of Justice Statistics. (2010, January 26). "Census of Jail Facilities, 2006." ICPSR26602-v1. Ann Arbor, MI: Inter-university Consortium for Political and Social Research [distributor]. http://doi.org/10.3886/ICPSR26602.v1

US Department of State. (2014, March). *Consular Notification and Access*. Washington, DC: Author.

US House of Representatives. (2006). *A Failure of Initiative: Final Report of the Select Bipartisan Committee to Investigate the Preparation for and Response to Hurricane Katrina, February 15, 2006*. Washington, DC: Author.

US Office of Juvenile Justice and Delinquency Prevention, US Department of Justice. (2011). *Emergency Planning for Juvenile Justice Residential Facilities*. Washington, DC: Author. Available at www.ncjrs.gov/pdffiles1/ojjdp/234936.pdf

Willis, G. (2013). "Rikers Island Inmates Costing NYC Millions in Lawsuits." *The Willis Report, Fox Business* [Website]. Available at http://www.foxbusiness.com/on-air/willis-report/blog/2013/01/14/rikers-island-inmates-costing-nyc-millions-lawsuits

Woods v. Edwards, 51 F .3d 577, 581 (5th Cir. 1995).

World Bank. (2003). *Building Safer Cities: The Future of Disaster Risk*. Washington, DC: Author.

6 A comparative analysis of proactive disaster mitigation measures in apartment buildings in Boston and Washington, DC

Seth H. Holmes

Introduction

Natural disasters such as hurricanes, earthquakes, droughts, and severe winter storms often produce power outages in urban and rural areas. Recent extreme weather events such as the 2003 blackout in the Eastern US and the 2012 Superstorm Sandy in the New York City metropolitan area illustrated the vulnerability of buildings to electrical power outages: many buildings were rendered uninhabitable due to their heating and/or cooling systems not functioning.[1]

During power outages, typical heating, ventilation, and air-conditioning (HVAC) systems in buildings cannot function due to various system electrical requirements; consequently, these buildings cannot provide adequate indoor thermal comfort and risk losing their livable conditions due to indoor overheating or overcooling depending on the season. Air conditioning systems utilize electric pumps and compressors to move heat out of buildings in order to provide cooling. Buildings that use a furnace or boiler for heating often still have a heating fuel source during a blackout (natural gas or oil); however, these systems are not functional without electricity to power the fans or pumps to move the heated air or water throughout the building.

Unfortunately, as climate change impacts increase, many parts of the world will experience more intense weather events, which may lead to a further increase in power outages.[2] One particular weather event that is causing concern is the increase in the frequency and duration of heat waves. Heat waves can cause electrical blackouts or brown-outs in various ways. Brown-outs, or temporary losses of power, occur when electrical demand surpasses supply due to excessive use of electrical air-conditioning equipment during a heat event. Electrical blackouts may also occur during a heat wave because power plants can be forced to shut down when surface water temperatures rise above safe levels for use in cooling plant equipment. The loss, or non-use, of air conditioning equipment during a blackout can lead to dangerously high indoor temperature conditions that can cause heat stress and potentially death.[3] This can be of particular concern in housing populations more vulnerable to heat stress, such as the elderly, young children, or chronically ill or unhealthy individuals.[4] Though many of these individuals may reside in hospitals or nursing facilities, more often than not they spend much of their time in residential settings such as multifamily or single-family homes. For instance, the majority of the heat-related deaths during the 1995 Chicago heat wave were elderly individuals living at home.[5]

Current building codes require residential buildings to have heating systems and, in some instances, cooling systems to maintain indoor thermal comfort; mandatory standards for operable windows exist to allow natural ventilation as a form of fresh air ventilation or to provide cooling when mechanical cooling is not required. For example, the state of Massachusetts, like many other US states, requires that occupied residential spaces have operable windows in exterior walls equal to 4 percent of the floor area for that space.[6] However, newer buildings with lower infiltration rates are also required to have mechanical ventilation to provide fresh air. Some larger residential buildings, much like commercial buildings, require 100 percent mechanical heating, cooling, and ventilation for building comfort and air quality needs. With the exception of the operable windows, all of these systems require electricity to function and consequently maintain the desired indoor conditions.

When buildings are without power, indoor air temperatures often "drift" up or down in relation to outdoor temperatures. However, this temperature drift is also influenced by the heat loss and/or gain through the building's envelope; design techniques that directly affect heat gain and loss include but are not limited to: windows, insulation, access to natural ventilation, and shading. If indoor thermal conditions rise too high or too low, indoor occupants have a higher risk of heat stress from hyperthermia or hypothermia. Though building codes have requirements for thermal comfort, they do not have requirements for indoor "passive habitability," or the ability for a building to maintain safe indoor conditions during long periods without power for mechanical systems.[7] Some cities, such as New York City, are developing building standards that require buildings to provide habitable conditions during power outages.[8]

The goal of this research paper is: (1) to evaluate a generic residential building and various envelope design techniques with regard to their ability to passively maintain indoor temperatures during a prolonged electrical blackout; and (2) to recommend a combination of these techniques for optimal residential passive habitability design in the US northeast.

Methodology

The method presented is based on four steps that are described in the following paragraphs. In the first step, an energy model of a baseline residential building is constructed and its indoor conditions during a four-day blackout are simulated. The second step involves the creation and simulation of a series of alterations to the baseline model that include modifications to the building's envelope. The third step compares the results of these individual modifications and simulates another series of combinations of the individual modifications in order to find optimal results. The final step compares the optimal building design for both the Boston, MA, and Washington, DC, climates.

Initially, an energy model of the baseline building was generated representing a typical residential housing unit. Whole-building energy models estimate heat flows in and out of a building and predict building energy use and interior thermal comfort conditions. For projects seeking LEED or BREEAM certification for high-performance buildings, energy models are required; similarly, building codes require energy models for projects seeking approval through performance-based analysis. For this paper, the whole-building energy modeling "Energy Plus" was utilized in simulating the baseline and variant models; this simulation engine was produced by the US Department of Energy.[9]

Weather data files were used in all the energy model simulations, including the most recent Typical Meteorological Year (TMY3) files for Boston and Washington, DC, produced by the US Department of Energy. Each weather file was analyzed to determine the hottest and coolest average day for each typical year with regard to dry-bulb temperature. A four-day summer blackout period was selected that includes the hottest day, as well as the day before and the two days after the hottest day. Similarly, a four-day winter blackout period was created around the average coldest day. Building simulations were conducted starting three days before each blackout period; the first three days of the simulation do not include blackout protocols, and thus all electrical systems in the building are functional including heating and cooling. By including functional days prior to the blackout, drift conditions can be properly estimated in relation to the previously controlled indoor conditions.

For this paper, a baseline energy model representing a multi-family building was created. A top-floor central unit was then selected from this model for evaluation because it was considered a "worst-case scenario" due to its greater heat gains and losses through the roof and due to its limited façade exposure for use in natural ventilation. The construction variables for the baseline building and top center unit included the following specifications:

- The interior dimensions for the residence were 26'-0" x 42'-0" (1,092 f²) with a 9'-0" interior ceiling height.
- Only one of the 26'-0" long walls was an exterior wall; this wall contained all the exterior windows in the space.
- The window-to-wall ratio was 40 percent, meaning 40 percent of the one exterior wall was modeled as a window.
- The window was unshaded.
- The window was inoperable.
- Wall construction was an insulated concrete block wall with brick facing. Insulation values were sized to meet the 1998 IECC energy code for the Boston climate zone.
- Roof construction was an insulated asphalt roof, similarly designed to meet the 1998 IECC energy code.
- Interior electric loads were scheduled to turn off during the electrical blackout period, thereby eliminating internal heat gain from equipment and lights.
- Interior occupant loads were considered 100 percent of the time during the blackout period.
- The energy model was designed as a two-zone residential unit: one zone represented the living space (including lounge, bed, and bath) and constituted 70 percent of the unit; the second zone represented the kitchen space and remaining square footage. The spaces were divided by a "virtual partition," meaning they were separate for calculations but there was no physical partition separating the two zones.

The baseline model was then simulated for both the winter and summer blackout periods. Indoor dry-bulb air temperatures were recorded for the three days leading up to each blackout, and for the four days during the blackout. Both the outdoor and simulated indoor temperatures were recorded for one-hour intervals, or 24 instances per day. The analysis for the baseline model and subsequent modified models focused on the living space zone within the residential unit.

For the second step, a series of envelope modifications were made to the baseline model and individually simulated for the winter and summer blackout periods. These modifications are collectively considered and referred to here as "Passive Habitability Measures" or PHMs. The modified models included the following PHMs for both the winter and summer analyses:

- Increasing the insulation in the walls and roof (insulation thickness tripled).
- Changing the window-to-wall ratio to 20 percent.
- Changing the window-to-wall ratio to 60 percent.
- Adding a 12″ deep external louvered shading system.

Additionally, the following PHMs were individually simulated for the summer analysis only:

- Modifying baseline windows to include an operable area equal to 4 percent of the unit floor area.
- Increasing the operable opening size to 8 percent of the unit floor area.
- Modifying the unit design to have cross ventilation. For this modification, both 26′-0″ long, opposing walls were considered external and had identically sized windows; these windows had total operable openings equal to 4 percent of the unit floor area.
- Modifying the unit design to have cross ventilation with openings equal to 8 percent of the unit floor area.

For the third step, the modified model results were compared to see which PHM option provided interior temperatures closer to the thermal comfort setpoint when compared to the baseline model. For the winter analysis, this meant finding the model that had the warmest interior temperatures throughout the four-day blackout; for the summer analysis, this meant finding the model with the coolest interior temperatures. In lieu of simulating all 49 possible PHM combinations, a categorical evaluation sequence was utilized to narrow the results more quickly. The first category included simulating for insulation and shading PHMs as well as the combination of the two; the resulting interior temperatures were compared to each other and to the baseline model. The second category simulated the 20 percent and 40 percent window-to-wall ratios and compared the results to the baseline. For the winter analysis only, the best design from the two previous categories was selected and an additional simulation of a combination of the two PHMs was conducted; all the results were then compared to determine the final optimized PHM design for winter conditions.

For the summer analysis, the third and fourth analysis categories were also considered. The third category included adding single-sided natural ventilation by including the 4 percent and 8 percent window-opening designs to the residential unit; the results of these two were compared to the baseline, which did not have operable windows. The fourth category included the cross-ventilation design, but these simulations utilized only the best opening size determined in the previous category. Similarly to the winter analysis, the best options from all four categories were then individually compared, as well as combined and compared in order to determine the optimal design combination for summer conditions.

Finally, the previously described simulation sequence was conducted for both the Boston and Washington, DC, climates to determine whether regional differences appeared. Final optimal results from both cities were compared to determine the following:

1   Are optimal PHM design combinations the same for each city?
2   Given the climate data, how different are the internal temperatures during the blackout for each city's optimal design?

Results

The results from the series of simulation analyses contained a variety of indoor temperature information as well as indicating the range of indoor temperatures for winter and summer conditions in the sample residential unit. For both the winter and summer blackout analysis in Boston, a combination of PHMs was better than the baseline design at keeping temperatures closer to indoor comfort levels; however, as expected, no combination of PHMs was able to keep temperatures in the comfort zone 100 percent of the time during the blackout.

For the Boston winter analysis, Figure 6.1 illustrates three items: (1) the outdoor temperatures during the seven-day analysis period; (2) the indoor temperature for the three-day period with powered HVAC equipment and the four-day blackout; and (3) the indoor temperature if there was no blackout during the seven-day analysis. These three items will be consistently shown in subsequent charts during the sequential analysis steps. Indoor temperature in the baseline design averages between 68°F and 77°F during the days with normal power, but then drops to as low as 51°F during the blackout.

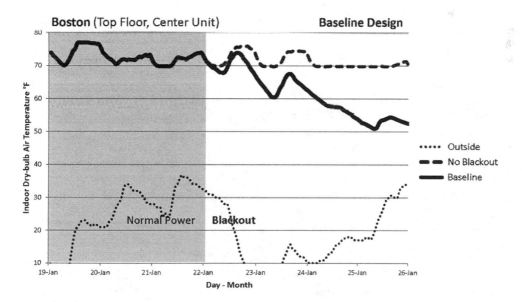

Figure 6.1 Winter indoor conditions – baseline building (Boston).

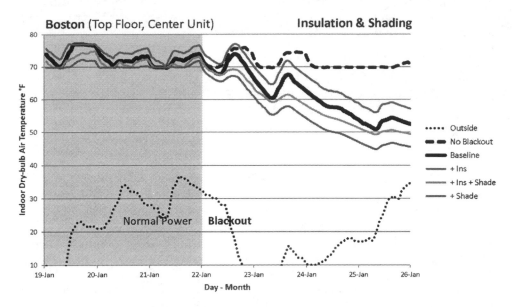

Figure 6.2 Winter indoor conditions, category 1 comparison – baseline building, insulation
PHM, shade PHM, and insulation and shade combination PHM (Boston).

Figure 6.2 illustrates the winter Category 1 analysis to evaluate and compare insulation and external shading PHMs. This comparison indicates that adding shading to
the design further reduced the indoor temperature to as low as 45°F; this is likely due
to the reduced solar heat gain in the space. On the other hand, adding insulation
increased the indoor blackout temperature by reducing heat loss through the envelope; the low temperature for this PHM was 56°F. The combination of shading and
insulation was not as successful as simply increasing insulation alone. Therefore,
adding insulation alone was considered the best option from Category 1 for the winter blackout.

Figure 6.3 illustrates the winter Category 2 analysis to evaluate and compare
window-to-wall ratios (WWR). This comparison indicates minor changes to indoor
temperatures during the blackout for both 20 percent and 60 percent WWR when
compared to the 40 percent WWR of the baseline. The 60 percent WWR did
increase temperatures over the baseline during the day, but reduced them further
at night due to extra heat loss through the glass. On the other hand, the 20 percent
WWR reduced daytime temperatures slightly compared to the baseline, but overall
increased the nighttime low temperature from 51°F up to 53°F. Therefore, a 20
percent WWR was considered the best option from Category 2 for the winter
blackout.

Figure 6.4 illustrates the final winter comparison analysis to evaluate and compare
the best combination of the previously selected PHMs. The Insulation and 20 percent
WWR PHMs, as well as a combination of the two, were compared to the baseline.
The combination Insulation + 20 percent WWR PHM was the most successful in that

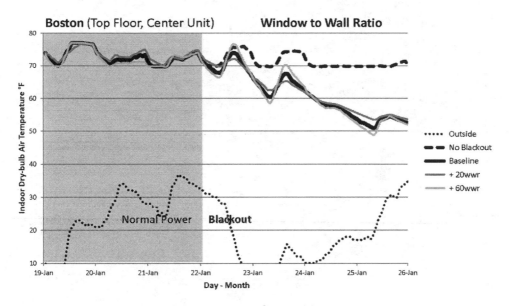

Figure 6.3 Winter indoor conditions, category 2 comparison – baseline building, 20 percent, and 60 percent window-to-wall ratio PHM (Boston).

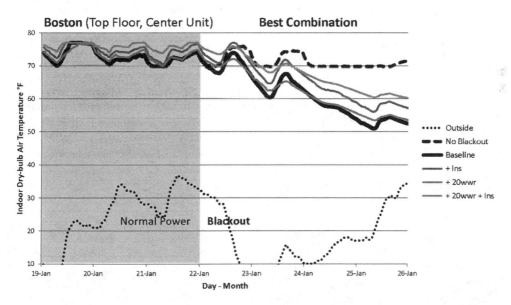

Figure 6.4 Winter indoor conditions, final comparison – baseline building, insulation PHM, 20 percent window-to-wall ratio PHM, and 20 percent/20 percent window-to-wall ratio and insulation PHM combination (Boston).

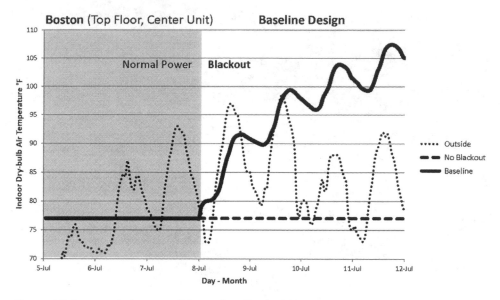

Figure 6.5 Summer indoor conditions – baseline building (Boston).

it increased the low blackout temperature from 51°F to 60°F. The insulation-only PHM did increase daytime temperatures slightly higher than the combination PHM for the first couple of days, but nighttime temperatures dropped further and by the third day, daytime temperatures were worse.

For the Boston Summer analysis, Figure 6.5 illustrates three items: (1) the outdoor temperatures during the seven-day analysis period; (2) the indoor temperature for the three-day period with powered HVAC equipment and the four-day blackout; and (3) the indoor temperature if there was no blackout during the seven-day analysis. These three items will be consistently shown in subsequent charts during the sequential analysis steps. The indoor temperature in the baseline design was consistently 77°F during the days with normal power, but then rose consistently each day to as high as 108°F during the fourth day of the blackout. Nighttime indoor temperatures did drop by between 3°F and 5°F in relation to daytime indoor high temperatures; however, daytime increases ranged from 7°F to 12°F and therefore continually built on previous day heat gains. The daytime highs for the four sequential blackout days were 92°F, 99°F, 104°F, and 108°F respectively.

Figure 6.6 illustrates the summer Category 1 analysis to evaluate and compare insulation and external shading PHMs. This comparison indicated that adding insulation marginally reduced the indoor temperature by one or two degrees F during the day; this was likely due to the reduced heat gain and loss through the envelope from day to day. On the other hand, adding external shading reduced the indoor blackout temperatures significantly; the maximum temperature fell from 108°F to 97°F. Combining shading and insulation was nearly, but not quite, as successful as simply external shading alone; the reduced maximum temperature for this PHM was 98°F. Therefore, adding external shading alone was considered the best option from Category 1 for the summer blackout.

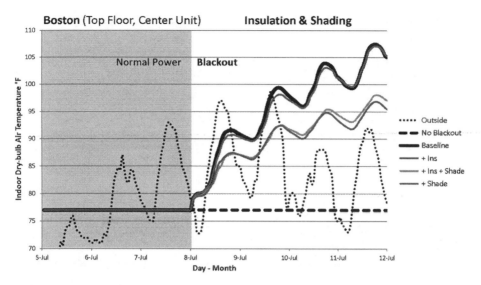

Figure 6.6 Summer indoor conditions, category 1 comparison – baseline building, insulation PHM, shade PHM and insulation, and shade combination PHM (Boston).

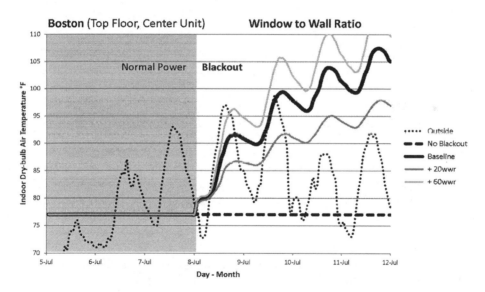

Figure 6.7 Summer indoor conditions, category 2 comparison – baseline building, 20 percent and 60 percent window-to-wall ratio PHM (Boston).

Figure 6.7 illustrates the summer Category 2 analysis to evaluate and compare window-to-wall ratios (WWR); this analysis does not include external shading. This comparison indicated larger changes to indoor temperatures due to WWR during the summer blackout when compared to the winter blackout analysis. The 60 percent WWR increased the maximum temperatures over the baseline from 108°F to 114°F; this was likely due to the increase in solar gain. The 20 percent WWR reduced the daytime

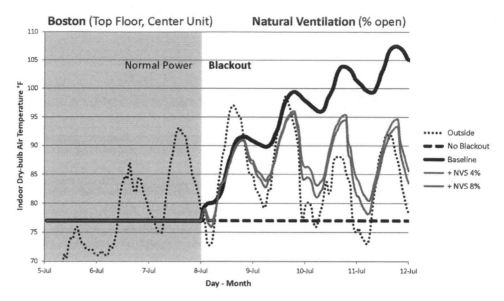

Figure 6.8 Summer indoor conditions, category 3 comparison – baseline building, 4 percent open single-sided ventilation PHM and 8 percent open single-sided ventilation PHM (Boston).

maximum temperature from 108°F to 98°F. Therefore, similar to the winter blackout analysis, a 20 percent WWR was considered the best option from Category 2 for the summer blackout.

Figure 6.8 illustrates the summer Category 3 analysis to evaluate natural ventilation design strategies in regard to the percentage of operable window area; this analysis does not include external shading and retains the 40 percent WWR baseline design. The single window in the baseline design was modified for two separate PHMs: (1) to have an operable area equal to 4 percent of the floor area (which is code minimum in many jurisdictions); and (2) to have an operable area equal to 8 percent of the floor area. Operable windows were simulated to be open all hours, with the exception of the hours between 10am and 8pm; these hours were selected for the windows to be closed because they are on average the hottest hours of the day. Both of these PHMs reduced the indoor temperatures for nearly all hours during the blackout, with the exception of the first few daytime hours of the first day. The maximum indoor temperature for both PHMs was 95°F; the minimum indoor temperatures for both the 4 percent and 8 percent opening PHMs were 81°F and 78°F respectively. During the hottest night of the blackout (day 1), the minimum indoor temperatures for both the 4 percent and 8 percent opening PHMs were 84°F and 83°F respectively. Therefore, an operable window equal to 8 percent of the floor area was considered the best option from Category 3 for the summer blackout.

Figure 6.9 illustrates the summer Category 4 analysis to evaluate and compare natural ventilation design strategies with regard to cross-ventilation and single-sided ventilation; this analysis does not include external shading and retains the 40 percent WWR baseline design. This analysis compares the previous category's single window with an

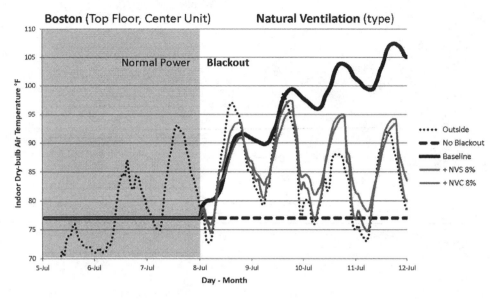

Figure 6.9 Summer indoor conditions, category 4 comparison – baseline building, 8 percent open single-sided ventilation PHM and 8 percent open cross ventilation PHM (Boston).

8 percent opening size (single-sided ventilation) to a design with the 8 percent opening area split between two windows on opposite walls (cross ventilation). Similarly to Category 3, operable windows were simulated to be open all hours, with the exception of the hours between 10am and 8pm. Both of these PHMs reduced the indoor temperatures for nearly all hours during the blackout compared to the baseline. The maximum and minimum indoor temperature for the single-sided PHM were 95°F and 78°F respectively; the maximum and minimum indoor temperature for the cross-ventilation PHM were 97°F and 75°F respectively. During the hottest night of the blackout (day 1), the minimum indoor temperatures for both the single-sided and cross-ventilation PHMs were 83°F and 80°F respectively. Though the cross-ventilation design had slightly higher daytime temperature results, it had substantially lower temperatures in relation to nighttime outdoor low temperatures; due to this result, cross-ventilation was considered the best option from Category 3 for the summer blackout.

Figure 6.10 illustrates the final summer comparison analysis to evaluate and compare the best combination of the previously selected PHMs. The external shading, 20 percent WWR, and 8 percent cross-ventilation PHMs, as well as a combination of the three, were compared to the baseline. The combination design of cross ventilation + 20 percent WWR + shading was the most successful in that it decreased the high blackout temperature from 108°F up to 87°F. Nighttime indoor low temperatures were all equal because all simulated combinations include cross-ventilation.

Figures 6.11 and 6.12 indicate the results for the winter and summer combination analyses for a Washington DC climate. Though specific indoor temperature results vary due to slight regional weather data differences, the same combination designs from the Boston analysis also produced the best results for the DC analysis.

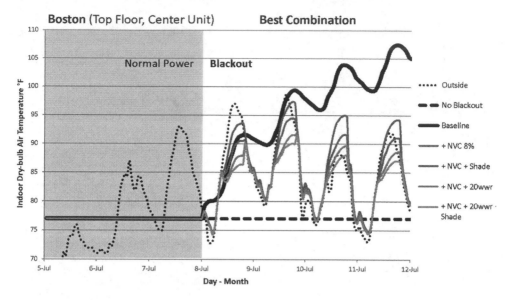

Figure 6.10 Summer indoor conditions, final comparison – baseline building, 8 percent open cross ventilation PHM, 8 percent cross ventilation + shade PHM, 8 percent cross ventilation and 20 percent window-to-wall ratio PHM and 8 percent cross ventilation + 20 percent window-to-wall ratio + shade PHM combination (Boston).

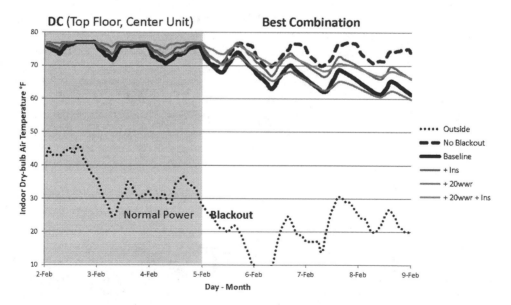

Figure 6.11 Winter indoor conditions, final comparison – baseline building, insulation PHM, 20 percent window-to-wall ratio PHM and 20 percent window to wall ratio + insulation PHM combination (Washington, DC).

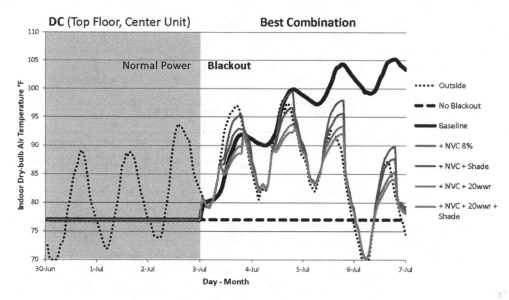

Figure 6.12 Summer indoor conditions, final comparison – baseline building, 8 percent open cross ventilation PHM, 8 percent cross ventilation + shade PHM, 8 percent cross ventilation + 20 percent window-to-wall ratio PHM and 8 percent cross ventilation + 20 percent window-to-wall ratio + shade PHM combination (Washington, DC).

Discussion

The results from the model comparison show significant changes to indoor winter and summer conditions based on the envelope design choices. For the winter analysis, it is clear that increased insulation is the primary PHM for reducing heat loss over the four-day analysis period. For the summer analysis, cross ventilation provided the most effective means for temperature reduction, particularly during the nighttime when outdoor temperatures subside somewhat.

Regarding the combination of PHMs to provide optimal design choices, the inclusion of the 20 percent WWR was successful in both the winter and summer conditions; the lower WWR reduced overheating in summer and heat loss in winter. This design measure is easier to include in new building designs, but may be difficult to implement in retrofitting existing buildings. If buildings are forced to retain higher WWR percentages, retrofit designers should investigate more insulated glass window units with a high-heat-gain low-e coating to help retain more heat in the winter. However, this should be paired with some form of summer shading in order to reduce excessive heat gain in the summer. Additionally, the 20 percent WWR percentage can sometimes compete with the need for glazed and operable window opening sizing, which is based on a percentage of the floor area; for residential units that do not have mechanical ventilation and cooling, this window sizing percentage will take precedence over the WWR percentage.

The winter analysis illustrated that increased insulation and no external shading was preferred, whereas the opposite was true during the summer analysis where shading was critical and insulation had a slight negative effect on indoor temperatures. These results indicate that external shading should be seasonally operable, such as removable exterior awnings or venetian blinds. Insulation is not as easily removable; however, the negative effect on internal temperatures in the summer was only slight, so buildings in the northeast may find greater annual benefits by including the extra insulation, as it is more critical for winter conditions.

The summer analysis illustrated that buildings should be designed for cross-ventilation with openings sized greater than the code minimum 4 percent of the floor area. Though single-sided ventilation was successful, cross-ventilation provided a quicker and larger drop to internal temperatures during the night. However, the timing of natural ventilation should also be considered by residents; properly timed natural ventilation is critical for night cooling during a summer blackout that happens during a heat wave. Because building designers have no say over the use of operable windows during a blackout, natural ventilation "best practice guidelines" should be developed for distribution to residents. If mechanically operated windows are installed, overrides should exist to allow for manual operation during blackout conditions.

The comparison between Boston and Washington saw little difference regarding extreme hot and cold four-day events. The Washington case study experienced slightly warmer conditions for both winter and summer; however, there was no change in the optimal PHM combination. This is likely to be due to the fact that only extreme conditions were analyzed and both cities experience fairly hot and cold conditions at least once per year. With that in mind, Boston experiences a higher frequency of cold events, while Washington experiences a higher frequency of hot events. Therefore, the risk associated with these types of events should be considered when making design choices; for instance, a Washington building owner may invest primarily in cross-ventilation opening sizing, whereas a Boston owner may primarily invest more in insulation.

This research is not extensive and further research is required in various related areas. Primarily, analyses should be conducted for a broader range of climate types in the US and internationally. Similarly, a cost-investment analysis could be associated with this type of study to determine the cost effectiveness of the various PHMs – i.e., which PHM provides the most effective habitability result for the least cost. A cost-effectiveness study could also investigate the energy savings each PHM provides to a building during normal building operating conditions. Additionally, this type of study should be done on other residential building and/or unit types. Finally, more research is needed to determine a proper indoor heat threshold and metric; though indoor air temperature significantly affects human comfort and homeothermy, it is only one of a variety of factors affecting human thermal regulation. The variables of humidity, air speed, radiant heat, metabolism, and clothing need to be considered in greater detail.

Conclusion

This study indicates that there is a great risk of extreme indoor temperatures if an extended winter or summer blackout happens during an extreme temperature event such as a cold snap or heat wave. Though not an extensive study, it is clear that there are certain combinations of typical building envelope design measures that can help reduce the risk of extreme indoor temperatures. For the northeast US, a combination of low window-to-wall ratios, external operable shading, and an increase to code minima for insulation and natural ventilation will be likely to provide higher indoor temperatures during winter blackouts and cooler indoor temperatures during summer blackouts. The use of these simple design features may lead to lower morbidity and mortality among vulnerable populations during prolonged blackouts during extreme outdoor temperature conditions.

Current US building codes require specific energy efficiency standards to be met; often these standards are met using the same building envelope design features analyzed here as passive habitability measures. However, current building codes do not have passive habitability standards and therefore do not cover how buildings should maintain adequate indoor conditions during periods of lost power or fuel. The measures analyzed here are not controversial, cutting-edge, or overly costly. Therefore, it seems reasonable for building codes to consider and address standards that require a basic level of passive habitability in buildings. Codes exist to provide basic levels of public safety; they should also consider passive thermal safety during disasters and extreme temperature situations.

Notes

1 D. Williams (2006), *Sustainable Design: Ecology, Architecture and Planning* (Hoboken, NJ: Wiley); PlaNYC (2013), *New York City Special Initiative for Rebuilding and Resiliency: A Stronger, More Resilient New York* (New York, NY: Author).
2 Intergovernmental Panel on Climate Change (2014), *Climate Change 2014: Impacts, Adaptation and Vulnerability IPCC Working Group II Contribution to AR5* (New York: United Nations, IPCC).
3 E. Klinenberg (2002), *Heat Wave: A Social Autopsy of Disaster in Chicago* (Chicago, IL: University of Chicago Press).
4 J. Kravchenko, A.P. Abernethy, M. Fawzy, and H.K. Lyerly (2012), "Minimization of Heat-wave Morbidity and Mortality," *American Journal of Preventive Medicine* 44, pp. 274–82; G. Barnett, M. Beaty, D. Chen, S. McFallan, J. Meyers, M. Nguyen, Z. Ren, A. Spinks, and X. Wang (2013), *Pathways to Climate Adapted and Healthy Low-Income Housing* (Gold Coast, Australia: National Climate Change Adaptation Research Facility).
5 Klinenberg (2002).
6 State Board of Building Regulations and Standards (2010), *780 CMR: Massachusetts Amendments to the International Building Code 2009* (Boston, MA: State of Massachusetts).
7 S. Holmes, T. Phillips, and A. Wilson (2015), "Overheating and Passive Habitability: Indoor Health and Heat Indices," *Building Research and Information*, doi:10.1080/09613218.201 5.1033875
8 Building Resiliency Task Force (2013), *Report to Mayor Michael R. Bloomberg and Speaker Christine Quinn* (New York, NY: Urban Green).
9 US Department of Energy (2014), *EnergyPlus Energy Simulation* [Software], retrieved May 6, 2014, from http://apps1.eere.energy.gov/buildings/energyplus/

Bibliography

Barnett, G., D. Chen, S. McFallon, J. Meyers, M. Nguyen, Z. Ren, et al. (2013). *Pathways to Climate Adapted and Healthy Low-Income Housing*. Gold Coast: National Climate Change Adaptation Research Facility.
Building Resiliency Task Force. (2013). *Report to Mayor Michael R. Bloomberg & Speaker Christine Quinn*. New York, NY: Urban Green.
Holmes, S., T. Phillips, and A. Wilson. (2015). "Overheating and Passive Habitability: Indoor Health and Heat Indices." *Building Research and Information*, doi:10.1080/09613218.201 5.1033875
Intergovernmental Panel on Climate Change. (2014). *Climate Change 2014: Impacts, Adaptation, and Vulnerability. IPCC Working Group II Contribution to AR5*. New York, NY: United Nations, Author.

Klinenberg, E. (2002). *Heat Wave: A Social Autopsy of Disaster in Chicago*. Chicago, IL: University of Chicago Press.

Kravchenko, J., A.P. Abernethy, M. Fawzy, and H.K. Lyerly. (2012). "Minimization of Heatwave Morbidity and Mortality." *American Journal of Preventative Medicine* 44, pp. 274–282.

Massachusetts State Board of Building Regulations and Standards. (2010). *780 CMR: Massachusetts Amendments to the International Building Code 2009*. Boston, MA: State of Massachusetts.

PlaNYC. (2013). *New York City Special Initiative for Rebuilding and Resiliency: A Stronger, More Resilient New York*. New York: Author.

US Department of Energy. (2014). *EnergyPlus Energy Simulation Software* [Software]. Retrieved May 6, 2014, from http://apps1.eere.energy.gov/buildings/energyplus/

Williams, D. (2006). *Sustainable Design: Ecology, Architecture, & Planning*. Hoboken, NJ: John Wiley & Sons.

7 Incorporating natural disasters into the undergraduate civil engineering curriculum

A case study of Hurricane Katrina and the Oso landslide

Thomas C. Cross and James Kaklamanos

Introduction

In recent years, there has been an increased focus on the role of the engineering and scientific communities in providing the public with adequate protection against the effects of natural disasters. The ability of engineers and scientists to develop methods by which natural disasters can be better understood, more effectively predicted, and more accurately quantified has been subjected to increased scrutiny. A keen awareness of the role these professions play in defining the quality of life and the responsibility inherent in this role is crucial to the survival of the engineering and scientific professions. To that end, engineering education can be enhanced with the addition of learning objectives involving natural disasters and associated engineering failures. For undergraduate civil engineering students, these curricular bridges can serve as real-world examples of various engineering concepts, as well as the broader impacts of design errors, engineering judgment, and public policy. This paper will focus on curricular enhancements within a junior-level course in geotechnical engineering, a subdiscipline of civil engineering that focuses on earth materials and their interaction with the built environment. Most students enrolled in ABET accredited civil engineering programs in the US are required to complete at least one course in geotechnical engineering at some point during their college career (typically in the junior year). As a case study, this paper will describe how natural disasters have been incorporated into the curriculum of *Civil Engineering 3020 – Geotechnical Engineering* at Merrimack College, a predominantly undergraduate, private institution in North Andover, MA, in the US. These curricular enhancements could readily be adopted in other geotechnical engineering courses, as well as in courses on related topics. This paper will focus upon the levee failures during Hurricane Katrina and the extent to which this disaster exemplifies the curricular concepts of a typical undergraduate geotechnical engineering course. A discussion of the Oso, Washington, landslide of 2014 will also be presented.

Geotechnical engineering curricula

A typical undergraduate geotechnical engineering course introduces students to fundamental soil mechanics concepts such as soil composition and classification, compaction, groundwater flow, subsurface stress, settlement, and shear strength. Subsequent courses build upon these basic concepts to address topics in geotechnical design, such as slope stability analyses, the design of foundations and earth-retaining structures, and earthquake engineering. At Merrimack College, the course outline of *Civil Engineering 3020 – Geotechnical Engineering* is presented as follows (the numbers in parentheses indicate the approximate number of 75-minute classes allotted to the topic):

1   Soil composition and classification (4)
2   Compaction (2)
3   Groundwater (3)
4   Stress (4)
5   Settlement (6)
6   Shear strength (5)
7   Applications of soil mechanics (3)

In this section of the paper, we provide a brief overview of each of these course topics, to guide the later discussion on Hurricane Katrina and the linkages between these course topics and the levee failures.

Soil composition and classification

The first topics typically covered in geotechnical engineering courses are the geologic origin of soils, physical properties for describing soils, and the soil classification systems used to determine the types of soil present at a site. In order of decreasing grain size, the four major soil types are gravels, sands, silts, and clays. Gravels and sands (coarse-grained soils) have larger grains than silts and clays (fine-grained soils). Gravels and sands derive much of their strength from the frictional forces that develop as the soil particles shear against one another when subjected to loading. In contrast, silts and clays derive much of their strength from cohesion, which can be caused by physical or chemical bonding. An additional fifth type of soil is organic soil. Organic soils contain a significant amount of organic material recently derived from decomposing plants or animals, are highly compressible, and are generally unsuitable for construction.[1]

Compaction

Compaction is the densification of soil via the removal of air voids and can be performed in the field using a variety of techniques and equipment. Most construction projects require some sort of earthwork, and the compaction of placed fill must meet specifications to ensure that the soil will have adequate stiffness and shear strength. The compaction specifications for earth dams and levees are among the most stringent of any type of earthwork, due to the potentially catastrophic consequences of failure.[2]

Groundwater

Groundwater is the term commonly used to describe all subsurface water within the ground, which is located within soil voids ("pores") and rock fissures.[3] *Permeability* is a soil property describing the rate at which groundwater is transmitted through a soil. Due to their larger grain sizes, gravels and sands are much more permeable than silts and clays, and therefore drain more quickly.[4]

Subsurface stress

Subsurface stresses in soil can be carried by both solid particles and pore water. *Effective stress* is the portion of the total stress carried by the solid particles. The effective stress in a soil is computed using the equation $\sigma' = \sigma - u$, where σ' = the effective stress, σ = the total stress, and u = the pore water pressure of the groundwater. Given this relationship, it is clear that effective stress decreases as the pore water pressure increases. Thus, the presence of groundwater reduces the effective stress in a soil. Effective stress is consequently affected by factors that influence groundwater such as rainfall, pumping practices, and irrigation activities. Furthermore, effective stress governs soil behavior and – perhaps most importantly – the shear strength of a soil.[5]

Settlement

Settlement is defined as the downward movement of soil due to induced vertical stresses. *Subsidence* is a term used to describe settlement that occurs over a large area. Settlement is induced in a soil whenever the vertical effective stress in that soil increases. In general, clays and silts experience larger amounts of settlement than sands and gravels, and this settlement occurs over a longer period of time.[6]

Shear strength

The *shear strength* of a soil is the greatest shear stress that the soil can sustain without failure. Shear stresses in soils are resisted by the interactions between the soil particles, primarily the rearrangement of those particles when sustaining stress. Using the Mohr-Coulomb failure criterion, which is the most common failure theory assumed for soil, the shear strength is defined by the equation $s = c' + \sigma' \tan \phi'$, where s = shear strength, c' = effective cohesion, σ' = effective stress on the failure plane, and ϕ' = effective friction angle. A soil's shear strength is therefore linearly related to its effective stress. When the applied shear stress exceeds the shear strength of that soil, the soil can no longer support the loading, and failure occurs.[7]

Failures can also occur due to the effects of groundwater. The *hydraulic gradient* is a ratio of the change in the total head (energy) of flowing water to a unit length in the direction of flow. Mathematically, the hydraulic gradient is defined as $i = \Delta h / \Delta l$, where i = the hydraulic gradient of the soil, h = the total head, and l = the distance the water travels.[8] The critical hydraulic gradient for water flowing through soil is defined as $i_{cr} = (\gamma_{sat} - \gamma_w)/\gamma_w$, where i_{cr} = the critical hydraulic gradient, γ_{sat} the saturated unit weight of the soil, and γ_w = the unit weight of water. The critical hydraulic gradient represents a state of zero effective stress ($\sigma' = 0$), which corresponds to zero shear strength in cohesionless soils. This indicates incipient failure.[9]

In the design of structures located adjacent to bodies of water, knowing the critical hydraulic gradient of the underlying soil is crucial to understanding the potential for soil failure. Underseepage is a form of groundwater motion that often precedes failure. When foundation soils underlying an earth structure such as a levee are sufficiently permeable, water may travel rapidly under the structure, thereby significantly increasing the pore pressure. The resulting reductions in effective stress and shear strength can initiate catastrophic instability, potentially resulting in piping failure. Piping failure occurs when seeping water erodes and removes soil, beginning from the exit point of the seepage path and advancing beneath the earth structure.[10] This process forms an enclosed, pipe-like channel through which water and eroded material may flow unimpeded. The occurrence of this phenomenon beneath a levee can be predicted by comparing the critical hydraulic gradient to the vertical exit gradient at the toe of the structure.[11] If the exit gradient is sufficiently raised due to the occurrence of underseepage, then there remains little resistance to soil erosion, and piping ensues.[12]

Hurricane Katrina: Disaster

Overview

On August 29, 2005, Hurricane Katrina made landfall as a Category 3 storm southeast of the city of New Orleans, LA.[13] The effects of wind-driven waves, high tide, precipitation, and reduced local atmospheric pressure resulted in storm surges ranging from 15 to 20 feet in the Gulf of Mexico and Lake Borgne to the east, and from 10 to 14 feet along the southern shore of Lake Pontchartrain to the north of New Orleans.[14] Levee breaches occurred at 50 locations along the city's hurricane protection system, resulting in significant flooding of approximately 80 percent of the city.[15] The storm was directly responsible for 1,200 fatalities and caused an estimated $108 billion in property damage.[16]

The catastrophic failure of the New Orleans hurricane protection system raised numerous questions regarding the underlying engineering and management practices employed in the development of the system. Several organizations conducted investigations, producing substantial evidence of poor engineering judgment and practice, disorganization of the construction and implementation of system components, and a failure to appreciate the risk to life and property should the system fail.

Levees and floodwalls in New Orleans

The New Orleans Flood Defense System (NOFDS) is an approximately 350-mile-long system composed predominantly of federally constructed and locally managed levees, floodwalls, and pumping stations that protect a series of separate basins.[17] The largest elements of this network were developed by the US Army Corps of Engineers (USACE) following Hurricane Betsy in 1965[18] and are often referred to as the Hurricane Protection System (HPS). The majority of the system is composed of earthen levees.[19] Prior to Hurricane Katrina, Hurricane Betsy was one of the most devastating hurricanes to affect New Orleans. The storm resulted in 75 deaths and over $1 billion in damage across southeast Louisiana and Florida, and caused catastrophic flooding in New Orleans' Lower Ninth Ward.[20] The purpose of the HPS was to prevent similar catastrophic flooding during future storms (such as Katrina).

Levees prevent the flow of water into populated or otherwise vulnerable areas by acting as a barrier against rising water and storm surge activity. After Hurricane Betsy, pre-existing levees were heightened to prevent flooding and maintain canal capacity.[21] Traditionally, the heightening of a levee is achieved through the addition of compacted material to the land side of the levee, which heightens the levee and increases the width of the levee's base.[22] Residential development alongside the existing levees, combined with the anticipated cost of compensating and removing residents, inhibited the implementation of the traditional heightening method. Consequently, the USACE pursued the construction of floodwalls atop the existing levees. These floodwalls – comprised of concrete walls supported by steel sheet piles driven into the existing earth levees – would provide additional height without widening the base of the levee.[23]

Three different styles of floodwalls were used, each named for its cross-sectional shape. *T-walls*, named for the similarity between their cross-sectional shape and an inverted letter "T," were supported by continuous steel sheet-pile walls extending from the above-ground concrete portion of the structure down into the earth below the levee, thereby reducing seepage beneath the wall. The T-walls were additionally supported by precast concrete or steel H-piles driven into the earth below. A slightly modified version of the T-wall, known as an *L-wall* due to its cross-sectional shape, was implemented in very limited quantities. *I-walls*, named for the similarity between their cross-sectional shape and the letter "I," were supported by continuous steel sheet-pile walls identical to those used for the T-walls. Unlike the T-walls, however, the I-walls did not include deep pile foundations to anchor the walls more strongly in place. The overwhelming majority of the floodwalls in the NOFDS/HPS were these simpler I-walls.[24]

Levee failures during Hurricane Katrina

The floodwalls were put to the test when Hurricane Katrina made landfall. Levee breaches occurred at 50 locations along the HPS, 46 of which were the result of the overtopping and subsequent erosion of the levees. The remaining four failures were the result of dramatic floodwall foundation failures along the 17th Street Canal, the Inner Harbor Navigation Canal (IHNC), and at two points along the London Avenue Canal. In addition to the breaches, approximately 220 miles of the levees and floodwalls experienced some sort of damage.[25]

Hurricane Katrina in the curriculum

Implementation at Merrimack College

The curricular concepts for geotechnical engineering are well illustrated by the failure of the New Orleans levee system during Hurricane Katrina. In *Civil Engineering 3020 – Geotechnical Engineering* at Merrimack College, Hurricane Katrina serves as an overarching theme throughout the entire semester. The first class of the semester is a case study of Hurricane Katrina that introduces students to the events leading up to Hurricane Katrina (including the history of the levee system); the mechanics behind the levee failures; and the failures in engineering, construction, maintenance, and public policy that ultimately contributed to the levee breaches. The students' first course assignment is to write a brief essay on some aspect of the disaster, focusing either on the past, the present (during the disaster), or the future, for example:

- The past: the pre-existing geological/geotechnical conditions that made New Orleans susceptible to hurricane devastation or the public policy decisions that contributed to New Orleans' vulnerability
- The present: why specific levee failures occurred during Hurricane Katrina
- The future: how the levees and hurricane protection system are being improved to protect New Orleans in the future.

During the semester, when new course topics are addressed, these topics are linked to the Hurricane Katrina levee failures. Once students understand the significance of the course content, they are more able to retain and appreciate the material. During the final class meeting of the semester, students reflect upon Hurricane Katrina through the lens of soil mechanics and geotechnical engineering, using the concepts learned throughout the semester. Hurricane Katrina serves as a poignant, real-world example of the importance of geotechnical engineering and the consequences of failure. In this section, specific recommendations are provided on how various aspects of the levee failures during Hurricane Katrina can be incorporated into courses in geotechnical engineering and related topics.

Importance of engineering geology

The performance of the HPS was heavily dependent upon the engineering properties of the underlying soils. To understand the engineering properties of the soils underlying the levees and floodwalls protecting New Orleans, knowledge of the geologic history of the area is required. New Orleans is located within the Mississippi River deltaic plain. River and lake activity have resulted in large quantities of fine-grained silts, clays, and organic materials, with 90 percent (excluding artificially filled areas) of New Orleans being underlain by swamp or marsh deposits.[26] One of the broad themes underlying the failures described in this chapter – and in many other geotechnical engineering failures throughout history – is the disconnect between engineering geology and geotechnical engineering. The geologic conditions under which soil is formed and deposited will ultimately influence how this soil behaves when subjected to loads.

Overtopping and erosion failures: Linkage to soil classification and compaction

The levee failures underlie the importance of proper soil classification: some soil types should be avoided in certain applications. Levees with a core constructed of well-compacted, low-permeability clay are typically more resistant to erosion than those constructed using coarse-grained soils (such as sands), which are susceptible to erosion.[27] Unfortunately, the use of erosion-susceptible soils as levee material was inexplicably common along some portions of the system. Some levees were composed primarily of material dredged from nearby waterways – a process known as hydraulic filling – that often consisted of erodible, lightweight sands that were poorly compacted.[28] Exacerbating the usage of inappropriate soils in levee construction, most of the levee system did not incorporate any form of protection against overtopping or erosion along the land sides of levees.[29] Relatively inexpensive measures, such as concrete splash-pads, could have been incorporated into the design to reduce erosion resulting from overtopping. Similarly, this erosion could have been reduced by using T-walls – which incorporate a horizontal concrete slab behind the vertical portion of the wall – more extensively, rather than opting for the less expensive, completely vertical I-walls.[30] The general lack of resilience to overtopping and erosion significantly increased the volume of flooding and the resulting devastation.[31] The selection of poor levee material and the general lack of system resiliency exemplify the tradeoff between technical adequacy and financial cost inherent in many engineering projects.

Settlement

Prior to Hurricane Katrina, the city of New Orleans was already an excellent case study for concepts involving settlement. The reasons that large areas of New Orleans lie below sea level can be explained by geotechnical engineering principles. Subsidence, combined with the slow rise in sea level, has left large portions of the city currently below sea level elevation.[32] This renders the city increasingly vulnerable to tropical storm events, particularly to flooding associated with a breaching of the levee system.[33] Several factors have led to the large-scale settlement occurring around New Orleans. Increases in the effective stress in the underlying soils – caused by structural loadings and by the pumping of groundwater – produces settlement. The slow drainage rates of the clays in the area result in persistent long-term settlement, which is exacerbated by the decomposition of organic matter. Furthermore, tectonic activity also contributes to the subsidence of the entire Gulf Coast region, including New Orleans.[34]

To accurately determine the elevations – relative to sea level – necessary to design and construct a flood protection system, the accuracy of the referenced benchmark datum is critical. When many of the floodwalls surrounding New Orleans were constructed, outdated variations of the NGVD29 datum, which is based on terrestrial reference points rather than on mean sea level, were referenced. The decision to use terrestrial benchmarks completely neglected the effects of subsidence and sea level rise on ground elevation. Therefore, the floodwalls were constructed to a lower protection height relative to sea level than originally mandated and authorized by Congress.[35] This led to an increase in overtopping during Katrina, and consequently contributed to the erosion-induced breaching of floodwalls and levees.[36] For example, the I-walls constructed along the 17th Street Canal levees in the 1990s were intended to extend 14.0 feet above sea level. Currently, these floodwalls lie 1.3 to 1.9 feet below that design elevation.[37] The peak storm surges caused by Katrina were commonly between 1 to 3 feet above the tops of the existing levees and floodwalls, only slightly exceeding the intended design elevation of those structures. If the levees and floodwalls had been constructed to the proper elevations, fewer overtopping-induced failures might have occurred.[38]

Groundwater flow

All of the four non-overtopping floodwall and levee foundation failures occurred without water levels reaching their design elevations.[39] The causes have been determined to consist of underseepage and piping, the development of a water-filled gap between the I-wall and the earth material of the surrounding levee, and the development of slip surfaces along weak strata in the soils beneath the levees.[40]

Underseepage played a role in the two breaches of levees along the London Avenue Canal. The I-walls of the levees were laterally supported solely by the cantilever action of their supporting steel sheet pile walls being embedded in the earth levees. Hydrostatic pressure due to rising water levels caused the portion of the I-wall atop the levees to deflect backwards (towards the land side), exposing the sheet piles and allowing water to fill the widening space. The gap widened until the water extended downward to the bottom tip of the sheet piles, allowing for water to flow more easily beneath the levee.[41] The groundwater exerted too large of a hydraulic gradient for the soil to remain intact, initiating piping failure. The formation of a water-filled gap was not considered in the design of the levees, despite the development of a similar gap during a 1985 I-wall test conducted by the USACE.[42] The likelihood of an underseepage-related failure was therefore inaccurately calculated and mistakenly ignored. This design omission underscores the importance of considering all possible failure modes in engineering design, an important concept for students to appreciate.

Shear strength failure

With the formation of the water-filled gap, the length of the surface providing shearing resistance against backward lateral translation was reduced, significantly increasing the potential for a shearing failure. This reduction in resisting forces is likely to have caused the lateral translation of the 17th Street Canal levee and floodwall.[43] The critical layer for the 17th Street Canal failure was a stratum of weak organic soil: the applied stresses exceeded the shear strengths of this layer, resulting in a catastrophic failure of the soil. This thin layer was composed of organic clayey silt and was discovered at varying depths between 8.3 and 11 feet below the ground surface in borings made under the supervision of the Independent Levee Investigation Team (ILIT) after Katrina. The water content for the layer was in excess of 270 percent, and physical evidence indicated that extremely high pore pressures likely developed as a result. This significantly reduced the shear strength along the layer, leading to the lateral translation and failure of the levee.[44] This layer had remained undetected despite drilling conducted during the construction of the levees prior to Katrina. The ILIT report deemed the failure to detect this layer to be the result of those drillings having been conducted without the supervision of an expert geological engineer with knowledge of the site's complexity and experience in detecting and analyzing such layers.[45] The failure of the 17th Street Canal levee can be incorporated in numerous course lessons throughout the semester, such as engineering geology, physical soil properties (and the implications of high water content), subsurface explorations, and shear strength.

The USACE also overestimated the strength of the foundation soils beneath and surrounding the levees. Directly beneath the centerline of a levee embankment, foundation soils experience greater effective stress due to the weight of the overlying levee material, and therefore exhibit greater shear strengths. The soils beneath the toe of a levee, however, are weaker because they are not subjected to the same amount of effective stress. During the design of the levees, tests were performed only on the soil directly beneath the centerline of the embankment, and shear strengths were estimated based on those tests. The strength of the soils outside the centerline of the levee and specifically along the toe – arguably the most critical point of the structure – were overestimated as a result. Furthermore, the factor of safety (FS) used for the design was only 1.3, an outdated figure originally selected for the design of levees providing protection for non-populous agricultural areas.[46] The target factor of safety was not reflective of the risk to human life posed by a potential failure of the levee system during a hurricane event; USACE standards call for target factors of safety of at least 1.4 to 1.5.[47] Overestimation of shear strength and usage of low factors of safety are unconservative practices that can lead to engineering failures, and this case study teaches students the importance of reasonable conservatism in engineering design.

The Oso, Washington, landslide

On March 22, 2014, an unstable slope east of Oso, Washington, catastrophically failed, resulting in the movement of approximately 7.6 million cubic yards of earth. Forty-three people were killed in the landslide, which now has the distinction of being the deadliest single landslide in American history. The landslide reached distances of 0.6 miles from the toe of the slope, destroying approximately 60 homes and other structures and temporarily damming the North Fork of the Stillaguamish River.[48] Like Hurricane Katrina, the Oso landslide illustrates a number of concepts that are taught in geotechnical engineering courses. Some of these concepts are briefly discussed in this section.

First, this disaster serves as an excellent example of the influence of groundwater on subsurface stresses and shear strengths. The 2014 slide took place along the edge of a plateau composed of glacially deposited sands on top of lake-deposited silts and clays.[49] In March 2014, Oso received near-record amounts of precipitation. The entire season had been especially wet, with precipitation in the area between February and March at 150–200 percent of the long-term average.[50] The sands at the site are very permeable, much more so than the underlying silts and clays. Thus, groundwater at the site drains through the sand layer, but the very low permeability of the saturated silts and clays beneath the sand causes groundwater to collect in the overlying sand.[51] When soil is saturated, the pore water reduces interlocking and shearing between the soil particles, which in turn reduces the soil's frictional strength (the primary source of strength in sands). The majority of the rainfall was most likely retained within the glacial sand layer along the top of the slope. The corresponding reduction in effective stress and soil strength was likely to have been a significant factor in the initiation of the slide.

The proximity of the Stillaguamish River to the toe of the hill was likely to have been another contributing factor for the landslide. Miller and Miller[52] concluded that there was a strong tendency for the Stillaguamish to preserve its position at the base of the slope, which results in regular slumping into the river. These gradual changes in geometry at the base of the slope ultimately reduce the stability of the entire slide mass, thereby increasing the potential for a large catastrophic failure.[53] The combination of soils present, historical geologic and groundwater activity, and previous slide activity should have served as a clear indicator that the area was at risk for further landslides.

In addition to the geological aspects of the disaster, policy decisions regarding land use in the Oso area have come under increased scrutiny. The permitting of residential development downslope of the Hazel slide, in spite of the site's history of landslides, has been widely questioned, as this ultimately placed more people in the path of the deadly landslide. Revelations of logging activity in the area (particularly behind the hill that failed) have raised questions regarding how effectively logging restrictions were enforced; logging in any area prone to landslide activity has been a point of engineering concern and is associated with a risk of increased groundwater recharge and consequent reduction of slope stability.[54]

There were many opportunities to learn from past failures in the Oso area; studies detailed the volatile combination of soil and groundwater conditions present at the site,[55] but the warning signs were not sufficiently acted upon. Prior to the catastrophic 2014 landslide, this specific hillside had experienced slope failures in 1949, 1951,

1964, 1967, 1988, and most recently (prior to the 2014 event) in 2006, when 300 yards of the hillside slid into – and temporarily dammed – the Stillaguamish River. Formal assessments for the probability of further landslide activity in the area were either never performed or never published.[56] Like Hurricane Betsy in New Orleans, the past landslides at Oso represented learning opportunities that were either missed or ignored. The Oso landslide, like Hurricane Katrina, serves as an important lesson for engineering students about the importance of engineering history and learning from the past.

Conclusions

The levee failures during Hurricane Katrina were one of the largest failures of a civil engineering system in American history. The inadequacies of the levee system, and the examples of its poor performance during Hurricane Katrina, are numerous. However, key failures in the engineering of the system illustrate the importance and relevance of fundamental geotechnical engineering concepts. The use of outdated and inappropriate data, the utilization of erodible material in the construction of levees, the lack of protection against overtopping, and the failure to design against underseepage make it apparent that poor engineering judgment is substantially to blame for the disastrous consequences of Hurricane Katrina. In simulations that modeled the effects of Hurricane Katrina for a scenario in which the floodwalls did not catastrophically fail and the pumping stations remained functional, the results indicated that nearly two-thirds of the deaths and more than half the property losses would not have occurred.[57]

Undergraduate civil engineering students represent the next generation of engineering professionals and will be responsible for addressing these types of disasters in the future. Events such as Hurricane Katrina and the Oso landslide provide real-world examples of concepts already being taught in undergraduate curricula. Incorporating natural disasters into the curriculum provides student engineers with a keener awareness of the impacts of engineering decisions and the applicability of technical concepts to real-world events. These curricular bridges highlight the importance and relevance of soil mechanics concepts such as soil classification, groundwater flow, settlement, subsurface stress, shear strength, and soil failures, as well as the broader implications of errors in design, maintenance, and public policy. Educating student engineers on the failures of the past is vital to preventing failures of this magnitude from occurring in the future, and in the wake of disasters such as Hurricane Katrina and the Oso landslide, preventing such events should be every engineer's priority.

Notes

1 D.P. Coduto, M.-C.R. Yeung, and W.A. Kitch (2011), *Geotechnical Engineering: Principles and Practices*, 2nd edn. (Upper Saddle River, NJ: Prentice Hall), pp. 122, 138–39, 157, 530–34, 545–46.
2 Ibid., pp. 190, 192–93, 207, 223.
3 Ibid., pp. 122, 253–54.
4 B.M. Das (2011), *Principles of Foundation Engineering*, 7th edn. (Stamford, CT: Cengage Learning), pp. 25–6.
5 Coduto et al. (2011), pp. 395–98, 409, 539.
6 Ibid., pp. 419, 422–23, 485–86.
7 Ibid., pp. 527–29, 533, 539.

8 Ibid., p. 261.
9 K.P. Terzaghi, R.B. Peck, and G. Mesri (1996), *Soil Mechanics in Engineering Practice*, 3rd edn. (New York: Wiley), pp. 85–86.
10 R.B. Seed, R.G. Bea, R.I. Abdelmalak, A.G. Athanasopoulos, G.P. Boutwell, J.D. Bray, J.-L. Briaud, C. Cheung, D. Cobos-Roa, J. Cohen-Waeber, B. D. Collins, L. Ehrensing, D. Farber, M. Hanemann, L. F. Harder, K. S. Inkabi, A. M. Kammerer, D. Karadeniz, R.E. Kayen, R. E. S. Moss, J. Nicks, S. Nimmala, J. M. Pestana, J. Porter, K. Rhee, M. F. Riemer, K. Roberts, J. D. Rogers, R. Storesund, A. V. Govindasamy, X. Vera-Grunauer, J. E. Wartman, C. M. Watkins, E. Wenk Jr., and S. C. Yim (2006), *Independent Levee Investigation Team Final Report: Investigation of the Performance of the New Orleans Flood Protection System in Hurricane Katrina on August 29, 2005* (Berkeley, CA: University of California, Berkeley), ch. 10, p. 3.
11 Terzaghi et al. (1996), pp. 222–23.
12 Seed et al. (2006), ch. 10, p. 3.
13 R.D. Knabb, J.R. Rhome, and D.P. Brown (2005, updated 2011), *Tropical Cyclone Report: Hurricane Katrina, 23–30 August 2005* (Miami, FL: National Hurricane Center), p. 3.
14 Knabb et al. (2005), p. 9; Seed et al. (2006), ch. 2 p. 23; L.E. Link, J.J. Jaeger, J. Stevenson, W. Stroupe, R.L. Mosher, D. Martin, J.K. Garster, D.B. Zilkoski, B.A. Ebersole, J.J. Westerink, D. T. Resio, R. G. Dean, M. K. Sharp, R. S. Steedman, J. M. Duncan, B. L. Moentenich, B. Howard, J. Harris, S. Fitzgerald, D. Moser, P. Canning, J. Foster, and B. Muller (2009), *Performance Evaluation of the New Orleans and Southeast Louisiana Hurricane Protection System: Final Report of the Interagency Performance Evaluation Task Force (IPET)* (Washington, DC: United States Army Corps of Engineers), pp. I-35–36.
15 Link et al. (2009), p. I-3.
16 E.S. Blake, C.W. Landsea, and E.J. Gibney (2011), *The Deadliest, Costliest and Most Intense United States Tropical Cyclones from 1851 to 2010 (and Other Frequently Requested Hurricane Facts)*, NOAA Technical Memorandum NWS NHC-6 (Miami, FL: National Oceanic and Atmospheric Administration / National Weather Service, National Hurricane Center), pp. 5, 7.
17 Seed et al. (2006), App. F, pp. 1–2; Link et al. (2009), p. I-29.
18 Seed et al. (2006), ch. 2, pp. 1–3.
19 Seed et al. (2006), ch. 10, p. 1.
20 Link et al. (2009), p. I-26; Blake et al. (2011), pp. 7, 9.
21 Seed et al. (2006), ch. 4, p. 11.
22 C.F. Andersen, J.A. Battjes, D.E. Daniel, B. Edge, W. Espey, Jr., R.B. Gilbert, T.L. Jackson, D. Kennedy, D.S. Mileti, J.K. Mitchell, P. Nicholson, C. A. Pugh, G. Tamaro, Jr., and R. Traver (2007), *The New Orleans Hurricane Protection System: What Went Wrong and Why: A Report by the American Society of Civil Engineers Hurricane Katrina External Review Panel* (Reston, VA: American Society of Civil Engineers), p. 21.
23 Seed et al. (2006), ch. 10, p. 1, ch. 11, pp. 16, 18.
24 Link et al. (2009), pp. I-29–30, I-118.
25 Ibid., pp. I-41, I-45–46.
26 Seed et al. (2006), ch. 3, pp. 3–5.
27 Ibid., ch. 2, pp. 6–7.
28 Ibid., ch. 6, pp. 3, 21, ch. 7, p. 11, ch. 8, p. 10.
29 Link et al. (2009), p. I-30.
30 Seed et al. (2006), ch. 8 pp. 7–8, ch. 11 p. 10; Link et al. (2009), p. I-30.
31 Ibid., pp. I-123–25.
32 Seed et al. (2006), ch. 4, p. 28.
33 Link et al. (2009), pp. I-25–26.
34 Seed et al. (2006), ch. 3, pp. 19–22, ch. 4, p. 28.
35 Ibid., ch. 4, p. 28.
36 Link et al. (2009), p. I-3.
37 Seed et al. (2006), ch. 4, p. 29.
38 Andersen et al. (2007), p. 67.
39 Link et al. (2009), p. I-3.
40 Seed et al. (2006), ch. 6, pp. 18–20, ch. 8., pp. 18–20, 39–40, 44; Link et al. (2009), pp. I-122–23.

41 Andersen et al. (2007), pp. 51–55.
42 Seed et al. (2006), ch. 8, p. 34, ch. 11, p. 17, ch. 12, pp. 14–15.
43 Andersen et al. (2007), pp. 50–51.
44 Seed et al. (2006), ch. 3, pp. 11–13, ch. 11, p. 19.
45 Seed et al. (2006), ch. 11, p. 19.
46 Seed et al. (2006), ch. 8, p. 34, ch.11, pp. 20–21.
47 Andersen et al. (2007), pp. 49–50.
48 J.R. Keaton, J. Wartman, S. Anderson, J. Benoit, J. de la Chapelle, R. Gilbert, and D.R. Montgomery (2014), *The 22 March 2014 Oso Landslide, Snohomish County, Washington*, Geotechnical Extreme Events Reconnaissance (GEER) Association Report, pp. 1, 33, 54, 159.
49 D.J. Miller and L.R. Miller (1999), *Hazel/Gold Basin Landslides: Geomorphic Review Draft Report* (Seattle, WA: M2 Environmental Services), p. 1.
50 J. Robertson (2014, April 2014), "Landslide in Washington State," *Science Features* [Blog], available at http://www.usgs.gov/blogs/features/usgs_top_story/landslide-in-washington-state/
51 Miller and Miller (1999), p. 1.
52 Ibid., p. 5.
53 D.J. Miller and J. Sias (1997), *Environmental Factors Affecting the Hazel Landslide* (Seattle, WA: M2 Environmental Services), pp. 3.9–10, 3.16; Miller and Miller (1999), p. 5.
54 M. Baker and J. Mayo (2014, March 26), "Logging OK'd in 2004 May Have Exceeded Approved Boundary," *The Seattle Times*; Miller and Sias (1997), pp. 1.1, 3.13–15.
55 Miller and Sias (1997); Miller and Miller (1999).
56 Keaton et al. (2014), pp. 54–55.
57 Andersen et al. (2007), p. 39.

Bibliography

Andersen, C.F., J.A. Battjes, D.E. Daniel, B. Edge, W. Espey, Jr., R.B. Gilbert, T.L. Jackson, D. Kennedy, D.S. Mileti, J.K. Mitchell, et al. (2007). *The New Orleans Hurricane Protection System: What Went Wrong and Why: A Report by the American Society of Civil Engineers Hurricane Katrina External Review Panel*. Reston, VA: American Society of Civil Engineers.

Baker, M., and J. Mayo. (2014, March 26). "Logging OK'd in 2004 May Have Exceeded Approved Boundary." *The Seattle Times*. Available at http://seattletimes.com/html/localnews/2023235343_mudslideovercutxml.html

Blake, E.S., C.W. Landsea, and E.J. Gibney. (2011). *The Deadliest, Costliest and Most Intense United States Tropical Cyclones from 1851 to 2010 (and Other Frequently Requested Hurricane Facts)*. NOAA Technical Memorandum NWS NHC-6. Miami, FL: National Oceanic and Atmospheric Administration / National Weather Service, National Hurricane Center.

Coduto, D.P., M.-C.R. Yeung, and W.A. Kitch. (2011). *Geotechnical Engineering: Principles and Practices*, 2nd edn. Upper Saddle River, NJ: Prentice Hall.

Das, B.M. (2011). *Principles of Foundation Engineering*, 7th edn. Stamford, CT: Cengage Learning.

Keaton, J.R., J. Wartman, S. Anderson, J. Benoit, J. deLaChapelle, R. Gilbert, and D.R. Montgomery. (2014). *The 22 March 2014 Oso Landslide, Snohomish County, Washington*. Geotechnical Extreme Events Reconnaissance (GEER) Association Report. Retrieved November 27, 2015, from http://www.geerassociation.org/index.html

Knabb, R.D., J.R. Rhome, and D.P. Brown. (2005, updated 2011). *Tropical Cyclone Report: Hurricane Katrina, 23–30 August 2005*. Miami, FL: National Hurricane Center. Retrieved November 27, 2015, from http://www.nhc.noaa.gov/pdf/TCR-AL122005_Katrina.pdf

Link, L.E., J.J. Jaeger, J. Stevenson, W. Stroupe, R.L. Mosher, D. Martin, J.K. Garster, D.B. Zilkoski, B.A. Ebersole, J.J. Westerink, et al. (2009). *Performance Evaluation of the New Orleans and Southeast Louisiana Hurricane Protection System: Final Report of the Interagency Performance Evaluation Task Force (IPET)*. Washington, DC: US Army Corps of Engineers.

Miller, D.J., and L.R. Miller. (1999). *Hazel/Gold Basin Landslides: Geomorphic Review Draft Report*. Seattle, WA: M2 Environmental Services. Retrieved November 27, 2015, from http://www.netmaptools.org/Pages/Hazel/Hazel_GoldBasin.pdf

Miller, D.J., and J. Sias. (1997). *Environmental Factors Affecting the Hazel Landslide*. Seattle, WA: M2 Environmental Services. Retrieved September 29, 2014, from http://www.netmaptools.org/Pages/Hazel/Hazel.pdf

Robertson, J. (2014, April 10). "Landslide in Washington State." *Science Features* [Blog]. Available at http://www.usgs.gov/blogs/features/usgs_top_story/landslide-in-washington-state/

Seed, R.B., R.G. Bea, R.I. Abdelmalak, A.G. Athanasopoulos, G.P. Boutwell, J.D. Bray, J.-L. Briaud, C. Cheung, D. Cobos-Roa, J. Cohen-Waeber, B. D. Collins, L. Ehrensing, D. Farber, M. Hanemann, L. F. Harder, K. S. Inkabi, A. M. Kammerer, D. Karadeniz, R.E. Kayen, R. E. S. Moss, J. Nicks, S. Nimmala, J. M. Pestana, J. Porter, K. Rhee, M. F. Riemer, K. Roberts, J. D. Rogers, R. Storesund, A. V. Govindasamy, X. Vera-Grunauer, J. E. Wartman, C. M. Watkins, E. Wenk Jr., and S. C. Yim (2006). *Independent Levee Investigation Team Final Report: Investigation of the Performance of the New Orleans Flood Protection System in Hurricane Katrina on August 29, 2005*. Berkeley, CA: University of California, Berkeley.

Terzaghi, K.P., R.B. Peck, and G. Mesri. (1996). *Soil Mechanics in Engineering Practice*, 3rd edn. New York: Wiley.

8 The social implications of disaster waste management on displaced residents

The case of Hurricane Katrina – a brief systematic review

Fatima M. Alfa

Background

This chapter is a systematic review and investigation of the social implications of disaster waste management on displaced citizens during the recovery phase of a natural disaster (a hurricane). The researcher will attempt to identify strategies for effective management of both disaster waste and citizen displacement. In the study, I will analyze the community-wide impact that cuts across a wide spectrum of infrastructures and services, such as clean water, healthcare, electricity, schools, access roads, safety, and so on.

The scope of this chapter is limited to the outcomes of Hurricane Katrina, which hit the Gulf Coast of the US in 2005, covering the southern states of Florida, Louisiana, Mississippi, and Alabama. The aftermath of Hurricane Katrina caused social problems such as the displacement of most residents, and the accumulation of natural waste from internally displaced persons.

Methods

I applied a realist synthesis methodology to existing studies on disaster waste management and the social impact of Hurricane Katrina from 2005 to 2014. I retrieved English-language publications from multiple databases and search engines such as the University of Maryland University College's (UMUC) OneSearch, Jstor, Elsevier, and Google.

I considered both a dependent variable (DV) and an independent variable (IV) in this research. The DV is displacement, while the IV is disaster waste management.

Introduction

In this study, I identify probable outcomes of evidence-based strategies in disaster recovery management. The research method that allows practitioners and management to modify strategies to fit the disaster is realist synthesis. The body of evidence in the research enabled the management and practitioners to monitor the dynamics closely and to coordinate them between the dependent and independent variables.

The scope of this research is limited to the effects of Hurricane Katrina. Documentation of evidence of the social impacts of Hurricane Katrina on the practitioners of evidence-based management (EBMgt) is provided. The review culminates in identified strategies for disaster waste management and will contribute to bringing management practitioners and researchers together. Evidence-based management practitioners, policy makers, and researchers will benefit from this research because it promotes good strategies for the disaster recovery process.

Evidence-based research synthesis

In the field of medicine, researchers pioneered evidence-based research (EBR) synthesis.[1] With the adoption of EBR came different nomenclature (e.g., evidence-based practice [EBP], EBMgt, evidence-based healthcare, etc.). Rousseau considered EBMgt to be "a knowledge-intensive, capacity-building way to think, act, organize and lead,"[2] which suggests that EBR/EBMgt could be essential to enhancing both management and leadership styles.

The systematic review

A systematic review is a comprehensive review process that identifies all relevant studies to answer a particular question.[3] Gough, Oliver, and Thomas defined systematic review as "a review of the research literature using systematic and accountable methods."[4] They assessed the validity of each study, taking the review process into account when reaching conclusions. This comprehensive process provides clarity.[5] The systematic review process for this research follows the review stages defined by Gough et al.,[6] as shown in Figure 8.1.

Research or review initiation

The research initiation emanates from my curiosity and interest in the management of natural disaster waste and displaced residents. Initially, my aim was to look at all kinds of natural disasters, but later I decided to narrow the focus to hurricanes – specifically, Hurricane Katrina.

Engaging stakeholders

The primary stakeholders for this review are Federal agencies (e.g., the Federal Emergency Management Agency [FEMA]; the US Environmental Protection Agency [EPA]; state, county, and city governments; humanitarian organizations such as the American Red Cross and Doctors Without Borders; policy makers; management practitioners; citizens in the affected areas; and the public). This research is independent because I am not in contact with FEMA or any of the identified stakeholders. However, I am familiar with FEMA's disaster recovery and waste management requirements and with several congressional reports on Hurricane Katrina.

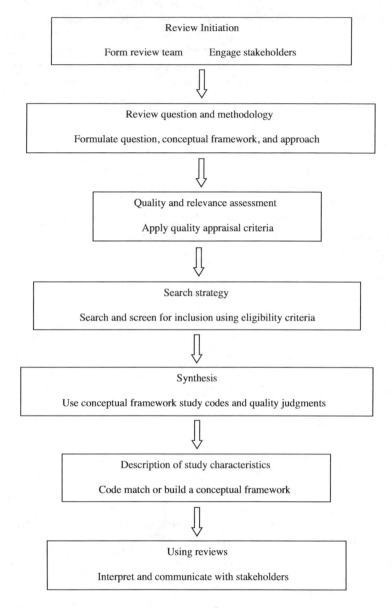

Figure 8.1 Common stages in a systematic review.[7]

Research question and methodology

The research question (RQ) is an important and critical part of any research. Formulating the RQ is the first step in a systematic review.[8] I have used two RQs:

RQ1: What is the social impact of disaster waste management on the residents?
RQ2: To what extent do disaster waste management and displacement affect each other?

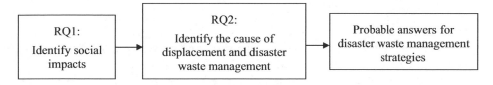

Figure 8.2 Research question mapping.

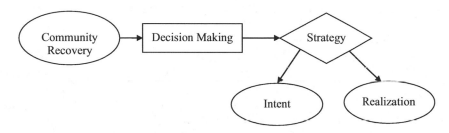

Figure 8.3 The adhocracy strategy.

With RQ1, I sought to identify the social impacts of disaster waste management and displacement. With RQ2, I sought to identify the cause of the phenomena being researched and the role of citizen participation.

Theoretical framework

I apply Mintzberg's theory of adhocracy strategy to the context, interventions, mechanisms, and outcomes (CIMO) of this study.[9] Mintzberg (2007) defined strategy as an action that emanates from decision-making.[10] This definition means that strategy takes place only after a decision-making process. The adhocracy strategy can facilitate the decision-making process for both management and the practitioner. I also postulate that the adhocracy strategy is a need- or situation-based strategy.[11] Mintzberg posited that an "adhocracy must hire and give power to experts, people whose knowledge and skills have been highly developed in training programs."[12] The adhocracy strategy is applicable to different situations. Mintzberg and McHugh argued that strategy concerns intent and realization, and that deliberate strategies (intentions realized) and emergent strategies (patterns realized despite the presence or absence of intentions – see Figure 8.3) must be used.[13] With deliberate strategy, the organization moves from intended strategy to realized strategy, while taking note of the unrealized strategy and knowing that an emergent strategy might also exist.

The context of disaster waste management and displaced residents fits into this theory because, as a social issue, different situations might arise that would warrant a different type of approach to resolution. Thus, adhocracy can address the flexibility of strategy, decision-making, and solutions that are much needed in disaster management.

Search strategy

Gough et al. recommended the development of an explicit search strategy for use over a spectrum of sources and for searching "databases of different disciplines."[14] The search strategy for the articles for inclusion and exclusion for this research follows these recommendations. The primary database used for the search was UMUC's OneSearch database; I also used Google Scholar, Jstor, and Elsevier. The DV (displacement) and IV (disaster waste management) were the main search criteria. I searched for scholarly peer-reviewed articles published in the English language between 2005 and 2014. The search terms reflected key words that boosted an exhaustive search for studies to help answer the research question. I have listed the search terms below.

- "Disaster waste management"
- "Hurricane Katrina" from 2005 to 2014
- "Hurricane Katrina" AND "recovery or cleanup" AND "social impact* or implication*"
- "Hurricane Katrina" AND "recovery or cleanup" AND "social impact* or implication*" from 2004 to 2014 (scholarly peer-reviewed articles)
- "Hurricane Katrina" AND "recovery or cleanup" AND "social impact* or implication*" from 2004 to 2014 (scholarly peer-reviewed articles, and SocIndex with full text)
- ("Hurricane Katrina") AND (recovery OR cleanup) AND (residents OR victims OR displacement) AND (implication*), 2005 to 2014 (scholarly peer-reviewed articles)
- Katrina AND (waste OR debris OR cleanup) AND manag* AND (resident* OR victim*) from 2005 to 2014 (scholarly peer-reviewed articles)

Inclusion-exclusion criteria

The primary inclusion and exclusion criterion for these research articles was relevance. As shown in Table 8.1, a search for Hurricane Katrina generated 76,269 articles (many of which were duplicates). I limited the search by year from 2005 to 2014, narrowing the list of results to 27,584 articles. I used a more specific search string ("Hurricane

Table 8.1 Inclusion/exclusion criteria

Parameter	Inclusion criteria	Exclusion criteria
Relevance to hurricane disaster waste management	Yes	No
Language	English	All other languages
Year of Publication	2005–2014	Before 2005
Publication Type	Full-text scholarly peer-reviewed papers and congressional reports	Trade journals, videos, etc.
Type of Disaster	Hurricanes	Other natural disasters
Location	Alabama, Florida, Louisiana, and Mississippi	Other locations in the US and other countries
Type of Study	Both primary and secondary	

Katrina" AND "recovery or cleanup" AND "social impact* or implication*") to narrow the results further to 3,109 articles. A restriction of the same search string to scholarly peer-reviewed articles published between 2005 and 2014 resulted in 1,150 articles. This number was still too high and contained many healthcare-related articles, so the search was narrowed again using only the SocIndex with a full text database, instead of searching all providers and academic journals. Only one article was located in this search. I then used this search string – Katrina AND (waste OR debris OR cleanup) AND manag* AND (resident* OR victim*) – to identify 249 articles.

Another search was conducted with the following string: ("Hurricane Katrina") AND ("recovery" OR "cleanup") AND ("residents" OR "victims" OR "displacement") AND ("implication"*) to identify ten articles, but only one article met the selection criterion of relevance. Finally, I searched the following string: "Disaster waste management" and identified five articles. I selected one article from this search result and identified two more articles from the reference lists of the selected article.

To ensure the integrity of the selection process, I used a published screening tool, the Critical Appraisal Skills Programme (CASP) International Network's (2014) manual review criteria (illustrated in Figure 8.4): a process of manual search (finding),

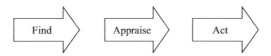

Figure 8.4 Manual review to build criteria standard data (modified).[15]

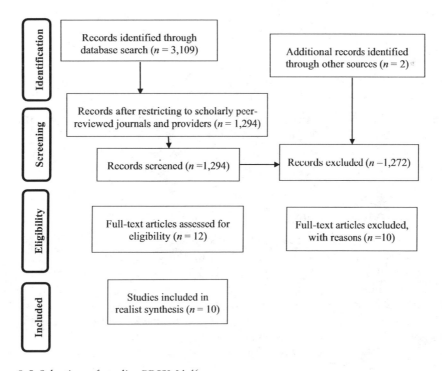

Figure 8.5 Selection of studies PRISMA.[16]

screening (appraisal), and selection (action). I browsed the abstracts of these articles to screen for articles relevant to disaster waste management, displaced residents, and the various social impacts.

Quality appraisal

Gough et al. posited that appraising the relevance of studies is an effective tool for reviewers to ensure the integrity and pertinence of studies used to develop the conclusions of a review.[17] I used the CASP tool (developed by the CASP International Network) to assess the quality of research articles for this research. The CASP tool was used to appraise quality in three categories, with a total of ten criteria that fell into the following three categories:

Assessment of the quality of the study reporting

This category consists of five criteria for appraising quality: the study purpose, methodology, design, sampling, and data. Any body of evidence that scored either "strong" or "medium" was selected for inclusion.

Assessment of the quality of the transparency and rigor of the study

Three criteria were used to assess transparency and rigor: bias, ethics, and rigor. In this category, all articles that were scored "strong" were selected, and articles that were scored "weak" or "medium" regarding the study population were all selected for inclusion because such relationships are difficult to determine.

Assessment of the quality of the study methods: Is there a clear statement of findings?

In this category, I assessed the study findings and value. All studies whose authors had clearly stated their findings and had a strong score were included. The global rating rubric was applied to the appraisal criteria as follows: Strong (S) = 1; Moderate (M) = 2; and Weak (W) = 3.

Data extraction and coding

Gough et al. defined data extraction or collection as "taking from the studies information about their focus of interest, methods, findings or authors' discussion."[18] Data were extracted using these key terms and phrases: disaster, disaster waste management, debris removal, recovery, relief, leadership, FEMA, strategy, displacements, social implications, hurricanes, Hurricane Katrina, decision-making, and citizen participation. Gough et al. defined coding as "any key wording, data extraction, or other annotation of studies to conduct a systematic review."[19] Saldana explained that a code does not necessarily have to reproduce the data. Saldana's view seemed to give researchers the choice of reproducing data or simply condensing it.[20] Gough et al. indicated that a researcher must have a detailed coding or data extraction system to describe the methods used in the study, the phenomenon under study, and its context, explaining that detailed coding also gives insight into the criteria for judging the study's trustworthiness and relevance to the review question.[21]

The coding of the body of evidence used in this review was done by category of themes, frameworks, and theories. The coding approach in this study is a synopsis and annotation of the main points of the study. First, I annotated the extracted data and stored them in Microsoft Excel. Data that suggested a waste management issue were then extracted.

Realist synthesis

A systematic review involves the synthesis of evidence gathered from EBR. Pawson et al. defined a realist synthesis as "an approach to reviewing research evidence on complex social interventions, which provides an explanatory analysis of how and why they work (or don't work) in particular contexts or settings."[22] This quality of the realist synthesis provides the management practitioner with the evidence for informed decisions.

The realist synthesis enabled me to identify the underlying causal mechanism in the variables of the study. The technique of identifying what works and what does not is very important to practitioners of EBR and management, as it is an identification tool that also proffers solutions to the problem of disaster waste management, displacement, and their social impacts.

Synthesis of findings using CIMO logic

The findings of the evidence used in this study show that a link exists between disaster waste management and displacement. The slow pace of disaster waste removal from access roads prolonged the displacement time, and the delayed removal of disaster waste posed a public hazard. I configured and analyzed the research evidence using a realist synthesis. The workings of the interventions were identified according to the premise that the realist synthesis would be used to explore the contextual workings of interventions.[23]

The discussion of the synthesis of findings is from the CIMO relationship found in the body of evidence for this research. The thrust of the CIMO logic is in finding what works in what context and what does not work. The implication is that several iterations of interventions can exist without the root cause of a problem being found.

Therefore, applying CIMO logic can potentially identify the intervention or mechanism that is not working in a particular situation. CIMO logic can be beneficial to practitioners because it lays out the mapping for identifying the workability of interventions. It has the potential to optimize the time spent on researching or looking for the root cause of a problem.

CIMO logic is also valuable to the practitioner because of its flexibility, for it can be applied to different situations. Therefore, one can classify the realist synthesis as a fit-for-purpose technique. Figure 8.6 illustrates CIMO logic.

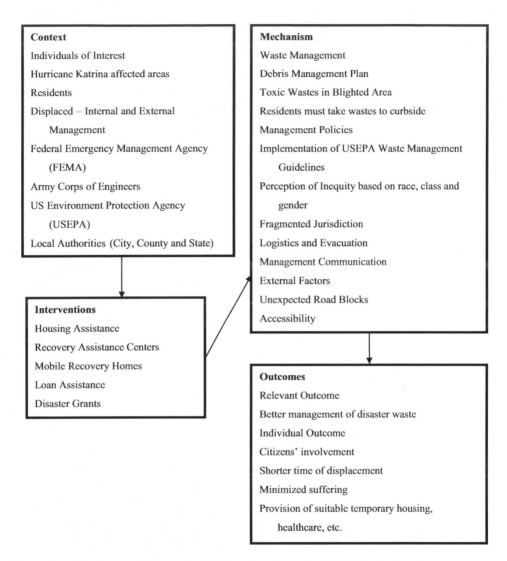

Figure 8.6 The Context-Intervention-Mechanism-Outcome (CIMO) relationship.
CIMO Logic

Context (C)

Areas, residents, and management that were affected by Hurricane Katrina are the contexts of this study. Internally displaced residents are those residents who have temporarily relocated to shelters or homes within their city. Management are the authorities responsible for all phases of disaster management, including FEMA, the Army Corps of Engineers, the EPA, local and municipal authorities such as city, county, and state agencies, and waste management contractors (see the context section of Figure 8.6, above).

Interventions (I)

Interventions refer to the various recovery programs such as housing, loans, and grants. It is expected to identify the interventions that worked and those that did not work. The identification of the conducive interventions through the application of Mintzberg's adhocracy theory[24] would point the practitioner to strategies that would lead to quicker disaster waste management and quicker return of displaced residents (see Figure 8.6).

Mechanisms (M)

The mechanisms of interest are disaster waste management, external factors, and management policies. I discuss and analyze these mechanisms in the three categories outlined in Figure 8.3.

Waste management

The mechanism applied here is the EPA's Guidelines for Disaster Waste Management, which also provide technical and management options on disaster waste. Individual municipal or state government is responsible for the specific guidelines for waste management in its jurisdiction.[25] The collaboration includes inter- and intra-agency experts who work together on the specific waste management option adopted.

Regarding the mechanism of the debris management plan, although guidelines for disaster waste management exist, it appears that no evidence of a debris management plan was in place for Hurricane Katrina. The lack of an existing plan for environmental hazards and debris removal created a challenge of catastrophic proportions. Hurricane Katrina generated a great amount of environmentally hazardous waste, thereby complicating the issue of debris removal.[26] The challenge of debris removal is that policies and procedures always hamper effective and efficient removal. The lesson from this challenge is that the US Department of Homeland Security should coordinate with the EPA in its efforts to remove debris during and after a disaster. In a congressional report on Hurricane Katrina, Luther identified the lack of a debris management plan as a challenge to disaster waste management.[27] Brown et al. also identified the disposal of large volumes, including toxic waste, as a challenge to waste managers.[28] A suitable strategy for this mechanism is a synergy of management, practitioners, and community members adopting a pre-disaster debris management plan. This synergistic approach will limit confusion and chaos and hasten debris removal.

Leadership style and policies

The mechanism of leadership style and policies was detrimental to disaster waste management. The perception among the residents was that some of the mechanisms of leadership or management policies promoted inequity according to class, race, and gender. Keithly and Rombough indicated that government policies (e.g., the refusal to support rebuilding projects and the lack of assistance for homeowners who wanted to rebuild) contributed to the displaced residents' decisions not to return home.[29] Pierre and Stephenson identified the government officials' inequity in the cleanup and recovery effort as a problem, asserting that the government authorized the disposal of toxic waste in landfills located in blighted neighborhoods despite protests from social justice activists.[30] Fragmented jurisdiction was another management policy applied. The lack of unified management, the unclear and overlapping roles and responsibilities of the command and control within the federal government, and the weakness and insufficiency of the regional planning and coordination structures undermined preparedness efforts.[31] The lessons learned from this challenge were that the federal government must revisit existing recovery plans and coordinate with regional offices to ensure that functional operational structures will ensure efficiency and effectiveness; the same must be done within the federal bureaucracy regarding the supply process.[32] Public involvement would enhance the proper management of disaster waste. Esworthy et al. identified this assurance as an important step in disaster recovery.[33]

Supporting the view of Esworthy et al., Brown et al. identified inadequate "guidance on the most effective way to incorporate communities into disaster waste decision making" as an impediment to disaster waste management following Hurricane Katrina.[34] They suggested that public consultation during the disaster waste management process might have increased public understanding of the actions necessary for efficient management of the waste, and that publicly unacceptable waste management options should have been identified before attempts were made to implement them. There was inadequate management communication between the authorities on Hurricane Katrina Waste Management and the residents. According to the United States Government, the hurricane destroyed much of the communication infrastructure, thereby impeding communication.[35] Supporting this view, Brown et al. identified the issues of "limited understanding of the impact of disaster waste management on community recovery" and "post-disaster communities' behavior on waste management programmes."[36] Therefore, communication was a huge challenge for disaster waste managers.

External factors

Lack of management communication also created a loophole for external factors to slow the process of recovering from Hurricane Katrina. Agencies that are not directly involved in the disaster management effort constitute external factors. A strategy of information dissemination to the public should be in place before a disaster hits. A Web database system for sharing information among agencies involved in recovery efforts would also enhance management communication.

Outcomes (O)

Disaster waste management

Outcomes are the results of the mechanisms applied in the context of Hurricane Katrina. The outcome of the EPA's disaster guidelines was ambiguous and lacked a sense of direction, thereby impeding the debris removal process. The guidelines did not work because of the lack of guidance on decision-making and option consideration in different disaster situations. The outcome of this lack of debris management was that authorities who were responsible for waste management appeared to have found the task overwhelming. Luther stated that a debris management plan would ensure the ability of property owners to return to an area and to assist with the cleanup, separating hazardous and non-hazardous waste and managing asbestos-contaminated waste.[37] The Boulder Valley District used an adhocracy strategy in community collaboration for disaster risk reduction and recovery that was dubbed "relocalizing."[38] The case of Boulder, CO, shows that accepting a synergy of informal and formal networks can yield successful outcomes.

The inability to separate wastes according to EPA guidelines slowed the process of leaving waste on the curbside for removal. This outcome shows that displaced residents and disaster waste management have direct causal effects on each other. Cooke indicated that many residents could not return to their homes until after the cleanup. Some residents were ready to return, but authorities disallowed them from returning because their neighborhoods had not been cleaned.[39] Most of the damaged buildings were rental properties, which made the residents less likely to return.[40] Road blockages not only prevented residents from returning to their homes, but they also posed a potential health risk, for they created pools of stagnant water and other contaminated wastes.[41]

The lack of a management plan or contingency plan also resulted in non-hurricane-related waste. Brown et al. stated that "the post-disaster waste generated from excessive unwanted donations, large amounts of healthcare wastes, [and] rotten food from power outages and emergency relief food packaging" added to the problems arising from the hurricane.[42] This new municipal waste also constituted a health hazard. The outcome of fragmented government policies was that they hindered the progress of waste management, thereby prolonging the recovery process.

Leadership style and policies

The outcome of management policies was the slow return of most residents to their homes. Green et al. identified other factors that contributed to the slow return of residents in the affected areas of New Orleans, for example, "shortage of labor, lack of insurance coverage, lack of capital, overwhelmed service sector, [and] limited access to schools and healthcare."[43] School enrollment dropped, and students did not have access to their schools.[44]

Community participation

The outcome of non-existent community participation in disaster waste management and decision-making resulted in a prolonged recovery process. Cooke recommended the return and participation of residents to hasten the cleanup process.[45] Involving the community in the decision-making process is therapeutic for the affected homeowners.

Management communication

One outcome of the lack of adequate communication mechanisms was the dumping of toxic waste in landfills sited in blighted neighborhoods by waste managers. Evidence shows that this management action resulted in public opposition to the use of construction and demolition landfills for mixed waste. Evidence also shows that management communication facilitates disaster waste management. This public opposition, in turn, led to a lawsuit and an eventual closing of the landfill sites.[46] Cooke gave examples of successful citizen participation from effective communication after the Lincoln, NE, flood of 1993.[47] Brown et al. indicated that communication was important to the successful and efficient debris removal after Hurricanes Frances and Jeanne.[48] It is not clear why the effective use of communication following these two hurricanes was not replicated after Katrina. It might be attributed to the level of devastation caused by Katrina. Esworthy et al. indicated that a critical part of disaster recovery is keeping the public well informed. Nevertheless, the authors caution that doing so could be difficult, especially in the early aftermath of storm events.[49]

External factors

The outcome of the external factors was that access to some sites was limited for both the waste management contractors and the public. This lack of access in turn slowed debris removal and affected the general disaster waste management process. An expedited disaster debris removal would have accelerated community recovery and the rebuilding process.

Conclusion

Through this research, I have demonstrated that a direct link exists between disaster waste management and the displacement of residents of disaster-affected areas.[50] As I have pointed out, it would have been easy to apply Mintzberg's adhocracy strategy[51] to address the confusion and chaos that ensued because of the multiplicity of strategies for humanitarian assistance. As mentioned, the adhocracy strategy is flexible, and it accommodates a mixture of different strategies with the aim of finding the one that works best in the particular circumstances, regardless of how diverse the viewpoints might be. The fact that experts in different areas of recovery can assemble and develop plans tailored to any community makes the strategy a viable option for the long-term issue of immediate attention to recovery following a natural disaster. For each of these outcomes, a concerted effort was made to identify the strategy that would be most applicable to achieve an effective outcome.

In addition, an effort was made to cite an example of where the strategy had been applied with the desired results. Evidence shows that some strategies were applied in the twentieth century, which worked with the clean-up of debris from disaster, but why those same strategies were not applied in the disasters of the twenty-first century is a subject for further research.

Evidence points to the fact that the government (at the federal, state, and local levels) has great resources to make recovery processes seamless; however, as I have demonstrated, the lack of proper coordination complicated the process.

If the adhocracy strategy were adopted, it would have led to a uniform guide that could have easily been adapted to any particular situation. As demonstrated in the research, using a prescriptive method to deal with disaster waste management and displaced citizens will not work to solve the issues of early removal of waste and the return of those citizens. The community must be involved in the decision-making process and the execution of the adopted strategy. Such involvement definitely creates a sense of ownership, as opposed to the feeling of displaced people who have others dictating decisions to them.

I have identified some strategies, such as a synergy of management, practitioners, and community members in adopting a pre-debris management plan, a synergy of formal and informal networks, diversity awareness, and the dissemination of a pre-disaster information plan, and a web-based database management system for sharing information among agencies. If applied, these strategies will contribute to the ongoing quest for immediate solutions to disaster debris removal and the return of displaced citizens.

Future study

Considering the lessons learned and published by different agencies, in congressional reports and hearings, it was evident that many difficulties and challenges were encountered in the Hurricane Katrina recovery efforts.[52] Therefore, the United States Congress required FEMA to develop a National Disaster Recovery Framework (NDRF) in recognition of the importance of improving disaster recovery in America. The intention was to have a framework on disaster relief that would benefit the whole nation in any kind of disaster, whatever its source. President Obama established the NDRF to look for ways to tackle the challenges of disaster relief efforts more effectively and efficiently. Previously, the approach had always been reactive – that is, communities looked for ways to address the issues after the disaster had struck. After several such failed experiences in properly and adequately bringing about the desired recovery efforts, it is now time to develop proactive measures that will address the same recovery efforts so that they are effective and efficient. The framework is an attempt to design a blueprint for disaster recovery, so that even if it does not meet a community's exact needs, adjustments can be made to make it work.

In the NDRF, the Federal Emergency Management Agency[53] addressed leadership in recovery efforts and detailed its roles and responsibilities in both the pre- and post-disaster eras. Given that the NDRF was developed almost six years after Hurricane Katrina, it is important to test its effectiveness. Therefore, a study comparing Hurricanes Katrina and Sandy might throw light on the effectiveness of the NDRF.

Notes

1 E. Barends, B. Janssen, W. ten Have, and S. ten Have (2013, January), "Effects of Change Interventions: What Kind of Evidence Do We Really Have?" *Journal of Applied Behavioral Science* [Online edition], doi:10.1177/0021886312473152, 3.
2 D.M. Rousseau (2012), "Envisioning Evidence-Based Management," in *The Oxford Handbook of Evidence-Based Management*, D.M. Rousseau, ed. (Oxford, UK: OUP), p. 1.
3 D. Denyer and D. Tranfield (2009), "Producing a Systematic Review," in *The Sage Handbook of Organizational Research Methods*, D.A. Buchanan and A. Bryman, eds. (Thousand Oaks, CA: Sage), p. 671.
4 Gough et al. (2012), 5.
5 M. Petticrew and H. Roberts (2006), *Systematic Reviews in the Social Sciences* (Malden, MA: Blackwell), pp. 28, 10.
6 Gough et al. (2012)
7 D. Gough, S. Oliver, and J. Thomas (2012), *An Introduction to Systematic Reviews* (Thousand Oaks, CA: Sage), p. 8.
8 Petticrew and Roberts (2006), p. 28; J.E. Squires, J.C. Valentine, and J.M. Grimshaw (2013), "Systematic Reviews of Complex Interventions: Framing the Review Question," *Journal of Clinical Epidemiology* 66(11), p. 1215.
9 H. Mintzberg (2007), *Tracking Strategies: Towards a General Theory* (Oxford, UK: OUP).
10 Ibid., p. 1.
11 Ibid., pp. 71–113.

12 Ibid., p. 199.
13 H. Mintzberg and A. McHugh (1985), "Strategy Formation in an Adhocracy," *Administrative Quarterly* 30(2), p. 161.
14 Gough et al. (2012), pp. 116, 118.
15 Critical Appraisal Skills Programme (2014), *CASP Website*, retrieved December 30, 2015, from http://www.casp-uk.net/#!internationally/c3cw
16 D. Moher, A. Liberati, J. Tetzlaff, D.G. Altman, and the PRISMA Group (2009), "Preferred Reporting Items for Systematic Reviews and Meta-Analyses: The PRISMA Statement," *PLoS Med* 6(6), p. e1000097.
17 Gough et al. (2012), p. 155.
18 Ibid., p. 138.
19 Ibid.
20 J. Saldana (2013), *The Coding Manual for Qualitative Researchers*, 2nd edn. (Los Angeles, CA: Sage).
21 Gough et al. (2012), p. 143.
22 R. Pawson, T. Greenhalgh, G. Harvey, and K. Walshe (2004), *Realist Synthesis: An Introduction* (London, UK: ESRC Research Methods Program), p. 5.
23 R. Pawson (2006), *Evidence-Based Policy: A Realist Perspective* (London: Sage), p. 74.
24 Mintzberg (2007).
25 C. Brown, M. Milke, and E. Seville (2011), "Disaster Waste Management: A Review Article," *Waste Management* 31, p. 1088.
26 US Government (2006), *The Federal Response to Hurricane Katrina: Lessons Learned* (Washington, DC: Author), p. 61.
27 L. Luther (2011), *Managing Disaster Debris: Overview of Regulatory Requirements, Agency Roles, and Selected Challenges* (Washington, DC: Congressional Research Service).
28 Brown et al. (2011), p. 1091.
29 D.C. Keithly and S. Rombough (2007), "The Differential Social Impact of Hurricane Katrina on the African American population of New Orleans," *Race, Gender and Class* 14(3–4), p. 150.
30 J.K. Pierre and G.S. Stephenson (2008), "After Katrina: A Critical Look at FEMA's Failure to Provide Housing for Victims of Natural Disasters," *Louisiana Law Review* 68(2), p. 447.
31 US Government (2006), p. 54.
32 Ibid., pp. 56, 59.
33 R. Esworthy, L.J. Schierow, C. Copeland, and L. Luther (2005, October), *Cleanup after Hurricane Katrina: Environmental Considerations* (Washington, DC: Congressional Research Service).
34 Esworthy et al. (2005); Brown et al. (2011), pp. 1093, 2012.
35 US Government (2006).
36 Brown et al. (2011), p. 1093.
37 Luther (2011), p. 9.
38 I. Kelman (2008), "Relocalising Disaster Risk Reduction for Urban Resilience," *Urban Design and Planning* 161(December), pp. 197–204.
39 T.J. Cooke (2009), "Cleaning up New Orleans: The Impact of a Missing Population on Disaster Debris Removal," *Journal of Emergency Management* 7(3), p. 24.
40 Ibid.
41 Brown et al. (2011), p. 1086.
42 Brown et al. (2011), p. 1089.
43 R. Green, L.K. Bates, and A. Smyth (2007), "Impediments to Recovery in New Orleans' Upper and Lower Ninth Ward: One Year after Hurricane Katrina," *Disasters* 3(4), p. 322.
44 Keithly and Rombough (2007), p. 150.
45 Cooke (2009), p. 21.
46 Luther (2011).
47 Cooke (2009), p. 23.
48 Brown et al. (2011), p. 1093.
49 Esworthy et al. (2005), p. 28.
50 Cooke (2009); Luther (2011); Brown et al. (2011).
51 Mintzberg (2007).

52 S.J. Czerwinski (2012, June), *Disaster Recovery: Selected Themes for Effective Long-Term Recovery*, GAO-12–813T (Washington, DC: Government Accountability Office), p. 1.
53 Federal Emergency Management Agency (2014), *National Disaster Recovery Framework: Strengthening Disaster Recovery for the Nation, 2011* (Washington, DC: Author).

Bibliography

Barends, E., B. Janssen, W. ten Have, and S. ten Have. (2013, January). "Effects of Change Interventions: What Kind of Evidence Do We Really Have?" *Journal of Applied Behavioral Science* [Online edition], 1–23, doi:10.1177/0021886312473152

Brown, C., M. Milke, and E. Seville. (2011). "Disaster Waste Management: A Review Article." *Waste Management* 31, pp. 1085–98.

Cooke, T.J. (2009). "Cleaning up New Orleans: The Impact of a Missing Population on Disaster Debris Removal." *Journal of Emergency Management* 7(3), pp. 21–31.

Critical Appraisal Skills Programme. (2014). *CASP Website*. Retrieved December 30, 2015, from http://www.casp-uk.net/#!internationally/c3cw

Czerwinski, S.J. (2012, June). *Disaster Recovery: Selected Themes for Effective Long-Term Recovery*. GAO-12–813T. Washington, DC: Government Accountability Office. Available at http://www.gao.gov/products/GAO-12–813T

Denyer, D., and D. Tranfield. (2009). "Producing a Systematic Review." In *The Sage Handbook of Organizational Research Methods*. D.A. Buchanan and A. Bryman, eds. pp. 671–89. Thousand Oaks, CA: Sage.

Esworthy, R., L.J. Schierow, C. Copeland, and L. Luther. (2005, October). *Cleanup after Hurricane Katrina: Environmental Considerations*. Washington, DC: Congressional Research Service. Available at http://assets.opencrs.com/rpts/RL33115_20051013.pdf

Federal Emergency Management Agency. (2014). *National Disaster Recovery Framework: Strengthening Disaster Recovery for the Nation, 2011*. Washington, DC: Author. Available at http://www.fema.gov/pdf/recoveryframework/ndrf.pdf

Gough, D., S. Oliver, and J. Thomas. (2012). *An Introduction to Systematic Reviews*. Thousand Oaks, CA: Sage.

Green, R., L.K. Bates, and A. Smyth. (2007). "Impediments to Recovery in New Orleans' Upper and Lower Ninth Ward: One Year after Hurricane Katrina." *Disasters* 3(4), pp. 311–35, doi:10.1111/j.1467–7717.2007.01011.x

Keithly, D.C., and S. Rombough. (2007). "The Differential Social Impact of Hurricane Katrina on the African American population of New Orleans." *Race, Gender and Class* 14(3–4), pp. 142–53. Available at http://www.jstor.org/discover/10.2307/41675296?uid=3739704&uid=2&uid=4&uid=3739256&sid=21104445294881

Kelman, I. (2008). "Relocalising Disaster Risk Reduction for Urban Resilience." *Urban Design and Planning* 161(December), pp. 197–204, doi:10.1680/udap.2008.161.4.197

Luther, L. (2011). *Managing Disaster Debris: Overview of Regulatory Requirements, Agency Roles, and Selected Challenges*. Washington, DC: Congressional Research Service. Available at http://assets.opencrs.com/rpts/RL34576_20110113.pdf

Mintzberg, H. (2007). *Tracking Strategies: Towards a General Theory*. Oxford, UK: OUP.

———, and A. McHugh. (1985). "Strategy Formation in an Adhocracy." *Administrative Quarterly* 30(2), pp. 160–97.

Moher, D., A. Liberati, J. Tetzlaff, D.G. Altman, and the PRISMA Group. (2009). "Preferred Reporting Items for Systematic Reviews and Meta-Analyses: The PRISMA Statement." *PLoS Med* 6(6), p. e1000097, doi:10.1371/journal.pmed.1000097

Pawson, R. (2006). *Evidence-Based Policy: A Realist Perspective*. London: Sage.

———, T. Greenhalgh, G. Harvey, and K. Walshe. (2004). *Realist Synthesis: An Introduction*. London, UK: ESRC Research Methods Program. Available at https://www.ccsr.ac.uk/methods/

Petticrew, M., and H. Roberts. (2006). *Systematic Reviews in the Social Sciences*. Malden, MA: Blackwell.

Pierre, J.K., and G.S. Stephenson. (2008). "After Katrina: A Critical Look at FEMA's Failure to Provide Housing for Victims of Natural Disasters." *Louisiana Law Review* 68(2), pp. 443–95. Available at http://digitalcommons.law.lsu.edu/cgi/viewcontent.cgi?article=6232&context=lalrev

Rousseau, D.M. (2012). "Envisioning Evidence-Based Management." In *The Oxford Handbook of Evidence-Based Management*. D.M. Rousseau, ed. pp. 3–24. Oxford, UK: OUP.

Saldana, J. (2013). *The Coding Manual for Qualitative Researchers*, 2nd edn. Los Angeles, CA: Sage.

Squires, J.E., J.C. Valentine, and J.M. Grimshaw. (2013). "Systematic Reviews of Complex Interventions: Framing the Review Question." *Journal of Clinical Epidemiology* 66(11), pp. 1215–22, doi:10.1016/j.jclinepi.2013.05.013

US Government. (2006). *The Federal Response to Hurricane Katrina: Lessons Learned*. Washington, DC: Author. Available at http://georgewbush-whitehouse.archives.gov/reports/katrina-lessons-learned/

9 An urban form response to disaster vulnerability

Examining tsunami evacuations in two Chilean cities

Jorge León and Alan March

Introduction

Rapid-onset urban disasters such as near-field tsunamis require prompt and usually autonomous responses from vulnerable populations, including appropriate decisions about critical activities such as evacuation and sheltering. The characteristics of urban forms can increase a community's capacity to deal with such crises as they unfold and help to achieve resilience to disasters; however, little research exists in this area,[1] and the actual changes that might bring about improvement are not commonly examined. In this respect, the majority of tsunami risk-reduction efforts analyze the built environment (e.g., land zoning) and emergency readiness actions (e.g., evacuation preparation) as separate areas of study and practice: the former focuses on long-term changes excluding evacuation analysis, while the latter examines the urban realm as a static context.[2] This chapter analyzes this gap, using research into the tsunami-prone cities of Iquique and Talcahuano in Chile. It includes a diagnosis of existing situations, a series of urban modifications to improve these conditions, and an appraisal of these changes using qualitative and quantitative tools.

The chapter is divided into four parts. First, a theoretical background regarding disasters (particularly tsunamis) and the role of urban form in coping strategies is set out. Second, the case studies of Iquique and Talcahuano are introduced, including the proposed research method. Third, the existing situation and the urban recommendations for improvement are examined with the aid of computer-based models and fieldwork. Fourth, the results and the critical issues for tsunami risk reduction in the urban built environment are discussed.

The role of urban form in coping with disasters

From emergency management to disaster risk reduction: An ongoing transition

The number of disasters recorded yearly has quadrupled since the 1960s[3] as a result of climate change, poor management of natural resources, and growing exposure of populations to natural hazards. Cities in the developing world in particular have become hot spots for disasters as urbanization gathers pace, including inappropriate characteristics such as high densities, poverty, and rapid expansion of informal settlements.[4] The rising disaster trend has been paralleled by a steady evolution in governmental coping tactics, from civil-defense-based response and relief approaches to risk-reduction (or mitigation) strategies.[5]

Urban form for increasing disaster resilience

Built environment disciplines (such as urban planning and design) can support mitigation strategies by managing and modifying the location, spatial arrangements, functions, and growth trends of cities and regions; they can also contribute to raising governmental and public awareness of the risk of natural hazards, and to achieving overall reductions in urban disaster vulnerability.[6] Less attention has been paid, however, to *how* these disciplines can foster appropriate urban forms capable of improving other essential disaster management activities, such as response and recovery, which might lead to increased community resilience to disasters. This can be defined as the ability to cope with such catastrophes by surviving them, minimizing their impacts, and recovering with minor social disruption.[7]

A critical case of rapid onset disasters: Near-field tsunamis

"Near-field" or "local" tsunamis are those generated within 1,000 km of an area of interest, which in turn implies arrival times between 30 minutes and approximately two hours.[8] In cases such as this, vulnerable populations (up to tens of thousands of people) have little time to make appropriate decisions about key disaster-response activities such as evacuation (considered "the most important and effective method to save human lives" in the case of a tsunami)[9] and sheltering, usually without any formal guidance. This context of crisis can be improved by an urban form capable of promoting resilience by supporting rapid and effective responses to the tsunami emergency.

There are two key features in any urban tsunami evacuation system: routes and safe assembly areas (either "horizontal," i.e., open public spaces, or "vertical," i.e., existing buildings or evacuation towers). Disciplines that foster appropriate urban forms comprising these elements, such as urban planning or urban design, can therefore have a strong influence on overall outcomes of emergency evacuations. Within tsunami-risk-reduction frameworks, urban form and evacuation should be strongly linked in a two-way relationship subject to ongoing review and improvement. Nevertheless, current approaches,[10] upon examination, typically exhibit a lack of connection between these two aspects. Urban-form-based tsunami risk-reduction approaches (such as zoning, relocation of population and activities, creation of buffer areas, and design or building codes) are long-term mitigation strategies; they do not normally involve a deep analysis of the actual population responses in the case of a tsunami emergency. In turn, evacuation analyses commonly examine the different aspects of recorded or potential crises (e.g., population behavior, departure/displacement times involved, risks, etc.), but with an exclusive focus on the description and/or diagnosis of the situation rather than a deeper examination of how the urban spaces themselves could be designed to improve the population's response (e.g., evacuation and sheltering). The urban form is addressed as an almost immutable context.

This chapter aims to bridge this gap between evacuation and the built environment by examining the role of urban form in achieving more effective and safer evacuations during near-field tsunamis, the paths to improving this role through appropriate urban design recommendations, and the possible quantitative and qualitative tools to appraise these. To achieve this, this chapter will introduce the case studies of two Chilean tsunami-prone and currently highly vulnerable cities: Iquique and Talcahuano. The chapter is focused on pedestrian evacuation, which is strongly encouraged over car-based evacuation in the case of a near-field tsunami emergency.[11]

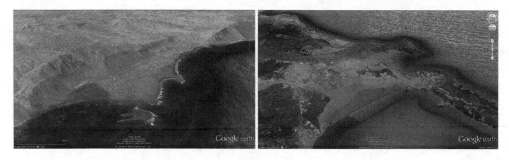

Figure 9.1 Satellite images of Iquique (left) and Talcahuano (right), overlapped with expected tsunami flood area (shaded in gray).[12]

Urban form and tsunami evacuation in two Chilean cities

Iquique and Talcahuano

Iquique (20°32'S, 70°11'W) and Talcahuano (36°43'S, 73°7'W) are two mid-size Chilean coastal cities located in tsunami-vulnerable coastal plains, with estimated populations of 284,539 and 172,000, respectively as of 2012 (see Figure 9.1).

These cities have been repeatedly affected by destructive tsunamis throughout recorded history: in 1604, 1715, 1868, and 1877 in the case of Iquique, and in 1570, 1657, 1730, 1751, 1835, and 2010 in the case of Talcahuano. The risk posed by tsunamis has not prevented urban growth occurring in vulnerable areas. The tsunami flood maps recently developed by the Chilean Navy[13] show that, as the outcome of historical development processes, Iquique and Talcahuano have large proportions of their urban areas in vulnerable conditions: 49.7 percent (around 770 hectares) and 66 percent (approximately 1,300 hectares), respectively. Currently, roughly 80,000 and 70,000 people live in tsunami-hazardous areas in Iquique and Talcahuano, respectively.

Research method and outcomes

Both case studies were examined in three sequential phases: (1) a diagnosis of the existing situation; (2) a proposal for strategic urban form changes; and (3) a critical examination of this modified scenario. In turn, each phase was developed at two different levels, each with critical importance: the macro-scale of the urban tsunami evacuation system, comprised of routes and safe assembly areas; and the micro-scale of the actual evacuees' experience of these routes and assembly areas during an emergency. The three phases were examined on these two scales by combining quantitative and qualitative methods, as described below.

Diagnosis of the existing situations

Macro-scale analyses were conducted with the aid of two computer-based models. The first one analyzed the urban spatial configuration between the evacuees' vulnerable locations and their safe destinations. In each city, secondary sources of

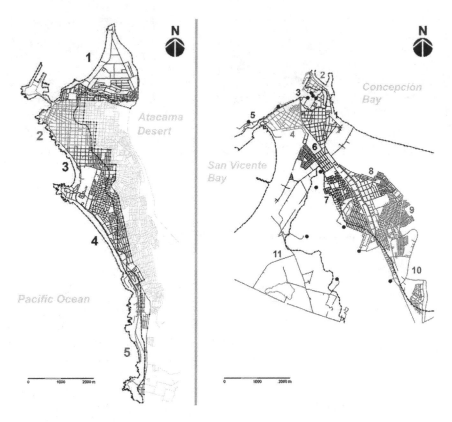

Figure 9.2 Existing urban network in vulnerable areas, evacuation zones (shaded in gray), assembly areas (dots), and expected tsunami flood limits (dotted line) in Iquique (left) and Talcahuano (right).[14]

information (maps and satellite images) were used to identify the closest safe destinations in case of a tsunami emergency. The shortest evacuation routes between these and each evacuee location were then calculated using the ArcGIS's Network Analyst extension (in this respect, it has been argued that pedestrians, especially evacuees, usually try to select the closest destination in terms of distance and time).[15] Tens of thousands of evacuation routes were mapped and their convergent spatial properties examined, with particular emphasis on overlaying densities and bottlenecks (see Figure 9.2 above).

The second model was an agent-based simulation, i.e., "a powerful modeling technique for simulating individual interactions and capturing group behavior resulting from individual interactions in a dynamic system,"[16] which was developed using Agent Analyst, an ArcGIS extension.[17] Every evacuee (i.e., "agent") in the tsunami-vulnerable areas of Iquique and Talcahuano was allocated a route to follow (obtained from the first model), a common pedestrian evacuation speed (1.4 m/s),[18] and a set of speed-decreasing rules (as a result of the interaction with the environment and with other agents).[19] The model iterated until every "agent" reached a safe destination and recorded the time required for that (see Tables 9.1–9.2 and Figures 9.3–9.4).

Both models required the spatial distributions of evacuees for daytime and nighttime scenarios as inputs. These were obtained from governmental data such as censuses and commuting studies. By spatially overlaying this information with the expected tsunami flood area in Iquique and Talcahuano, respective vulnerable populations of 108,881 and 72,675 (daytime) and of 83,331 and 69,514 (nighttime), were identified.

Table 9.1 Terrain slope and pedestrian speed conservation[20]

Slope (degrees)	Speed conservation (%)
0°	100%
0° – 5°	90%
5° – 15°	80%
15° – 30°	40%
30° – 45°	15%
More than 45°	5%

Table 9.2 Estimated pedestrian speed conservation factors[21]

Road classification					
Width_type	1	1	1	2	2
Traffic volume_type	1	2	3	1	2
Speed Conservation (%)	50%	65%	80%	55%	70%
Width_type	2	3	3	3	4
Traffic volume_type	3	1	2	3	1
Speed Conservation (%)	85%	60%	75%	90%	75%
Width_type	4	4	5	5	5
Traffic volume_type	2	3	1	2	3
Speed Conservation (%)	80%	95%	85%	90%	100%

Legend		
Width_type	1	One lane (small)
		One lane (wide)
	3	Two lanes (small)
	4	Two lanes (wide)
	5	Four lanes
Traffic volume	1	High
	2	Medium
	3	Low

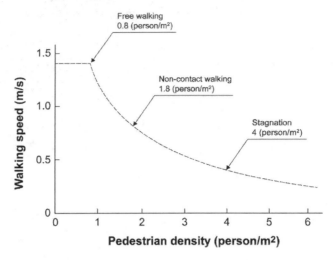

Figure 9.3 Uni-directional walking speeds as a function of pedestrian density.[22]

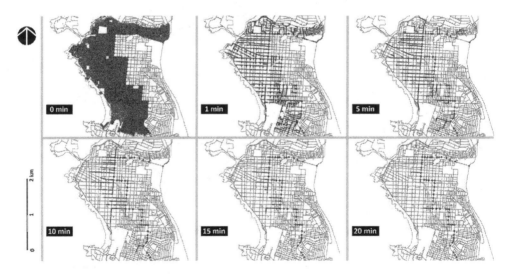

Figure 9.4 Snapshots from the Iquique agent-based tsunami evacuation model. The evacuees are shown as dark points, the urban network as gray lines, the assembly areas as light points, and the expected flooding area with a light gray line.[23]

To minimize the excessive consumption of computing resources and processing times, each case study was divided into different evacuation zones that were examined separately, according to the approach proposed by Imamura et al.[24] Iquique was divided into five zones (as defined in the city emergency response plan), whilst 11 zones were identified in Talcahuano, based on the location of the safe assembly areas.

Two different scenarios were tested in each case study: an "optimistic" one, where all the agents started simultaneously, and a "pessimistic" scenario (closer to observed

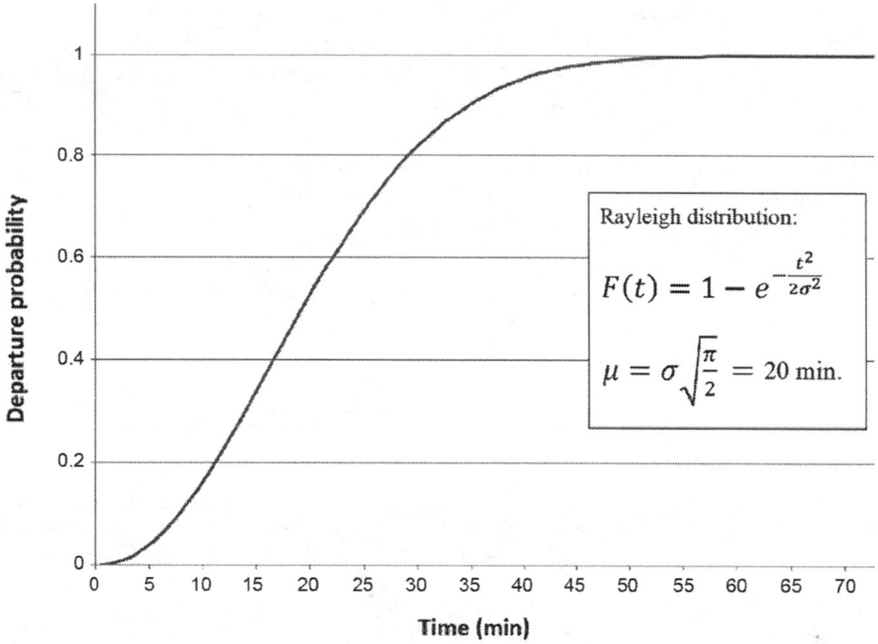
Departure probability
1
0.8
0.6
0.4
0.2
0
0 5 10 15 20 25 30 35 40 45 50 55 60 65 70
Time (min)
Rayleigh distribution:
$F(t) = 1 - e^{-\frac{t^2}{2\sigma^2}}$
$\mu = \sigma\sqrt{\frac{\pi}{2}} = 20$ min.

Figure 9.5 Rayleigh cumulative distribution of evacuees' departure times, with a mean value (μ) of 20 minutes.[25]

behaviors during past emergencies around the world), where few evacuees departed at the beginning of the simulation and the number gradually increased before slowly decreasing until it ceased. These starting times were assigned according to a Rayleigh cumulative distribution, with an average departure time of 20 minutes (coinciding with the estimated arrival of the first tsunami wave in these cities).[26] For this stochastic scenario, 50 simulation rounds were conducted to obtain an average value of the required times of evacuation for every zone (see Figure 9.5).

The existing micro-scale conditions in both cities in turn were examined during fieldwork studies conducted in April and May 2013. These studies included a thorough survey of the priority evacuation routes and safe assembly areas previously identified during the modeling phase, according to the urban safety parameters identified through an extensive literature review.[27] These parameters were enhanced with the information collected by examiners during two tsunami evacuation drills conducted in these cities in 2013.[28]

Urban form changes for tsunami evacuation

The diagnosis phase allowed the identification of critical parts in Iquique and Talcahuano's urban realms, where strategic urban form changes could be developed to improve the existing conditions for a tsunami evacuation. The interventions focused on the areas that the modeling and fieldwork identified as the most vulnerable: the middle and northern parts of Iquique and the southeastern area of Talcahuano. At the macro-scale, the Iquique proposal includes the creation of vertical evacuation points (e.g., tsunami-resistant towers), the pedestrianization of sections of priority evacuation routes, and the creation of a new pedestrian passage through a former airport lot. In the case of Talcahuano, interventions are focused on reconnecting the urban network across existing disruptive ground-level features, such as a railway and marshy areas; two vertical evacuation facilities are also proposed. Figure 9.6 below presents a summary of the proposals.

At the micro-scale level, in turn, the proposals aim to mitigate the abundant evacuation-related problems of the priority routes and assembly areas, as identified during fieldwork in Iquique and Talcahuano. These problems can be classified into three categories: (1) poor physical conditions of the built environment (e.g., critical street sections were seismically vulnerable, with high facades located too close to the sidewalk); (2) poor evacuation-related design of public spaces (e.g., narrow sidewalks, hazardous features such as electricity wires and gas tanks, wrongly placed urban furniture elements, etc.);

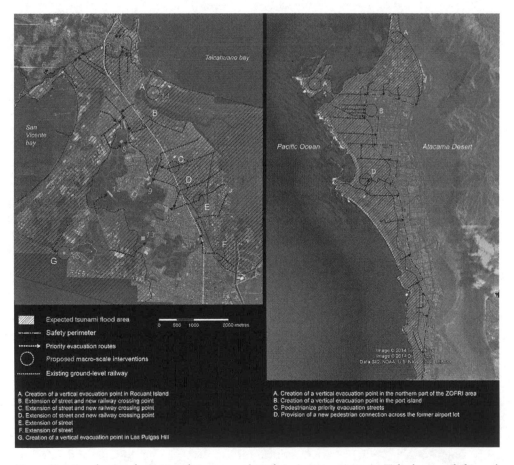

Figure 9.6 Synthesis of suggested macro-scale urban interventions in Talcahuano (left) and Iquique (right).[29]

and (3) inappropriate use of the evacuation space (e.g., cars parked obstructing sidewalks, street vendors on sidewalks, etc.).

In addition to mitigating these problems, the proposals aim to enhance the evacuation-related capacities of these spaces. To achieve this, wayfinding and information-provision systems should be incorporated, such as solar-powered street lighting, tsunami signage, and remotely operated information and communication points.[30] Emergency-relief features such as solar-powered electricity and fresh water provisions are similarly necessary, whilst Japan's experience underlines the importance of reinforcing the role of these places as memorials of past catastrophes or warnings for future disasters, for instance by including sculptures or other pieces of art.[31] When combined with macro-scale strategies, micro-scale changes such as those described above could lead to an overall improvement of the urban realm, not only for evacuation purposes.

Examination of the modified situations

The proposed macro-scale modifications in Iquique and Talcahuano were examined with the aid of the urban configuration and agent-based models introduced above. The modified evacuation zones were 1, 2, 3, and 4 in Iquique and 6, 8, 9, 10, and 11 in Talcahuano (see Figure 9.2 above). The changes in the total required times for evacuation and in the rate of safe evacuees in time are summarized in Tables 9.3–9.5 and Figures 9.7–9.8 below.

The urban configuration model's results show that a more dense and "connected" street pattern (i.e., one that includes more nodes and links) can lead to shorter evacuation routes (see, for example, Zones 1 and 3 in Iquique, with populations located at similar average distances from safe areas). The model also shows that the maximum length of the evacuation routes can be significantly reduced (up to around 43 percent of their original extent) as a result of the proposed changes, especially in the zones where new vertical evacuation facilities are included (Zones 1 and 2 in Iquique and 6 and 11 in Talcahuano). In turn, average lengths can also be considerably reduced (up to 12 percent) by minor changes in the street pattern arrangements, such as adding new crossing points to the existing ground-level railway in Talcahuano.

The agent-based model, in turn, shows that in an "optimistic" scenario, the safe areas can be reached within 20 minutes (i.e., the expected arrival time of a near-field tsunami) by roughly 90 percent of the daytime and nighttime populations in Iquique and by around 60 percent of them in Talcahuano. In a "pessimistic" scenario with delayed departures, these percentages diminish to around 25 percent and 10 percent in Iquique and Talcahuano, respectively. The suggested urban form modifications could in turn significantly reduce the total evacuation times. In an "optimistic" scenario, Iquique would reduce its total evacuation times from 42 to 33 minutes during the daytime and nighttime, whilst in Talcahuano the reduction would be from 48 to 37 minutes (daytime) and from 39 to 37 minutes (nighttime). In turn, in the "pessimistic" scenario Iquique would reduce its total evacuation time from 88.1 to 84.7 minutes (daytime) and show no noticeable impact on the nighttime total evacuation time (85.6 minutes).

Table 9.3 Comparison between existing and modified scenarios, urban configuration model, Iquique and Talcahuano[32]

Urban configuration analysis

Case study	Zone	Population		Average distance to the safe assembly area (km)		Longest evacuation route (km)				Average aevacuation route (km)			
						Existing		Modified		Existing		Modified	
		Day	*Night*	*Day*	*Night*	*Day*	*Night*	*Day*	*Night*	*Day*	*Night*	*Day*	*Night*
Iquique	1	16840	14352	0.43	0.25	2.63	2.63	1.74	1.14	0.64	0.39	0.60	0.39
	2	34543	20483	0.55	0.49	2.69	2.42	1.50	1.50	0.64	0.57	0.61	0.57
	3	31015	25873	0.42	0.31	1.924	1.924	1.63	1.63	0.55	0.41	0.53	0.40
	4	22556	18960	0.46	0.45	2.06	2.06	2.06	2.06	0.62	0.59	0.61	0.59
	5	3927	3663	0.24	0.22	1.07	1.06	N/A	N/A	0.48	0.44	N/A	N/A
Talcahuano	1	7627	7655	0.65	0.71	1.53	1.69	N/A	N/A	0.83	0.91	N/A	N/A
	2	1877	1003	0.30	0.46	1.15	1.14	N/A	N/A	0.42	0.52	N/A	N/A
	3	560	880	0.54	0.47	1.23	1.23	N/A	N/A	0.77	0.65	N/A	N/A
	4	4535	5387	0.52	0.47	1.37	1.38	N/A	N/A	0.73	0.69	N/A	N/A
	5	1292	77	0.60	0.44	1.38	1.29	N/A	N/A	0.55	0.78	N/A	N/A
	6	11180	11170	0.73	0.56	2.1	1.89	1.78	1.39	0.8	0.73	0.73	0.72
	7	4219	6252	0.69	0.70	1.10	1.36	N/A	N/A	0.63	0.72	N/A	N/A
	8	10589	16655	0.93	1.06	2.76	2.94	2.53	2.54	1.29	1.48	1.18	1.3
	9	11436	15572	1.29	1.16	2.82	2.84	2.47	2.47	1.72	1.59	1.51	1.39
	10	6877	8019	1.02	0.91	2.36	2.36	2.28	2.19	1.52	1.29	1.39	1.23
	11	9844	5	1.16	0.46	3.81	0.66	2.68	0.66	1.66	0.42	1.53	0.42

Table 9.4 Total evacuation times for existing and modified "optimistic" scenarios, Iquique and Talcahuano[33]

Existing/Modified and Day/Night "Optimistic" Evacuation Scenarios, Iquique and Talcahuano

Case study	Zone	Main land use	Scenario	Population	Total evacuation times (min)	
					Existing	*Modified*
Iquique	1	ZOFRI + Residential	Day	16,840	42	28
			Night	14,352	42	19
	2	Port + CBD	Day	34,543	36	23
			Night	20,483	33	23
	3	Residential + Services + Cavancha Beach	Day	31,015	28	22
			Night	25,873	25	21
	4	Residential + Services	Day	22,556	33	33
			Night	18,960	33	33
	5	Residential	Day	3,927	17	N/A
			Night	3,663	17	N/A
Talcahuano	1	Residential – Commercial	Daytime	7,627	23	N/A
			Night	7,655	24	N/A
	2	CBD	Day	1,877	19	N/A
			Night	1,003	22	N/A
	3	Residential – Commercial	Day	560	21	N/A
			Night	880	21	N/A
	4	Residential – Commercial	Day	4,535	22	N/A
			Night	5,387	24	N/A
	5	Port – Industrial	Day	1,292	27	N/A
			Night	77	25	N/A
	6	Residential – Industrial	Day	11,180	26	23
			Night	11,170	27	21
	7	Residential	Day	3,697	16	N/A
			Night	6,252	16	N/A
	8	Residential	Day	10,589	38	35
			Night	16,655	38	35
	9	Residential	Day	11,436	39	37
			Night	15,572	39	37
	10	Residential – Industrial	Day	6,877	37	35
			Night	8,019	37	35
	11	Industrial	Day	9,844	48	37
			Night	5	17	17

Table 9.5 Total evacuation times for existing and modified "pessimistic" scenarios, Iquique and Talcahuano[34]

Existing/Modified and Day/Night "Pessimistic" Evacuation Scenarios, Iquique and Talcahuano

Case study	Zone	Main land use	Scenario	Population	Total evacuation times (min)	
					Existing	Modified
Iquique	1	ZOFRI + Residential	Day	16,840	88.1	79.2
			Night	14,352	69.8	68.3
	2	Port + CBD	Day	34,543	81.8	75.6
			Night	20,483	79.2	75.5
	3	Residential + Services + Cavancha Beach	Day	31,015	80.8	74.9
			Night	25,873	76.9	72.4
	4	Residential + Services	Day	22,556	84.9	84.7
			Night	18,960	85.6	85.6
	5	Residential	Day	3,927	68.5	N/A
			Night	3,663	68.1	N/A
Talcahuano	1	Residential – Commercial	Day	7,627	74.1	N/A
			Night	7,655	75.6	N/A
	2	CBD	Day	1,877	65.1	N/A
			Night	1,003	66.4	N/A
	3	Residential – Commercial	Day	560	67.9	N/A
			Night	880	68.6	N/A
	4	Residential – Commercial	Day	4,535	72.1	N/A
			Night	5,387	73.3	N/A
	5	Port – Industrial	Day	1,292	72.4	N/A
			Night	77	61.4	N/A
	6	Residential – Industrial	Day	11,180	78.2	75.7
			Night	11,170	75.3	71.7
	7	Residential	Day	3,697	67.1	N/A
			Night	6,252	68.4	N/A
	8	Residential	Day	10,589	87.1	85.1
			Night	16,655	87	85
	9	Residential	Day	11,436	88.1	85
			Night	15,572	88.6	84
	10	Residential – Industrial	Day	6,877	85.3	84.8
			Night	8,019	85.5	84.5
	11	Industrial	Day	9,844	95.8	84.5
			Night	5	46.7	46.7

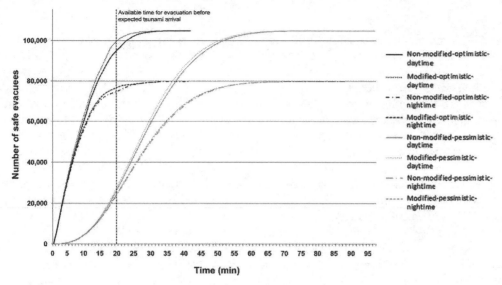

Figure 9.7 Total number of safe evacuees in time zones 1 to 4, Iquique, for non-modified/modified, daytime/nighttime.[35]

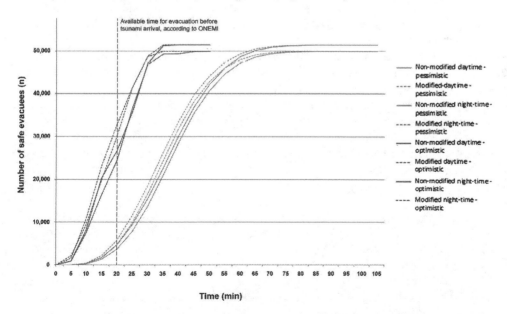

Figure 9.8 Total number of safe evacuees in time zones 6 and 8–11, Talcahuano, for non-modified/modified, daytime/nighttime.[36]

Discussion

Results and critical issues

Due to its proximity (around 100 km) to a source of large-magnitude earthquakes (the fault line between the Nazca and the South American tectonic plates), the Chilean coast can be affected by a tsunami in 20 minutes or even less after a major tremor. The results of this chapter show that despite increasing risk-reduction efforts (e.g., evacuation training, early warning systems) this condition implies significant ongoing vulnerability for Chilean coastal communities. Macro-scale models confirm that a complete evacuation of the two case studies examined could not be achieved within a 20-minute window, even in a very "optimistic" scenario (i.e., assuming an immediate departure of all the vulnerable populations after the perceived earthquake and following the shortest routes between their locations and the safe areas). More "pessimistic" (or realistic) conditions (such as delayed departure times among the evacuees) would in turn significantly increase the total times required for evacuation (and therefore the vulnerability). In turn, fieldwork examination showed overall poor evacuation-related conditions in the urban built environments of Iquique and Talcahuano. This is the result of deterioration, poor design, and the inappropriate use of public space.

As a first step to improving these issues, a series of macro and micro-urban scale changes were proposed and tested in these cities. The results of the models show that the length of the evacuation routes and the total evacuation times could be significantly reduced by interventions such as the creation of new vertical evacuation points and new connections in existing street patterns. However, the improvement in the rate of safe evacuees in time is critically restricted by the actual number of people using the proposed facilities; this quantity is likely to be dependent upon evacuees' initial locations during an emergency. Moreover, the pessimistic scenario shows that, as evacuation times are extended due to delayed departures, the benefits (in terms of the rate of safe evacuations) diminishes.

This paper argues that the existing tsunami evacuation vulnerabilities in Iquique and Talcahuano are the historical outcome of four interrelated critical issues. Although based on a case study method in Chile, these issues could also be examined in other tsunami-prone contexts with different socio-cultural characteristics. The first key topic is the *long gaps between recurrences* of tsunamis. For a given coastal location, destructive tsunamis can occur repeatedly over centuries or even decades. This characteristic can undermine both long-term risk reduction efforts and everyday tsunami awareness among the population and authorities (especially in developing contexts with multiple competing needs), leading to urban growth patterns that increase vulnerability. For instance, between the major earthquakes of 1877 and 2014, Iquique increased its urban area from approximately 70 to 1,500 hectares; of these, around 770 (51.3 percent) are located in tsunami-vulnerable zones along the coast. In the case of Talcahuano, between the 1835 and 2010 events, the urban area expanded from roughly 15 to 1,900 hectares, with 1,300 of these (68.4 percent) in vulnerable zones.

The second critical issue is related to *siting decisions* and *inertia*. Coastlines have always attracted human settlements for housing, maritime facilities, and resort developments, a tendency fostered by the long gaps between devastating tsunamis.[37] Once coastal towns start to develop specific location-related characteristics as the result of multiple social and environmental factors, "the urbanization process tends to be definitive and in general terms irreversible."[38] Although relocation has been suggested as a possible response to this problem, it has proven difficult in several contexts due to the tight historical bonds between the population and its site.[39] Iquique and Talcahuano's development models have historically been based on taking advantage of their coastal locations, first as fishing coves and afterwards as mining/industrial ports and tourist hubs. Progress has boosted population growth, and nowadays around 40 percent of Iquique's and Talcahuano's populations live in tsunami-vulnerable locations.

A third key issue is the *urban form characteristics* of the city. On a macro scale, certain street patterns are better suited for rapid evacuation than others. For instance, the dense orthogonal grid, a result of the urban guidelines embodied in the "Leyes de Indias" enacted in 1573 by King Felipe II of Spain[40] is common in Latin American cities. For evacuation purposes, the grid provides straight, nearby routes with multiple redundancies; this is an important advantage over other urban patterns with more "hierarchical" layouts, such as those with curvilinear distributor roads, branching patterns, and many cul-de-sacs (common in recent urbanization).[41] On a micro-scale level, in turn, certain characteristics of the urban form might seriously exacerbate the vulnerability of evacuees. For instance, Iquique's narrow streets (typically 10–11 meters wide) (a common attribute of Spanish cities located in low-population areas) might lead to bottlenecks during an emergency. Moreover, several of these streets cross informally developed areas with very poor seismic qualities. As discussed above, the streetscape of the case studies examined is also characterized by an overall poor evacuation-related design of urban furniture elements.

Finally, the fourth critical issue is the *actual response* of the city's population and emergency systems during a near-field tsunami emergency. The analysis of past emergencies in Iquique, Talcahuano, and other tsunami-prone cases allows the identification of common harmful characteristics in these types of situation: (i) an overall collapse across the interrelated elements in human settlements (e.g., lifeline infrastructures, communication networks, etc.) due to "cascading consequences" or "sequential failures" usually provoked by a large-magnitude earthquake;[42] (ii) little or no guidance provided by authorities (including the provision of information), due to sequential failures, the lack of enough personnel to manage tens of thousands of evacuees simultaneously, and the institutional need to save critical equipment for the post-disaster stage; (iii) evacuation-hindering behaviors among the population, such as incorrect routing, late departure, use of cars, entering vulnerable areas instead of leaving them (for instance, to pick up children from school), and returning home before a warning is cancelled.

Notes

1 P. Allan and J. Roberts (2009), "Urban Resilience and the Open Space Network," *Tephra* 22, pp. 55–59; P. Allan, M. Bryant, C. Wirsching, D. Garcia, and M.T. Rodriguez (2013), "The Influence of Urban Morphology on the Resilience of Cities Following an Earthquake," *Journal of Urban Design* 18(2), pp. 242–62.
2 Drawing on the work of E.N. Bernard (1995), *Tsunami Hazard Mitigation: A Report to the Senate Appropriations Committee* (Seattle, WA: National Oceanic and Atmospheric Administration); S. Murata, F. Imamura, K. Katoh, Y. Kawata, S. Takahashi, and T. Takayama (2010), *Tsunami: To Survive from Tsunami*, Advanced Series on Ocean Engineering Vol. 32 (Singapore: World Scientific Publishing); J. Preuss (1988), *Planning for Risk: Comprehensive Planning for Tsunami Hazard Areas* (Arlington, VA: Urban Regional Research, National Science Foundation); S. Scheer, V. Varela, and G. Eftychidis (2012), "A Generic Framework

for Tsunami Evacuation Planning," *Physics and Chemistry of the Earth* 49, pp. 79–91; N. Shuto and K. Fujima (2009), "A Short History of Tsunami Research and Countermeasures in Japan," in *Proceedings of the Japan Academy, Series B* 85(8), pp. 267–75.

3 International Disaster Database (2011), *Disaster Trends*, retrieved March 20, 2012, from http://www.emdat.be/disaster-trends

4 J. Joerin and R. Shaw (2010), "Climate Change Adaptation and Urban Risk Management," in *Climate Change Adaptation and Disaster Risk Reduction: Issues and Challenges (Community, Environment and Disaster Risk Management)*, R. Shaw, J. Pulhin, and J. Pereira, eds., pp. 195–215 (Bingley, UK: Emerald); M. Pelling (2003), *The Vulnerability of Cities: Natural Disasters and Social Resilience* (Sterling, VA: Earthscan); C. Wamsler (2014), *Cities, Disaster Risk and Adaptation* (New York, NY: Routledge).

5 L. Pearce (2003), "Disaster Management and Community Planning, and Public Participation: How to Achieve Sustainable Hazard Mitigation," *Natural Hazards* 28(2–3), pp. 211–28; M. Tarrant (2006), "Risk and Emergency Management," *The Australian Journal of Emergency Management* 21(1), pp. 9–14.

6 R.J. Burby, ed. (1998), *Cooperating with Nature: Confronting Natural Hazards with Land Use Planning for Sustainable Communities, Natural Hazards and Disasters* (Washington, DC: Joseph Henry Press); Wamsler (2014).

7 S. Cutter, L. Barnes, M. Berry, C. Burton, E. Evans, E. Tate, and J. Webb (2008), "A Place-Based Model for Understanding Community Resilience to Natural Disasters," *Global Environmental Change* 18(4), pp. 598–606.

8 The International Oceanographic Commission and the United Nations Educational, Scientific, and Cultural Organization (2008), *Tsunami Preparedness–Information Guide for Disaster Planners*, IOC Manuals and Guides No. 49 (Paris, France: UNESCO); National Tsunami Hazard Mitigation Program. (2005). Tsunami [0][0]Terminology. Retrieved June 9, 2014, from http://nthmp-history.pmel.noaa.gov/terms.html.

9 N. Shuto (2005), "Tsunamis: Their Coastal Effects and Defense Works," Scientific Forum on the Tsunami, Its Impact and Recovery, Asian Institute of Technology, Thailand, p. 8.

10 Bernard (1995); S. Kazusa (2004), *Tsunami and Storm Surge Hazard Map Manual* (Tokyo, Japan: Japanese Office of Disaster Management); Murata et al. (2010); Preuss (1988); Scheer, Varela, and Eftychidis (2012); N. Shuto and K. Fujima (2009), "A Short History of Tsunami Research and Countermeasures in Japan," in *Proceedings of the Japan Academy, Series B* 85(8), pp. 267–75.

11 IOC and UNESCO (2008); L.D. Samant, L.T. Tobin, and B. Tucker (2008), *Preparing Your Community for Tsunamis: A Guidebook for Local Advocates* (Palo Alto, CA: GeoHazards International); S. Scheer, A. Gardi, R. Guillande, G. Eftichidis, V. Varela, B. De Vannsay, and L. Colbeau-Justin (2011), *Handbook of Tsunami Evacuation Planning* (Luxembourg: European Union).

12 Image produced by the Servicio Hidrográfico y Oceanográfico de Armada (SHOA) and Google Earth (2012).

13 SHOA (2012), *Proyecto CITSU*, retrieved June 8, 2012, from http://www.shoa.cl/index.htm

14 Image provided by the authors. Reprinted with permission.

15 J. Gehl (2010), *Life between Buildings: Using Public Space*, 6th edn. (Copenhagen, Denmark: The Danish Architectural Press); J. Stroehle (2008), *How Do Pedestrian Crowds React When They Are in an Emergency Situation? Models and Software* (Urbana, IL: University of Illinois at Urbana-Champaign).

16 X. Chen and F.B. Zhan (2008), "Agent-Based Modelling and Simulation of Urban Evacuation: Relative Effectiveness of Simultaneous and Staged Evacuation Strategies," *Journal of the Operational Research Society* 59(1), p. 25.

17 K.M. Johnston (2013), Agent Analyst: Agent-Based Modeling in ArcGIS (Redlands, CA: Esri Press).

18 Ando et al. (1988), as cited in R.A. Smith (1995), "Density, Velocity and Flow Relationships for Closely Packed Crowds," *Safety Science* 18(4), pp. 321–27.

19 M. Mück (2008, June), "Tsunami Evacuation Modelling. Development and Application of a Spatial Information System Supporting Tsunami Evacuation Planning in South-West Bali," Geography Diploma Thesis, Institut für Geographie Universität Regensburg; J. Post, S. Wegscheider, M. Mück, K. Zosseder, R. Kiefl, T. Steinmetz, and G. Strunz (2009), "Assessment

of Human Immediate Response Capability Related to Tsunami Threats in Indonesia at a Sub-National Scale," *Natural Hazards and Earth System Sciences* 9(4), pp. 1075–86.

20 Adapted from Post et al. (2009).

21 Adapted from Mück (2008).

22 Adapted from Ando et al. (1988), as cited in Smith (1995).

23 Data gathered and image produced by the authors. Reprinted by permission.

24 F. Imamura, A. Muhari, E. Mas, M.H. Pradono, J. Post, and M. Sugimoto (2012), "Tsunami Disaster Mitigation by Integrating Comprehensive Countermeasures in Padang City, Indonesia," *Journal of Disaster Research* 7(1), pp. 48–64.

25 Adapted from E. Mas, A. Suppasri, F. Imamura, and S. Koshimura (2012), "Agent-Based Simulation of the 2011 Great East Japan Earthquake/Tsunami Evacuation: An Integrated Model of Tsunami Inundation and Evacuation," *Journal of Natural Disaster Science* 34(1), pp. 41–57, and Mas et al. (2013).

26 E. Mas, B. Adriano, and S. Koshimura (2013), "An Integrated Simulation of Tsunami Hazard and Human Evacuation in La Punta, Peru," *Journal of Disaster Research* 8(2), pp. 285–95.

27 A. Ciborowski (1982), "Physical Development Planning and Urban Design in Earthquake-Prone Areas," *Engineering Structures* 4(3), pp. 153–60; R.A. Davidson and H.C. Shah (1997), *An Urban Earthquake Disaster Risk Index* (Stanford, CA: Department of Civil and Environmental Engineering, Stanford University); M. Erdik (1994), "Developing a Comprehensive Earthquake Disaster Master Plan for Istanbul," in *Issues in Urban Earthquake Risk*, B.E. Tucker, M. Erdik, and C.N. Hwang, eds., pp. 125–66 (Dordrecht, the Netherlands: Springer Science + Business Media); B.A. Nadel (2004), "Home and Business Security, Disaster Planning, Response, and Recovery," in *Building Security: Handbook for Architectural Planning and Design*, B.A. Nadel, ed, p. 12.1 (New York, NY: McGraw-Hill); J.M. Ercolano (2008), "Pedestrian Disaster Preparedness and Emergency Management of Mass Evacuations on Foot: State-of-the-Art and Best Practices," *Journal of Applied Security Research* 3(3–4), pp. 389–405; L. He and S. Xu (2012), "Urban Design Strategies of Public Space Based on Danger Stress Response," *Advanced Materials Research* 450–451, pp. 1026–31; New York State Department of Transportation (2013), "Pedestrian Facility Design," in *Highway Design Manual No. 71* (New York, NY: Author), https://www.dot.ny.gov/divisions/engineering/design/dqab/hdm/chapter-18; National Fire Protection Association (2012), *NFPA 101: Life Safety Code* (Quincy, MA: Author); Scheer, Varela, and Eftychidis (2012).

28 J.-M. Walker (2013, August 8), *Informe Técnico de Evaluación. Simulacro Macrozona de Terremoto y Tsunami, Evacuación del Borde Costero. Regiones de Arica y Parinacota, Tarapacá, Antofagasta y Atacama* (Santiago, Chile: ONEMI).

29 Image produced by the authors. Reprinted by permission.

30 C. Rabaçal, C. Ferreira, R. Levy Salvador, S. Silva, and N.M. Sousa (2014), "Enhancement of Urban Security through Community Empowerment – A Local Perspective," paper presented at the Fifth International Disaster and Risk Conference, August 24–28, Davos, Switzerland.

31 A. Suppasri, N. Shuto, F. Imamura, S. Koshimura, E. Mas, and A.C. Yalciner (2012), "Lessons Learned from the 2011 Great East Japan Tsunami: Performance of Tsunami Countermeasures, Coastal Buildings, and Tsunami Evacuation in Japan," *Pure and Applied Geophysics* 170(6–8), pp. 993–1018.

32 Data gathered by the authors.

33 Ibid.

34 Ibid.

35 Ibid.

36 Ibid.

37 National Tsunami Hazard Mitigation Program (2001), *Designing for Tsunamis: Seven Principles for Planning and Designing for Tsunami Hazards* (Washington, DC: Author).

38 S. Menoni and G. Pesaro (2008), "Is Relocation a Good Answer to Prevent Risk? Criteria to Help Decision Makers Choose Candidates for Relocation in Areas Exposed to High Hydrogeological Hazards," *Disaster Prevention and Management* 17(1), p. 34.

39 Ibid.; A. Oliver-Smith (1991), "Successes and Failures in Post-Disaster Resettlement," *Disasters* 15(1), pp. 12–23.

40 A. Wyrobisz (1980), "La ordenanza de Felipe II del año 1573 y la construcción de ciudades coloniales españolas en la América," *Estudios Latinoamericanos* 7, pp. 11–34.
41 S. Marshall (2005), *Streets and Patterns* (New York, NY: Spon Press).
42 D.J. Alesch and W. Siembieda (2012), "The Role of the Built Environment in the Recovery of Cities and Communities from Extreme Events," *International Journal of Mass Emergencies & Disasters* 30(2), pp. 197–211; R.G. Little (2002), "Controlling Cascading Failure: Understanding the Vulnerabilities of Interconnected Infrastructures," *Journal of Urban Technology* 9(1), pp. 109–23.

Bibliography

Alesch, D.J., and W. Siembieda. (2012). "The Role of the Built Environment in the Recovery of Cities and Communities from Extreme Events." *International Journal of Mass Emergencies & Disasters* 30(2), pp. 197–211.

Allan, P., M. Bryant, C. Wirsching, D. Garcia, and M.T. Rodriguez. (2013). "The Influence of Urban Morphology on the Resilience of Cities Following an Earthquake." *Journal of Urban Design* 18(2), pp. 242–62.

———, and J. Roberts. (2009). "Urban Resilience and the Open Space Network." *Tephra* 22, pp. 55–59.

Bernard, E.N. (1995). *Tsunami Hazard Mitigation: A Report to the Senate Appropriations Committee*. Seattle, WA: National Oceanic and Atmospheric Administration.

Burby, R.J., ed. (1998). *Cooperating with Nature: Confronting Natural Hazards with Land Use Planning for Sustainable Communities, Natural Hazards and Disasters*. Washington, DC: Joseph Henry Press.

Chen, X., and F.B. Zhan. (2008). "Agent-Based Modelling and Simulation of Urban Evacuation: Relative Effectiveness of Simultaneous and Staged Evacuation Strategies." *Journal of the Operational Research Society* 59(1), pp. 25–33.

Ciborowski, A. (1982). "Physical Development Planning and Urban Design in Earthquake-Prone Areas." *Engineering Structures* 4(3), pp. 153–60.

Cutter, S., L. Barnes, M. Berry, C. Burton, E. Evans, E. Tate, and J. Webb. (2008). "A Place-Based Model for Understanding Community Resilience to Natural Disasters." *Global Environmental Change* 18(4), pp. 598–606.

Davidson, R.A., and H.C. Shah. (1997). *An Urban Earthquake Disaster Risk Index*. Stanford, CA: Department of Civil and Environmental Engineering, Stanford University.

Ercolano, J.M. (2008). "Pedestrian Disaster Preparedness and Emergency Management of Mass Evacuations on Foot: State-of-the-Art and Best Practices." *Journal of Applied Security Research* 3(3–4), pp. 389–405.

Erdik, M. (1994). "Developing a Comprehensive Earthquake Disaster Master Plan for Istanbul." In *Issues in Urban Earthquake Risk*. B.E. Tucker, M. Erdik, and C.N. Hwang, eds. pp. 125–66. Dordrecht, the Netherlands: Springer Science + Business Media.

Gehl, J. (2010). *Life between Buildings: Using Public Space*. 6th edn. Copenhagen, Denmark: The Danish Architectural Press.

He, L., and S. Xu. (2012). "Urban Design Strategies of Public Space Based on Danger Stress Response." *Advanced Materials Research* 450–451, pp. 1026–31.

Imamura, F., A. Muhari, E. Mas, M.H. Pradono, J. Post, and M. Sugimoto. (2012). "Tsunami Disaster Mitigation by Integrating Comprehensive Countermeasures in Padang City, Indonesia." *Journal of Disaster Research* 7(1), pp. 48–64.

International Disaster Database. (2011). *Disaster Trends*. Retrieved March 20, 2012, from http://www.emdat.be/disaster-trends

International Oceanographic Commission, the, and the United Nations Educational, Scientific, and Cultural Organization. (2008). *Tsunami Preparedness–Information Guide for Disaster Planners*. IOC Manuals and Guides No. 49. Paris, France: UNESCO.

———. (2013). *Tsunami Glossary*. Rev. ed. Paris, France: UNESCO.

Joerin, J., and R. Shaw. (2010). "Climate Change Adaptation and Urban Risk Management." In *Climate Change Adaptation and Disaster Risk Reduction: Issues and Challenges (Community, Environment and Disaster Risk Management)*. R. Shaw, J. Pulhin, and J. Pereira, eds. pp. 195–215. Bingley, UK: Emerald.

Johnston, K.M., ed., D.G. Brown, N. Collier, H.R. Ekbia, M.J. Fraley, E.R. Groff, M.A. Gudorf, N. Li, A. Ligmann-Zielinska, M.J. North, et al. (n.d.). *Agent Analyst: Agent-Based Modeling in ArcGIS* [Software]. Retrieved December 30, 2015, from http://resources.arcgis.com/en/help/agent-analyst/

Kazusa, S. (2004). *Tsunami and Storm Surge Hazard Map Manual*. Tokyo, Japan: Japanese Office of Disaster Management.

Little, R.G. (2002). "Controlling Cascading Failure: Understanding the Vulnerabilities of Interconnected Infrastructures." *Journal of Urban Technology* 9(1), pp. 109–23.

Marshall, S. (2005). *Streets and Patterns*. New York, NY: Spon Press.

Mas, E., A. Suppasri, F. Imamura, and S. Koshimura. (2012). "Agent-Based Simulation of the 2011 Great East Japan Earthquake/Tsunami Evacuation: An Integrated Model of Tsunami Inundation and Evacuation." *Journal of Natural Disaster Science* 34(1), pp. 41–57.

Mas, E., B. Adriano, and S. Koshimura. (2013). "An Integrated Simulation of Tsunami Hazard and Human Evacuation in La Punta, Peru." *Journal of Disaster Research* 8(2), pp. 285–95.

Menoni, S., and G. Pesaro. (2008). "Is Relocation a Good Answer to Prevent Risk?: Criteria to Help Decision Makers Choose Candidates for Relocation in Areas Exposed to High Hydrogeological Hazards." *Disaster Prevention and Management* 17(1), pp. 33–53.

Mück, M. (2008, June). "Tsunami Evacuation Modelling. Development and Application of a Spatial Information System Supporting Tsunami Evacuation Planning in South-West Bali." Geography Diploma Thesis, Institut für Geographie Universität Regensburg. Available at http://www.gitews.org/tsunami-kit/id/E4/sumber_lainnya/rencana_evakuasi/bali_badung/Tsunami%20Evacuation%20Modelling%20by%20Matthias%20Mueck%20ENG.pdf

Murata, S., F. Imamura, K. Katoh, Y. Kawata, S. Takahashi, and T. Takayama. (2010). *Tsunami: To Survive from Tsunami*. Advanced Series on Ocean Engineering, Vol. 32. Singapore: World Scientific Publishing.

Nadel, B.A. (2004). "Home and Business Security, Disaster Planning, Response, and Recovery." In *Building Security: Handbook for Architectural Planning and Design*. B.A. Nadel, ed. p. 12.1. New York, NY: McGraw-Hill.

National Fire Protection Association. (2012). *NFPA 101: Life Safety Code*. Quincy, MA: Author.

National Tsunami Hazard Mitigation Program. (2001). *Designing for Tsunamis: Seven Principles for Planning and Designing for Tsunami Hazards*. Washington, DC: Author.

———. (2005). *Tsunami Terminology*. Retrieved June 9, 2014, from http://nthmp-history.pmel.noaa.gov/terms.html

New York State Department of Transportation. (2013). "Pedestrian Facility Design." In *Highway Design Manual No. 71*. New York, NY: Author. https://www.dot.ny.gov/divisions/engineering/design/dqab/hdm/chapter-18.

Oliver-Smith, A. (1991). "Successes and Failures in Post-Disaster Resettlement." *Disasters* 15(1), pp. 12–23.

Pearce, L. (2003). "Disaster Management and Community Planning, and Public Participation: How to Achieve Sustainable Hazard Mitigation." *Natural Hazards* 28(2–3), pp. 211–28.

Pelling, M. (2003). *The Vulnerability of Cities: Natural Disasters and Social Resilience*. Sterling, VA: Earthscan.

Post, J., S. Wegscheider, M. Mück, K. Zosseder, R. Kiefl, T. Steinmetz, and G. Strunz. (2009). "Assessment of Human Immediate Response Capability Related to Tsunami Threats in Indonesia at a Sub-National Scale." *Natural Hazards and Earth System Sciences* 9(4), pp. 1075–86.

Preuss, J. (1988). *Planning for Risk: Comprehensive Planning for Tsunami Hazard Areas*. Arlington, VA: Urban Regional Research, National Science Foundation.

Rabaçal, C., C. Ferreira, R. Levy Salvador, S. Silva, and N.M. Sousa. (2014). "Enhancement of Urban Security through Community Empowerment – A Local Perspective." Paper presented at the Fifth International Disaster and Risk Conference, August 24–28, Davos, Switzerland.

Samant, L.D., L.T. Tobin, and B. Tucker. 2008. *Preparing Your Community for Tsunamis: A Guidebook for Local Advocates.* Palo Alto, CA: GeoHazards International.

Scheer, S., A. Gardi, R. Guillande, G. Eftichidis, V. Varela, B. De Vannsay, and L. Colbeau-Justin. (2011). *Handbook of Tsunami Evacuation Planning.* Luxembourg: European Union.

Scheer, S., V. Varela, and G. Eftychidis. (2012). "A Generic Framework for Tsunami Evacuation Planning." *Physics and Chemistry of the Earth* 49, pp. 79–91.

Servicio Hidrográfico y Oceanográfico de Armada. (2012). *Proyecto CITSU.* Retrieved June 8, 2012, from http://www.shoa.cl/index.htm

Shuto, N. (2005). "Tsunamis: Their Coastal Effects and Defense Works." Scientific Forum on the Tsunami, Its Impact and Recovery, Asian Institute of Technology, Thailand.

Shuto, N., and K. Fujima. (2009). "A Short History of Tsunami Research and Countermeasures in Japan." *Proceedings of the Japan Academy, Series B* 85(8), pp. 267–75.

Smith, R.A. (1995). "Density, Velocity and Flow Relationships for Closely Packed Crowds." *Safety Science* 18(4), pp. 321–27.

Stroehle, J. (2008). *How Do Pedestrian Crowds React When they Are in an Emergency Situation? Models and Software.* Urbana, IL: University of Illinois at Urbana-Champaign.

Suppasri, A., N. Shuto, F. Imamura, S. Koshimura, E. Mas, and A.C. Yalciner. (2012). "Lessons Learned from the 2011 Great East Japan Tsunami: Performance of Tsunami Countermeasures, Coastal Buildings, and Tsunami Evacuation in Japan." *Pure and Applied Geophysics* 170(6–8), pp. 993–1018.

Tarrant, M. (2006). "Risk and Emergency Management." *The Australian Journal of Emergency Management* 21(1), pp. 9–14.

Walker, J.-M. (2013, August 8). *Informe Técnico de Evaluación: Simulacro Macrozona de Terremoto y Tsunami, Evacuación del Borde Costero. Regiones de Arica y Parinacota, Tarapacá, Antofagasta y Atacama.* Santiago, Chile: ONEMI.

———. (2013, October 15). *Informe Técnico de Evaluación. Simulacro Macrozona de Terremoto y Tsunami, Evacuación del Borde Costero, Regiones del Biobío, La Araucanía, Los Lagos y Aysén.* Santiago, Chile: ONEMI.

Wamsler, C. (2014). *Cities, Disaster Risk and Adaptation.* New York, NY: Routledge.

Wyrobisz, A. (1980). "La ordenanza de Felipe II del año 1573 y la construcción de ciudades coloniales españolas en la América." *Estudios Latinoamericanos* 7, pp. 11–34.

10 Can laws and policies foster disaster risk reduction?

The case study of Brazil, with particular focus on housing laws, policies, and programs

Karen Da Costa and Paulina Pospieszna

Introduction

In this paper, we consider what role laws and policies can play in disaster risk reduction. The background for this discussion is provided in Section 2, which covers the Hyogo Framework for Action (HFA) and its references to the adoption of laws and policies by states in their commitment to reduce the risk of disasters. Section 3 considers how far different states have progressed in their commitment to adopt laws and policies designed to reinforce disaster risk reduction. We chose Brazil as a case study, and Section 4 outlines the main features relevant to disasters in this country. It is suggested that one of the key areas in which disaster risk reduction needs to advance in Brazil is housing for low-income families. Section 5, therefore, covers not only the general legal and policy framework for disasters in the country but also a novel public initiative on housing. Section 6 provides an evaluation of the progress made on disaster risk reduction in Brazil to date, focusing on the improvement of housing for low-income families, the population most vulnerable to disasters. Finally, the paper provides conclusions and recommendations.

The methodology adopted consists in the review of the pertinent literature published both in English and in Portuguese. Although it was not possible to conduct a field study, from the analysis of the literature we identified a certain pattern of issues in relation to the novel housing program adopted in Brazil, which thus informed our assessment and conclusion.

The Hyogo framework for action, with a focus on adopting laws with a strong implementation framework at the national level

Disaster risk reduction (DRR) is understood as the concept and practice of reducing disaster risks through systematic efforts to establish and manage the causal factors of disasters, such as reducing exposure to hazards, lessening the vulnerability of people and property, wisely managing the land and the environment, and improving preparedness for adverse events.[1] The International Strategy for Disaster Reduction (ISDR) provides a channel for cooperation among governments, organizations, and civil society actors to assist in implementing the framework for disaster risk reduction. An expected product of this DRR initiative is the reduction of disaster losses, through improvement in lives and in the social, economic, and environmental assets of communities and countries that are prone to disasters.

Recognizing the need for a comprehensive approach to DRR, at the 2005 World Conference on Disaster Reduction in Kobe, Japan, 168 governments adopted the United

Nations-endorsed Hyogo Framework for Action (HFA) as a ten-year plan to make the world safer from natural hazards. Since then the HFA has served as a road map for DRR and sets out five priorities for action, each elaborated into a number of specific areas.[2] National laws and regulations are covered in Priority for Action 1, indicating the need for states to "ensure that disaster risk reduction is a national and a local priority with a strong institutional basis for implementation."[3] Furthermore, states are encouraged to develop policy, legislative, and institutional frameworks for disaster risk reduction and to "adopt, or modify where necessary, legislation to support disaster risk reduction, including regulations and mechanisms that encourage compliance and that promote incentives for undertaking risk reduction and mitigation activities."[4]

Although the HFA is not a binding treaty, the initiative led many national legislative bodies to draft or to revise laws on disaster risk reduction. An analysis of government reports gathered using the HFA Monitor[5] indicates that progress is taking place in DRR in the passing of national legislation. However, the implementation and enforcement of laws passed following the adoption of the HFA proved to be problematic given a specificity of national legal contexts. According to the Mid-Term Review of the HFA,[6] the implementation of policies requires the combination of knowledge of the risk factors, the strong involvement of well-informed public groups (including local communities), and, above all, well-drafted laws and regulations. These laws and regulations should clearly define responsibilities for all levels of government, tailored to the specific circumstances of the disaster-prone countries. Moreover, new laws and regulations should be harmonized with legislative frameworks in all sectors that have a direct impact on how disaster risk is managed. In 2015, the third UN World Conference on DRR took place in Sendai, Japan, with the aim of adopting a post-HFA document. This new document should assist states and other partners in their efforts to achieve more effective DRR.[7]

Overview of the global picture on national laws and disaster risk reduction

The defective or absent implementation of the HFA at the national level, together with the inadequate enforcement of laws on DRR, poses a significant challenge. Against this background, actors such as the International Federation of Red Cross and Red Crescent Societies (IFRC) and the United Nations Development Programme (UNDP) reviewed legislative frameworks in various jurisdictions in order to identify key gap areas. Different countries were considered as case studies. Difficulties and successes in the adoption of laws on DRR and their implementation were identified, and recommendations were made.[8]

Aspects covered in this exercise included an assessment of whether adopted national frameworks make DRR a priority for community-level action in a given country; whether the laws promote the involvement of communities and their access to information about DRR; and whether there is adequate funding for DRR activities at the community level. Building codes, land-use regulations, and other legal incentives were also analyzed. Finally, attention was paid to whether and how laws take into account local variabilities in hazard, exposure, and vulnerability, as well as the country's general circumstances.

As a result of this exercise, the IFRC identified four different categories of national legal frameworks addressing natural hazards: (1) No specific law on natural disasters;

(2) Response focus: legislation on natural disasters that is in force, but focused on response and usually limited to specific types of natural hazards; (3) Disaster management; and (4) DRR focus: a holistic approach, regulating a range of relevant matters directly related to the reduction of disaster risk.[9] Applying this typology, the IFRC considered Brazil an example of category 3. The country has laws on natural disasters in force, but these mostly cover disaster management, including different phases of disasters, together with general aspects of prevention.

Since this assessment, Brazil has adopted important legal and policy developments, which may strengthen the reduction of disaster risk. This is the case especially in relation to new social housing initiatives and how these may impact DRR. We consider that taking Brazil as a case study may serve as a comparison to be used in further studies covering other countries facing similar vulnerabilities and social outlooks.

Overview of the Brazilian case: General disaster risks, outline of disasters, and the housing situation

Brazil is subject to natural hazards, especially floods, landslides, and droughts. Droughts affect the greatest number of people in the country, but floods lead most commonly to disasters, with a great number of deaths and economic losses.[10] Floods and landslides in the Brazilian states of Santa Catarina in 2008 and Rio de Janeiro in 2011 resulted in thousands of deaths and hundreds of thousands left homeless.

These events gave rise to much debate on how the country can best cope with future disasters and how it can reduce vulnerability to natural disasters. Although Brazil is a country with moderate economic growth[11] and relatively good human development indicators (the country is 79th out of 187 countries, with a rating of 0.744), inequality between the rich and the poor is substantial.[12] The poor are much more likely to be affected by disasters than wealthier communities, for low-income families are not only more likely to live in disaster-prone areas but also to live in inadequate houses.

Brazilian laws on civil defense are those that most directly address DRR, but they have been frequently amended in recent years. Brazil's National Civil Defense System (SINDEC) was organized in 1988, reorganized in 1993, updated in 2005, and amended by later legislation. Civil defense matters are currently assigned mainly to the Ministry of National Integration (MNI), through its National Secretariat of Civil Defense (NSCD).[13] The MNI is the main body directly assigned to tasks relating to DRR, being the overall coordinator of the SINDEC. Another organ that covers DRR issues, especially those pertaining to the right to housing, is the Ministry of the Cities (MoC), created in 2003. The MoC addresses, among other things, risk in urban areas, especially when associated with precarious housing. One initiative in this context is the "Action to Support the Prevention and Eradication of Risks in Settlements," part of the MoC's "Urbanization, Regularization, and Integration of Precarious Settlements" program. The program supports municipal projects for upgrading slums, including the reduction of disaster risks associated with landslides and floods.[14] Actions range from training municipal technicians on dealing with risk; supporting municipalities to produce risk-reduction plans; and financing corrective structural measures addressing risk, such as engineering interventions to contain slopes.[15]

The right to housing is covered by the 1988 Brazilian Constitution,[16] as well as by later legislation. The recent establishment of the MoC is certainly a landmark development in the country in terms of housing, reflecting a lengthy popular struggle aiming to make progress in strengthening and guaranteeing both citizenship and dignity, especially for those living in slum areas. One of the most acute social problems Brazil faces today is its housing deficit. Rapid urbanization and population growth in big cities such as Rio de Janeiro have caused tremendous housing challenges. As a result, those who cannot afford adequate housing tend to build makeshift dwellings in poor areas, known as *favelas*. Most such shelters do not meet minimal conditions to be considered habitable. As a general rule, the makeshift houses that form *favelas* are built without observance to any laws or safety regulations and with inadequate materials. They often have precarious or no access to basic services such as water, electricity, and sewage systems.

In Brazil, shantytowns are often located in areas subject to landslides, floods, and fires, and thus their inhabitants are constantly at risk of being killed by natural hazards and domestic accidents (most households use gas as energy for cooking). For many decades, favela inhabitants did not have formal title either to the land they occupied or to the houses they built and lived in.[17] The inhabitants of such informal communities were not recognized as homeless, but they did not have access to ways of legally securing their shelter and the area where they lived.

Tackling the housing situation in Brazil is a complex task that requires, among other things, clear regulations. The United Nations Special Rapporteur on the right to adequate housing, a procedure of the Human Rights Council (resolution 15/8), provides some practical directives towards the implementation of the right to adequate housing. The Special Rapporteur regularly publishes reports, covering issues such as practical guiding principles aiming to assist states in addressing housing issues across the world. The 2013 Report of the Special Rapporteur to the UN General Assembly emphasizes possession rights and recommends to states that

> the legal recognition of the rights of those occupying public, private or community land and housing for a prescribed period, through adverse possession of land and housing, above the rights of absentee owners or the State, is an important measure to ensure that land and housing is being used in the most socially productive manner and to fulfill the right to adequate housing for all.[18]

At the legislative level, Brazil has laws in line with these recommendations, including providing formal title to dwellings. Its 1988 Constitution contains a specific chapter on Urban Policy,[19] which, in article 183, recognizes adverse possession (*usucapião*) of urban land used for a home after five years of possession without interruption or opposition, provided that the possessor does not own any other property.[20]

The issue of regularizing titles in Brazilian cities facing inadequate housing such as Rio de Janeiro pertains not only to the right to *moradia* (that is, the right to housing) but also to the "right to the city." According to scholars, the right to the city is a "right to urban life" that involves the right to occupy physical space, "a right not to be expelled, a right not to suffer segregation" and a right to participate in all levels of decision-making.[21] Thus it involves both habitation and participation.[22] Fernandes suggested that the Brazilian legal system incorporated the right to the city, in particular through the 2001 federal legislation known as the Statute of the City,[23] as well as the drafting of the World Charter on the Right to the City.[24] The Statute of the City further developed the urban policy based on the constitution, thus covering new "legal forms for transferring ownership of private land (that urban squatter-specific form of *usucapião*, adverse possession) and a legal form for the concession of a real right to use public land (a specialized form of lease)."[25]

These recent legislative developments articulate the social function of the city and of property and have led to important institutional reforms. Examples include the creation of the Ministry of Cities and the establishment of the National Council of Cities, the National Program to Support Sustainable Urban Land Regularization, and the National Campaign for Participatory Municipal Master Plans. A separate but related initiative resulted in a constitutional amendment providing a social right to housing in 2000, with a law enacted in 2005 to implement this right.[26]

Another important initiative designed to improve urbanization in Brazil is the adoption of a framework supporting low-income families to purchase their own houses. The "My House, My Life" (*Minha Casa, Minha Vida*, or MCMV) program facilitates the purchases of houses and apartments by low-income families so that they can live in their own property. The program aims for compatibility with other initiatives, such as the avoidance of disaster risks in any future housing development. A later section of this paper assesses how this initiative has so far been implemented.

In terms of participation, in the last decade, Brazil has adopted more ways to facilitate popular consultation and participation, including on matters relating to natural hazards and housing. One example relating to civil defense is that there is greater public consultation and participation in the body responsible for proposing civil defense policies, the *Conselho Nacional de Defesa Civil* (CONDEC). However, there is a need for greater transparency in the way that civil society representatives are chosen.[27] Another problem is that the law covering the general aspects and structure of civil defense is far from clear, making its application rather confused, particularly because it does not accurately indicate the role of the different actors in a disaster situation. The majority of actions are still undertaken for the most part at the governmental level; community-level work needs to be further strengthened, with more direct involvement and empowerment of community members in order to make them less vulnerable to disasters.[28]

In its 2011 country report, the IFRC pointed out that Brazil's legal framework on DRR presented no comprehensive national law covering DRR. Consequently, there was some ambiguity of roles among the various levels of government concerning specific DRR activities. The gaps also suggested the lack of legislation clearly providing for early warning, weak coordination among different sections of the government, and lack of precision in specifying the role of different actors in a disaster situation. After the publication of this report, Brazil adopted new legislation in an attempt to give a more prominent role to DRR. This law includes the National Policy on Protection and Civil Defense setting obligations for different entities, including municipalities, to take a series of preventive measures aiming to minimize the risk of disasters.[29] This also relates to the housing situation in the country, and the legislative framework concerning both areas will be reviewed in the following sections.

Overview of recent laws and policies on the right to housing and how they address DRR

Against the background of the recent disasters affecting the country, Brazil adopted a number of steps with a view to improving the legal and institutional framework addressing disasters and the housing situation, especially that of the people living in the most precarious conditions.

The key legislation on DRR currently in force is Law 12.608 of 2012, which instituted a National Policy on Protection and Civil Defense together with disaster management bodies and which incorporated ideas contained in the HFA.[30] This law paved the way for the creation of a system for disaster information and monitoring. Municipalities are in charge of identifying and mapping disaster risk areas within their territory and are responsible for ensuring that high-risk areas are not used for human habitation.[31]

This legislation emphasizes the importance of DRR, thus following the international trend set by the HFA. It indicates that all levels of government (national, regional, and municipal administrations) shall adopt DRR measures and that they should strive to collaborate with each other.[32]

Law 12.608/2012 updated the legislative framework on disasters, especially Law 12.340 of December 1, 2010 and its Regulatory Decree 7.257 of August 4, 2010. However, Law 12.608/2012 still needs further regulation. This shall take place through the adoption of a decree on the subject, indicating how exactly preventive action and general disaster policies are to be implemented.

As already suggested, in Brazil, people living in precarious conditions are among the most vulnerable in the event of disasters. The housing situation in Brazil is far from ideal, with about 6 percent of the national population (over 11 million people) living in precarious conditions in slum areas. Not only do they face inadequate access to public services, but also precarious houses are often built upon areas that are permanently inundated or are built insecurely on the slopes of hills.[33] As disaster is an issue affecting all parts of the country, and because Brazilian national policy on disasters currently recommends that disaster awareness should be taken into account throughout the state apparatus, it is important to make clear what main developments are taking place in the housing sector and how disaster risks are being taken into account.[34]

Recent legislative changes, notably the Statute of the City, included disaster considerations in matters of urban housing in Brazil. Accordingly, the use of land and urban space avoids exposing the population to disaster risks.[35] Another new legislative development is that municipalities registered in the National Record on Municipalities as being highly susceptible to disaster risk (especially severe landslides and flash floods) are obliged to develop a Director Plan for their territory. This should clearly indicate disaster risk areas and measures for disaster prevention, together with areas that may need to relocate some residents due to their high disaster risk.[36]

The disaster risk maps should indicate which municipalities contain risky areas in order for them to be monitored by the National Center for the Monitoring and Alert of Natural Disasters (*Centro Nacional de Monitoramento e Alertas de Desastres Naturais*, or CEMADEN),[37] a recently created federal body that monitors and informs registered municipalities about the likelihood of disasters in their respective territories.

The assessment of laws, programs, and their implementation: An example to be followed?

In relation to laws addressing disasters, despite legislative changes mandating municipalities with specific disaster risks to develop a Disaster Plan for their territory, to date, very few municipalities in Brazil have implemented disaster risk maps. Currently federal funds are available for municipalities upon request from the central government for financial and technical support to map disaster risks. Due to the federal organization of the state, the federal government cannot impose the mapping exercise at the municipal level; instead, it must wait to receive municipal requests for federal support for the mapping exercise. Municipal requests may be delayed due to a lack of information or interest on the part of municipalities to engage in this initiative. Brazil has over 5,000 municipalities; as of 2014, a total of 795 municipalities throughout the country had conducted disaster risk mapping and were being monitored by CEMADEN for disaster risks.[38],[39] CEMADEN's aim is to monitor over 1,000 cities in the next few years.[40] This suggests that a multi-sectoral approach is needed for tangible results to be achieved in disaster prevention. The initiative seems to be a good start in creating more awareness in society and a better-informed public for coping with disaster risks.

An assessment of the most significant recent initiative on the right to housing, the low-income housing program MCMV, suggests the following. First of all, the program was created at the same time as the federal Statute of the City was adopted. This represented a legal landmark for the country, allowing for the elaboration of a national housing strategy with greater participation of different sectors of the population through the creation of a participatory council. Secondly, the MCMV program itself was designed to address sectors of the population that have historically been excluded from official housing programs, namely the poorest section of the population. Part of this group corresponds to households earning between three and ten times the minimum wage officially set by federal authorities of the country. In Brazil, 91 percent of the national housing deficit relates to families earning less than three times the minimum wage.[41]

The national housing deficit is estimated to be six million houses, and the MCMV program aims to tackle this in the following way.[42] In 2013, about 2.7 million houses were contracted under the MCMV program; approximately 1.2 million have been delivered, with about R\$160 billion in government subsidies.[43] The government is in charge of framing the program and financially supporting people's access to subsidies, but the construction work is left to the private sector.[44] The fundamental role played by the private sector has posed some challenges, which will be discussed later.

The project is framed by the national government, but local (municipal) governments are involved to a considerable extent throughout the program, especially in relation to families with the lowest incomes. Local governments both elaborate concrete project proposals in order to secure financial support from the national government and provide complementary land and infrastructure to initiatives within the MCMV program taking place in their own territory. They also keep records of all those participating in the program.[45] However, the MoC offers no general financial incentives to municipalities to purchase land for the MCMV program, and thus the private sector alone is left to buy land and build houses within the program.[46] The municipal bodies tend to allow real estate developers to have prospective areas recognized as suitable for participating in the MCMV program. This is done especially in their classification as "Areas of Special Social Interest" (*Zona Especial de Interesse Social*, or ZEIS), which identifies certain unoccupied areas to be used for the resettlement of persons who previously lived in inadequate areas.[47] The ZEIS labeling of an area can also occur in relation to areas that were occupied by low-income residents who had no title to the property. These people can then be given legal title to the property and can benefit from services, including engineering improvements.

Having the private sector as the leading force in the MCMV program has led to difficulties, namely the construction of houses in isolated areas located on the outskirts of cities, which are cheaper and thus more affordable for beneficiaries of the MCMV program to purchase. The flipside of this development is that, with less access to public transport and public services (notably health centers and public schools), together with fewer job opportunities, the MCMV program tends to reinforce systems of exclusion from the city.[48] This makes it more difficult, if not completely impossible, to implement the "right to the city," which was the central initiative of the legislation in the Statute of the City.

Another criticism of the MCMV program is that it has only partially allocated money to those most in need – those households living on under less than three times the minimum wage. The MCMV program initially allocated 40 percent of its resources to these families (who comprise about 91 percent of the housing deficit in Brazil). The budget allocation for this category of beneficiaries was later increased to 60 percent.[49] Moreover, a few private groups participating in the program have benefitted enormously from tax incentives and further official financial support, which led to higher profits and land prices. Various authors have highlighted the poor quality of the newly built houses as an issue of concern.[50] There has also been criticism that projects built in the MCMV program, especially those carried out by building companies operating nationwide, tend to adopt a single model for different areas of the country. Thus no attention has been paid to variations such as climate, which should have been taken into account, bearing in mind the importance of environmental concerns in DRR.[51] Similarly, there have been instances of violations of municipal regulations by the failure of builders to comply with the requirements to provide green leisure areas within or near MCMV-project areas.[52]

To date, the program seems to have prioritized quantity over the quality of the houses built so far. This may be a consequence of the poverty and the lack of power of most of the beneficiaries of the program, namely families earning less than three times the minimum wage per month. It is suggested that the program is failing to deliver a more far-reaching structural change in the country's urban space and land markets.[53]

Beyond this, Brazil is once again facing financial and macroeconomic difficulties with a combination of fiscal constraints, tighter external credit conditions, and increasing interest rates affecting its economy. With the state less able to keep subsidizing this ambitious housing program, it is very likely that low-income households may not be able to count on state support for them to buy their houses for long, and the project itself might not be able to live up to its promises.[54] However, the picture is not altogether grim in relation to this far-reaching program. To date, MCMV has delivered a considerable number of houses to the poorest sectors of the population, though it has been pointed out that success stories within the project are closely linked to the involvement of the local government and its technical bodies, together with pressure exerted by social movements to see that effective progress takes place.[55]

The nexus between the MCMV housing program and disaster risks is important, because an improved housing situation for low-income families suggests that the population facing the greatest risk of disaster would become more resilient and able to cope with natural hazards such as floods and landslides. In this context the effectiveness of the program depends both on the funding authorities identifying the correct areas to carry out the program and on the delivery of good-quality housing. It is clear that other factors play an equally important role, such as education and drilling exercises for the population living with the risk of disasters, but these further areas deserve separate investigations, which is beyond the scope of this paper.

Conclusion and recommendations

Laws, policies, and programs are very important for DRR. The HFA has paved the way for legislative developments to take place in different jurisdictions. The example of Brazil suggests that this framework was taken into account in new legislation on disasters, though more remains to be done. In relation to disaster policies, the risk mapping exercise is a commendable initiative that requires better integration and collaboration among different levels of government.

On the right to housing, the ambitious MCMV program represents an important initiative that may inspire other countries. However, to date, there is some disconnect between theory and practice, especially regarding the contrast between the notion of the right to the city and what is being currently delivered under this program. The reality is that the right to the city seems to have faded in practice, because houses have been built in remote areas that have little access to public services and transport and provide few job opportunities. Part of this picture relates to the problematic role currently played by the private sector, marked by the delivery of poor-quality housing units which are not adapted to local needs, including the weather conditions.

This study suggests that in Brazil, legislation may have facilitated important developments with a view to improving housing and thus reducing disaster risks for poor people, who are also sometimes relocated to less-risky areas. Despite the notable concept of the MCMV program, however, the main recommendation is that it be further adjusted so that the housing situation can be effectively improved and disaster risks to be considerably reduced.

Notes

1 United Nations International Strategy for Disaster Reduction (2009), *UNISDR Terminology on Disaster Risk Reduction* (Geneva, Switzerland: Author), pp. 10–11. See also United Nations Office for Disaster Risk Reduction (2004), *Living with Risk: A Global Review of Disaster Reduction Initiatives* (Geneva, Switzerland: Author), p. 23.
2 United Nations International Strategy for Disaster Reduction (2007), *Words into Action: A Guide for Implementing the Hyogo Framework, Annex III: Indicators for Assessing Progress in Implementing Hyogo Framework for Action Priorities 1–5* (Geneva, Switzerland: Author), pp. 151–52.
3 United Nations International Strategy for Disaster Reduction (2005), *Hyogo Framework for Action: Priorities for Action* (Geneva, Switzerland: Author), Priority for Action 1, p. 6.
4 Ibid., Priority for Action 1 (i)(c).
5 See "National HFA Monitor, 2013–2015," (n.d.), *PreventionWeb*, retrieved November 21, 2015, from http://www.preventionweb.net/english/hyogo/hfa-monitoring/national/
6 UNISDR (2011), *Hyogo Framework for Action 2005–2015: Building the Resilience of Nations and Communities to Disasters, Mid-Term Review 2010–2011* (Geneva, Switzerland: Author).
7 For more on the conference, see UNISDR (2013), "Towards a Post-2015 Framework for Disaster Risk Reduction," *PreventionWeb*, retrieved November 21, 2015, from http://www.preventionweb.net/posthfa/; UNISDR (n.d.), "Post-2015 Framework for Disaster Risk Reduction," *UNISDR Website*, retrieved November 21, 2015, from http://www.unisdr.org/we/coordinate/hfa-post2015
8 See the 2011 IFRC country case studies on Brazil, the Dominican Republic, Nepal, and South Africa, retrieved November 21, 2015, from http://www.ifrc.org/en/what-we-do/idrl/research-tools-and-publications/disaster-law-publications/; E. Mosquini (n.d.), "Disaster Laws," discussion paper presented at the 31st International Conference of the Red Cross Red Crescent, Geneva, Switzerland, November 28–December 1, 2011, retrieved November 21, 2015, from http://www.ifrc.org/PageFiles/132698/ENGLISH%20Disaster%20

Laws%20discussion%20paper%20FINAL.pdf; United Nations Development Programme (2007), *A Global Review: UNDP Support to Institutional and Legislative Systems for Disaster Risk Management* (New York, NY: Author); International Federation of the Red Cross and Red Crescent Societies (2010), *Hyogo Framework for Action: Red Cross Red Crescent Mid-Term Review October 2010* (Washington, DC: Author); UNISDR (2011); Global Network of Civil Society Organisations for Disaster Reduction (GNDR) (2011), *If We Do Not Join Hands: Views from the Frontline* (London, UK: Author); GNDR (2009), *Clouds but Little Rain: Views from the Frontline – A Local Perspective of Progress Towards Implementation of the Hyogo Framework for Action* (London, UK: Author).

9   International Federation of the Red Cross and Red Crescent Societies (n.d.), *Legislation for Disaster Risk Reduction* [Web page], retrieved November 21, 2015, from http://www.ifrc. org/en/what-we-do/idrl/about-disaster-law/legislation-for-disaster-risk-reduction/

10 IFRC (2011), *Analysis of Legislation Related to Disaster Risk Reduction in Brazil* (Geneva, Switzerland: Author), p. 18.

11 The annual percentage of GDP growth was 7.6 percent in 2010, 3.9 percent in 2011, 1.8 percent in 2012, and 2.7 percent in 2013. World Bank (n.d.), *World Development Indicators* [Web page], retrieved December 1, 2015, from http://data.worldbank.org/data-catalog/ world-development-indicators

12 According to the World Bank's World Development Indicators, the Gini coefficient for Brazil was 52.9 in 2013. A Gini index is a measure of statistical dispersion of the income distribution of a country's residents; a low Gini coefficient indicates a more equal distribution, with 0 corresponding to complete equality. According to new data, Brazil is now ranks 79th in the world, with a rating of 0.744. World Bank (n.d.); United Nations Development Programme, *Human Development Reports: Brazil – Human Development Indicators* [Web page], retrieved December 1, 2015, from http://hdr.undp.org/en/countries/profiles/BRA

13 Universidade Federal de Santa Catarina and Centro Universitário de Estudos e Pesquisas sobre Desastres (2012), *Capacitação Básica em Defesa Civil* (Florianópolis, Santa Catarina: Author), p. 21.

14 IFRC (2011), p. 23.

15 Ibid.

16 Housing figures are among the social rights listed in Articles 5–17 of the Brazilian Constitution. For a non-official English translation of the 1998 Brazilian Constitution, see "Brazil Constitution" [Web page], available at http://www.v-brazil.com/government/laws/constitution.html

17 M.P. Poirier (2012), "Brazilian Regularization of Title in Light of Moradia, Compared to the United States' Understandings of Homeownership and Homelessness," *University of Miami Inter-American Law Review* 44, p. 268.

18 R. Rolnik (2013, December 30), *Report of the Special Rapporteur to the UN General Assembly* (New York, NY: UN).

19 Constitution of the Federative Republic of Brazil October 5, 1988, tit. VI, ch. II. (This section consists of Articles 182 and 183.)

20 For more about the strengths of the 1988 Constitution, see E. Fernandes (1998), "Law and Urban Change in Brazil," in *Illegal Cities: Law and Urban Change in Developing Countries*, E. Fernandes and A. Varley, eds., pp. 140–56 (London, UK: Zed Books).

21 C. Butler (2009), "Critical Legal Studies and the Politics of Space," *Sociological and Legal Studies* 18, pp. 325–26.

22 Poirier (2012), p. 272.

23 Brazil Law 10.257, July 10, 2001.

24 E. Fernandes (2007), "Constructing the 'Right to the City' in Brazil," *Social and Legal Studies* 16, pp. 201–02.

25 Poirier (2012), p. 275.

26 Ibid., p. 273.

27 For example, see the following comparative study, which points out the need for greater transparency in disaster management structures in Brazil, especially at the local level: E.L. Tompkins, M.C. Lemos, and E. Boyd (2008), "A Less Disastrous Disaster: Managing Response to Climate-Driven Hazards in the Cayman Islands and NE Brazil," *Global Environmental Change* 18(4), pp. 742–43.

28 IFRC (2011), p. 5.

29　See Federal law no. 12.608, April 10, 2012.
30　Ibid.
31　Ibid., Article 8 (IV), (VII).
32　Ibid., Articles 2 and 4 (I).
33　Data gathered from a 2011 study conducted by the Brazilian Institute of Geography and Statistics (Instituto Brasileiro de Geografia e Estatística, or IBGE). See J. Garcia (2011, December 21), "Mais de 11 milhões vivem em favelas no Brasil, diz IBGE; maioria está na região Sudeste," *Universo Online* [Website], available at http://noticias.uol.com.br/cotidiano/ultimas-noticias/2011/12/21/mais-de-11-milhoes-vivem-em-favelas-no-brasil-diz-ibge-maioria-esta-na-regiao-sudeste.htm
34　Law 12.608, April 10, 2012, Article 3.
35　Ibid., Article 24, which modifies Law 10.257 of July 10, 2001, Article 2(VI).
36　Ibid., Articles 42-A and 42-B.
37　Centro Nacional de Monitoramento e Alertas de Desastres Naturais (n.d. a), CEMADEN Website, retrieved November 21, 2015, from http://www.cemaden.gov.br/index.php
38　Interview with Carlos Nobre, representative of the Ministry of Science, Technology, and Innovation, which is in charge of supporting municipalities in mapping their disaster risk areas. See "País terá centro de alerta para prever desastres naturais: Entrevista com Carlos Nobre" (2011, October 6), *Instituto Humanitas Unisinos*, retrieved November 21, 2015, from http://www.ihu.unisinos.br/noticias/501600-pais-tera-centro-de-alerta-para-prever-desastres-naturais-entrevista-com-carlos-nobre
39　CEMADEN (n.d. b), *Municípios Monitorados* [Web page], retrieved November 21, 2015, from http://www.cemaden.gov.br/municipiosprio.php
40　"País terá centro. . ." (2011).
41　As of January 1, 2015 the minimum wage in Brazil is R$788.00 per month. This represents around €250.00, £200.00, or US$300.00 per month – see M. Izaguirre (2014), "Decreto fixa salário mínimo em R$ 788 a partir de 1º de janeiro," *Valor* [Online edition], retrieved November 21, 2015, from http://www.valor.com.br/brasil/3839830/decreto-fixa-salario-minimo-em-r-788-partir-de-1. See also J. Klink and R. Denaldi (2014), "On Financialization and State Spatial Fixes in Brazil: A Geographical and Historical Interpretation of the Housing Program My House, My Life," *Habitat International* 44, pp. 223–24; M.A. Buzzar, M.C. Barros, C. Teixeira, C. Boggi, R. Urnahi, and M. Viana (2014), "Minha Casa Minha Vida Entidades e as Possibilidades de Renovação da Política Habitacional," paper presented at the XV Encontro Nacional de Tecnologia do Ambiente Construído, November 12–14, 2014, Maceió, Alagoas, Brazil, 1589.
42　J. Berger, N. Medvedovski, and L. Mörschbächer (2014), "Avaliação da Gestão Pós-Ocupação, Estudo de Caso: Condomínios do Programa Minha Casa Minha Vida em Pelotas-RS," paper presented at the XV Encontro Nacional de Tecnologia do Ambiente Construído, November 12–14, 2014, Maceió, Alagoas, Brazil, p. 1874.
43　Klink and Denaldi (2014), p. 224.
44　Ibid.
45　Ibid.
46　Buzzar et al. (2014), p. 1592, addressing two MCMV projects built in São Paulo.
47　See the "Statute of the City," Article 4, V, f.
48　Buzzar et al. (2014), pp. 1590–91; Berger et al. (2014), pp. 1876–77.
49　Klink and Denaldi (2014), p. 224.
50　This can be identified in the literature covering different regions in Brazil. See, for example, Buzzar et al. (2014); L. Logsdon and D. Campos, "O 'PMCMV' em Cuiabá-MT uma análise da qualidade dos projetos destinados às familias de baixa renda, 2012–2014," paper presented at the XV Encontro Nacional de Tecnologia do Ambiente Construído, November 12–14, 2014, Maceió, Alagoas, Brazil (covering locations in Cuiabá/Mato Grosso state); J.M.B.M. Hybiner, A.C.G. Tibiriçá, A.W.B. de Carvalho, M.G. Murat, and C. Hosken (2014), "Uso da NBR15575: 2013 Na avaliação técnico-construtiva de um conjunto habitacional," paper presented at the XV Encontro Nacional de Tecnologia do Ambiente Construído, November 12–14, 2014, Maceió, Alagoas, Brazil (covering locations in Viçosa, Minas Gerais state).
51　Hybiner et al. (2014), p. 2363.

52 Berger et al. (2014), pp. 1876, 1880–81, on the situation in Pelotas, Rio Grande do Sul state.
53 Klink and Denaldi (2014), p. 225.
54 Ibid., p. 226.
55 Ibid., p. 225.

Bibliography

Berger, J., N. Medvedovski, and L. Mörschbächer. (2014). "Avaliação da Gestão Pós-Ocupação, Estudo de Caso: Condomínios do Programa Minha Casa Minha Vida em Pelotas-RS." Paper presented at the XV Encontro Nacional de Tecnologia do Ambiente Construído, November 12–14, 2014, Maceió, Alagoas, Brazil.

Butler, C. (2009). "Critical Legal Studies and the Politics of Space." *Sociological and Legal Studies* 18, pp. 325–26.

Buzzar, M.A., M.C. Barros, C. Teixeira, C. Boggi, R. Urnahi, and M. Viana. (2014). "Minha Casa Minha Vida Entidades e as Possibilidades de Renovação da Política Habitacional." Paper presented at the XV Encontro Nacional de Tecnologia do Ambiente Construído, November 12–14, 2014, Maceió, Alagoas, Brazil.

Centro Nacional de Monitoramento e Alertas de Desastres Naturais. (n.d. a). CEMADEN Website. Retrieved November 21, 2015, from http://www.cemaden.gov.br/index.php

———. (n.d. b). *Municípios Monitorados* [Web page]. Retrieved November 21, 2015, from http://www.cemaden.gov.br/municipiosprio.php

Fernandes, E. (1998). "Law and Urban Change in Brazil." In *Illegal Cities: Law and Urban Change in Developing Countries*. E. Fernandes and A. Varley, eds. London, UK: Zed Books.

———. (2007). "Constructing the 'Right to the City' in Brazil." *Social and Legal Studies* 16, pp. 201–2.

Garcia, J. (2011, December 21). "Mais de 11 milhões vivem em favelas no Brasil, diz IBGE; maioria está na região Sudeste." *Universo Online* [Website]. Available at http://noticias.uol.com.br/cotidiano/ultimas-noticias/2011/12/21/mais-de-11-milhoes-vivem-em-favelas-no-brasil-diz-ibge-maioria-esta-na-regiao-sudeste.htm

Global Network of Civil Society Organizations for Disaster Reduction (GNDR). (2009). *Clouds but Little Rain: Views from the Frontline – A Local Perspective of Progress Towards Implementation of the Hyogo Framework for Action*. London, UK: Author.

———. (2011). *If We Do Not Join Hands: Views from the Frontline*. London, UK: Author.

Hybiner, J.M.B.M., A.C.G. Tibiriçá, A.W.B. de Carvalho, M.G. Murat, and C. Hosken. (2014). "Uso da NBR15575: 2013 Na avaliação técnico-construtiva de um conjunto habitacional." Paper presented at the XV Encontro Nacional de Tecnologia do Ambiente Construído, November 12–14, 2014, Maceió, Alagoas, Brazil.

International Federation of the Red Cross and Red Crescent Societies. (n.d.). *Legislation for Disaster Risk Reduction* [Web page]. Retrieved November 21, 2015, from http://www.ifrc.org/en/what-we-do/idrl/about-disaster-law/legislation-for-disaster-risk-reduction/

———. (2010). *Hyogo Framework for Action: Red Cross Red Crescent Mid-Term Review October 2010*. Washington, DC: Author.

———. (2011). *Analysis of Legislation Related to Disaster Risk Reduction in Brazil*. Geneva, Switzerland: Author.

Izaguirre, M. (2014, December 30). "Decreto fixa salário mínimo em R$ 788 a partir de 1º de janeiro." *Valor* [Online edition]. Retrieved November 21, 2015, from http://www.valor.com.br/brasil/3839830/decreto-fixa-salario-minimo-em-r-788-partir-de-1

Klink, J., and R. Denaldi. (2014). "On Financialization and State Spatial Fixes in Brazil: A Geographical and Historical Interpretation of the Housing Program My House, My Life." *Habitat International* 44, pp. 223–24.

Logsdon, L., and D. Campos. "O 'PMCMV' em Cuiabá-MT uma análise da qualidade dos projetos destinados às familias de baixa renda, 2012–2014." Paper presented at the XV

Encontro Nacional de Tecnologia do Ambiente Construído, November 12–14, 2014, Maceió, Alagoas, Brazil.

Mosquini, E. (n.d.). "Disaster Laws." Paper presented at the 31st International Conference of the Red Cross Red Crescent, November 28–December 1, 2011, Geneva, Switzerland. Retrieved November 21, 2015, from http://www.ifrc.org/PageFiles/132698/ENGLISH%20 Disaster%20Laws%20discussion%20paper%20FINAL.pdf

"National HFA Monitor, 2013–2015." (n.d.). *PreventionWeb*. Retrieved November 21, 2015, from http://www.preventionweb.net/english/hyogo/hfa-monitoring/national/

"País terá centro de alerta para prever desastres naturais: Entrevista com Carlos Nobre." (2011, October 6). *Instituto Humanitas Unisinos*. Retrieved November 21, 2015, from http://www. ihu.unisinos.br/noticias/501600-pais-tera-centro-de-alerta-para-prever-desastres-naturais-entrevista-com-carlos-nobre

Poirier, M.P. (2012). "Brazilian Regularization of Title in Light of Moradia, Compared to the United States' Understandings of Homeownership and Homelessness." *University of Miami Inter-American Law Review* 44, pp. 259–312.

Rolnik, R. (2013, December 30). *Report of the Special Rapporteur to the UN General Assembly*. New York, NY: UN.

Tompkins, E.L., M.C. Lemos, and E. Boyd. (2008). "A Less Disastrous Disaster: Managing Response to Climate-Driven Hazards in the Cayman Islands and NE Brazil." *Global Environmental Change* 18(4), pp. 742–43.

United Nations Development Programme. *Human Development Reports: Brazil – Human Development Indicators* [Web page]. Retrieved December 1, 2015, from http://hdr.undp.org/ en/countries/profiles/BRA

———. (2007). *A Global Review: UNDP Support to Institutional and Legislative Systems for Disaster Risk Management*. New York, NY: Author.

United Nations International Strategy for Disaster Reduction. (n.d.). "Post-2015 Framework for Disaster Risk Reduction." *UNISDR Website*. Retrieved November 21, 2015, from http:// www.unisdr.org/we/coordinate/hfa-post2015

———. (2005). *Hyogo Framework for Action: Priorities for Action*. Geneva, Switzerland: Author.

———. (2007). *Words into Action: A Guide for Implementing the Hyogo Framework, Annex III: Indicators for Assessing Progress in Implementing Hyogo Framework for Action Priorities 1–5*. Geneva, Switzerland: Author.

———. (2009). *UNISDR Terminology on Disaster Risk Reduction*. Geneva, Switzerland: Author.

———. (2011). *Hyogo Framework for Action 2005–2015, Building the Resilience of Nations and Communities to Disasters, Mid-Term Review 2010–2011*. Geneva, Switzerland: Author.

———. (2013). "Towards a Post-2015 Framework for Disaster Risk Reduction." *PreventionWeb*. Retrieved November 21, 2015, from http://www.preventionweb.net/posthfa/

UN Office for Disaster Risk Reduction. (2004). *Living with Risk: A Global Review of Disaster Reduction Initiatives*. Geneva, Switzerland: Author.

Universidade Federal de Santa Catarina and Centro Universitário de Estudos e Pesquisas sobre Desastres. (2012). *Capacitação Básica em Defesa Civil*. Florianópolis, Brazil: Author.

World Bank. (n.d.). *World Development Indicators* [Web page]. Retrieved December 1, 2015, from http://data.worldbank.org/data-catalog/world-development-indicators

Part 3
Asia

Introduction

Adenrele Awotona

Asia occupies 30 percent of the world's land mass. However, 40 percent of the world's disasters between 2004 and 2013 occurred there, resulting in 80 percent of the world's disaster-related deaths,[1] the vast majority of whom were the poor, the vulnerable, and those living in remote areas. This makes Asia the continent most susceptible to disaster in the world. Indeed, the 2004 Indian Ocean tsunami, which, according to the United Nations, killed over 220,000 people and made millions homeless, was one of the deadliest natural disasters in human history. The economic losses have also been enormous: in 2012 alone, Asia lost US$15 billion due to natural disasters, according to data released by the Center for Research on the Epidemiology of Disasters (CRED), a drop from 2011, when the region recorded a staggering $300 billion loss, mostly due to Japan's tsunami and Thailand's floods. In 2013, Asia again suffered the greatest damage in the world (49.3 percent of global disaster damage).[2]

Furthermore, about 40 percent of the earth's land area is threatened by desertification, which is intensified by climate change and human activities. The UN Convention to Combat Desertification has estimated that by 2030, "water scarcity in some places will displace up to 700 million people. Nine Asian Development Bank member countries including China, India and Pakistan, have large land areas within the arid, semi-arid and dry sub-humid zones and therefore, remain most vulnerable to desertification."[3] The Asian Development Bank has noted that Asia's economic progress will be challenged by the increasing quantity of floods, landslides, and other disasters.[4]

It is against this background that the various authors of the chapters in this section examine an array of approaches to disaster risk reduction, relief, and recovery in different Asian communities.

Chapters 11 and 12 report case studies from India, a subcontinent whose unique geo-climatic conditions make it vulnerable to natural disasters. Approximately 57 percent of India is vulnerable to earthquakes, ranging from magnitude 1.7 to 9.3 on the Richter scale; 12 percent and 8 percent are vulnerable to floods and cyclones, respectively. In terms of losses of human lives and national assets, the Super Cyclone of October 1999 caused more than 9,000 deaths; the Bhuj earthquake of January 2001 resulted in 14,000 deaths; and the Indian Ocean tsunami of December 2004 accounted for the deaths of nearly 15,000 persons.

In Chapter 11, Sweta Byahut discusses the salient features of the proposed legal and administrative framework for developing regulations to enable a resilient built environment for the city of Delhi, India. This was part of a comprehensive reform proposed for 2004–06 to streamline Delhi's development regulation system. Development regulation not only shapes a city's built environment and plays a crucial role in determining its quality of life, image, and efficiency, but also determines its safety and resiliency in the face of disasters and climate impact. Byahut notes that most Indian cities are growing rapidly under the *license raj*[5] regime, and widespread unplanned development has resulted in unattractive cities and low-quality building stock, often in violation of building regulations and vulnerable to natural disasters. Delhi's development regulation system, the author continues, has long been perceived as being complicated, non-transparent, unfair, and costly, resulting in low levels of compliance and a large stock of poor-quality buildings that can be considered unauthorized or illegal. Byahut then proposes that interventions are needed at all levels to address this: from improving primary building legislation to establishing more appropriate monitoring systems, ensuring the certification of professionals and the rationalization of regulations. According to Byahut, the proposals, which were prepared with the guidance and support of the Municipal Corporation of Delhi (MCD) and overseen by a multi-organization steering committee, consisted of several stages: reviewing the legal and institutional frameworks for development regulation in Delhi; framing a legislative reform agenda; drafting new procedure bylaws; formulating performance and planning regulations; establishing a process of preparing local area plans; developing a framework to involve professionals in ensuring compliance; and legislative amendments to enable all of the above. Byahut states that the broader objective was to increase the scope of development regulation in order to develop a resilient and robust building stock and implement a culture of compliance.

In Chapter 12, Hem Chandra, Ankita Pandey, Sarika Sharma, Nitin Bharadwaj, Leela Masih, and Supriya Trivedi examine the degree of success of a medical preparedness disaster plan to treat and save the maximum number of lives at or very near a disaster site. The recent cloudburst in June 2013 in the Kedarnath temple area of Uttarakhand in India was a case study. Medical management at the disaster site was problematic due to the loss of logistics, but possible at the nearest hospitals. Uttarakhand's hilly terrain and seismic zone made the task of efficient medical delivery difficult, leading to a loss of more than 3,000 human lives and assets worth US$670 million. Hem Chandra and his colleagues concluded that the geographic location of Uttarkhand requires a special medical preparedness disaster plan to be formulated and that the armed forces as well as paramilitary staff should play a significant role in both the formulation of the process and its implementation.

In 1915, the Rockefeller Foundation (RF) and the closely related International Health Board (IHB) decided that malaria control would be one of their major public health emphases. Beginning in the southern US and expanding globally in the 1920s, these organizations engaged in community-based demonstration projects that influenced other international organizations as well as national and imperial projects. At the same time, the RF and IHB conducted national anti-malaria campaigns that were judged at the time, and in historical perspective, as less successful than the localized, community-based projects. Both approaches emphasized education, including establishing institutes of public health, supporting national and colonial public health training programs and the training of paraprofessionals at the local level. Chapter 13, by Darwin H. Stapleton, examines several iterations of the Rockefeller and IHB programs, focusing most closely on the contemporary and long-term results of community-based programs in China, Indonesia, and Sri Lanka, which were the most effective and durable responses to malaria as a public health disaster. He draws on the archives of the Rockefeller Foundation at the Rockefeller Archive Center (Sleepy Hollow, NY), as well as published work by a range of international scholars.

Turkey has two important active faults, the Anatolian Fault and the Northern Anatolian Fault, which span the northern and eastern part of the country, and constitute a 90-percent coverage in earthquake hazard zones. The Northern Anatolian fault is a strike-slip fault moving towards the west at a rate of 3 cm per year, and has historically ruptured from the eastern rural region of the country approaching the west; it ruptured in the western industrial and populated region of Kocaeli in 1999. In Chapter 14, Derin N. Ural scrutinizes the outcomes of a large effort to improve Turkey's disaster management system, including disaster policies and enforcement of regulations that were undertaken by the government, nongovernmental organizations, and universities in the years following the Kocaeli earthquake. The evaluation of the efficacy of those efforts is imperative because, according to Ural, the scientific prediction for the next rupture is expected to be to the west of Kocaeli, which coincides with the Istanbul Metropolitan area, inhabited by over 13 million residents. Ural observes that the aftermath of the Kocaeli earthquake required coordinated and rapid local and national disaster management responses and recovery. However, she notes that Turkey's disaster management system in 1999 was not prepared to cope with the large disaster, which resulted in over 17,000 casualties and damaged 80,000 buildings.

In Chapter 15, Mahmood Hosseini and Yasamin O. Izadkhah examine risk reduction programs in developing countries that are located in earthquake-prone areas, with a focus on Iran. The authors begin by outlining the necessity of seismic risk reduction in earthquake-prone countries such as Iran. They then present a simple formulation for risk evaluation and reduction by introducing and using the concept of "reduction indices." This is followed by their examination of some factors that intensify seismic risk in developing countries, such as the low quality of construction, lack of sufficient code enforcement and regulations, poverty, insufficient attention of authorities to disaster threats, and the lack of awareness and preparedness in the community at risk. Finally, Hosseini and Izadkhah address the role of management in risk reduction with regard to some suggested main lines of actions and proposed risk reduction formulation.

Chapter 16, by Xuepeng Qian, Weisheng Zhou, and Kenichi Nakagami, starts by reviewing the efficacy of the idea of the pairing aid system which was introduced in China for the economic development of the minority-inhabited border areas in the 1950s and later established as an institutional framework for national policies to meet three different objectives: economic development of the minority-inhabited border areas, grand infrastructure construction projects, and disaster relief and recovery. The pairing aid system of Tangshan Earthquake in 1976 was the first trial, where it proved to be effective in rescue and reconstruction work. It also proved successful in disaster relief and recovery after the Wenchuan Earthquake in 2008, in Japan after the Niigata Earthquake in 2004 and the Great East Japan Earthquake in 2011. This chapter then compares the pairing aid systems from several perspectives of formation, organization, and operation in different social and cultural contexts of China and Japan. It examines the improvement and institutionalization of the pairing aid system at different scales and forms of disaster management in order to identify a sustainable, mutual aid system among provinces and cities all over the two countries, which will be necessary when large-scale disasters happen in the future.

According to Yutaka Sho in Chapter 17, all of the 13 nuclear power plants in Tohoku, Japan, were creating energy for Tokyo's benefit when the 2011 Tohoku earthquake struck. The high visibility of the long-term effects of the Fukushima No. 1 meltdown divided opinion around the world. The majority of the citizens of Japan, Germany, and Italy took their anti-nuclear stance to the streets and to their parliaments. Ignoring the citizens' consensus, however, Sho notes that the Japanese government in Tokyo continues to fund and export nuclear technology as a way out of the recession. On the ground, established professionals and non-architects alike rushed in to the reconstruction effort with varying methods and ideologies. Some designers published papers addressing the urgent issues that emergency shelters face, while others took the disaster as an opportunity for tabula rasa planning schemes to revamp the poor and depopulated Tohoku region. The latest scheme, according to Sho, is dark tourism – turning the Fukushima No. 1 plant into a theme park in order to create jobs and to memorialize the disaster; fundraising efforts in Tokyo for the "Disneyfication" of Fukushima is well underway. In this chapter, Sho reviews, analyzes, and critiques architectural and planning projects in post-earthquake Tohoku in order to contemplate ways to build connections between affected rural regions such as Tohoku and unscathed cities such as Tokyo.

In Chapter 18, Yu Wang and Hans Skotte investigate the reconstruction activities undertaken following the 2008 Sichuan earthquake and how the mode of rebuilding of the village of Taoping left the traditional Qiang settlement still vulnerable, but in new ways. They also raise the question as to how one should embed the principal methods of disaster risk reduction (DRR) into the prevailing approach to lived-in cultural heritage reconstruction. For their analysis, Wang and Skotte employ two modes: the Disaster Risk Assessment methodology and the Pressure and Release (PAR) model. A risk assessment was conducted to identify and measure the underlying disaster risks and current vulnerability of (the reconstructed) Taoping, while the Pressure and Release model was used to analyze the path-generating vulnerability. Using these models, the authors were able to uncover the effects of the recent reconstruction on Taoping's present vulnerability as a lived-in cultural heritage settlement. Wang and Skotte note in their research that those two modes of analysis encounter a complex situation in which the heritage of Taoping intimately bonds with the "lived-in" quality of the local community's relationship to its heritage. This exposes the social dimension as a crucial factor when identifying present-day vulnerabilities. The risk assessment reveals that the lived-in cultural heritage village of Taoping still remains at high-level risk with regard to earthquakes and at mid-level risk with regard to weather-caused landslides and floods. According to the PAR model analysis, the consequence of the interaction between community and post-disaster reconstruction could be defined as a state of "dynamic pressure" pushing the village towards an "unsafe state." This is because of the community's material reaction to the official reconstruction, which in effect weakened some of the rebuilt structures, and because the official reconstruction required that the livelihood of the community change from farming to tourism due to the building of a new settlement on the land used for farming, leaving the original village a "tourist village." Wang and Skotte conclude that embedding a risk assessment approach into future reconstruction efforts of damaged lived-in cultural heritage sites requires that all the relevant stakeholders address the engagement and contribution of the people who constitute the "lived-in" dimension of the cultural site.

In Chapter 19, Fariha Tariq discusses a few of the efforts being made to improve the conditions of slums in Pakistan and describes the process undertaken by professionals, NGOs, and educational institutions to envision safe, healthy houses made of good-quality, inexpensive local materials, and designed for a community's unique environmental and social conditions. In Pakistan, a vast stretch of dock area on the Ravi bank in Lahore remained under chest-deep water for about eight weeks following heavy showers during the monsoon of July 2013. About 50,000 slum-dwellers in this part of the city were severely affected by rain and water-logging. There are 227 slum settlements in Lahore today, which face housing deprivations during severe environmental conditions. Tariq notes that there is a dearth of financial and technical resources to build/renovate damaged and destroyed houses, especially for low-income households. Recently there have been development efforts in Pakistan that have promoted architectural intervention as the key to helping local people overcome vulnerability and poverty, improve their lives, and change their communities. Mobilizing and educating them to build safe, healthy, and environmentally sound houses, however, has proven beneficial. Tariq concludes by proposing what she describes as "progressive, participatory ways to design, develop, and deliver shelter facilities in adverse environments."

Notes

1 D. Guha-Sapir, P. Hoyois, and R. Below (2014), *Annual Disaster Statistical Review 2013: The Numbers and Trends* (Brussels, Belgium: Centre for Research on the Epidemiology of Disasters).
2 Ibid.
3 "Asia's Disaster Toll" (2013), *Climate Change in Asia* [Website], Konrad-Adenauer-Stiftung Media Programme Asia, retrieved January 20, 2016, from http://ejap.org/environmental-issues-in-asia/natural-disasters-asia.html
4 Ibid.
5 *License Raj* refers to the elaborate system of licenses, regulations, and red tape that was required to establish and run businesses in India up until 1990. The system is currently being dismantled as part of the liberalization reforms initiated by the Indian government in 1991.

Bibliography

"Asia's Disaster Toll." (2013). *Climate Change in Asia* [Website]. Konrad-Adenauer-Stiftung Media Programme Asia. Retrieved January 20, 2016, from http://ejap.org/environmental-issues-in-asia/natural-disasters-asia.html

Guha-Sapir, D., P. Hoyois, and R. Below. (2014). *Annual Disaster Statistical Review 2013: The Numbers and Trends*. Brussels, Belgium: Centre for Research on the Epidemiology of Disasters. http://www.cred.be/sites/default/files/ADSR_2013.pdf

11 An effective development regulation system for a resilient built environment

A reform project for Delhi, India

Sweta Byahut

Background

Disaster risk reduction in cities of the developing world stresses the importance of a resilient and robust building stock to address urban vulnerability. Recent building collapses have resulted in massive losses of lives and property and have led authorities to investigate, improve, and streamline building regulations. Building safety became a national priority in India after the January 2001 Gujarat earthquake, in which approximately 14,000 people lost their lives. Subsequent investigations showed that this earthquake was within the expected seismicity of the region, but that the buildings had not been constructed to seismic standards and had used substandard materials, resulting in massive building collapses when the earthquake struck. In 2005, massive floods in Mumbai resulted in over 1,000 casualties; again, investigations revealed that substandard buildings had been constructed in vulnerable and low-lying areas. A string of isolated building failures also highlighted the need for building regulation: in November 2011 a building collapsed in east Delhi, killing 71 people and injuring 100 others, and in April 2013 another seven-story building collapsed in Thane, killing 74 people. This building was constructed illegally and occupied by construction workers and their families. About 20 people were taken into custody, including senior municipal officials who were accused of taking bribes and turning a blind eye to illegal and inexpensive construction, believed to be part of the powerful nexus of builders, politicians, municipal officers, and police. The most horrific collapse, that of the Rana Plaza building in Dhaka, Bangladesh, which housed several garment manufacturing factories and killed over 1,100 people in April 2013, is still fresh in people's memories. This was also blamed on substandard construction materials and a blatant disregard for building safety codes. In all of these incidents, subsequent investigations exposed substandard and/or illegal construction as the main reason for the buildings' failures and the loss of lives.

We understand that unplanned urbanization – unregulated growth that is haphazard and unregulated – is a major source of vulnerability in India, and that reforming the building regulation system is critical to disaster risk reduction. In the wake of recent disasters, the Indian government has made a commitment to develop a better system for preventing such devastating damage in its cities.[1] Integrating disaster risk reduction into development policies is more cost-effective than the relief and rehabilitation necessary after disasters. India has developed recommendations and a road map for improving the current regulatory system in order to achieve a culture of compliance in several ways. The first step is the preparing of updated Model Town and Country Planning

Legislation, Zoning Regulations, Development Control, and Building Regulations/ Bylaws. It suggests including hazard maps and appropriate policies for mitigating hazards, updating the current zoning system, establishing a registration system for architects, structural engineers, and construction engineers, and a certified inspection system for new developments.[2] Furthermore, the 73rd and 74th constitutional amendments give rural and urban local bodies increased local self-governance, which enables them to initiate preparedness, mitigation, recovery, and rehabilitation in a more decentralized manner responsive to local needs. While local governments are responsible for reducing urban vulnerability, they typically lack the capacity to regulate or enforce regulations. There is also a lack of willingness or political support to implement systemic or institutional reforms.[3]

Menoni et al. discussed the separate but related concepts of resilience and vulnerability.[4] Resilience is characterized by the capability of a community to absorb a disturbance, such as a natural disaster, without compromising its structure and function. Vulnerability, on the other hand, is how prone a community is to damage after an extreme event. Reducing a community's vulnerability to a natural disaster does not necessarily increase its resilience in the event of a disaster, and vice versa. Furthermore, resilience comprises a region's capacity to store knowledge and experience, plan for future change, reorganize without falling into decline, and keep crucial facilities functioning. Resilience should be systemic and integrated into all aspects of development regulation, and "implementation does not come automatically from regulation."[5] Therefore, policies that provide incentives for compliance need to be developed.

Earthquake vulnerability results from poor building quality rather than the earthquake itself: earthquakes are often within the expected levels of regional seismicity but buildings are not constructed to withstand them, resulting in widespread damage and casualties. India has had scientific knowledge of seismic structural design since the 1960s, but many buildings still fell in the Gujarat 2001 earthquake, resulting in very high human losses. The post-earthquake experience in Gujarat demonstrated that most casualties occurred because the collapsed buildings were not built (nor were they required to be built) to the safety standards stipulated in the National Building Code of India, because at the time the Code was rarely enforced at the local level. Thiruppugazh[6] identified non-compliance with building codes, substandard construction, lack of regulation and enforcement, poor construction, the use of poor-quality materials, aging building stock, lack of professional training in seismic design and construction, and not having a professional engineering association for oversight as major causes for high seismic vulnerability. In post-earthquake Gujarat, an entirely new set of safety-related building regulations were uniformly applied and enforced across the state. However, the enabling legislation, the Gujarat Town Planning and Urban Development Act of 1976, still failed to clearly delineate responsibilities for compliance and has not prevented violations.[7]

This culture of non-compliance is ubiquitous in Indian cities because a system of self-regulation of compliance with building codes simply does not exist. There is agreement that site engineers, real estate developers, and builders must be held more directly accountable for compliance; however, at present the market actually penalizes builders for complying with building regulations. Citizens are unaware of urban vulnerability and the need for building safety provisions, so are unwilling to pay more for a building that meets safety codes. At the same time, the cost of developing a compliant building is higher, so developers suffer financially. It is interesting to note that because developers typically comply with building codes for public buildings, no government buildings collapsed in the Gujarat 2001 earthquake because government construction is funded by the government, while private citizens try to cut corners when trying to cover the costs associated with compliance.[8]

Yates[9] points to the moral dilemma that can occur when building structures to withstand earthquakes. Developers are aware that the government and NGOs will bail out victims of earthquake damage, so they are willing to compromise buildings' quality and safety and do not invest in appropriate insurance. In most Indian states, compliance with Indian Standard (IS) codes that define building design standards for seismic considerations (IS 1893 and IS 4326) is usually not even required for private construction. When it is, enforcement is deficient, and structural engineers do not assume legal responsibility for code violations or building failures: "having a well-designed [regulatory] system does not guarantee compliance."[10] Builders and owners do not comply with regulations because they are complex, out of date, inappropriate, and expensive to implement or monitor. Furthermore, the inadequate enforcement system makes violations difficult to detect or punish. These widespread deficiencies must be addressed in a comprehensive manner in order to establish a culture of compliance.

Leaf[11] highlights high enforcement costs in the context of rapidly urbanizing countries. Informal housing systems may actually provide a solution to the high cost of housing, but are often unsafe and vulnerable. More affordable housing discourages informal and unauthorized housing systems and increases compliance. In Indian cities, development tends to be more closely monitored in central city areas than it is in suburban areas. As cities expand, land-use controls such as FAR or FSI (Floor Area Ratio or Floor Space Index) and building heights are restricted within central city areas to limit density, encouraging people to suburbanize and the growth of informal and unregulated development on the outskirts, which is usually substandard and unsafe.[12] If FAR limits were relaxed and more vertical and denser growth in central city areas encouraged, construction could be closely monitored, leading to better-quality building stock. When speaking of a regularization of informal or unauthorized settlements, Menoni et al. assert that "in order to reduce their physical and systemic vulnerability, policies aimed at their legislation and integration into urban areas should be accompanied by actions to retrofit buildings or to provide basic infrastructures and services."[13] Risk reduction needs to be mainstreamed within local urban governance and planning and implementation processes.[14] Legal and institutional arrangements with improved capacities and accountability are important for disaster risk management at the local level, and these capacities should be integrated in land-use planning practices.[15] Other recommendations include raising the standard of work ethics, developing a culture of risk reduction, and enabling local institutions to train and certify professionals who value investing in a resilient built environment.

Non-compliance with regulations cannot be solved through additional regulations or increased enforcement, but instead must be dealt with through the integration of development with vulnerability reduction, good governance, awareness, and capacity building. In Gujarat after the 2001 earthquake, the responsibilities of everyone involved in development were defined and processes were established to make them accountable. Loopholes in the existing development regulation system were closed, and efforts were made to educate the public about seismic safety and to train civil engineers for disaster mitigation. As a result, people are now generally more willing to pay for safer buildings.[16] It is clear that developing a culture of compliance goes beyond codes, to reforming entire legal, economic, and social systems. A well-functioning development regulation system is recognized as an effective tool to improve public safety, develop resilient building stock, ensure a safer built environment, and enforce a culture of compliance.

Weaknesses in Delhi's existing regulation system

Delhi is a historic city (it is actually seven different cities established over the centuries in layers), with the old city of Shahjahanabad being more than 500 years old and the capital city of New Delhi built a century ago. Major areas developed as organic, unplanned settlements. Today, the greater Delhi region, known as the National Capital Territory of India, is one of the largest and fastest growing mega-cities. It has expanded into neighboring states, and now includes a population of almost 27 million. The present building regulations in Delhi are drawn with too broad a brush, and are not suitably nuanced to deal with local contexts or ground realities. Making these practically enforceable would require a more detailed level of planning and the developing of area-specific building bylaws which could take into account Delhi's diversities.[17]

The development regulation system in Delhi is under stress: about 40 percent of existing buildings are unauthorized in some manner, making them vulnerable and unsafe. Much of the current construction continues to be unauthorized. Building regulations are widely perceived to be outdated, rigid, and complicated, while the roles of professionals involved in the building industry are not clearly defined. Enforcement is a challenge for authorities, and procedures are lengthy and complicated. In addition, existing regulations are perceived to be irrelevant and insufficiently nuanced, paying little regard to areas that need special accommodation such as urban villages, the old Walled City, and slums. Delhi lies in the Zone IV seismic hazard zone, which is a High Damage Risk Zone and encompasses areas predisposed to up to MSK-VIII earthquake risks. The failure to regulate development in Indian cities puts them at serious risk of natural and man-made disasters. Over the years, many plaintiffs have filed suits against the Municipal Corporation of Delhi (MCD) in an attempt to regularize unauthorized construction, and several such court cases are pending. Critical issues such as seismic safety, heritage conservation, access for the disabled, urban design, and redevelopment have been mostly ignored in Delhi.

Enforcement is an important issue with regard to current regulations because authorities themselves perceive most building regulations to be unenforceable. Procedures to be followed before, during, and after construction are not clear, and all associated processes are complicated and lengthy. By putting the onus for compliance on the regulating authority, the existing system leaves room for corruption and does not adequately involve the professionals or share with them any responsibility for ensuring compliance. All these weaknesses exacerbate the issue of jurisdiction: several agencies, including the Municipal Corporation of Delhi, the New Delhi Municipal Council, and the Delhi Development Authority, are all involved in regulating development, but their roles are unclear and their jurisdictions often overlap.

Recognizing all of these issues, the Municipal Corporation of Delhi (MCD) embarked on an ambitious reform process in 2004. The objectives were to improve Delhi's severely stressed building regulation system by revising the Delhi Municipal Corporation (DMC) Act of 1957, which is the enabling legislation for building regulations. In addition, new procedures and building performance bylaws were to be framed. Another objective was to decrease the pervasive rent-seeking behavior by encouraging self-regulation through professional engagement and providing incentives to foster a culture of compliance and to develop resilience and a safer built environment. A decade later, the proposals have been only partially implemented, but there are many important lessons to be learnt from this experience.

The reform process

To address all of these concerns, the reform process for Delhi's building regulation system was carried out in three phases from 2003 to 2006. The project was sponsored by the Indo-USAID FIRE D II Project and the National Institute of Urban Affairs and was overseen by a multi-stakeholder Steering Committee. Phase I, *Policy Agenda and Legislative Intentions*, was formulated by an expert group[18] after examining key inadequacies in Delhi's existing building bylaws, enforcement model, and enabling legislation. It proposed amending the Delhi Municipal Corporation Act of 1957 and framing new building bylaws for Delhi. Specifically, it recommended framing new building regulations for Delhi with emphasis on streamlining procedure regulations; clearly defining roles and responsibilities for professionals involved in construction compliance; establishing an agency for chartering professionals; and formulating a process for initiating Local Area Plans (LAPs) and area-specific building bylaws to be prepared at the neighborhood level by the MCD. In addition, it recommended amending the Delhi Municipal Corporation Act of 1957 to enable all of the above and to mandate clearly and empower the MCD to regulate development and prepare LAPs. Table 11.1 outlines these recommendations.

Table 11.1 Policy agenda and legislative intentions reform framework[19]

Policy agenda and legislative intentions			
1 New building bylaws	2 Amendments to the DMC Act	3 SPV (PASPRA)	4 Local area plans
Review current bylaws	Draft amendment to Act	Conceptualize SPV	Prepare guidelines and select consultants
Draft new building bylaws	Amend Act	Prepare business plan	Prepare pilot local area plans
Notify bylaws		Incorporate SPV	Prepare area-specific planning bylaws
		Operationalize SPV	Adopt area-specific building bylaws
Operationalize new system for regulating development			

In Phase II, the team undertook the drafting of the proposed building bylaws and amendments to the DMC Act 1957 and the formulation of draft guidelines to prepare the LAP in an intensive process. Building regulations were restructured to simplify and clearly distinguish between procedure, building performance, and planning bylaws. This represents a radical departure from existing legislative and regulatory traditions, and attributed much greater responsibility to professionals, along with commensurate powers, autonomy, and privilege. An agency for registering professionals engaged in development regulation called the Professional and Service Providers Rating Agency (PASPRA) was conceptualized. In Phase III, the MCD initiated the preparation of five model LAPs and prepared area-specific building bylaws in pilot neighborhoods with the intention of making them implementable in practice, taking into account Delhi's vast diversities. The next section describes the salient features of the proposed new building regulation system.

Salient features of the new proposed development regulation system

Unlike the existing building bylaws, which lack clarity and are perceived to be rigid, the proposed bylaws were developed following a clear structure and organized distinctly into procedure bylaws, building bylaws, and planning regulations. The procedure bylaws would include procedures for obtaining a building permit, for monitoring construction, and for obtaining a building use permit. The building performance bylaws would include minimum standards that a building has to be built to (structural strength, fire safety, parking layout, etc.) and refer to the National Building Code of India and additional bylaws. The planning bylaws would specify the maximum density, height, and bulk of buildings, parking, permissible and non-permissible uses, and would refer to the Delhi Master Plan for additional bylaws.

This proposed regulatory system was not designed to be set in stone, but is meant to be multi-tiered and have graded flexibility. The Delhi Municipal Corporation Act 1957 is the enabling legislation for development regulations and empowers the MCD, and is therefore designed to be fairly rigid and difficult to modify. On the other hand, the building bylaws specify procedures for regulation, building performance standards, and planning standards, and are developed more like a rigid skeletal framework that is less difficult to modify. Lastly, the schedules and forms include supporting details and formats that may require periodic revisions, and the MCD was given more flexibility to modify them when needed.

Streamlining procedure bylaws and professional self-regulation

Procedures to obtain a building permit before construction begins, to monitor construction, and to obtain a building use permit after construction was completed were clarified and streamlined. The proposed bylaws require developers to engage qualified professionals (such as architects, advocates, and structural engineers), and most importantly, the bylaws clearly lay out the roles and responsibilities of various participants involved at all stages of construction, as shown in Table 11.2 below.

Table 11.2 Clarifying roles and responsibilities of professionals[20]

Professional	Roles and responsibility
Owner	Primary responsibility for complying to building bylaws
Advocate	Responsible for verifying ownership of applicant for building permit
Architect	Responsible for ensuring that architectural design meets with bylaws Responsible for meeting procedural requirements of MCD
Structural Engineer	Responsible for ensuring that structural design meets with bylaws
Construction Engineer	Responsible for ensuring that construction complies with approved plans
Owner	Responsibility for maintenance

It was proposed that all professionals involved in construction, including advocates, architects, construction engineers, and structural engineers, must be listed on record (registered) with the MCD to ensure that they possess the required skills and qualifications to obtain and maintain their licenses based on their minimum qualification and experience requirements and periodic competency tests. This would ensure that professionals are willing to self-regulate, abide by rules defined in the building bylaws, and submit to the purview of the Professional Oversight Committee in the case of disputes. In the long term, PASPRA was proposed to list and grade the professionals based on qualifications, experience, and professional abilities.

It was proposed that the procedure for obtaining building permits would be differentiated based on the type of building to be constructed. The architect was empowered to issue a *Deemed Building Permit* for *simple buildings* (as defined in the schedule) after scrutinizing and certifying plans and would register such buildings with the MCD, which would then undertake random scrutiny of approved plans and construction to ensure compliance. A regular Municipal Building Permit from the MCD would be required for all *complex buildings*, and the architect would scrutinize and certify plans. Construction could begin only after a building permit had been granted, and the owner and architect would be required to notify MCD when it was granted. The building permit would lapse if construction did not begin within one year of the permit being obtained, and lapsed permits would have to be revalidated in order to move forward with construction.

Procedures were also streamlined for various construction phases to ensure that adequate safety procedures are followed on the construction site, that debris is not dumped on any public right-of-way, and construction does not cause undue nuisance or danger. The building permit and prescribed information must be prominently displayed on site. The construction engineer was made responsible for compliance with approved plans, and the architect for notifying the MCD of progress. The MCD will still inspect construction at the plinth, first floor, middle floor (in the case of high-rises), and topmost floor levels in the case of complex buildings, and will grant a permit to proceed with further construction only after inspection. In the case of change in any of the professionals involved or change in ownership, the building permit will lapse automatically and new professionals will be appointed, who must then revalidate the building permit. For complex buildings, a building use permit has to be obtained from the MCD upon the completion of construction; in the case of simple buildings, this can be issued by the architect.

Specifying building performance bylaws

Performance bylaws define the minimum standards that a building has to be built to for structural strength, fire safety, parking layout, etc. It was proposed that older prescriptive specifications should be reduced and that mandatory standards should be limited to issues that affect the public domain or safety. In most cases, these originate from national standards, such as the Indian Standard Codes and the National Building Code of India, while some originate from other legislation.[21] Mandatory standards were specified for clearances from trunk infrastructure, parking, lifts and escalators, rainwater management, solid waste management, air pollution, water and noise pollution, structural safety, fire prevention and safety, maintenance of buildings, and a few architectural elements (railings, compound walls, minimum floor height, etc.). It was proposed that several standards that had previously been specified should be left to the discretion of the architect, including lighting, ventilation, heating, air conditioning, water supply, sanitation, drainage, electrical infrastructure, and accessibility.

It was proposed that hazard and seismic safety should be incorporated into performance bylaws based on the "Model Town and Country Planning Legislation, Zoning Regulations, Development Control and Building Bylaws for Hazard Zones of India 2004" by the Ministry of Home Affairs. In addition, for all important complex buildings, the structural engineer was required to submit a Structural Design Basis Report, along with a third-party verification of the structural design. For fire safety, the Delhi Fire Prevention and Fire Safety Act, 1986, was made applicable to buildings 15 meters or higher and performance-based fire standards were mandated for low-rise buildings. Accessibility was recommended for non-public areas based on the manual published by the Office of the Chief Commissioner for Persons with Disabilities, Government of India. These were not mandated, as it is difficult to ensure compliance, but in complex buildings the architect was made responsible for incorporating accessibility standards.

Relevant provisions on heritage conservation were proposed for incorporation in the legislative amendments and the bylaws that provide for the constitution of a Heritage Conservation Committee empowered to designate heritage buildings, precincts and natural features and to define guidelines and requirements for the designated areas. Additional clearance from the Heritage Conservation Committee would be needed prior to issuing a building permit near these.

Introducing local area planning and area-specific planning regulations

The Delhi Master Plan for the horizon year 2021 defines the requirements for floor area ratio (FAR), building setbacks and margins, use, and parking, through its development control regulations. These are currently uniform for building use/typology and do not consider the location of the building. The consequence of this is that they are unresponsive to market forces and sensitive to existing local conditions on the ground, and therefore discourage compliance. Currently, the proposed bylaws include a brief reference to the Master Plan prepared and enforced by the Delhi Development Authority (DDA). In many areas, stringent development controls need to be more realistic and relaxed, so it was proposed that the MCD institute a process for preparing and implementing Local Area Plans (LAPs) wherein area-specific building bylaws and development controls would be formulated as part of a comprehensive and integrated LAP prepared at the neighborhood scale. These would require in-depth study, analysis, and planning at the micro-level in a context-sensitive approach.

The proposed amendments to the DMC Act would empower the MCD to prepare LAPs and area-specific bylaws. The Master Plan does not present a clear strategy for dealing with widespread unauthorized development, but it does include the provision for local area planning within it and recognizes the need for it in Delhi neighborhoods. To initiate this process, five pilot LAPs were initiated for selected neighborhoods in Delhi. There would be at least 150 such LAPs needed for all of Delhi, so they need to be expedited. In addition, provisions would need to be made in the Delhi Master Plan to accommodate LAPs and area-specific bylaws under the mandates of the Master Plans and Zonal Plans as prepared and enforced by the DDA. The LAPs would be based on detailed analyses of local conditions and development pressures, and provide micro-level detail of neighborhoods within the broader macro-level Master Plan and metropolitan-level Zonal Plans.

Establishing a system for granting variances

A transparent and fair procedure to grant variances (essentially, permits to not comply with certain regulations) was introduced in the proposed legislative amendments. While this is a common occurrence in Western development regulation systems where Board of Zoning Appeals routinely grant variances, such a procedure is not clearly defined in Indian cities. Currently, in Delhi ad-hoc *condonations* are sometimes granted by the Municipal Commissioner or the Central Government. However, this can be very time-consuming, difficult, and contentious. An amendment was therefore proposed that a Variance Committee be established with clearly defined powers and limits. The variance procedure would be a quasi-judicial, transparent, and systematic process and would focus on the technical merit of an application for a variance to address unforeseen situations or technological innovations.

Dealing with unauthorized construction

One of the most important recommendations of the systemic reform is the empowering of the MCD to deal with unauthorized construction, cases where construction on site deviates from approved plans, and cases where a building is occupied without a having a building use permit or is being used in some way not in accordance with the permit. In the proposed system, wherever necessary, the MCD would be empowered: to give notice to stop unauthorized construction; to revoke a building permit or a building use permit; to forcibly to stop construction and seal a construction site or a building; to demolish unauthorized construction; and to penalize professionals responsible for violations. In addition, the Professional Oversight Committee was empowered to affix responsibility and penalties on defaulting professionals, and violations of this nature were also made cognizable offenses.

The biggest challenge is dealing with the existing stock of unauthorized buildings, as large-scale demolitions are impractical, undesirable, and unnecessary. There needs to be a practical yet sensible procedure to bring into the mainstream a large majority of unauthorized buildings through regularization/legalization. There is a widespread perception that existing regulations are unrealistic and the standards set are too high to be achievable, and thereby contribute to illegality in the first place. Regulations, therefore, need to be more realistic and take ground realities into consideration. The proposed building performance bylaws make it easier to deal with the existing stock of unauthorized buildings because many prescriptive bylaws were removed and left to the discretion of professionals. However, planning bylaws are more challenging because they are determined by the Delhi Master Plan and disregard the ground realities throughout the city. The regulatory reforms, therefore, recommend the speedy preparation of LAPs and area-specific bylaws.

Making the developers responsible

Because the legislation does not define "real estate developers" or describe their role, it is difficult to make them directly accountable for compliance. However, developers have been made indirectly accountable in several ways: first, they are held accountable by making it mandatory for them to hire listed professionals; secondly, landowners are held directly accountable in the streamlined procedures, which make the developers accountable as well; and, finally, unauthorized construction has been made a cognizable offense.

The focus of Delhi's proposed building bylaws is on delivering buildings that are legal, safe, and built to standards by engaging the right professionals, ensuring professional self-regulation, and streamlining the procedure regulations. Under the framework, all consumer-related issues can be solved by consumer courts, and all other building-related disputes can be resolved by proposed Municipal Building Tribunals (MBTs, appellate tribunals), proposed to be newly constituted under the proposed amendments to the DMC Act and meant to empower neighbors and owners to approach the MBT to swiftly resolve any individual disputes pertaining to LAPs on a case-by-case basis. Currently, all complaints escalate to public interest litigation (PIL) in a regular court, whereas the MBT has the potential to provide an alternate forum for citizens seeking redress of building-related issues.

Implementation

Several years after the reform process was undertaken, there has been mixed success. Although there is greater accountability on the part of the professionals, Delhi was not able to amend the enabling municipal legislation, which is key to systemic reform. Consequently, many recommendations were implemented in a piecemeal manner that has not resulted in overall systemic reform. However, although the ambitious legislative changes have not been implemented, the MCD has since initiated a simplified procedure in Delhi and has specified the role of structural engineers in the building plan approval process.[22] This has helped clarify the roles and responsibilities of professionals such as architects, structural engineers, and construction supervisors, but the responsibilities of owners and developers are still muddled and complex. MCD also took steps to establish a techno-legal regime for ensuring building compliance based on the PASPRA's recommendations. Furthermore, the DDA has endorsed the proposal to prepare LAPs at the municipal level, subordinate to the Master/Zonal Plans. The consultants who developed the five initial pilot LAPs faced difficulties in dealing with unauthorized construction in extremely dense, old neighborhoods, and future LAP initiatives will need to learn from these experiences and complexities. These demonstrated that creating an LAP can be an extremely tedious and convoluted process, but they are still more effective than blanket amnesty/regularization schemes for unauthorized buildings.

In many ways, Delhi's prominence as India's capital is another reason contributing to limited implementation. Unlike other major cities in India that enjoy considerable autonomy and local self-governance, Delhi has to defer to India's central government. Even though Delhi is a state, both the crucial land and police departments fall under the jurisdiction of the Lieutenant Governor of Delhi, not the Chief Minister. Important planning matters, such as those relating to the Delhi Master Plan or development regulation reform, need approval from the Indian Government's Ministry of Urban Development. Periodic changes in the bureaucratic leadership have also caused the reform process to lose steam over time. Because these political circumstances and this administrative framework are unique to Delhi, these proposed regulation reforms may be considerably less difficult to implement in other major Indian cities.

Discussion

There are immense challenges in trying to introduce a culture of compliance and to strengthen regulatory systems where none have existed before. Reforming Delhi's system for regulating development is a process aimed at the long-term emergence of a culture of compliance, as well as strengthening the MCD's role as a professional and efficient regulating authority and decreasing the rent-seeking behavior prevalent among the regulating authorities. However, it is important to consider whether such systemic changes in Delhi are possible in a complex (and sometimes dysfunctional) democracy such as India. The entire process relies on the belief that altering the structure within which people operate is the key to successful implementation and will bring about change in attitudes and behaviors, rather than just imposing additional layers of stringent regulation. It relies on clearly defined roles, responsibilities, and procedures, self-regulating market-type mechanisms, and reduced opportunities for profit-seeking. Most importantly, it relies on engaging civil society in governance, creating economic incentives for compliance, and enforcing harsh penalties for non-compliance. To enable all of the above, the legislation and bylaws that define the structure of the system need to be modified.

Success relies on dismantling the old *license raj* in cities and introducing a mature and streamlined method for development regulation. The recommendations have the potential to increase the scope of regulations and refocus them onto issues of public concern and safety. This overarching, overhauling, and ambitious set of reforms will not be easy to implement and will need to undergo many changes based on experience. The main threat is from vested interests that prefer the status quo and resist any efforts for systemic changes. The new regulation system essentially aims to infuse a new synergy between the new building bylaws and the Master Plan for Delhi, and it will the job of the reformers to garner support for these objectives.

These regulatory issues are not unique to Delhi and are prevalent in several cities in the developing world that have grown organically over centuries with little planning intervention. Lessons learnt from these comprehensive reform proposals to streamline the development regulation system in Delhi provide valuable lessons for similar cities in the rapidly urbanizing developing world. Specifically, the project detailed in this paper provides insights into systemic improvements that are required for a robust built environment as well as for improving planning practices in developing countries.

Properly regulated urban development can be a powerful tool to reduce urban vulnerability, but it requires large resources and systemic reforms.

Acknowledgments

From 2004–2006 the author was a core member of the EPC-TRF team of planners, lawyers, and specialist consultants that drafted the new proposed Delhi building bylaws and proposed amendments to the Delhi Municipal Corporation Act, 1957. The project was funded by the USAID and prepared under the guidance of the Steering Committee and the Delhi Municipal Corporation.

Notes

1 Ministry of Home Affairs, India (2008), "Building a New Techno-Legal Regime for Safer India" (New Delhi, India: Government of India).
2 Ibid.
3 I. Davis (2000), "Cities of Chaos," in *India Disasters Report: Towards a Policy Initiative*, P.V. Unnikrishnan and S. Parasuraman, eds., pp. 77–80 (New Delhi, India: OUP).
4 S. Menoni and C. Margottini, eds. (2011), "Shift in Thinking," in *Inside Risk: A Strategy for Sustainable Risk Mitigation*, pp. 287–327 (Milan, Italy: Springer-Verlag Italia).
5 Ibid., p. 319.
6 V. Thiruppugazh (2007), "Urban Vulnerability Reduction: Regulations and Beyond," in *The Indian Economy Sixty Years after Independence*, R. Jha, ed., pp. 200–12 (London, UK: Palgrave Macmillan).
7 Ibid.; V. Thiruppugazh and S. Kumar (2010), "Lessons from the Gujarat Experience: Disaster Mitigation and Management," in *Recovering from Earthquakes: Response, Reconstruction and Impact Mitigation in India*, A. Revi and S. Patel, eds., pp. 223–37 (New Delhi, India: Routledge).
8 Ibid.
9 A. Yates (2002), "The Nexus between Regulation Enforcement and Catastrophic Engineering Failures," paper presented at the Current Issues in Regulation: Enforcement and Compliance Conference, University of South Australia, September 2–3, 2002, Melbourne, Australia.
10 Ibid.
11 M. Leaf (1994), "Legal Authority in an Extralegal Setting: The Case of Land Rights in Jakarta, Indonesia," *Journal of Planning Education and Research* 14(1), pp. 12–18.
12 K.S. Sridhar (2010), "Impact of Land Use Regulations: Evidence from India's Cities," *Urban Studies* 47(7), pp. 1541–69.
13 Menoni and Margottini (2011), p. 298.
14 C. Kenny (2012), "Disaster Risk Reduction in Developing Countries: Costs, Benefits and Institutions," *Disasters* (36)4, 559–88; S. Byahut and D. Parikh (2005), "Integrating Disaster Mitigation Practices in Urban Planning Practices in India," ProVention Consortium Applied Research Grant Report.
15 Davis (2000).
16 Thiruppugazh (2007); Thiruppugazh and Kumar (2010).
17 US Agency for International Development (2008), "Preparation of Local Area Plans: Pilot Projects for Delhi, India. Final Report," *Indo-USAID FIRE (D) Project*, retrieved June 1, 2014, from http://pdf.usaid.gov/pdf_docs/pnaea781.pdf
18 This expert group consisted of the Kolkata-based Times Research Foundation (TRF) and Environmental Planning Collaborative (EPC), a not-for-profit urban planning firm based in Ahmedabad, India.
19 Data Prepared by the Environmental Planning Collaborative and Times Research Foundation Team for the Municipal Corporation of Delhi.
20 Municipal Corporation of Delhi (2004), *Policy Agenda and Legislative Intentions*, consensus paper prepared by the Environmental Planning Collaborative and Times Research Foundation; Municipal Corporation of Delhi (2005a), *Proposed New Building Bylaws for Delhi*,

prepared by the Environmental Planning Collaborative and Times Research Foundation; Municipal Corporation of Delhi (2005b), *Proposed Amendments to the Delhi Municipal Corporation Act 1957*, prepared by the Environmental Planning Collaborative and Times Research Foundation.

21 These include a host of legislation, including: the Indian Electricity Rules, 1956; the Petroleum Pipelines Act, 1962; the Delhi Fire Prevention and Fire Safety Act, 1986; the Specifications for Rain Water Harvesting by the Central Ground Water Authority, the Solid Waste (Management and Handling) Rules by the Ministry of Environment and Forests; the Air Pollution Control Act, 1981; the Water Act, 1974; and the Noise Pollution (Regulation and Control) Rules, 2000.

22 Municipal Corporation of Delhi (2009).

Bibliography

Byahut, S. (2006). "Greater Role for Professionals in the Development Regulation System." *GICEA News* 70(1), pp. 18–19.

———, and D. Parikh. (2005). "Integrating Disaster Mitigation Practices in Urban Planning Practices in India." ProVention Consortium Applied Research Grant Report.

Davis, I. (2000). "Cities of Chaos." In *India Disasters Report: Towards a Policy Initiative*. P.V. Unnikrishnan and S. Parasuraman, eds. pp. 77–80. New Delhi, India: OUP.

Delhi Development Authority. (2007). *Draft Master Plan for Delhi – 2021*. New Delhi, India: Author. Retrieved June 1, 2014, from http://dda.org.in/planning/draft_master_plans.htm

Kenny, C. (2012). "Disaster Risk Reduction in Developing Countries: Costs, Benefits and Institutions." *Disasters* 36(4), pp. 559–88.

Leaf, M. (1994). "Legal Authority in an Extralegal Setting: The Case of Land Rights in Jakarta, Indonesia." *Journal of Planning Education and Research* 14(1), pp. 12–18.

Menoni, S., and C. Margottini, eds. (2011). "Shift in Thinking." In *Inside Risk: A Strategy for Sustainable Risk Mitigation*. pp. 287–327. Milan, Italy: Springer-Verlag Italia.

Municipal Corporation of Delhi. (2005a). *Proposed New Building Bylaws for Delhi*. Prepared by the Environmental Planning Collaborative and Times Research Foundation. Delhi, India: Municipal Corporation of Delhi.

———. (2005b). *Proposed Amendments to the Delhi Municipal Corporation Act 1957*. Prepared by the Environmental Planning Collaborative and Times Research Foundation.

———. (2008). "Building a New Techno-Legal Regime for Safer India." New Delhi, India: Government of India.

———. (2009, October 22). *Simplified Procedures for Building Permits for Residential Plotted Development in Approved Layout Plans*. Office order dated October 22, 2009. New Delhi, India: Author.

Municipal Corporation of Delhi. (2004). *Policy Agenda and Legislative Intentions*. Consensus Paper Prepared by the Environmental Planning Collaborative and Times Research Foundation, Submitted to the Municipal Corporation of Delhi.

Roy, A. (2005). "Urban Informality." *Journal of the American Planning Association* 71(2), pp. 147–58.

Sridhar, K.S. (2010). "Impact of Land Use Regulations: Evidence from India's Cities." *Urban Studies* 47(7), pp. 1541–69.

Thiruppugazh, V. (2007). "Urban Vulnerability Reduction: Regulations and Beyond." In *The Indian Economy Sixty Years after Independence*. R. Jha, ed. pp. 200–12. London, UK: Palgrave Macmillan.

———, and S. Kumar. (2010). "Lessons from the Gujarat Experience: Disaster Mitigation and Management." In *Recovering from Earthquakes: Response, Reconstruction and Impact Mitigation in India*. A. Revi and S. Patel, eds. pp. 223–37. New Delhi, India: Routledge.

US Agency for International Development. (2008). "Preparation of Local Area Plans: Pilot Projects for Delhi, India. Final Report." *Indo-USAID FIRE (D) Project*. Retrieved June 1, 2014, from http://pdf.usaid.gov/pdf_docs/pnaea781.pdf

Yates, A. (2002). "The Nexus between Regulation Enforcement and Catastrophic Engineering Failures." Paper presented at the Current Issues in Regulation: Enforcement and Compliance Conference. University of South Australia, September 2–3, 2002, Melbourne, Australia.

12 Medical preparedness for natural disasters in India

Perspectives and future measures

Hem Chandra, Ankita Pandey,
Supriya Trivedi, Sarika Sharma,
Nitin Bharadwaj, and Leela Masih

Introduction

A disaster is an event or series of events giving rise to casualties and damage or loss of property, infrastructure, environment, essential services, or means of livelihood on a scale that is beyond the normal coping capacity of the affected community. Disasters disrupt progress and destroy developmental efforts, often pushing nations back several decades in their quest for progress. Thus, efficient management of disasters rather than mere response to their occurrence has received increased attention, both within India and abroad.[1] The *Oxford Dictionary of English Etymology* defines a disaster as a "sudden or great calamity." Similarly, the *Collins Dictionary* describes disaster as "an occurrence that causes great distress or destruction." The United Nations defines disaster as "the occurrence of sudden or major misfortune which disrupts the basic fabric and normal functioning of the society or community."[2] A disaster is a sudden, calamitous event that seriously disrupts the functioning of a community or society and causes human, material, and economic or environmental losses that exceed the community's or society's ability to cope using its own resources.

In the Disaster Management Act of 2005, a disaster is defined as

> a catastrophe, mishap, calamity, or grave occurrence in any area, arising from natural or man-made cause, or by accident or negligence which results in substantial loss of life or human suffering or damage to, and destruction of property, or damage to, or degradation of, environment, and is of such a nature or magnitude as to be beyond the coping capacity of the community of the affected area.[3]

Most major disasters occur largely as unforeseen events. Thus a disaster has the following main features: (1) unpredictability; (2) unfamiliarity; (3) speed; (4) urgency; (5) uncertainty; and (6) threat.

Types of disaster

All disasters can be broadly divided into two categories – natural and human-made disasters. Natural disasters are uncontrollable, as indeed nature is uncontrollable; such events are often termed "acts of God." The occurrence of human-made disasters, on the other hand, is rapidly increasing in our present-day technological society, and is a result of human failure, error, or malfunction of some structure or system designed by humans. Similarly, while there is sometimes an element of warning in natural disaster, there is generally none in human-made disasters, and this lack of warning makes avoidance difficult. The following are examples of the different types of disasters under the above-mentioned categories:

1 **Natural Disasters** can be of the following types: (1) earthquakes; (2) floods; (3) storms; (4) cyclones; (5) epidemics; (6) famines; (7) landslides.
2 **Man-Made Disasters** include: (1) air, rail, and sea disasters; (2) fires; (3) explosions; (4) building collapses; (5) industrial accidents; (6) Riots and stampedes; (7) terrorism and mass shootings; and (8) mass suicides.

A brief profile of India

India is one of the oldest civilizations in the world, with a kaleidoscopic and rich cultural heritage. It has achieved all-round socio-economic progress during the 67 years since its independence. The country has become self-sufficient in agricultural production and is now one of the top industrialized countries in the world, and one of the few nations to have gone into outer space to conquer nature for the benefit of the people. It covers an area of 3,287,590 km^2, extending from the snow-covered Himalayan heights to the tropical rain forests of the south. As the seventh-largest country in the world, India stands apart from the rest of Asia, marked off by mountains and the sea, making the country a distinct geographical entity. Bounded by the great Himalayas in the north, it stretches southwards past the Tropic of Cancer, where it tapers off into the Indian Ocean between the Bay of Bengal on the east and the Arabian Sea on the west.[4]

Vulnerability to disasters

India is one of the most disaster-prone countries in the world. This is largely due to its geo-climatic conditions, combined with its high population density and other socio-economic factors. India is vulnerable, in varying degrees, to a large number of natural as well as human-made disasters due to its unique geo-climatic and socio-economic conditions, particularly floods, droughts, cyclones, earthquakes, landslides, avalanches, and forest fires. These factors, accelerating the intensity and frequency of disasters, are responsible for the heavy toll of human lives and disruption of the life support system in the country. The natural geological setting of the country is the primary reason for its increased vulnerability. The geo-tectonic features of the Himalayan region and the adjacent alluvial plains make the region susceptible to earthquakes, landslides, water erosion, etc. Though peninsular India is considered to be the most stable area, occasional earthquakes in the region show that geo-tectonic movements are still occurring within its depths. The tectonic features, characteristic of the Himalayas, are also prevalent in the alluvial plains of the Indus, the Ganga, and the Brahmaputra, as the rocks lying below the alluvial plains are extensions of the Himalayas. Thus this region is also quite prone to seismic activities. Along with these natural factors, various human activities, such as increasing demographic pressure, deteriorating environmental conditions, deforestation, unscientific development, faulty agricultural practices and grazing, unplanned urbanization, construction of large dams on river channels, etc., are also responsible for the accelerated impact and increase in frequency of disasters in the country.[5]

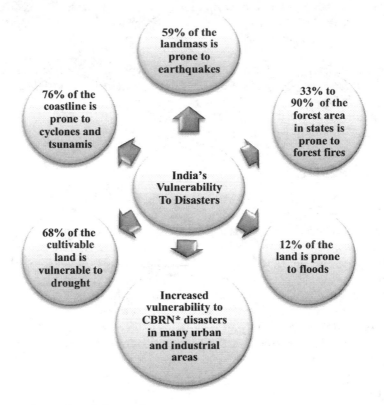

Figure 12.1 India's vulnerability to disasters.
CBRN: Chemical, Biological, Radiological, Nuclear

Common types of disasters in India

The High Power Committee on Disaster Management has identified various types of disasters. Tsunamis were added to this list in 2005.

1   Water- and climate-related disasters:

    a   Floods and drainage management;
    b   Cyclones;
    c   Tornadoes and hurricanes;
    d   Hailstorms:
    e   Cloudbursts;
    f   Heat and cold waves;
    g   Snow avalanches;
    h   Droughts;
    i   Sea erosion;
    j   Thunderstorms; and
    k   Tsunamis.

2 Geological disasters:

    a   Landslides and mudflows;
    b   Earthquakes;
    c   Dam failures/bursts; and
    d   Mine disasters.

3 Accidental disasters:

    a   Chemical, industrial, and nuclear disasters;
    b   Urban fires;
    c   Mine flooding;
    d   Major building collapses;
    e   Festival-related disasters;
    f   Electrical disasters and fires;
    g   Air, road, and rail accidents;
    h   Boats capsizing; and
    i   Village fires.

4 Biological disasters:

    a   Epidemics;
    b   Pest attacks;
    c   Cattle epidemics; and
    d   Mass food poisoning incidents.[6]

The 2013 Uttarakhand floods

Series of events

Uttarakhand is well known for the pilgrimage circuit of Chota Char Dham: Kedarnath, Badrinath, Gangotri, and Yamunotri. It is known as the "Land of the Gods." Kedarnath is one of the ancient and famous pilgrimage sites in Uttarakhand, India. It is located in the snow-covered area of the Himalayan region at the height of approximately 3,583 meters above sea level in the Mandakini valley of Rudraprayag District, Uttarakhand. Due to extreme weather conditions, it is not possible to visit this holy place at any time of the year; it is only safe from May to October. There is conflict between the tourism industry, population growth, and the several hydroelectric projects underway in the Uttarakhand district. Since Uttarakhand became a state, its population has increased approximately 141 percent. This region is seismically and ecologically very sensitive and delicate, so even a minute change can create a dangerous disaster.

Information provided by the Indian Meteorological Department shows that rainfall in Uttarakhand from June 1–18, 2013, was 385.1 mm, whereas the normal rainfall for this time of year is 71.3 mm – an increase of440 percent. Heavy precipitation augmented the river discharge, and almost all of the region's major rivers rose above the danger level. This caused severe damage to life and property in different parts of the state.

On June 16 and 17, 2013, heavy rains and the bursting of the moraine-dammed Chorabari Lake caused the flooding of the Saraswati and Mandakini Rivers in the

Table 12.1 Major natural disasters in India over the last ten years

Name of disaster	Year	State and area	Effect on human life
Gujarat Earthquake	2001	Bhuj, Bhachau, Anjar, Ahmadabad, and Surat in Gujarat State	25,000 deaths and 6.3 million people affected
Tsunami	2004	Coastline of Tamil Nadu, Kerala, Andhra Pradesh, Pondicherry, and Andaman	10,749 deaths, 5,640 missing, 2.79 million persons affected
Kashmir Earthquake	2005	Kashmir and the surrounding Himalayan region	86,000 deaths
Maharashtra Floods	2005	Maharashtra	1,094 deaths, 167 injured, and 54 missing
Kosi Floods	2008	North Bihar	527 deaths, and 3.33 million persons affected
Cyclone Nisha	2008	Tamil Nadu	245 deaths
Drought	2009	252 districts in ten states	–
Leh Cloudburst	2010	Leh in Jammu and Ladakh in Kashmir	–
Sikkim Earthquake	2011	Northeastern India, epicenter near the Nepal border and Sikkim	–
Uttarakhand Floods (Heavy rains, flash floods, landslides, and a sustained cloudburst)	2013	Uttarakhand	Approx. 5,700 deaths

Rudraprayag district of Uttarakhand. The prolonged heavy downpour resembled a cloudburst event in the Kedarnath valley and surrounding areas, which damaged 18 km of the banks of the Mandakini River between Kedarnath and Sonprayag, and completely washed away the towns of Gaurikund, Rambara, and Kedarnath. The India Meteorological Department reported that this heavy to very heavy rainfall in the higher Uttarakhand, Himachal, and Nepal Himalaya region converged with the Southwest Monsoon trough and westerly disturbances, which led to the formation of dense clouds over the Uttarakhand Himalaya region.[7]

The Himalayan flash floods in the state of Uttarakhand

The Himalayan state of Uttarakhand faced an unprecedented natural disaster on the night of June16, 2013, due to a combination of factors, namely early heavy rainfall, movement of southwest monsoon winds, and the formation of a temporary lake.[8] A sudden gush of water engulfed the centuries-old Kedarnath temple and washed away everything in its vicinity in a matter of minutes. The June 2013 flash floods in Uttarakhand took a tremendous toll, with an estimated 4,200 villages affected and more than 5,000 citizens going missing. It resulted in severe destruction of property, including heritage and religious structures such as the ancient Kedarnath temple. Government agencies and the priests of the Kedarnath temple planned mass cremations of the hundreds of victims one week after the tragedy. The disaster was caused in part by a model of development that has become all too common across the country: the use of dynamite for road construction, indiscriminate mining, unsustainable dam construction, dumping of debris into rivers, rampant deforestation, and large-scale encroachments have increased the probability of flash floods and landslides and weakened the Himalayan ecosystem overall, making such occurrences a question of when rather than if. Haphazard cheek-by-jowl construction and unplanned growth increasingly characterize the process of urbanization.

The second event occurred on June 17, 2013, at 6:45 a.m., after the overflow and collapse of the moraine-dammed Chorabari Lake, which released a large volume of water that caused another flash flood in the town of Kedarnath, leading to widespread devastation downstream.[9]

On June 18, more than 12,000 pilgrims were stranded at Badrinath, the popular pilgrimage center located on the banks of the Alaknanda River. Rescuers at the Hindu pilgrimage town of Haridwar on the river Ganga recovered the bodies of 40 victims washed down by the flooded rivers by June 21, 2013. The search for the bodies of those who died during the June disaster in the Kedar valley continued for several months; even as late as September 2013, about 556 bodies were found during the fourth round of search operations, of which 166 bodies were severely decomposed.[10] Although the Kedarnath Temple itself was not damaged, its base was inundated with water, mud, and boulders from the landslide, damaging its perimeter. The temple was flooded with water, resulting in several deaths due to drowning and panic-driven stampedes.[11] Even after a week, dead bodies had not been removed from Kedarnath, resulting in water contamination in the Kedarnath valley; villagers who depended on spring water suffered various health problems, such as fever and diarrhea.[12] When the floods receded, satellite images showed a new stream at the town of Kedarnath. The sacred portals of the Kedarnath temple were reopened to devotees on May 4, 2014, about a year after the flash floods that had left hundreds of people dead and many others stranded.

Recently, the risk of natural disasters has increased in the area as a result of increasing anthropogenic activities. This trend is likely to continue to increase in the future along with activities such as pilgrimages, tourism, etc. The natural flow paths of the channels are obstructed by the construction of man-made structures, resulting in the deviation of the flow of the river from its natural course.

Possible causes of the Uttarakhand disaster

The National Institute of Disaster Management (NIDM), in one of its first reports on the Uttarakhand floods, blamed "climatic conditions combined with haphazard human intervention" in the hills for the disaster. Ageomorphologic study of the area indicated that the surface slopes consist mostly of glacial, fluvio-glacial, or fluvial materials, which are mostly unconsolidated and loose in nature. The drainage studies indicate a migratory or shifting nature of the river systems. They are also very erosive, especially when loaded with sediments (the erosive power of river water carrying sediments is exponentially greater than that of clean water). The loss of green cover due to deforestation for road construction and other activities such as building construction, mining, and hydroelectric projects also resulted in increased surface flow and the rise of the riverbed due to the disposal of debris in the rivers. The abnormally high amount of rain (more than 400 percent above normal) in the hill state was caused by the conjunction of Westerlies with the monsoonal cloud system. Heavy precipitation swelled rivers, both upstream and downstream. Besides the rainwater, a huge quantity of water was probably released by the melting of ice and glaciers due to high temperatures during May and June. The water not only filled up the lakes and rivers that overflowed, but also may have caused the breaching of moraine-dammed lakes in the upper reaches of the valley. The Alaknanda and Mandakini rivers (both tributaries of the Ganga) caused much destruction because they returned to their old courses, where buildings had been constructed over a period of time.[13]

It is an open secret that unplanned urbanization, industrialization, and unscrupulous development plans are responsible for most of the human deaths and destruction of private and public property. The Himalayan region is a sensitive zone, full of water bodies, forests, and glaciers; common sense dictates that the local ecology be treated with care and caution. The mania for energy has led to damming of practically all water bodies, small and large. Every activity done in the name of development destroys trees and disturbs the local ecology. This has been going on for a long time and has accelerated in the past decades after the so-called economic reforms and liberalization. The increasing influx of religious tourists and their activities have further aggravated the situation.

The consequences of the disaster

The scale of the disaster

The extent of damage as reported by the state government was as follows:

1   Number of affected districts = 13
2   Number of villages affected (habitations) = 1,603
3   Number of cremated bodies and the missing = approx. 5,700

Search, rescue, and relief operations

In any disaster, the first responders are state government officials. As soon as the matter came to the notice of the Government of India, all the required central ministries were mobilized. On June 16, 2013, the NDRF was asked by the Ministry of Home Affairs (MHA) to move their teams urgently to Uttarakhand to augment their existing deployment in the state.

Evacuation

The NDRF deployed 14 teams for the Uttarakhand operation and rescued 9,657 people. The ITBP deployed about 1,200 personnel for the operation and rescued more than 33,000 people. The IAF deployed about 45 helicopters for the operation and rescued more than 23,500 people. The Indian Army deployed 8,000 personnel, including 150 Special Forces, and rescued more than 38,500 people. Twelve army helicopters were deployed and 20 civil aircraft were utilized by the state government in its operations, and evacuated approximately 12,000 people. More than 135,000 persons were evacuated from the affected areas in the shortest possible time, notwithstanding the widespread destruction of roads, difficult terrain, and extremely challenging weather conditions.[14]

Relief and supplies

Essential supplies such as food, drinking water, medicines, kerosene, blankets, etc., were continuously provided through airdrops as well as via road transportation (vehicles, mules, etc.) wherever the connections were restored. Sixty-nine relief camps were set up to care for 151,629 pilgrims and local residents. Two thousand tons each of wheat and rice were allocated to Uttarakhand at economic cost by the Union of India.

Communications

Point-to-point communication by telephone was established at Kedarnath, Badrinath, Barkot, and Harshil. One hundred and five satellite phones were distributed by the Government of India to various central and state agencies for the purpose of urgent disaster management duties in Uttarakhand and Himachal Pradesh, in order to facilitate communication across the state.

Financial assistance to the state

The Prime Minister of India undertook an aerial survey of the affected areas and announced a US$150 million aid package for disaster relief efforts in the state.[15] Several state governments announced financial assistance, with Uttar Pradesh pledging US$3.8 million, the governments of Haryana, Maharashtra and Delhi US$1.5 million each, the governments of Tamil Nadu, Odisha, Gujarat, Madhya Pradesh, and Chhattisgarh US$750,000 each (Hindustantimes.com). The US Ambassador to India extended financial help of US$150,000 through the United States Agency for International Development (USAID) to the NGOs working in the area, and provided further financial aid of USD $75,000.[16]

The health consequences of disasters

Health consequences vary according to types of disaster. At one end of the spectrum, vector-borne epidemics and waterborne diseases are the short- and long-term health impacts of floods, respectively; at the other, repeated cycles or eruptions of violence may actually be fallout from ethnic conflicts, according to mental health experts. The main direct public health impacts of a flood are drowning and injuries. Indirect impacts include waterborne diseases, diarrhea, dysentery, hepatitis and poliomyelitis, respiratory infections, snakebites, skin infections, and conjunctivitis. It is a well-established fact that diarrheal diseases are a major cause of mortality and morbidity in emergencies, and studies have shown that they contribute to between 25 and 50 percent of all deaths.

Disasters strike at the very roots of human emotions and invariably cause uncontrollable psychological reactions. Psychosocial coping depends on the ability of the victims to adjust psychologically, the capacity of community structures to adapt to crises, and the individual and institutional help available. At the individual level, post-disaster psychosocial trauma includes anxiety, neuroses, sleeplessness, and lethargy. At the community level, psychosocial trauma manifests itself in the form of school dropouts among children and high rates of alcoholism, divorce, and suicide among adults.[17]

Disaster management and medical preparedness

The Disaster Management Act of 2005 makes clear that "disaster management" means a continuous and integrated process of planning, organizing, coordinating, and implementing measures which are necessary or expedient for: (i) preventing danger or the threat of any disaster; (ii) mitigating or reducing the risk, severity, or consequences of any disaster; (iii) capacity-building; (iv) preparedness to deal with any disaster; (v) prompt response to any threatening disaster situation or disaster; (vi) assessing the severity or magnitude of effects of any disaster; evacuation, rescue and relief; and (vii) rehabilitation and reconstruction. Disaster management can be defined as the organization and management of resources and responsibilities for dealing with all humanitarian aspects of emergencies, particularly preparedness, response, and recovery in order to lessen the impact of disasters.

Disaster management involves all levels of government. Nongovernmental and community-based organizations play a vital role in the process. Successful disaster management planning must encompass the situations that occur before, during, and after disasters. Prevention and mitigation – reducing the risk of disasters – involves activities that either reduce or modify the scale and intensity of the threat faced, or, by improving conditions, reducing or modifying the elements causing the risk.

Preparedness

Preparedness brings us to the all-important issue of disaster preparedness. The process embraces measures that enable governments, communities, and individuals to respond rapidly to disaster situations in order to cope with them effectively. Preparedness includes the formulation of viable emergency plans, the development of warning systems, the maintenance of inventories, public awareness and education, and the training of personnel. It may also embrace search-and-rescue measures as well as evacuation plans for areas that may be at risk from a recurring disaster. Medical preparedness for disasters is aimed at creating an institutional mechanism and systems that would result in the coordinated working of emergency responders, hospital managers, and local and regional officials.

Early warning

Early warning is the process of monitoring the situation in communities or areas known to be vulnerable to slow-onset hazards and passing the knowledge of the pending hazard onto the people affected. To be effective, warnings must be related to mass education and training of the population, who would then know what actions they must take when warned.

Risk reduction

It is self-evident that an area located in a seismic zone is always susceptible to earthquakes. So, if such an area were to become a hub of social activity centered on tourism, extensive loss of life and property would be bound to happen. Hence, it is suggested that such areas should be promoted only for tourism, not for extensive urbanization or as a base for human dwellings. It is also recommended that hotels and other buildings built in such areas be designed to withstand seismic activity and built within safety parameters.

Risk management

Adequate steps must be taken for disaster preparedness. First of all, special task force units, under the supervision of the armed forces, should be developed and in place to carry out early evacuation when necessary in vulnerable areas. First-Referral Units (FRU) under the Armed Forces Medical Corps should be established in different locations in hilly terrain. The geographical conditions of such areas mean that only army personnel have the required training and capacity to overcome such difficult terrain. The FRU should also help with the upgrading of CHCs and PHCs in such areas for disaster preparedness and in developing a disaster plan. The district-level hospitals need to be upgraded into tertiary-level treatment centers and level-one trauma centers. Finally, the evacuation plan to take victims from the FRUs/PHCs and CHCs to the nearest military/district hospital needs to be coordinated with surface and air transport.

Institutional arrangements for medical preparedness

The programs and procedures suited to the needs of the people in the state are formulated and implemented by the state governments. At the national level, health programs are implemented by the Ministry of Health and Family Welfare, which also plays a key role in augmenting capacities at all levels, including extending necessary help during disasters and emergencies.

Hospitals are also prone to seismic hazards. Thus, alternative modalities must be in place for prompt and effective disaster response. One such modality is a mobile hospital – a prefabricated, self-contained, container-based hospital that can be deployed by road, rail, or air. This can be rapidly deployed to provide medical care to disaster victims. Training physicians to serve effectively in emergency rooms is a prerequisite for managing trauma cases brought to hospital emergency departments. Hospital disaster management provides the opportunity to plan, prepare, and, when needed, enable a rational response in the case of disasters and mass casualty events.

The main objective of a hospital emergency/disaster plan is to optimally prepare the staff and institutional resources of the hospital for effective performance in different disaster situations. The plan should be designed to enable the survival of as many patients as possible. Patients must be distributed proportionally to other healthcare facilities. Hospitals provide full-time emergency services 24 hours per day, seven days a week. The initial chaos of any disaster scenario in a hospital can be minimized by properly training the staff of the hospital to execute their part of the plan and the specific role(s) each staff member will play in the case of a disaster. The hospital administration must appoint an incident commander to carry out commands and coordinate the overall disaster response. The medical superintendent or doctor on call should confirm the emergency with the relevant administrative state office(s) or police station. If possible, he or she should obtain information about the precise location and time of the event, the type of incident, the estimated number of causalities, the potential risks, and the exposed population; the hospital's supply of the required drugs and other medical items should also be assessed to ensure that extra supplies of medical items are obtained as soon as possible. Health facilities and treatment centers must be established at the disaster site, along with vaccination services, site operation camps, and a unit to dispose of animal carcasses. All ambulatory patients whose release does not pose a health risk should be discharged – if possible, they should be transported to their home areas. An efficient ambulance service is an essential part of the casualty service, so that casualties can be transported from the scene of the disaster to first-aid posts and hospitals. (Van-based) mobile teams for first aid and referral should be designated according to district; these teams must be mobile enough to rush to the affected area in minimal time and should be equipped to deal with accidents. An adequate number of mobile units with trained personnel, testing facilities, communication systems, and emergency treatment facilities must be set up.[18]

New strategies for a safer future

In disaster situations, a quick rescue and relief mission is expected; however, damage can be considerably reduced if adequate preparedness levels are achieved. Indeed, it has been noticed in the past that as and when attention has been given to adequate preparedness measures, the loss of life and property has been considerably reduced. Going along with this trend, the disaster management setup in India has, in recent years, been oriented towards a strong focus on preventive approaches, mainly through administrative reforms and participatory methods.

Preparedness measures such as the training of the community, the development of advanced forecasting systems and effective communications, and, above all, a sound and well-networked institutional structure involving government organizations, academic and research institutions, the armed forces, and NGOs have greatly contributed to the overall management of disasters in the country. This can clearly be seen from the various instances of reduced damage from disasters due to better preparedness and coordinated inter-agency response. Preparedness is key to breaking the disaster cycle. Good practices are a result of the heightened awareness and sensitivity towards communities at risk, and the approach of reducing community vulnerability to disasters has paid rich dividends. The first step in this direction is to identify vulnerable communities. Those communities periodically exposed to natural hazards, and, of these, the communities with poorly developed coping powers (such as economically weak communities), are the first focus of preparedness efforts.

The National Disaster Management Authority (NDMA) is chaired by the Prime Minister of India and has an overarching presence in the field of disaster management; it is responsible for formulating the National Act and related policies, and for preparing the National Plan. To consolidate the efforts already made for disaster preparedness, it is essential that the NDMA effectively discharges its statutory responsibilities and that the roles and responsibilities of other entities are clearly demarcated, documented, disseminated, and monitored.

Lessons learnt

The recent events in Uttarakhand have shown, more than ever, that we need a development strategy for the Himalayas that takes into account the vulnerability of the region and the need to protect the environment. We need to think about a pan-Himalayan strategy so that states can evolve common policies based on the region's natural resources – forests, water, biodiversity, organic and specialty foods, and nature tourism – but without generating further adverse effects on the environment.

Forest preservation

Forests are particularly important for the survival of people in the high Himalayan regions. They depend on the forests for their animals' fodder and water for agriculture. Forests also preserve biodiversity and prevent soil erosion, which is vital for the local ecosystem. Thus, hydropower and other development projects must not be allowed without compensatory afforestation.

Stopping the damming of rivers

Water is another key resource of the region, flowing from high glaciers and mountains to the plains. This resource is both an opportunity and a threat to its ecology and economy. Currently, there is a mad rush to build run-of-the-river projects and dams across the region. All Himalayan states are awarding hydroelectric projects to private companies at a breakneck speed. In the Ganga basin alone, Uttarakhand has identified projects totaling nearly 10,000 MW of power and plans for approximately 70 more. The impact of these projects on the ecology and hydrology of the region must be seriously evaluated, particularly in the light of increasing extreme climatic events. The funny thing is that all the power will be sold to states on the plain, while the local people are put at risk.

Promoting local agriculture

Himalayan states try to boost their economies using their unique products. They also recognize the need to keep their agriculture organic. The soil on the hill slopes is deficient in nutrients, so farmers often shift cultivation and make considerable efforts to obtain manure as fertilizer; needless to say, the returns are meager and the labor almost completely discounted and undervalued. Thus, there is a need to start a dialogue on the future of agriculture in this region. Adventure and nature tourism, alongside religious pilgrimages, are the most obvious route to economic development in the Himalayan states, and so should be promoted, but with inbuilt safeguards to protect the fragile ecology of the region.[19] Some common safeguards in this regard are as follows:

1   Similar to sanctuaries and national parks, create buffer areas within 5–10 km surrounding the pilgrimage sites where development is restricted.
2   Promote homestead tourism rather than hotel/motel tourism, based on policy incentives.
3   Promote the reuse and recycling of waste, energy efficiency, and renewable energy sources at all hill tourist spots.
4   Ensure that credible environmental compliance mechanisms are in place for each project in which local people have a key role. Today, we have no credible environmental compliance in place.
5   No projects should be cleared until and unless a credible cumulative impact assessment for all affected river basins and sub-basins has been conducted, which includes a carrying capacity study. Certain rivers and high-risk zones should be declared "no project areas" in each basin.

6   Implement an early warning system to forecast and disseminate information regarding all kinds of disasters, particularly those related to rainfall and landslides.
7   Protect and conserve rivers, riverbeds, and flood plains, including aquatic biodiversity.
8   The number of pilgrims admitted to a particular site must be limited to the number within the carrying capacity of the site. There should be no permanent construction at pilgrimage destinations for accommodating visitors. If accommodation is provided, it should be tented so that there is no excavation for building materials, no site clearance, no felling of trees, and no construction on river banks.
9   In the short and medium terms, we will have to maintain well-trained and well-equipped teams of rescuers and persons who can handle natural calamities, with the location of each team being carefully designated so that personnel and machines can be activated very quickly in the event of any natural disaster.

Conclusion

Disasters disrupt normal life and can also result in significant losses of infrastructure, population, and government facilities. The importance of disaster preparedness, particularly disaster mitigation and prevention efforts, cannot be overstated in such a scenario.

Disaster management has to be a multi-disciplinary and proactive approach. Besides various measures for implementing institutional and policy frameworks, the disaster prevention, mitigation, and preparedness initiatives undertaken by the central and state governments, the community, civil society organizations, and the media also play a key role in achieving our goal of moving together towards a safer India. Disaster management should be based on the three Rs of Rescue, Relief, and Rehabilitation. It has a preventive dimension which should be a part of long-term developmental planning.[20]

The message being put across is that, in order to move towards safer and sustainable national development, development projects should be sensitive towards disaster mitigation. Our mission is to reduce vulnerability to all types of hazards, be they natural or human-made. This is not an easy task, particularly in light of India's vast population and the multiple natural hazards it is exposed to. However, if we are firm in our conviction and resolve that the government and the people of this country are not prepared to pay the price in terms of mass casualties and economic losses, the task, though difficult, is achievable, and we shall achieve it.

Our vision for 2020 is to build a safer and more secure India through sustained collective effort, the synergy of national capacities, and people's participation. What looks like a dream today will be transformed into reality in the next two decades. This is our goal, and we shall strive to achieve it with a missionary zeal. The path ahead, which looks difficult today, will become much easier as we move along together.

Notes

1   Ministry of Home Affairs (India) (2013), *Annual Report 2013* (New Delhi, India: Author).
2   The *Oxford Dictionary of English Etymology* defines a disaster as a "sudden or great calamity"; similarly, the *Collins Dictionary* describes disaster as "an occurrence that causes great distress or destruction." The UN defines disaster as "the occurrence of sudden or major misfortune which disrupts the basic fabric and normal functioning of the society or community." (*Oxford Dictionary of English Etymology*, 1st edn., s.v. "disaster"; Collins Dictionary, 7th edn., s.v. "disaster"; United Nations Disaster Management Training Program, 1992).

3 The Disaster Management Act (2005, December 23), Law No. 53 of 2005, *The Gazette of India*, Part II, Section 1, No. 64.
4 *Report of the Comptroller and Auditor General (CAG Report) of India on Performance Audit of Disaster Preparedness in India* (2013), Union Government (Civil) Ministry of Home Affairs Report No. 5. (New Delhi, India: Author).
5 Ibid.
6 National Centre of Disaster Management (2001), *High-Powered Committee Report* (New Delhi, India: Author, Indian Institute of Public Administration).
7 National Institute of Disaster Management (2013), *Brief Report on a Visit to Alaknanda Valley, Uttarakhand, Himalaya, 22–24 June, 2013* (New Delhi, India: Author).
8 A. Kumar (2013), "Demystifying A Himalayan Tragedy: Study of 2013 Uttarakhand Disaster," *Journal of Indian Research* 1, pp. 106–16.
9 D.P. Dobhal, A.K. Gupta, M. Mehta, and D.D. Khadelwal (2013, July 25), "Kedarnath Disaster: Facts and Plausible Causes," *Current Science* 105(2), pp. 171–74.
10 N. Gokhale (2013, June 27), "Uttarakhand: Army Commander Walks with 500 People Out of Badrinath," NDTV.com, retrieved December 23, 2015, from http://www.ndtv.com/india-news/uttarakhand-army-commander-walks-with-500-people-out-of-badrinath-526670
11 "Kedarnath Temple Stays Intact; Its Surroundings Have Gone with [the] Flow," (2013, June 18), *Hindustan Times* [Online edition].
12 D. Tiwary and V. Mohan, "Uttarakhand Floods: Epidemic Looms as People Complain of Fever, Diarrhea," (2013, June 26), *The Times of India* [Online edition].
13 National Institute of Disaster Management (n.d.), *Home Page* [Website], Ministry of Home Affairs, New Delhi, retrieved December 23, 2015, from www.nidm.gov.in
14 "Uttarakhand Floods: Bodies of Chopper Crash Victims Recovered" (2013, June 26), *Mid-Day* [Online edition].
15 S. Iyer (2013, June 27), "Narendra Modi Sends Rs 3 cr More, Renews Offer to Help in Uttarakhand," *Hindustan Times* [Online edition].
16 "U.S. Ambassador Announces Relief to Flood Victims in Uttarakhand," (2013, June 24), *The Hindustan Times* [Online edition].
17 The World Health Organization (1992), *The Psychosocial Consequences of Disasters: Prevention and Management* (Geneva, Switzerland: Author).
18 Disaster Mitigation and Management Centre, Uttarakhand Secretariat (n.d.), *State Disaster Management Action Plan for the State of Uttarakhand* (Dehradun, India: Author).
19 "A Critical Look at India's Disaster Management: Lessons from the Uttarakhand Disaster" (2014), *Issues of India* [Blog], retrieved June 11, 2014, from https://socialissuesindia.files.wordpress.com/2013/10/disaster-management-in-india.pdf
20 S. Parasuraman and P.V. Unnikrishnan, *Disaster Response in India: An Overview*, retrieved December 23, 2015, from http://www.indianstrategicknowledgeonline.com/index.php?t=Disaster%20Management&start=75

Bibliography

"A Critical Look at India's Disaster Management: Lessons from the Uttarakhand Disaster." (2014). *Issues of India* [Blog]. Retrieved June 11, 2014, from https://socialissuesindia.files.wordpress.com/2013/10/disaster-management-in-india.pdf

"Capacity Building in Asia Using Information Technology Applications (CASITA)." *Asian Disaster Preparedness Center (ADPC)*. Retrieved December 23, 2015, from http://www.adpc.net/casita/Prog.html

"Char Dham and Hemkund Sahib Yatra to Restart from May 2014." (2014, April 24). *Biharprabha News* [Web site]. Retrieved April 24, 2014, from http://news.biharprabha.com/2014/04/char-dham-yatra-restarts-in-may-uttarakhand-assures-safety/

"Disaster." (n.d.). *Oxford Dictionaries* [Online edition]. Retrieved December 23, 2015, from http://www.oxforddictionaries.com/definition/english/disaster

Disaster Management Act, The. (2005, December 23). Law No. 53 of 2005. *The Gazette of India*, Part II, Section 1, No. 64.

Disaster Mitigation and Management Centre, Uttarakhand Secretariat. (n.d.). *State Disaster Management Action Plan for the State of Uttarakhand*. Dehradun, India: Author.

Dobhal, D.P., A.K. Gupta, M. Mehta, and D.D. Khadelwal. (2013, July 25). "Kedarnath Disaster: Facts and Plausible Causes." *Current Science* 105(2), pp. 171–74.

Gokhale, N. (2013, June 27). "Uttarakhand: Army Commander Walks with 500 People Out of Badrinath." *NDTV.com*. Retrieved December 23, 2015, from http://www.ndtv.com/india-news/uttarakhand-army-commander-walks-with-500-people-out-of-badrinath-526670

Iyer, S. (2013, June 27). "Narendra Modi Sends Rs 3 cr More, Renews Offer to Help in Uttarakhand." *Hindustan Times* [Online edition]. Retrieved June 27, 2013, from http://www.hindustantimes.com/india/narendra-modi-sends-rs-3-cr-more-renews-offer-to-help-in-uttarakhand/story-Yl4JUF7kBn1MrZjkgN6lMN.html

"Kedarnath Temple Stays Intact; Its Surroundings Have Gone with [the] Flow." (2013, June 18). *Hindustan Times* [Online edition]. Retrieved June 19, 2013, from http://www.hindustan-times.com/india/kedarnath-temple-stays-intact-its-surroundings-have-gone-with-flow/story-usmJ2UmB9Rc8ZQCrLTR4kN.html

Kumar, A. (2013). "Demystifying A Himalayan Tragedy: Study of 2013 Uttarakhand Disaster." *Journal of Indian Research* 1, pp. 106–16.

Ministry of Home Affairs (India). (2004). *Disaster Management in India: A Status Report*. New Delhi, India: Author.

———. (2013). *Annual Report 2013*. New Delhi, India: Author.

National Centre of Disaster Management. (2001). *High-Powered Committee Report*. New Delhi, India: Author, Indian Institute of Public Administration.

National Institute of Disaster Management. (n.d.). *Home Page* [Website]. Ministry of Home Affairs, New Delhi. Retrieved December 23, 2015, from www.nidm.gov.in

———. (2013). *Brief Report on a Visit to Alaknanda Valley, Uttarakhand, Himalaya, 22–24 June, 2013*. New Delhi, India: Author.

Parasuraman, S., and P.V. Unnikrishnan. *Disaster Response in India: An Overview*. Retrieved December 23, 2015, from http://www.indianstrategicknowledgeonline.com/index.php?t=Disaster%20Management&start=75

Report of the Comptroller and Auditor General (CAG Report) of India on Performance Audit of Disaster Preparedness in India. (2013). Union Government (Civil) Ministry of Home Affairs Report No. 5. New Delhi, India: Author.

"Rs 1000 cr Relief for Flood Hit Uttarakhand Announced." (2013, June 19). *Hindustan Times* [Online edition]. Retrieved June 19, 2013, from http://www.hindustantimes.com/india/rs-1000-cr-relief-for-flood-hit-uttarakhand-announced/story-gfXj6WDPclmMKkhBzpjodN.html

Tiwary, D., and V. Mohan. "Uttarakhand Floods: Epidemic Looms as People Complain of Fever, Diarrhea." (2013, June 26). *The Times of India* [Online edition]. Retrieved June 26, 2013, from http://timesofindia.indiatimes.com/india/Uttarakhand-floods-Epidemic-looms-as-people-complain-of-fever-diarrhoea/articleshow/20770540.cms

"U.S. Ambassador Announces Relief to Flood Victims in Uttarakhand." (2013, June 24). *The Hindustan Times* [Online edition]. Retrieved June 24, 2013, from http://www.hindustantimes.com/india/us-ambassador-announces-relief-to-flood-victims-in-uttarakhand/story-oxgkeI9BC4pZbsaO8UjDgO.html

"UP Govt Announces Financial Help to Uttarakhand." (2013, June 19). *Zee News* [Web site]. Retrieved December 18, 2015, from http://zeenews.india.com/news/uttar-pradesh/up-govt-announces-financial-help-to-uttarakhand_856082.html

"Uttarakhand Floods: Bodies of Chopper Crash Victims Recovered." (2013, June 26). *Mid-Day* [Online edition]. Retrieved June 26, 2013, from http://www.mid-day.com/articles/uttarakhand-floods-bodies-of-chopper-crash-victims-recovered/219888

World Health Organization, The. (1992). *The Psychosocial Consequences of Disasters: Prevention and Management*. Geneva, Switzerland: Author. Available at http:www.who.int/iris/handle/10665/58986

13 Community-based responses to epidemic diseases as a potential public health disaster

The Rockefeller experience, 1915–1950

Darwin H. Stapleton

This is a case study of long-term disaster mitigation planning by an international philanthropic organization, the Rockefeller Foundation. In the first half of the twentieth century the foundation engaged in disaster planning through its programs to control epidemic diseases and through the creation of community-based local health units. This study views those programs through the lens of disaster mitigation, and focuses on the health units and the means by which they served to mitigate public health disasters.

It is clear that effective disaster mitigation requires a "rear-view mirror" for effective preparation and planning. Two important volumes of disaster-mitigation studies rely on historical reviews as underpinnings for the development of effective plans and policies, and this examination – which goes further back into history than those studies – is intended to provide additional historical perspective for disaster planners and policy-makers.[1]

The Rockefeller Foundation was created in 1913 as one of the final philanthropic acts of John D. Rockefeller. He took a great interest in medicine and public health, establishing the Rockefeller Commission for the Eradication of Hookworm Disease in 1909 and the International Health Board in 1913. In the latter year he also established the Rockefeller Foundation, by far the largest philanthropic organization the world had seen and which remained the largest until 1950. The International Health Board always coordinated its operations with the Foundation, and the two were merged in 1928.

The Rockefeller Foundation's motto has always been "for the well-being of mankind throughout the world." Until 1950, its major focus was the amelioration of health conditions, and its work in that field generally had two prongs: the development of strategies to control major diseases, centering on yellow fever, malaria, and hookworm (especially in epidemic settings and conditions) but also devoting some attention to typhus, tuberculosis, and influenza; and the encouraging of nations, including their colonies, to establish comprehensive public health programs centered on community-based health services.

Historically, much of the attention on disaster preparedness and disaster recovery has focused on the role of government – which in virtually all situations has had the most resources for comprehensive planning and response. However, the role of non-profit organizations, particularly philanthropic foundations, needs to be considered by disaster policy-makers. One study of foundations interested in health matters noted that "the role of foundations in policy making has been largely ignored in public policy literature." Yet in the field of health, in particular, foundations play a significant role because "much of foundation funding goes to affect public policy, health professions education, health planning, facilities construction, and research, significantly in public health, an area critical to disaster planning."[2]

The best-known element of the Rockefeller strategy for epidemic disease control was to find effective ways to kill the insect vectors: mosquitoes in the case of malaria and yellow fever, and the human body louse in the case of typhus. The identification of the specific vectors and the search for effective insecticides have been covered extensively by historians, in part because the Rockefeller Foundation played a major role in promoting the use of DDT during World War II and immediately afterward. The astounding effectiveness of DDT – which was recognized by General Eisenhower and Winston Churchill as an important addition to the tools of warfare – led to a radical change in disease-control strategies and to its use throughout the world. It took some years before its long-term environmental effects were fully understood. DDT's use was severely limited after 1970 (it is still used in some countries to control insect-borne disease epidemics) but our planet and its human population are still coping with the aftermath of DDT saturation.[3]

The lesser-known element of Rockefeller public health work has also had long-lasting effects, but for the betterment of humankind. This is the health unit strategy that the Rockefeller Foundation did not invent but adopted from its experience in the southern United States, where it funded pilot programs at the county level, particularly for malaria control.[4] The Rockefeller philanthropies demonstrated an early interest in community-based health programs. The trustees and officers were aware of health programs based administratively at the county level that had been instituted in Kentucky and North Carolina at the turn of the twentieth century. Such work included a full-time health office, a veterinarian (the health of animals in rural districts being almost as important as human health), a clerk, and "a chemical and bacteriological laboratory."[5] This mode of health delivery caught the attention of the Rockefeller operatives and it became central to their plans for upgrading health conditions not only in the American south but also globally.[6] The first annual report of the International Health Board stated optimistically that

> The plan of work adopted for each country makes provision for beginning operations on a small scale. . . . The opening of work in each new country must be in the nature of an experiment. . . . When the effective working unit for these conditions has been ascertained this unit can be multiplied at will.[7]

The first significant "excellent example" of the application of this strategy outside the United States was in Brazil.[8] In 1916, International Health Board (IHB) officer Lewis Hackett was sent to Brazil after several of that country's states invited the IHB to set up hookworm control projects. Hackett arrived on the heels of the IHB Yellow Fever Commission's visit to Brazil, after which the Commission had urged that special attention be paid to the possibility of yellow fever epidemics there. Thus, Hackett had a divided mandate: to deal immediately with the endemic hookworm problem but also to create early warning mechanisms for epidemic disease. He worked with all levels of government administration and community leaders to implement these programs. At the local level, Hackett established the first anti-hookworm health unit in 1917 (and several more within the next five years), and at the federal level he saw the creation of a Yellow Fever bureau for monitoring and controlling that disease.[9]

Moreover, the local health units were quickly mobilized not only for anti-hookworm work but also for the control of epidemic malaria and (with the support of the Yellow Fever bureau) for Yellow Fever control as well. One important outcome of Hackett's work was the coordination of federal, state, and local health programs, which had been absent in Brazil before the IHB's intervention.[10] By 1923, six years later, it was reported that the local health units in Brazil were engaged with the control not only of hookworm and malaria but also potential epidemics of typhoid, trachoma, syphilis, meningitis, and diphtheria.[11] One episode was recounted by Hackett's successor in Brazil, John H. Janney, to demonstrate the disaster-control possibilities of community-based health units:

> In October 1922 the value of the permanent [community-based health] post was . . . demonstrated by the manner in which a small outbreak of cerebro-spinal meningitis was handled. On the 15th a family of three, all ill, came to Morro Agudo for medical care. The local physician made a diagnosis of meningitis which was verified by an examination of spinal fluid in the post laboratory. . . . They were found to have come from a neighboring county. . . . The district involved was then divided into an infected, a suspected and a free zone and immediate steps were taken to stop all traffic between the zones and between the two counties. An emergency isolation hospital was installed at Morro Agudo and the existing cases transferred to this hospital at once. This was all done so quickly and effectively that when the nearest representative of the State Sanitary Service arrived he found everything arranged and being handled to his complete satisfaction.[12]

With this kind of success, it is not surprising that by the mid-1920s the Foundation had convinced other nations of the efficacy of local health units, spreading the practice to various places in South America and Europe, and by the 1930s to south and east Asia. It is the latter region, south and east Asia, and specifically the projects in Sri Lanka, Indonesia, and China, to which this discussion now turns.

The IHB's first involvement in Sri Lanka, then the British colony of Ceylon, occurred in 1915 when it attempted to initiate a program to control hookworm among the workers on the large agricultural plantations. It was estimated that 90 percent of the workers were infected with hookworm. Facing stiff resistance from the plantation owners, who saw that they would incur substantial costs in building sanitary facilities, the hookworm campaign was unsuccessful. Nonetheless, it allowed the Board "to acquire first-hand knowledge about the socio-economic conditions" in Sri Lanka.[13] The failure of the hookworm campaign "convinced the [Board's] officials" of the need to work

> in the context of broad public health requirements . . . however . . . this realization did not come about as a natural progression . . . it was born out of a dynamic response to the local people, who cooperated with the hookworm campaign while voicing their criticisms of its limited scope. . . . The people of the villages and towns demanded attention to more immediate health problems. They questioned the rationale of giving treatment for hookworm infection in the context of numerous other diseases such as typhoid, smallpox, dysentery and malaria, which were more serious and debilitating.[14]

In this situation, the Rockefeller officials rethought their work in Sri Lanka and opted for the more fundamental and longer-term response by creating an "integrated public health campaign" based on local health units. The units had a broad mandate that ranged from the collection of vital statistics to health education to anti-malaria projects, with a continuing focus on hookworm control as an integrated program.

By 1929, there were five health units serving approximately 5 percent of the Sri Lankan population. Throughout the 1930s, additional units were created, such that there were nearly 25 by the end of the decade. Local support for the units included the funding of vaccination campaigns by local donations and support from Social Service Leagues that promoted child welfare and improved sanitation. "Private citizens donated buildings and lands to set up [the units] and provided water pumps and other equipment"[15] and "the medical officers in charge of the health units were mainly native Sri Lankans or other south Asians."[16] The health units clearly drew on local resources as well as Rockefeller Foundation support.

In the view of one historian, the health units "were largely responsible for reducing the maternal and infant mortality rates in the country." In regard to epidemic disease, the post-World War II anti-malaria campaign based on DDT "developed by the health units . . . in accordance with the unique ecological conditions in the island turned out to be remarkably cost effective" and contributed significantly "to the dramatic fall of the mortality rate in the country." According to the World Health Organization, the mitigation of disease in Sri Lanka is attributable to the nation's "community-based" approach to long-term health care.[17]

More briefly, a similar story emerged in Indonesia, beginning when it was the Netherlands East Indies. The Rockefeller officer assigned to that colony established the first health unit there in the late 1920s, and a health unit focused on training health workers was created in 1936.[18] According to anthropologist Eric Stein, who studied the colonial health regime extensively, "Although many Dutch physicians remained skeptical of . . . [the] preventive approach . . . Javanese doctors often approached the rural hygiene projects with enthusiasm, extending the model into their own administrative domains as heads of regional health services."[19] The first health demonstration unit, which became the model for the other units, survived the Japanese occupation and the War of Independence afterward. Again, according to Stein, "the key features of the Rockefeller . . . rural hygiene model were retained and further developed" in the 1950s, as new health demonstration units were created throughout Indonesia. They were staffed by trained local health educators and health workers who knew the socio-economic resources of the nation.[20]

In China, the health units were initiated with a more top-down approach. The first health unit, the Peking (Beijing) Health Demonstration Station, established in 1925, was regarded as a training facility for the Chinese students in the Department of Hygiene and the Nursing School at the Rockefeller-funded Peking Union Medical College. After graduation many of those students were involved in the establishment of other health units in China.

The Peking Health Station was one of the few urban community-based health units sponsored by Rockefeller philanthropy. Because it was integrated with city government functions (initially the police department), the Station carried out a wider range of functions than most local health units: inspections of health conditions in factories and restaurants were among its responsibilities, for example. Nonetheless, the Station's primary role was disaster mitigation, with its primary mission initially described as "control of communicable diseases such as smallpox, typhus, diphtheria, typhoid, plague and trachoma."[21]

Perhaps the most important consequence of the training function of the Peking Station was that a number of the students and graduates of the College subsequently moved to the Mass Education Movement's experimental work in the rural district of Dingxian, south of Beijing. The public health work of the Movement was founded on the idea that basic public health services could be provided by trained locals who would regularly visit the villages in the area. Fundamentally, this was an attempt at grass-roots control of disease to avoid epidemic outbreaks.

Observations of the success of the work at Dingxian influenced a generation of Chinese public health officials, and some writers have found in the public health work at Dingxian the roots of the "barefoot doctor" mode of rural health delivery during the early years of the People's Republic.[22] A recent scholar has described Dingxian as "unique and successful" in improving rural health in China.[23]

Inspired by the idea of preventing, or at least mitigating, the effect of epidemic disease on rural populations, the Rockefeller Foundation in the first half of the twentieth century aided in the creation of local health units that were either inspired by resident citizens or drew on them for the effective functioning of the units, or both. Recently the writer of an article on disaster-mitigation planning referred to the importance of a "model [that is] community- and region-based: it is foremost a tool for local communities and regional governments. It is based upon local knowledge supplemented by experts and it is to be used by both large and small communities."[24] This describes the health unit model that was pioneered by the Rockefeller Foundation throughout the world. Disaster mitigation planners should look back at that experience, as well as its long-term effects, in order to understand the matrices and determinants of its success.[25]

Notes

1 C.E. Haque, ed. (2005), *Mitigation of Natural Hazards and Disasters: International Perspectives* (Dortrecht, Netherlands: Springer); C.E. Haque and D. Etkin, eds. (2012), *Disaster Risk and Vulnerability: Mitigation through Mobilizing Communities and Partnerships* (Montreal: McGill-Queen's University Press).

2 C.S. Weissert and J.H. Knott (1995, Winter), "Foundations' Impact on Policy Making: Results from a Pilot Study," *Health Affairs* 14, pp. 275–86.

3 D.H. Stapleton (2000, Winter), "The Short-Lived Miracle of DDT," *American Heritage of Invention and Technology* 15, pp. 34–41; D.H. Stapleton (2005, July), "A Lost Chapter in the Early History of DDT: The Development of Anti-Typhus Technologies by the Rockefeller Foundation's Louse Laboratory, 1941–1944," *Technology and Culture* 46, pp. 513–40; B. Eskenazi, J. Chevrier, L. Goldman Rosas, H.A. Anderson, M.S. Bornman, H. Bouwman, A. Chen, B.A. Cohn, C. de Jager, D.S. Henshel, F. Leipzig, J.S. Leipzig, E.C. Lorenz, S.M. Snedeker, and D. Stapleton (2009, September), "The Pine River Statement: Human Health Consequences of DDT Use," *Environmental Health Perspectives* 117, pp. 1359–67.

4 D.H. Stapleton (2014), "Technological Solutions: The Rockefeller Insecticidal Approach to Malaria Control, 1920–1950," in *The Global Challenge of Malaria: Past Lessons and Future Prospects*, F.M. Snowden and R. Bucala, eds., pp. 21–26 (Singapore: World Scientific Publishing).

5 "The First All-Time County Health Unit" (n.d.) and L.L. Lumsden to J.A. Ferrell, January 29, 1924, folder 11, box 2, RG 5.2, Rockefeller Foundation Archives (hereafter RFA), Rockefeller Archive Center, Sleepy Hollow, New York, U.S.

6 An early review of Rockefeller community-based health work can be found in the Rockefeller Foundation's (1920) *Annual Report for 1919* (New York, NY: Author), pp. 74–80.

7   The Rockefeller Foundation and the International Health Commission (1915), *First Annual Report* (New York: IHC), p. 24.

8   Rockefeller Foundation (1918), *Annual Report, 1917* (New York, NY: Author), p. 157.

9   D.H. Stapleton (2005), "Lewis W. Hackett and the Early Years of the International Health Board's Yellow Fever Program in Brazil, 1917–1924," *Parassitologia* 47, pp. 353–60.

10 W. Rose, "Observations on [the] Public Health Situation and Work of the International Health Board in Brazil," October 25, 1920, folder 152, box 25, RG 5.2, RFA.

11 J.H. Janney, "The Plan and Progress of County Health Work in Brazil," September 5, 1923, folder 154, box 25, RG 5.2, RFA.

12 Ibid.

13 S. Hewa, (2012), "The Alma-Ata Declaration, Rockefeller Foundation and the Development of Primary Health Care in Sri Lanka: A Model for Health Promotion," in Science, Public Health and the State in Modern Asia, L. Bu, D.H. Stapleton, and K.-C. Yip, eds., pp. 88–100 (London, UK and New York, NY: Routledge).

14 Ibid., pp. 78–79.

15 Ibid., p. 87.

16 Ibid.

17 Ibid., p. 88.

18 Rockefeller Foundation (1937), *Annual Report for 1936* (New York, NY: Rockefeller Foundation), p. 121.

19 E.A. Stein (2012), "Hygiene and Decolonization: The Rockefeller Foundation and Indonesian Nationalism, 1933–1958," in *Science, Public Health and the State in Modern Asia*, L. Bu, D.H. Stapleton, and K.-C. Yip, eds., p. 55 (London, UK: Routledge).

20 Ibid., pp. 60–61.

21 L. Bu (2012), "Beijing's First Health Station: Innovative Public Health Education and Influence on China's Health Profession," in *Science, Public Health and the State in Modern Asia*, L. Bu, D.H. Stapleton, and K.-C. Yip, eds., p. 129–143 (London, UK and New York, NY: Routledge).

22 This section draws on D.H. Stapleton (2012), "'Removing the Obstacles to Public Health Work': Context for the Rockefeller Philanthropies and Public Health in China and Japan, 1920–1940," in *Science, Public Health and the State in Modern Asia*, L. Bu, D.H. Stapleton, and K.-C. Yip, eds., pp. 93–112 (London, UK and New York, NY: Routledge), and (2014), "Connecting Philanthropy with Innovation: China in the First Half of the Twentieth Century," in *Philanthropy for Health in China*, J. Ryan, L.C. Chen, and T. Saich, eds., pp. 120–36 (Bloomington and Indianapolis, IN: Indiana University Press).

23 L. Bu (2014), "John B. Grant: Public Health and State Medicine," in *Medical Transitions in Twentieth-Century China*, B. Andrews and M. Brown Bullock, eds., pp. 212–26 (Bloomington, IN: Indiana University Press), p. 219.

24 L. Pearce (2005), "The Value of Public Participation during a Hazard, Impact, Risk and Vulnerability (HIRV) Analysis," in *Mitigation of Natural Hazards and Disasters*, C.E. Haque, ed., pp. 79–109 (Dortrecht, Netherlands: Springer), p. 83.

25 Recommended readings regarding health units and malaria, especially in south and east Asia are: S. Hewa (1995), *Colonialism, Tropical Medicine, and Imperial Medicine* (Lanham, MD: University Press of America); J. Farley (2004), *To Cast Out Disease: A History of the International Health Division of the Rockefeller Foundation (1913–1951)* (New York, NY: OUP); K.-C. Yip, ed. (2009), *Disease, Colonialism, and the State: Malaria in Modern East Asian History* (Hong Kong: Hong Kong University Press); Bu, Stapleton, and Yip, eds., *Science, Public Health and the State in Modern Asia*, op. cit.; Snowden and Bucala, eds., *The Global Challenge of Malaria: Past Lessons and Future Prospects*, op. cit.; and Ryan, Chen, and Saich, eds., *Philanthropy for Health in China*, op. cit.

Bibliography

Bu, L. (2012). "Beijing's First Health Station: Innovative Public Health Education and Influence on China's Health Profession." In *Science, Public Health and the State in Modern Asia*. L. Bu, D.H. Stapleton, and K.-C. Yip, eds. London, UK and New York, NY: Routledge.

———. (2014). "John B. Grant: Public Health and State Medicine." In *Medical Transitions in Twentieth-Century China*. B. Andrews and M. Brown Bullock, eds. pp. 212–26. Bloomington, IN: Indiana University Press.

Eskenazi, B., J. Chevrier, L. Goldman Rosas, H.A. Anderson, M.S. Bornman, H. Bouwman, A. Chen, B.A. Cohn, C. de Jager, D.S. Henshel, et al. (2009, September). "The Pine River Statement: Human Health Consequences of DDT Use." *Environmental Health Perspectives* 117, pp. 1359–67.

Farley, J. (2004). *To Cast Out Disease: A History of the International Health Division of the Rockefeller Foundation (1913–1951)*. New York, NY: OUP.

Haque, C.E., ed. (2005). *Mitigation of Natural Hazards and Disasters: International Perspectives*. Dortrecht, Netherlands: Springer.

———, and D. Etkin, eds. (2012). *Disaster Risk and Vulnerability: Mitigation through Mobilizing Communities and Partnerships*. Montreal, Canada: McGill-Queen's University Press.

Hewa, S. (1995). *Colonialism, Tropical Medicine, and Imperial Medicine*. Lanham, MD: University Press of America.

———. (2012). "The Alma-Ata Declaration, Rockefeller Foundation and the Development of Primary Health Care in Sri Lanka: A Model for Health Promotion." In *Science, Public Health and the State in Modern Asia*. L. Bu, D.H. Stapleton, and K.-C. Yip, eds. London, UK and New York, NY: Routledge.

Pearce, L. (2005). "The Value of Public Participation during a Hazard, Impact, Risk and Vulnerability (HIRV) Analysis." In *Mitigation of Natural Hazards and Disasters*. C.E. Haque, ed. pp. 79–109. Dortrecht, Netherlands: Springer.

Rockefeller Foundation, The. (1918). *Annual Report, 1917*. New York, NY: Author.

———. (1920). *Annual Report for 1919*. New York, NY: Author.

———. (1937). *Annual Report for 1936*. New York, NY: Author.

———, and the International Health Commission. (1915). *First Annual Report*. New York, NY: IHC, 1915.

Stapleton, D.H. (2000, Winter). "The Short-Lived Miracle of DDT." *American Heritage of Invention and Technology* 15, pp. 34–41.

———. (2005). "Lewis W. Hackett and the Early Years of the International Health Board's Yellow Fever Program in Brazil, 1917–1924." *Parassitologia* 47, pp. 353–60.

———. (2005, July). "A Lost Chapter in the Early History of DDT: The Development of Anti-Typhus Technologies by the Rockefeller Foundation's Louse Laboratory, 1941–1944." *Technology and Culture* 46, pp. 513–40.

———. (2014). "Connecting Philanthropy with Innovation: China in the First Half of the Twentieth Century." In *Philanthropy for Health in China*. J. Ryan, L.C. Chen, and T. Saich, eds., pp. 120–136. Bloomington and Indianapolis, IN: Indiana University Press.

———. (2014). "'Removing the Obstacles to Public Health Work': Context for the Rockefeller Philanthropies and Public Health in China and Japan, 1920–1940." In *Science, Public Health and the State in Modern Asia*. L. Bu, D.H. Stapleton, and K.-C. Yip, eds., pp. 93–112. London, UK and New York, NY: Routledge.

———. (2014). "Technological Solutions: The Rockefeller Insecticidal Approach to Malaria Control, 1920–1950." In *The Global Challenge of Malaria: Past Lessons and Future Prospects*. F.M. Snowden and R. Bucala, eds., pp. 21–26. Singapore: World Scientific Publishing.

Stein, E.A. (2012). "Hygiene and Decolonization: The Rockefeller Foundation and Indonesian Nationalism, 1933–1958." In *Science, Public Health and the State in Modern Asia*. L. Bu, D.H. Stapleton, and K.-C. Yip, eds. pp. 51–70. London, UK and New York, NY: Routledge.

Weissert, C.S., and J.H. Knott. (1995, Winter). "Foundations' Impact on Policy Making: Results from a Pilot Study." *Health Affairs* 14, pp. 275–86.

Yip, K.-C., ed. (2009). *Disease, Colonialism, and the State: Malaria in Modern East Asian History*. Hong Kong: Hong Kong University Press.

14 The disaster life cycle in Turkey
Planning, policy, and regulation changes

Derin N. Ural

Introduction

Turkey is a nation with active faults spanning the land from east to west and north to south. Between 1999 and 2014, 287 recorded earthquakes caused damage in Turkey, resulting in 100,000 deaths, 170,000 injuries, and 700,000 structural failures. The economic consequences of these earthquakes have impacted Turkey's GDP at a rate higher than any other country: 66 percent of Turkey' land lies in first- and second-degree earthquake risk zones.[1] The country's total population is 70 million, and 11 of its largest cities are situated in high-risk earthquake zones: 70 percent of the population is at risk of being impacted by future earthquakes and 75 percent of its largest industries are situated in high-risk areas. Therefore, the challenge for Turkey is to be prepared for these earthquakes and to minimize potential losses with proper engineering design, insurance, code enforcement, and construction. Proper enforcement of seismic design codes is critical, as well as having knowledgeable and professional practicing engineers. There are lessons to be learned from the recent devastating earthquakes that have occurred in Turkey, specifically the 1999 Kocaeli (M_s 7.4) and Duzce (M_s 7.2) earthquakes.

The Kocaeli earthquake

A magnitude M_s 7.4 earthquake occurred in the northwestern Marmara region of Turkey on August 17, 1999 at 3:02 am local time. The earthquake lasted a lengthy 45 seconds, causing thousands of buildings to collapse and sustain heavy, moderate, or light damage due to the strong tremors. The epicenter was located along the Northern Anatolian Fault, a right lateral strike-slip fault, at 40.70 N and 29.91 E and at a depth of 15.9 km. The earthquake affected one of the most densely populated regions of the country (approximately 15 million residents), leaving over 600,000 people homeless. The official losses reported included 15,226 deaths and over 23,000 injured. The affected region is home to over half of the country's industry. The direct economic losses caused by this earthquake were reported as USD$1.3 billion, and indirect economic losses caused by disruption in production and industry as USD$2.4 billion.[2] One hundred and twenty-five large-scale Turkish companies on the stock exchange reported losses of USD$600 million due to short- and medium-term production interruptions and delays. Turkey's GDP growth fell by 3 percent as a consequence of the earthquakes in 1999.

The damage during and following the Kocaeli earthquake included 27,634 heavily damaged or collapsed buildings, as well as over 27,000 moderately damaged and over

31,000 lightly damaged buildings. The majority of the buildings that sustained damage were multi-story reinforced concrete structures. Each building housed up to 12 units (or 12 families).

According the UN OCHA situation report following the earthquake, an assessment by the IBS Marketing Research Services found that a striking 17 percent of the buildings in the affected region were designated as safe to be inhabited.[3] It is troubling that 83 percent of all the buildings that sustained damage were deemed to be structurally unsafe to inhabit without retrofitting and remediation work.

As described above, 55,062 buildings sustained moderate or heavy damage. Assuming that each building comprised 10 units, reconnaissance efforts indicated that 550,620 families needed immediate assistance. Unfortunately, some families were among the dead and injured. Of the remaining population, many migrated outside the disaster-impacted region to live with relatives. The large number of damaged buildings led to the immediate need for emergency temporary shelters, where tent units were set up for those families in need of shelter. The agencies providing the tents and the number of tents set up in the 121 tent cities designated throughout the region are listed in Table 14.1. Temporary and permanent housing projects were to follow for these displaced families.

Transportation arteries and lifelines also sustained heavy damage due to the earthquakes. The Trans-European Motorway (TEM) between the country's capital, Ankara, and Kocaeli was damaged due to the collapse of an overpass on the motorway in Arifiye, and the differential settlement affecting the surviving parts of the structure led to its closure for two days. Delivery of relief from the capital was therefore not possible for 48 hours and was further delayed afterwards due to road failure. Traffic was allowed at a moderate pace after six days. Telephone communication lines were severely damaged during the earthquake, and communication was restored only after 48 hours.

The earthquake caused main power lines to fail, resulting in power outages throughout the entire country. Water and sewage pipelines sustained heavy damage, cutting off the region's access to clean water. The damage to these lifeline systems totaled USD$12 billion. The Ministry of Finance announced it would cover USD$7 billion of these losses, and tax revenue would pay for the remainder. As a consequence of the loss of power and water to the region, industry came to a standstill, even though some facilities did not sustain structural damage from the earthquake.

The failure of the main transportation artery and all phone communication delayed the initial damage survey of the region. The Prime Minister visited the region to assess

Table 14.1 The number of tents set up in 121 tent cities following the earthquake[4]

Donors	Number
Red Crescent	40,680
Armed Forces	2,122
International Donors	54,841
Private Sector	7,970
Total Number of Tents	105,613

the damage and called on his colleagues in the capital, Ankara, to assist by addressing them through the only means of communication available, a live national television broadcast, which caused public chaos.

As a secondary effect of the earthquake, fires broke out in the oil tanks of one of the country's largest refineries, located in Izmit, immediately following the earthquake. These fires continued for six days. Assistance from private and international organizations was required to contain the fire. Additionally, a petrochemical plant located south of the Marmara Bay had a large chemical spill into the bay and contaminated the local soil; as a result, the agriculture sector sustained setbacks and losses of produce which constituted serious long-term environmental consequences that required remediation through updated mitigation policies.

The foreign monetary contributions to Turkey following the earthquake (reported by the UN OCHA)[5] included direct assistance of USD$300 million and a loan with a long-term repayment plan for USD$757 million from the World Bank. Turkey also received an additional loan from the International Monetary Fund (IMF) of USD$4 billion. Of the World Bank loan of USD$757 million, USD$252 million was specified as an Earthquake Emergency Recovery Loan, and USD$505 million was allocated for the Marmara Earthquake Emergency Reconstruction (MEER) Project. The USD$4 billion IMF loan was utilized to alleviate the negative impact of the earthquake on the economy and to assist in measures taken to fight inflation, as the new taxes following the earthquake were likely to increase the inflation rate. According to Turkish officials, the cost of rebuilding residential buildings, industrial facilities, public services, and infrastructure amounted to over USD$25 billion.

The following section outlines the policies set forth in Turkey following the compound disaster and the impact of the change brought about by the 1999 Kocaeli earthquake.

Minimizing risk through policies on disaster resilience

Disaster management prior to the establishment of the Turkish republic

The disasters that have occurred in Turkey date back to the year 553, during the Gokturk ("Celestial Turk") Empire, when records indicate that the Istanbul earthquake in that year caused the city walls to collapse. The second recorded disaster was another earthquake in Istanbul in the year 865 during the Byzantine Empire: historical documents indicate that one-third of the structures in Istanbul collapsed. The third large disaster recorded in the history of the country occurred in 1500, during Ottoman rule. It caused 1,300 deaths and extensive damage to 1,070 homes. The scale of the response to this earthquake led to the formation of the first disaster-management system recorded in the country. The Ottoman Sultan ruled that each family who had lost a home would be given 20 gold coins in order to rebuild. This first documented empire-led assistance initiated a "state insurance program" that lasted up until 1999 in the modern Republic of Turkey.

Disaster management following the establishment of the Turkish republic

Upon the foundation of the Republic of Turkey in 1923, information was recorded and documented regarding significant earthquakes that occurred. The policy changes due to these disasters included the establishment of a Ministry of Development and

Housing and a law requiring each municipality to incorporate guidelines for construction and roadways.

Fourteen major earthquakes occurred in the third decade of the Republic (1943–1953). These disasters caused 8,000 deaths and damaged over 60,000 homes. The severe and repeated losses in this decade led to the most significant policy change in disasters as the government established its first "Turkish National Earthquake Design Code," enacted in 1945 along with the preparation of the first earthquake risk map of Turkey. The Office of Earthquakes within the Ministry of Public Works and the office of the State Hydraulic Works were established during this decade.

Significant policy changes were made during the fourth decade of the Republic of Turkey (1953–1963), including the establishment of the Ministry of Housing and Settlements and Civil Defense. The most important Disaster Law (No. 1051, later known as 7269), which included specific guidelines on disaster preparedness, response, and recovery, was initiated in 1959. It stated that the government would be responsible for providing eligible homeowners with permanent housing units without charge in the event of severe damage or collapse. This "government insurance" program deterred homeowners from purchasing disaster insurance for their property, as there was a state guarantee that they would be given a safe new home if severe damage was sustained.

The fifth decade of the Republic of Turkey brought a change in policy-making: The Prime Ministerial State Planning Organization (recently renamed the Ministry of Development) developed policy and initiated official five-year development plans. Under the auspices of the Ministry, think tanks were established and experts in each field, including public authorities, members of academia, and non-governmental groups convened to initiate new policies. The first five-year plan was initiated in 1963, and this system has continued uninterrupted until today. The significance of the five-year plans in terms of disaster policies following the Kocaeli earthquake is outlined below.

Significance of the seventh five-year development plan (1996–2000)

Turkey experienced eight severe earthquakes during the seventh planning period (Table 14.2), which caused a total of 18,398 deaths and 147,232 structural collapses.

Table 14.2 Earthquakes during the seventh planning period

Earthquake location and date	Magnitude	Deaths	Number of structural failures
Mecitozu-Amasya, 1996	5.6	1	2,606
Antakya, 1997	5.4	1	1,841
Karliova-Bingol, 1998	5.0	–	148
Ceyhan-Adana, 1998	6.2	146	31,463
Kocaeli, 1999	7.8	17,480	73,342
Duzce, 1999	7.5	763	35,519
Cankiri, 2000	6.1	1	1,766
Sultandagi-Afyon, 2000	5.8	6	547

This planning period entailed important policy changes regarding planning and preparations for metropolitan areas in Turkey. The two largest earthquakes, the Kocaeli and Duzce events, prompted the process of policy-making for disaster-resilient communities.

The Turkish Government established strong programs with international support and cooperation. One successful program was the establishment of the "Earthquake Mitigation Research Center" in collaboration with the Japan International Cooperation Agency. In addition, the United Nations Development Program (UNDP) initiated the "Improving the Disaster Management System in Turkey" project in Turkey in 1997.

Another success was the establishment of the "Center of Excellence for Disaster Management" at the Istanbul Technical University with support from the US Federal Emergency Management Agency (FEMA) and funding from the Office of Foreign Disaster Assistance at the US State Department. The Center offered a pioneering Master's degree program in Disaster Management in Turkey, as well as publishing the first disaster management books in Turkish with funding from USAID. Certificate programs for local disaster stakeholders soon followed.

A mandatory earthquake insurance program was enacted in 1999, as the losses from the earthquakes proved to be larger than the budget had planned for. The details of the insurance program and the participation rates will be presented in the following sections.

Another important new policy was the initiation of the first Turkish Emergency Management Directorate (TAY) in November 1999. The establishment of TAY was a step forward for Turkey for disaster preparedness at a national level. The law specified that the Director of TAY would report to the Deputy Prime Minister, who is in charge of disaster affairs. The law did not deal with affairs at the local level, which were left for discussion as part of a future policy change. Therefore, the main strength of the new TAY office was to coordinate all stakeholder agencies and ministries in preparing for and responding to a natural or man-made disaster at the national level.

The significance of the eighth five-year development plan (2001–2005)

Policy changes continued through the eighth development plan period as the impact of the 1999 Kocaeli and Duzce earthquakes continued to remind policymakers of the imminent danger of future earthquakes. Six major earthquakes occurred during this period; the greatest losses were caused by the magnitude 6.4 Bingol earthquake of 2003, which resulted in 176 deaths and 6,000 collapsed structures.

The eighth development plan preparations included for the first time a task force on disaster management, which prepared guidelines for future planning and policy making and were influential in forming plans. The policies implemented during this planning period included an updated Seismic Design Code for Turkey in 2002. Regulations regarding the Civil Defense Search and Rescue teams were also passed in 2002, enhancing the response preparations for future emergencies and disasters.

An important area that needed policy changes was the environmental impact of disasters. During the eighth planning period, a regulation on the assessment of the environmental impact of projects in areas of high seismic risk was passed. This was a major improvement for the preparedness of secondary emergencies following large disasters.

The retrofit and rebuilding of schools and hospitals at risk in Istanbul began in 2004 as a major component of the Marmara Earthquake Emergency Reconstruction (MEER) project financed by the World Bank. The plan discussed the need for a national professional engineering licensing system in Turkey. This system currently allows all graduates of civil engineering programs to act as professional engineers without any testing procedure. The levels of education in Turkey vary greatly, and thus the collapse of structures is proportional to the adequacy of the practicing engineers.

Major national initiatives following the 1999 Kocaeli earthquake

Disasters are extraordinary events that result in severe destruction of property, loss of life, and economic consequences. The losses following the 287 earthquakes that occurred in Turkey between 1900 and 2014, combined with the impact of the 4,067 floods that occurred in during the same time period, amounted to over 100,000 deaths and losses in 30,000 communities.

Turkey's initial policy for disaster management, passed in 1960, proved insufficient with regard to the response and recovery efforts following the 1999 Kocaeli earthquake. The law included response and recovery plans, but neglected disaster mitigation and planning. The large losses led Turkey to revisit its strategies for disaster preparedness. A task force, including local and national leaders, academics, and NGOs worked together to identify areas for improvement in the national disaster management system. The main challenges were identified as:

1   The lack of coordination of disaster-management-related offices throughout the nation at the local and national levels.
2   The need to allocate funds to mitigation-related projects. According to the law of 1960, disaster funds were allocated for rebuilding after major disasters. Therefore, homeowners relied on the State for rebuilding instead of investing in disaster insurance. The government's policy for rebuilding was to grant families who had lost their homes in a disaster a new home without charge. Considering the extent of the damage to over 50,000 buildings during and after the Kocaeli earthquake, the funds were not sufficient to continue with this policy. A national disaster insurance program that allowed homeowners to insure their property was needed.
3   The need for a national professional licensing system for practicing civil engineers in Turkey. The number of structures without adequate engineering and which failed to comply with design codes is greater than those that comply with the code. Accountability following structural damage is an area that needs attention from policy makers.

The policy improvement in 2009 brought with it the establishment of the following offices within the Disaster Management Directorate: Planning and Mitigation, Response, Recovery, Civil Defense, Earthquake Preparedness, Strategic Development, Intelligent Computer Systems, and Communication and Legal Services. The policy change prompted a new national focus on the mitigation and planning procedures that were needed based on the evidence of past losses.

Structural retrofit and reconstruction program

The MEER–World Bank Project (2002–2014) had significant goals, including the provision of post-disaster permanent housing (including 14,644 units in a two-year time frame); the upgrading of trauma and healthcare facilities; the strengthening of the emergency management capacity by the establishment of the Turkish Emergency Management Directorate (similar to FEMA in the US); and the implementation of a seismic risk mitigation and remediation program for priority public buildings.

The European Investment Bank, the European Commission Development Bank, and the Islamic Development Bank joined the World Bank in financing this major retrofit and reconstruction project. The project included the retrofitting and reconstruction of schools, hospitals, and critical government buildings. The total budget for this (primarily mitigation) project was USD$1.6 billion. As the expected direct economic losses in the event of a severe earthquake in Istanbul are expected to reach USD$50 billion, this mitigation project is an excellent way to decrease the country's current disaster vulnerability and reduce expected future losses.

The MEER project had a subcomponent called the Istanbul Seismic Risk Mitigation and Emergency Preparedness Project (ISMEP). ISMEP's initial project was to increase the capacity of the Disaster Management Directorate in Istanbul. Thus, the construction of two new disaster management centers in Istanbul, one on the European side of the city in Hasdal, and a second on the Asian side of the city in Akfirat, were completed. These two centers replace the single center currently in Cagaloglu. The data and coordination efforts are mirrored on the two sites, as the population of Istanbul exceeds 10 million and a major disaster in the city would require emergency operations on the two continents that the city straddles.

The ISMEP project also included an assessment of the vulnerability of all public schools, which found that of the city's 1,084 public school buildings, 66 percent required seismic retrofitting, 33 percent had to be demolished and rebuilt, and 1 percent were safe and required no remediation. Vulnerability assessments were also carried out on government buildings: 108 were assessed in detail, of which 50 percent required a seismic retrofit, 48 percent were not built adequately to sustain earthquake loading and were therefore demolished and rebuilt, and the remaining 2 percent were safe and required no intervention. The third group of critical facilities included in the ISMEP Project were the hospitals in Istanbul. The detailed vulnerability assessment of the 45 hospitals concluded that 47 percent required a seismic retrofit, while 53 percent were found not to meet the necessary seismic design code and were thus were demolished and rebuilt. It was noted that no hospital building met the required seismic design codes and all required remediation.

Istanbul has a very rich historic and cultural heritage and is home to many significant structures. As part of the ISMEP project, 176 historically and culturally important structures were analyzed, including the Ottoman palaces, Byzantine and Roman churches, and museums. The ISMEP project funding did not extend to cover the possible remediation projects for these cultural and historic structures, so the projects have been prepared and will be implemented with future funding.

Another important component of the project included public education, including an explanation of the retrofits and rebuilding efforts as well as public preparedness for a future emergency or disaster.

The national earthquake insurance program

Protecting property against possible disasters by insurance is a method to prevent losses. However, when possible, the public tends to opt for federal aid programs instead of private insurance. Therefore, the federal government should ensure that insurance rates are properly based on the disaster risks of the area of the property being insured. It is known from previous disasters that building codes are frequently not enforced in regions at high risk for hazards.[6] In the US, the Insurance Research Council reported that a quarter of the insured losses caused by Hurricane Andrew in Florida in 1992 could have been prevented. This is a significant amount, as the total losses amounted to USD$34.5 billion. Enforcement of current building codes and checking code compliance is crucial in mitigating potential losses.

The Kocaeli earthquake, as reported by MunichRe and the NatCatService, was the eighth-largest earthquake to strike Turkey in terms of loss of life between the years 1980 and 2013. According to MunichRe, the overall losses in Kocaeli amounted to USD$12 billion.[7] The amount of the losses that were insured was limited to USD$600 million, only 5 percent of total losses. The remaining 95 percent had to be paid for by the government, which exceeded its capabilities, and thus a new Mandatory Earthquake Insurance program was initiated.

According to SwissRe, among the 91 metropolitan areas in Europe, Istanbul has the most people endangered by a single disaster scenario.[8] The forecast is that 6.4 million residents of Istanbul would be negatively impacted by an earthquake, followed by 2.7 million residents of Paris being impacted by a river flood and 2.2 million residents of London being impacted by a winter storm.

The natural hazard insurance policies of Turkey, Iceland, and Spain are outlined in Table 14.3. The similarities and differences in the programs are shown and the public's understanding of the policies' "compulsory" nature is compared.

The initiation of the compulsory earthquake insurance program in Turkey in 1999 was a step forward in preparation for future losses. However, as can be seen in Table 14.3, the rate of penetration of the program is only 27 percent. Iceland and Spain, with similar programs, have 100 percent penetration. Therefore, additional policies and initiatives are needed to increase the penetration rate in Turkey.

Table 14.3 National natural hazard insurance programs[9]

Country	Turkey	Iceland	Spain
Nature of program	Legal public entity	Public entity	Public business institution
Insured	Homeowners	Private and commercial asset owners	Private and commercial asset owners
Coverage	Earthquakes and any fires, explosions, landslides, or tsunamis caused by an earthquake	Earthquake, flood, volcanic eruption, avalanches, landslides	Earthquakes, floods, extraordinary storms, tornadoes, tsunamis, volcanic eruptions, meteorites, terrorism
Compulsion	Compulsory for every residential building registered in a state registry	Compulsory for all buildings	Compulsory within a property or personal insurance cover
Penetration (2012)	27%	100%	100%
State Guaranteed	Yes	Yes	Yes

The national professional engineering licensing program

The eighth plan of the Ministry of Development included discussion of the need for professional engineering licensing in Turkey. The number of structures with inadequate structural safety is alarming, at less than 20 percent. A component of the World Bank project therefore addressed the issue.

The project included the voluntary training of civil engineers under the authority of the Ministry of Public Works and Settlement. Professional training sessions were carried out in order to inform and educate engineers on seismic design codes. The training was offered in coordination with universities and local Civil Engineering society members. The training materials and programs were prepared locally, and a train-the-trainers program was completed for 3,631 civil engineers throughout Turkey.

This initiative was a local and unsustainable solution to a problem that requires national and sustainable results. The professional licensing systems that are successful around the world need to be examined and Turkey needs to initiate its own national system. A system is needed to verify that licensed engineers are competent, have engineering degrees that are recognized, are regulated, and are held accountable for their work.

The systems implemented in the US, Canada, United Kingdom, India, Pakistan, the European Union, and Turkey are outlined in Table 14.4. Each country has a system in place that includes criteria on the quality of the engineering program the candidate is a graduate of, the number of years of work experience the candidate has (preferably working under the guidance of a professional engineer), and the examination process.

The system currently in place in Turkey, as seen in Table 14.4, does not require candidates be graduates of accredited programs; nor is there a period of mandatory work experience prior to obtaining the title of Professional Engineer. The title is granted by the Chamber of Engineers, upon graduating with a Bachelor's degree from

Table 14.4 Outline of national professional engineering licensing programs

Country	USA	Canada	United Kingdom
Education	Graduate of an ABET-accredited university	Graduate of CEAB-Accredited University	Approved Education (MA level)
Mandatory Work Experience	Accumulate at least four years' engineering experience	1 Work under direction of a PE-certified supervisor for at least four years 2 Work experience reviewed by the Association	None
Exam Requirement	1 Complete the FE Exam and become an EIT 2 Complete a written Principles and Practice in Engineering (PE) Exam to become PE-certified	Complete the Professional Practice Exam	No exam, charter engineer qualification regulated by the Engineering Council, commercial leadership and management competencies needed

Table 14.4 (Continued)

Country	India	Pakistan	Europe	Turkey
Education	BA or MA degree	BA degree from a Pakistan-accredited university	BA from an accredited university	Bachelor's degree from any Turkish university
Mandatory Work Experience	Seven years' work experience, followed by registration with municipalities	Five years' work experience	None	None
Exam Requirement	No exam, Institution of Engineers offers PE	Complete Engineering Practice Examination, administered by the Pakistan Engineering Council	No exam, apply to the European Federation of National Engineering Association (FEANI).	No exam, register with the Chamber of Engineers

any engineering program in Turkey; there is no exam requirement. In order to reduce future losses and to initiate an accountability system, new policies and initiatives are needed to improve the national professional engineering licensing system.

Conclusions

The disaster life cycle in Turkey has room for growth. This paper outlined disaster occurrences dating back to the year 553, with information regarding human and economic losses. The policy changes following disasters regarding disaster management, including planning, mitigation, response, and recovery were presented in detail. The first disaster management policy dates back to 1500, during the Ottoman Empire, when the Sultan passed a law to assist households that sustained losses with compensation in the form of 20 gold coins. Upon the establishment of the modern Republic of Turkey, the state assistance and insurance program continued to support households that had sustained losses, in the form of providing new homes to replace those that had sustained heavy damage, at no cost to the owner. This policy, part of Law no. 1051, later known as Law 7269, gave the public the understanding that they were not required to prepare for disasters by having their own plan and insurance.

The 1999 earthquake response and recovery proved to policymakers that this government insurance plan needed to be changed, and therefore a compulsory national earthquake insurance program was implemented. The program continues to develop, however: there is a critical need to increase the rate of penetration among the public, perhaps by introducing new policies and initiatives. Another lesson learned included the fact that the quality of the building stock needs improvement. Assessments regarding the current building stock concluded that 17 percent of the building stock could be expected to withstand a future earthquake, but the remaining 83 percent requires retrofit or reconstruction. Therefore, a policy regarding an affordable grant or loan system for homeowners to rehabilitate their homes needs to be considered. Perhaps more importantly, new construction projects need proper engineering and seismic code enforcement. To achieve this, a proper professional engineering licensing system needs to be adopted and implemented in Turkey. This paper outlined professional licensing programs in countries that have fewer losses following earthquake events of similar magnitudes. The 1999 Kocaeli earthquake resulted in 55,062 reinforced concrete buildings sustaining major damage or collapse. The institution of a new professional engineering licensing program will bring accountability and quality control and will minimize losses.

This paper has also outlined all the major events and policy changes in Turkey. The 1999 Kocaeli and Duzce events served to prompt major policy initiatives, but these changes have lost momentum. This case has shown us that nations need to create change and initiate policies as part of mitigation initiatives rather than as a response to a disaster. If a nation selects the latter choice, the losses will continue to grow and the public will continue to be victims instead of survivors. Turkey has shown that when the public sector, private sector, NGOs, and academe come together, positive policy changes can rapidly become reality. This trend needs to be sustained and its achievements applauded in order to ensure a safer future with fewer losses to society as a whole.

Notes

1 Ministry of Development, Turkey (2014), *10th Development Plan for Disaster Management*, Report # KB2888-OIK.731 (Ankara, Turkey: Author).
2 European Parliament (1999), "Economic Consequences of the Recent Earthquakes in Turkey and Greece," European Parliament Document 8594 (Brussels, Belgium: Author).
3 United Nations Office for the Coordination of Humanitarian Affairs (1999), *OCHA Turkey Earthquake Situation Report 23* (New York, NY: Author).
4 Ibid.
5 Ibid.
6 H. Kunreuther (1996), "Mitigating Disaster Losses through Insurance," *Journal of Risk and Uncertainty* 12, pp. 171–87.
7 Munich Re (2014), NatCatSERVICE [Web page], *MunichRe Website*, retrieved January 2, 2016, from http://www.munichre.com/us/weather-resilience-and-protection/rise-weather/natcat-service/index.html
8 Swiss Re (2013), *Mind the Risk: Europe* (Zurich, Switzerland: Author).
9 Ibid.

Bibliography

American Society for Testing and Materials. (2010). *10 Standard Guides for Emergency Operations Center (EOC) Development*. West Conshohocken, PA: Author.
Coburn, A., and R. Spence. (1992). *Earthquake Protection*. West Sussex, UK: Wiley.
———, R. Spence, and A. Pomonis. (1991). *Vulnerability and Risk Assessment*. New York, NY: UNDP Disaster Management Training Program.
Comfort, L., ed. (1988). *Managing Disaster*. Durham, NC: Duke University Press.
———, B. Wisner, S. Cutter, R. Pulwarty, K. Hewitt, A. Oliver-Smith, J. Wiener, M. Fordham, W. Peacock, and F. Krimgold. (1999). "Reframing Disaster Policy: The Global Evolution of Vulnerable Communities." *Environmental Hazards* 1(1), pp. 39–44.
Emergency Management Accreditation Program. (n.d.). *EMAP Website*. Retrieved January 2, 2016, from http://www.emaponline.org/
European Parliament. (1999). "Economic Consequences of the Recent Earthquakes in Turkey and Greece." European Parliament Document 8594. Brussels, Belgium: Author.
Fischer, H.W. (1994). *Response to Disaster*. Lanham, MD: University Press of America.
Foster, H.D. (1980). *Disaster Planning – The Preservation of Life and Property*. New York, NY: Springer.

Ganapati, N.E., and A. Mukherji. (2014). "Out of Sync: World Bank Funding for Housing Recovery, Post-Disaster Planning, and Participation." *Natural Hazards Review*, ASCE 15(1), pp. 58–73.

Gigliotti, R., and J. Ronald. (1991). *Emergency Planning for Maximum Protection*. Newton, MA: Butterworth-Heinemann.

Godschalk, D.R. (2003). "Urban Hazard Mitigation: Creating Resilient Cities." *Natural Hazards Review*, ASCE 4(3), pp. 136–43.

Grand National Assembly of Turkey, the (TBMM). (1999, December 23). *Report of the Research Commission of the Turkish Parliament*, The Grand National Assembly of Turkey, Report No. 10/66, 67, 68, 69, 70.

Gülkan, P. (1998). *The Ceyhan-Mises Earthquake of 27 June 1998: A Preliminary Reconnaissance Report*. Middle East Technical University Disaster Management Implementation and Research Center Report No. METU/DMC 98–01. Ankara, Turkey: METU.

Herman, R.E. (1982). *Disaster Planning for Local Government*. New York, NY: Universe Books.

ISO/PAS. (2007). 22399:2007 Societal Security – Guideline for Incident Preparedness and Operational Continuity Management, International Organization for Standardizations, Geneva, Switzerland.

Joffe, H., T. Rossetto, C. Solberg, and C. O'Connor. (2013). "Social Representations of Earthquakes: A Study of People Living in Three Highly Seismic Areas." *Earthquake Spectra* 29(2), pp. 367–97.

Kasuba, R., and P. Vohra. (2004). "International Mobility and the Licensing of Professional Engineers." *World Transactions on Engineering and Technology Education* 3(1), pp. 43–46.

King, A., D. Middleton, C. Brown, D. Johnston, and S. Johal. (2014). "Insurance: Its Role in Recovery from the 2010–2011 Canterbury Earthquake Sequence." *Earthquake Spectra* 30(1), pp. 475–91.

Kohler, N., J.U. Peter, P. Hausmann, and L. Muller. (2013). *Compulsory Insurance against Natural Hazards in Germany – An Overview of Positions*. Munich, Germany: Swiss Re.

Kunreuther, H. (1996). "Mitigating Disaster Losses through Insurance." *Journal of Risk and Uncertainty* 12, pp. 171–87.

Lindell, M., and C. Prater. (2003). "Assessing Community Impacts of Natural Disasters." *Natural Hazards Review*, ASCE 4(4), pp. 176–85.

Ministry of Development, Turkey. (2014). *10th Development Plan for Disaster Management*. Report # KB2888-OIK.731. Ankara, Turkey: Author.

Munasinghe, M., and C. Clarke. (1994). *Disaster Prevention for Sustainable Development*. Washington, DC: The World Bank.

Munich Re. (2014). NatCatSERVICE [Web page]. *MunichRe Website*. Retrieved January 2, 2016, from http://www.munichre.com/us/weather-resilience-and-protection/rise-weather/natcat-service/index.html

National Council of Examiners for Engineering and Surveying, the. (n.d.). *NCEES Website*. Retrieved May 2014, from www.ncees.org

National Fire Protection Association. (2013). *1600 Standard on Disaster/Emergency Management and Business Continuity Programs*. Quincy, MA: Author. Retrieved January 2, 2016, from http://www.nfpa.org/assets/files/AboutTheCodes/1600/1600–13-PDF.pdf

Nelson, A.C., and S.P. French. (2002). "Plan Quality and Mitigating Damage from Natural Disasters: A Case Study of the Northridge Earthquake with Planning Policy Considerations." *Journal of the American Planning Association* 68(2), pp. 194–207.

Raphael, B. (1986). *When Disaster Strikes*. New York, NY: Basic Books.

Şener, S.M., A. Tezer, M. Kadıoğlu, I. Helvacıoğlu, and L. Trabzon. (2002). *Development of a National Emergency Management Model*. Istanbul, Turkey: ITU Press.

Stephenson, R.S. (1991). *Disasters and Development*. New York, NY: UNDP Disaster Management Training Program.

Swiss Re. (2013). *Mind the Risk: Europe*. Zurich, Switzerland: Author. Retrieved January 2, 2016, from http://media.swissre.com/documents/Swiss_Re_Mind_the_risk.pdf

Turkish Statistical Institute. (n.d.). *Statistics for Turkey*. Ankara, Turkey: Author. Retrieved July 2014, from http://www.turkstat.gov.tr

UDSEP 2023 Turkish Prime Ministry Earthquake Preparedness Strategy. (2013, February). Ankara, Turkey: Prime Minister's Office. Retrieved July 2014, from http://www.deprem.gov.tr/sarbis/Doc/Belgeler/UDSEP2012–2023.pdf

UK Civil Contingencies Act 2004 (c 36). http://www.legislation.gov.uk/ukpga/2004/36/contents.

United Nations Office for the Coordination of Humanitarian Affairs. (1999). *OCHA Turkey Earthquake Situation Report 23*. New York, NY: Author.

Ural, D. (1999). "Disaster Policies and the Economic Impact of Disasters: A Case Study for Turkey." METU Disaster Management Implementation and Research Center Report No. METU/DMC 99–01. Ankara, Turkey: METU.

———. (2006). "International Disaster Management Cooperation: A Case Study for Turkey and the USA." Washington, DC: The Future of Emergency Management, Department of Homeland Security, Federal Emergency Management Agency, and the Emergency Management Institute.

———. (2007a). "The Disaster Management Perspective of the November 20, 2004 Event in Istanbul", Emergency Management Lessons for Homeland Security – Policing Response to Terrorism NATO Security through Science Series E. *Human and Societal Dynamics* 19, pp. 375–82.

———. (2007b). "Disaster Management Education and Policies in Turkey." Integration of Information for Security NATO Science for Peace and Security Series C: Environmental Security, 397–403. Dordrecht, the Netherlands: Kluwer.

US Department of Defense. (2008, July). *United Facilities Criteria 4-141-04*. Washington, DC: Author.

US Federal Emergency Management Agency. (1998). *Project Impact Report*. Release No. 1293–71. Washington, DC: Author.

Wallace, W.A. (1990). "On Managing Disasters." In *Nothing to Fear: Risks and Hazards in American Society*. A. Kirby, ed., pp. 261–280. Tucson, AZ: University of Arizona.

Wright, J., and P. Rossi, eds. (1981). *Social Science and Natural Hazards*. Cambridge, MA: ABT Associates.

15 Main lines of action for seismic risk reduction in developing countries

A case study of Iran

Mahmood Hosseini and Yasamin O. Izadkhah

Introduction

Countries with a high exposure to natural hazards, such as Iran, are more likely to experience disasters in terms of both casualties and economic losses. Many countries well known for seismic hazards, such as Armenia, China, India, Iran, Nepal, Pakistan, and Turkey, simply do not have the infrastructure necessary to reduce vulnerability and harm, making them very vulnerable to natural disasters. A few examples are given in chronological order in the first part of this chapter to reveal the vulnerability of some of these developing countries, including Iran, to natural hazards.

The massive earthquake that struck Spitak, Armenia, on December 7, 1988, took at least 25,000 lives and left 50,000 injured and around 500,000 homeless. Many poorly constructed Soviet buildings across the region sustained serious damage or collapsed. The city of Spitak was severely damaged and the nearby cities of Leninakan (later renamed Gyumri) and Kirovakan (later renamed Vanadzor) sustained substantial damage as well. The tremor also damaged many surrounding villages.[1] In that earthquake, around 314 buildings collapsed and 1,264 required repair or strengthening at a reconstruction cost of 15 billion rubles (at 1989 currency value).[2]

The 1999 Izmit earthquake, also known as the Kocaeli earthquake, struck northwestern Turkey on August 17, 1999, killing around 17,000 people and leaving approximately half a million people homeless.[3] Many, possibly hundreds, of buildings that straddled the fault collapsed because their foundations were torn apart, undoubtedly causing hundreds of casualties or more. In the aftermath of the earthquake, it was found that most of the buildings had not met the design requirements of the country's building code and included details that were not earthquake-resistant. The lack of construction supervision by engineers had allowed for on-the-spot field design modifications and other measures (i.e., no checks or oversight). Many of the buildings had been built with poor and inappropriate construction materials and showed poor workmanship. In addition, many buildings had knowingly been built on active faults and in areas of high liquefaction potential, and most of them were not engineered, but had been built merely following past experience.[4]

In the last two decades, two major earthquakes have also occurred in Iran, including the events in Manjil-Roudbar in June 1990 and Bam in December 2003. The country is located in the Alpine-Himalayan seismic belt, one of the most active tectonic regions in the world, and these two events demonstrated the country's high vulnerability to earthquakes. The Manjil-Roudbar earthquake occurred in Gilan province, claiming about 15,000 lives and injuring more than 30,000.[5] The earthquake also left more than 500,000 people homeless and affected three cities: Roudbar, Manjil, and Loshan.

Almost 700 villages were damaged in the populated areas of the country in the northern and western parts of the Alborz Mountains. The 2003 Bam earthquake occurred in Kerman province, in southeastern Iran. It destroyed around 85 percent of the cities of Bam and Baravat, as well as neighboring villages, with a total population of approximately 143,000 people.[6] It also claimed 33,000 lives and left about 10,000 injured and more than 75,000 homeless (see Figure 15.1).

Figure 15.1(a, b) Bam after the earthquake in 2003.[7]

All these examples show the vulnerability to natural disasters of many developing countries, including Iran. It is therefore necessary to develop and follow risk-reduction programs in these countries, for which the main lines of action are discussed in this chapter. In the next section, the elements of risk reduction and the fundamental concepts of disaster risk management are addressed, and a simple formulation for risk evaluation and reduction is presented by introducing and using the concept of "reduction indices." Then some of the factors that intensify seismic risk in developing countries, including poor quality of construction, lack of sufficient code enforcement and regulations, poverty, insufficient attention of authorities to disaster threats, and finally lack of sufficient awareness and preparedness, are briefly explained, with a particular emphasis on Iran. The main lines of action for seismic risk reduction are then discussed, with special attention to the situation in developing countries. To conclude, the role of management in risk reduction is addressed based on the proposed lines of action and the proposed risk-reduction formulation. The suggested ideas presented in this chapter regarding the main lines of action for seismic risk reduction can help developing countries achieve more effective disaster reduction programs for a safer future.

Elements of seismic risk reduction

Risk (R) is generally defined as the combination of hazard (H) and vulnerability (V):[8]

$$R = H \times V \tag{1}$$

in which H and V, and therefore, R, are all probabilities. This equation can be applied to any system subjected to any kind of hazard on any scale. For example, the system can be a human body subjected to a disease that could easily infect almost all human beings. In order to reduce the risk of human illness caused by that disease, the virus should either be destroyed, to decrease the value of H to zero, or humans should be vaccinated against it, so that the value of V becomes zero. Reducing the risk value to zero, which is the ideal case, is not always possible; accepting some level of risk is inevitable. This is the case for several diseases for which there is no vaccine. In such cases, minimizing the level of exposure to the hazard, for example the exposure of people to the disease, is a good way to reduce the value of H. Fortifying systems which are subjected to hazards, for example strengthening people's bodies against the virus or bacteria by some medical measures, can also reduce the value of V.

Another example to consider is that of a building in an earthquake-prone area. The hazard in this case is the intense ground shaking, which the building may not withstand if its structural system is not strong enough (i.e., is vulnerable). In this case, as the value of H cannot be essentially reduced, the building's structural system should be strengthened to reduce its vulnerability. The last example to consider is that of Earth, which is subjected to the hazard of global warming. Some parts of the Earth have low-latitude coastal areas and are vulnerable to rising sea levels, and thus have higher risk in this respect. In order to reduce the risk, the temperature rise should be prevented or the vulnerable areas should be protected by some coastal structures that can stop water inundating the land.

Based on Equation (1) and the examples presented, it is evident that, to reduce the R value, it is necessary to reduce either the H or the V value. Clearly, these reductions require actions that should be taken by following certain procedures, and these actions and procedures need proper management. Some scholars have considered the effect of management in risk reduction using the following formula:

$$R = \frac{H \times V}{M} \qquad (2)$$

As can be seen in Equation (2), management has been contributed to the risk formula in the denominator to show that better management leads to more reduction in risk. However, Equation (2) has a mathematical deficiency. In fact, in the case of no management, the M value will be 0, which means a mathematically meaningless infinite R value. In order to overcome this problem, it is suggested that, by using the concepts of existing and reduced values for hazard, vulnerability, and risk, as well as the concept of reduction indices, the risk formula is used in a modified form in which the role of management can be included with no mathematical deficiency, as described in the following section.

Formulating risk reduction based on the concept of reduction indices

To overcome the mathematical deficiency of Equation (2) it is suggested that the following fundamental concepts usually used in disaster risk management should be used with some revision, as follows:[9]

- **The Existing Risk Level** (R_e): the level of risk if no risk-reduction measure is employed.
- **The Reduced Risk Level** (R_r): the level of risk after applying risk-reduction measures.
- **The Existing Hazard Level** (H_e): the level of hazard in an area if no hazard-reduction measure is employed.
- **The Reduced Hazard Level** (H_r): the level of hazard in an area after applying hazard-reduction measures.
- **The Existing Vulnerability Level** (V_e): the level of vulnerability of properties (buildings and infrastructures) if no vulnerability-reduction measure is employed.
- **The Reduced Vulnerability Level** (V_r): the level of vulnerability of properties (buildings and infrastructures) after applying vulnerability-reduction measures.
- **Hazard Reduction Index** (I_{Hr}): an index that shows the level of hazard reduction. Its value varies between 0 (no reduction) and 1 (full reduction).
- **Vulnerability Reduction Index** (I_{Vr}): an index that shows the level of vulnerability reduction. Its value varies between 0 (no reduction) and 1 (full reduction).
- **Risk Reduction Index** (I_{Rr}): an index that shows the level of risk reduction. Its value varies between 0 (no reduction) and 1 (full reduction).
- **Reduction Measure**: every action, practice or service which results in reducing the level of hazard, vulnerability, or risk in a community or any part of that community, provided that the "reduction potential" exists for that purpose.
- **Reduction Potential** (P_r): the possibility level of reducing the amount or the probability of any specific hazard or vulnerability level in an area or in a community.
- **Reduction Execution Level** (E_r): the amount of execution of reduction acts by responsible people (either authorities or users) in the community to benefit from

the possible measures of reducing the level of hazard, vulnerability, or risk, and can be defined as the ratio of "the utilized or active capacity" to "the existing capacity."

Based on the above-mentioned definitions, Equation (1) can be rewritten in following forms to state two concepts of "Existing Risk" and "Reduced Risk," which are:

$$R_e = H_e \times V_e \tag{3}$$

and

$$R_r = H_r \times V_r \tag{4}$$

In Equation (3), R_e represents the existing risk level that threatens the community if no risk management program is implemented, which means that there is no reduction in hazard or vulnerability. Alternatively, in Equation (4), R_r represents the reduced risk level that threatens the community after the employment of the "Risk Management Program." Apparently, if risk reduction measures are implemented in the community, the risk can be reduced to some extent, depending on how those measures are executed.

Based on the above definitions, the existing and reduced risk levels can be related to each other using the following simple formula:

$$R_r = R_e \, (1 - I_{Rr}) \tag{5}$$

in which I_{Rr} is the Risk Reduction Index," the value of which can vary between 0 and 1. As described earlier, management is a key factor in order to achieve R_r. In fact, good management is a prerequisite to move from the state of $R_r{=}R_e$ (no risk reduction) to $R_r{=}0$ (full risk reduction). Hence, in the case of "no management" $I_{Rr}{=}0$, and in the "ideal case," $I_{Rr}{=}1$. Similarly, the following equations can be written for relations between the "Reduced Hazard Level" and the "Existing Hazard Level," as well as the "Reduced Vulnerability Level" and the "Existing Vulnerability Level":

$$H_r = H_e \, (1 - I_{Hr}) \tag{6}$$

$$V_r = V_e \, (1 - I_{Vr}) \tag{7}$$

where I_{Hr} and I_{Vr} are, respectively, the "Hazard Reduction Index" and the "Vulnerability Reduction Index" obtained as the product of the "Reduction Potential" and the "Reduction Enforcement Level" or "Reduction Execution Level" in the community, namely:

$$I_{Hr} = P_{Hr} \times E_{Hr} \tag{8}$$

$$I_{Vr} = P_{Vr} \times E_{Vr} \tag{9}$$

In Equation (8), P_{Hr} and *her* are, respectively, the "Hazard Reduction Potential" and the "Hazard Reduction Execution or Implementation Level," and similarly in

Equation (9) P_{Vr} and E_{Vr} are, respectively, the "Vulnerability Reduction Potential" and the "Vulnerability Reduction Execution or Implementation Level" in the community. The concepts of "hazard reduction potential" and "vulnerability reduction potential," whose values are between 0 and 1, are considered here to acknowledge the fact that, in many cases, it is not actually possible to reduce the hazard or the vulnerability.

The values of "Hazard Reduction Execution or Implementation Level" and "Vulnerability Reduction Execution or Implementation Level" are also between 0 and 1, where 0 means the case in which no reduction measure is implemented in the community at all and 1means the situation in which the reduction measures are fully implemented. Combining Equations (3) to (7), one can write:

$$I_{Rr} = I_{Hr} + I_{Vr-} I_{Hr} \times I_{Vr} \tag{10}$$

in which the values of I_{Hr} and I_{Vr} are given by Equations (8) and (9) respectively. With the increase in management level, the indices' values start to move from 0 toward 1. Clearly, in order to have a value larger than 0 for either I_{Hr} or I_{Hr}, both elements of potential and implementation or execution should exist. If the reduction potential mentioned in Equations (8) and (9) is not used, there is actually no management implemented, and therefore there is no difference between the existing risk and the reduced risk. This concept can also be stated using the "Management Efficiency Index" (MEI), which can be defined as the ratio of the "Utilized Capacity" (C_U) to the "Existing Capacity" (C_E):

$$MEI = \frac{C_U}{C_E} \tag{11}$$

In the next section, factors that intensify the seismic risk level in developing countries, with emphasis on Iran, are presented based on which of the above formulations can be more appropriately used in "Disaster Risk Management." As mentioned above, this will serve as a case study.

Main factors intensifying the seismic risk level in developing countries

Factors that intensify the seismic risk level in developing countries are addressed separately below, with emphasis on Iran.

The poor quality of construction

The poor quality of construction is a two-fold issue: (1) the use of poor-quality construction materials; and (2) the improper use of good-quality materials. Poor materials include masonry such as mud, brick, and stone, which have been used in many developing countries. Bam, in southeast Iran, is a good example in this regard (see Figure 15.2).

Inappropriate use of good-quality materials can be due either to inadequately skilled workers or the lack of proper supervision in construction, as another example from a neighboring country shows. Most of the buildings damaged in the Izmit, Turkey, earthquake did not meet the design requirements of the building code and included

Figure 15.2 Using low-quality materials resulted in the vast destruction in the Bam, Iran, earth-
       quake of December 2003.[10]

details that were not earthquake-resistant, including inadequate vertical and horizontal reinforcing steel bars and the widespread use of mildsteelbars.[11]

Failure to sufficiently enforce codes and regulations

With regard to compliance with the regulations and standards of construction, two
aspects should be considered: enforcement by authorities and the compliance of the
community in following regulations. In developing countries, there is evidence of
deficiencies in these two aspects. Unfortunately, the codes and regulations are not
sufficiently inclusive and detailed, and therefore create some ambiguities for users.
The deficiency with regard to the second aspect relates to the lack of people's awareness of regulations and rules due to insufficient education. Examples of both of these
problems can be found in Iran. In developing countries, enforcement and community
acceptance of rules is generally strong in the capital but gradually diminishes when one
examines smaller cities, towns, and villages. For example, the construction deficiencies
illustrated in Figure 15.3 found in the cities of Kashan and Qom in central Iran, both
of which are close to the capital, show that in spite of the experience of destruction in
past earthquakes, drastic mistakes are still made in the construction of buildings which
clearly leave them quite vulnerable to damage in future earthquakes.

Figure 15.3 Examples of deficiencies in building construction in the cities of Kashan and Qom, Iran, 2010.[12]

Poverty

The capacity to survive and recover from the effects of a natural disaster is the result of two factors, the physical magnitude of the disaster in a given area and the socio-economic conditions of individuals and social groups in that area. Vulnerability is differentiated by social groups in almost all natural disasters. Altogether, it is estimated that 90 percent of victims and 75 percent of all economic damage accrues to developing countries.[13] According to Thouret and D'Ercole:

> The relation between socio-economic conditions and the impact of natural disasters can be expressed in this way: due to economic constraints, the poor are forced to live in precarious homes, made of flimsy, non-durable materials, on the least-valued plots of land. They build their shacks on steep hillsides, on floodplains, in fragile ecosystems and watersheds, on contaminated land, right-of-ways and even in the vicinity of active faults as well as other inappropriate areas. Inappropriate location evidently invites serious social and environmental problems. By comparison, the homes of the upper and middle classes are better constructed, built with hardier materials on more stable terrain and their residents enjoy better services; furthermore, they have more resources with which to rebound from disasters.[14]

Insufficient attention of authorities to disaster threats

Another factor that intensifies the impact of disasters in developing countries is the lack of proper governance. This is happening when the number of deaths from earthquakes is increasing as a percentage of total worldwide deaths (10 percent in 50 years).[15] It is clear that governments need to do more to protect their citizens by choosing higher levels of seismic zonation than those currently in use, which are based on 475-year hazard maps.

In fact, in most developing countries where democratic governing systems are relatively new, the governments tend to work on short-term projects that can be completed in a period of a few years. Most risk-reduction programs, however, need plans and programs that require decades to be developed and implemented. As a result, governments usually do not show interest in such long-term plans and activities.

It is also very important for authorities to assign adequate time to address various disasters threatening the community, particularly those disasters that occur rarely, such as major earthquakes and floods. However, sometimes the authorities are mostly absorbed in their daily jobs and are therefore distracted from issues related to disaster management. Furthermore, their daily engagements make it difficult to find appropriate time for training. In this regard, it is imperative that the authorities should be prepared to allocate sufficient time within their routine work to provide training in disaster risk reduction issues, for example, one month in a year.[16]

Lack of enough awareness and preparedness

A disaster risk reduction strategy involving a "seismic safety culture" will not be successful unless it raises public awareness of the earthquake threat and its associated risks. In this regard, it is assumed that in societies where there is a lack of disaster awareness and education, people are less resilient to the consequences of disasters.

In addition, public awareness education training and preparedness is required for poverty reduction and environmental protection. In this regard, efforts towards drawing up national and regional disaster risk management strategies have encountered significant challenges. The problem lies not with the use and adoption of technology in itself, but with the more entrenched culture of resistance to sharing information. Ensuring appropriate technical capacity at the national and local levels should be addressed through public awareness, education, and training in the use of global technical training. For example, although public education about earthquakes began in Iran more than 20 years ago, the level of awareness is still not satisfactory.

The main lines of actions for seismic risk reduction

Based on the issues discussed in the previous sections of this chapter, the main lines of action for seismic risk reduction can be described as follows.

Figure 15.4 depicts the needs and main stages of seismic risk reduction, which include convincing the community, risk evaluation, identifying risk mitigation

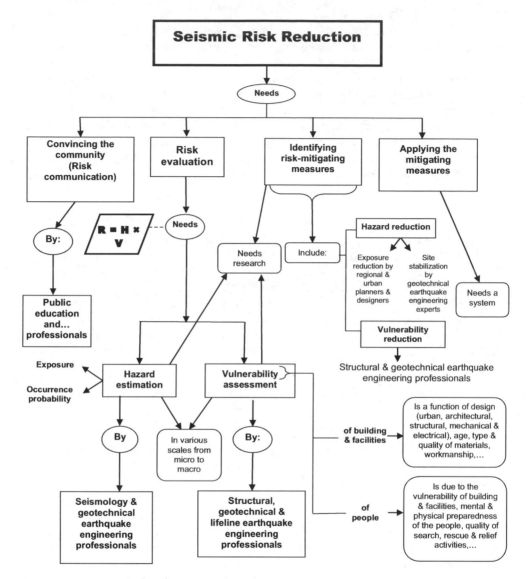

Figure 15.4 Seismic risk reduction needs and stages.

measures, and, finally, applying the measures. Each of these stages and related activities should be done by a group of experts and specialists, as illustrated in the figure.

Furthermore, Figure 15.5 shows the disaster mitigation system proposed by the authors, which is applicable to all communities, including developing countries.

In brief, based on the figures, three main lines of action for seismic risk reduction can be identified as development, retrofitting, and preparedness. These are explained separately below.

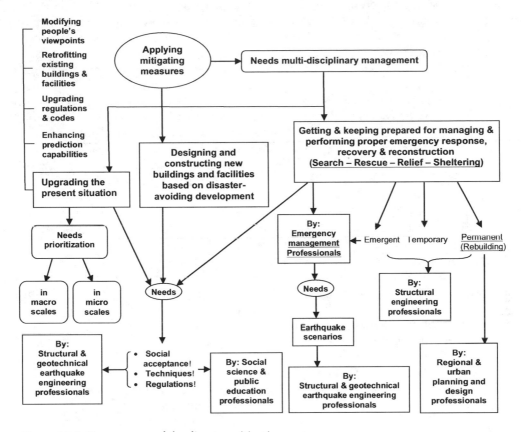

Figure 15.5 Components of the disaster mitigation system.

Disaster-avoiding development

Any development program requires five planning and design stages, including regional (land-use) planning, urban planning and design, architectural design, structural design, and, finally, non-structural design. In this regard, built environments can be categorized as:

- Buildings (with various functions, such as residences, businesses, educational facilities, health and rescue centers, and facilities for social activities);
- Special structures (such as dams, bridges, cooling towers, and industrial structures); and
- Lifelines (including water and wastewater, energy, transportation and communication systems).

For buildings and special structures, seismic design codes suggest calculating seismic hazards using microzonation maps, developed based on the maximum hazard created by all seismic faults. However, for lifeline systems that have extended structures, such as pipelines that extend for kilometers in different areas with varying hazard values, design codes usually do not provide a specified procedure for seismic hazard calculation. Therefore, there is a need for more specific hazard analysis for lifeline systems. Specifically, source-based seismic hazard maps should be developed rather than general maps in order to achieve scenario-based disaster risk reduction. Additionally, another important issue with regard to seismic risk reduction of lifeline systems is looking at them based on their service area[17] rather than provincial division.

Any planning and design task should be based on standards and codes. Most developing countries have adopted codes and standards of developed countries that are not necessarily compatible with the conditions of their countries. On the other hand, all standards and codes refer to some hazard maps in macro and micro scales. Macrozonation maps have been developed for most developing countries for various hazards. However, in developing countries such as Iran, microzonation maps have been prepared for only a few major cities, if any.

Three other important points with regard to disaster-avoiding development are the quality of construction materials, the level of skill in the work force, and the level of supervision. Unfortunately, developing countries, including Iran, suffer from certain shortcomings with regard to all of these issues. In most developing countries, construction is still done using traditional methods, which are time-consuming, costly, and produce low-quality results. Therefore, the industrialization of construction is a must for these countries. Construction techniques and quality have not been improved to keep pace with the development of codes and standards in Iran.

Another issue is the need for reconstruction plans in existing disaster-prone rural and urban areas. In this regard, differences between the reconstruction of an existing damaged city and the development of a new one should be taken into consideration. There are some main issues that should be considered for this purpose, including: present and future population studies; land ownership identification after the event; debris removal planning; identification of and differentiation between those components which should be rehabilitated or rebuilt in their previous locations, those which should be relocated, and those which should be added to the city.[18]

Seismic retrofitting of the built environment

In order to upgrade the built environment, it is first necessary to perform a risk evaluation of vulnerable existing buildings and facilities in various scales, which in turn needs more precise hazard analysis and, accordingly, more reliable vulnerability assessment. Seismic evaluation of more essential facilities requires non-linear dynamic analysis, which is time-consuming and costly, as well as system identification works for more important facilities, which in turn requires both instrumentation and signal processing. The required equipment and tools for this work are usually high-tech and are not easily accessible in developing countries. Also, implementing retrofitting techniques requires high-level expertise, which is usually unavailable in most developing countries.

In addition to financial and technological limitations for seismic upgrading, there is also a time limitation, as the next earthquake could happen at any time. Therefore, it is necessary to perform a prioritization of the existing facilities before making decisions on any retrofit program. Various techniques for quick seismic evaluation, such as using fragility curves, would also help the prioritization process. However, in developing countries, these curves are rarely available because of a lack of the relevant statistics. Therefore, these curves have to be developed based on numerical simulations and non-linear dynamic analysis, which are time-consuming, costly, and require advanced expertise.

The quality of retrofit design and construction is of great importance from both reliability and economy aspects. For some essential facilities such as petrochemical plants, the retrofit should be implemented only in the short time period of a routine plant overhaul. For this reason, both the retrofit method and its implementation technique must be compatible with the speed required by the industrial facility. However, this compatibility needs high levels of construction expertise and technology, which are not available in most developing countries such as Iran.

Upgrading the community's preparedness level

For any disaster threatening a community, people can be divided into three main categories. The first category includes the authorities, which have the services necessary for the implementation and management of the programs. The second group consists of experts and skilled people who have the knowledge to understand the hazard and the capability to plan and implement the reduction measures. The last group includes those people who are affected by the disaster threat and who can be trained for self-help, and at the same time can benefit from the guidance and services of the two other groups. Each of these three categories needs training in order to upgrade the preparedness level necessary to face a disastrous event. For the first group, training emergency managers is very important.[19] For those experts who have knowledge about a specific hazard, usually there is no need for specific training, except for those who have little experience and rarely interact with other groups in the community. In this regard, their interaction with the authorities and with the people needs to be facilitated properly. More focus should be put on the last group – the people – who comprise the largest part of the community.

Each of these groups bears certain responsibilities with regard to disaster preparedness. For the first category, the related responsibilities include:

- Preparing a sufficient number of disaster scenarios with the help of and consultation with the related experts;
- Creating a structured local emergency management in which links between the key persons and between these individuals and government officials are clearly defined;
- Preparing the required detailed action plan for each scenario;
- Providing financial support.

It should be mentioned that for disaster scenarios, the source-specific hazard maps are necessary, and they should be prepared independently for each seismic source.

For the expert groups, the following responsibilities apply:

- Providing the required consultancy to the authorities for preparing the necessary scenarios and corresponding action plans;
- Preparing updated training materials for upgrading the preparedness level for different groups of the community;
- Evaluating the trend of the upgrading process due to changes which may arise in the governing body as well as the population.

For the last group, the responsibilities that can be addressed include:[20]

- Assigning the roles of all individuals in the neighborhood according to their capabilities;
- Participating in neighborhood planning;
- Linking with neighborhood organizaions;
- Planning and performing drills to maintain preparedness; and
- Implementing the allocated tasks and responsibilities during and after the disasters.

It should also be noted that neighborhood councils could be activated in order to achieve successful neighborhood self-help planning.

Pursuing the three main lines of action in parallel, including development, retrofitting, and preparedness, is essential for successful risk management in each country; however, the weight assigned to each of these lines of action may vary from country to country. In developing countries, such as Iran, more weight should be placed on the first and third priorities, which are development and preparedness issues. However, in relatively developed countries, the second line of action should be given priority.

It is also worth mentioning that in developing countries, such as Iran, modifying people's perspectives, which is necessary to convince the community of the need to take action to reduce risk, is much more difficult than in developed countries.

The role of management in risk reduction

The role of management in risk reduction can be clearly seen in Equations (5) to (11) in the above section "Formulating Risk Reduction." The formulas use the concepts of "Reduction Index," "Reduction Potential," and "Reduction Implementation or Execution Level." On this basis, management should be applied in two areas to achieve successful risk reduction:

- Realizing whether there is any "reduction potential" in the case of both hazard and vulnerability, and how this potential can be increased; and
- Planning to increase the "reduction execution level" by considering various scientific, technological, social/cultural, and economic aspects.

The following two examples may help the reader to better understand the intended concepts and formulation.

1 As the first case, consider a road subject to a landslide hazard. If the sliding area is not very wide, it is usually possible to build a retaining wall at the foot of the hill in order to prevent slippage. This means that the potential for reducing the hazard does exist ($P_{hr} = 1$). Additionally, if the retaining wall is constructed with enough strength and resistance against soil slippage, then the reduction measure has been completely executed ($E_{hr} = 1$). Therefore, by using Equation (8), I_{Hr} is equal to 1, and this in turn results in a value of 1 for I_{Rr} by using Equation (10), and finally, the value of R_r will be 0 using Equation (5). This means that, in this case, the ideal or perfect risk reduction has happened.

2   As the second example, think of a community in which 80 percent of its school
    children have been well trained and educated in how to take shelter in the event
    of a disaster, and those children have fully transferred this knowledge to their
    families. This means that 80 percent of the community knows how to take
    shelter. If it is assumed that all building structures in this community are
    earthquake-resistant (only non-structural hazards exist), and that correct shelter-
    ing can reduce the number of casualties by up to 90 percent, the potential for
    reducing the number of casualties is $0.8 \times 0.9 = 0.72$. Now, if only 70 percent
    of the trained people really take shelter correctly, which means that the execu-
    tion level is 0.7, or 70 percent, then the "Risk Reduction Index" for life-loss
    in this community will be $0.72 \times 0.7 = 0.504$, or 50.4 percent. Of course, better
    management of disaster education and public awareness will increase the per-
    centage of trained people and the number of those who find appropriate shelters,
    and thus the risk reduction index will be increased as well.

Obviously, it is not possible to reduce the level of natural hazards in any region on
a large scale. For example, the level of seismicity in a region depends on the activities
of existing seismic faults in that region, and as long as the level of seismicity of those
active faults is high, the level of seismic hazard will not change, and therefore there is
no way to reduce it. However, with the development of new cities and the expansion
of facilities in existing cities, it is always possible to avoid the hazardous areas, at least
for essential facilities. In fact, by selecting appropriate sites for new facilities, it is possible to reduce the level of exposure to seismic hazards, such as ground shaking, fault-
ing, landslide, liquefaction, rock fall, large settlements, tsunami occurrence, and so on.
This may be called "Hazard-Avoidance-Oriented Development and Renovation."

Furthermore, if a distinction is considered between the concepts of "hazard" and "hazard exposure," the usage of a building can contribute to hazard exposure and consequently to "hazard avoidance." For example, consider a critical facility such as a public
center in which a large number of people work and which is located in a zone of high
seismic activity, such as the metropolises of Tehran, the capital of Iran, or Islamabad in
Pakistan. The level of exposure can be decreased in these types of buildings by changing
them from public centers into residential buildings, because, in the latter, fewer people
will be exposed to the hazard. In the case of existing buildings and facilities that do not
undergo a change of function, the reduction potential for hazard or hazard exposure level
is zero. However, in the case of future buildings, the potential can be more than zero, and
even 1.0, if an appropriate development plan is followed based on good management.

With regard to the second area of management application (vulnerability reduction),
it is necessary to upgrade the knowledge level of various groups of people in the community and to provide the required facilities for implementing that knowledge. For
example, in order to realize the role of management in reducing vulnerability, it must
be mentioned that vulnerability reduction deals with issues such as quality of:

- Architectural and structural design of the buildings;
- Design of mechanical and electrical facilities of the building;
- Construction materials;
- Workmanship;
- Technical supervision and inspection;
- Maintenance, and so on.

In most developing countries, including Iran, these aspects are usually not fully considered at all levels of management, and this fact can be one of the main deficiencies hindering the achievement of successful risk reduction programs in these countries.

Concluding remarks

To conclude, it is worth mentioning that governments should provide the opportunity for experts in the social sciences and in psychology to communicate directly and widely with people and to talk to them about disaster issues. Informing people about hazards will encourage them to identify potential hazards and to reduce their social, economic, physical, and environmental vulnerabilities, which in turn will improve their capacities. Furthermore, implementing techniques and facilities also requires advanced expertise that is usually unavailable in many developing countries. To resolve this deficiency, technology transfer from developed countries to developing ones is crucial. Finding the appropriate risk mitigation measures for various natural hazards is a multidisciplinary task which needs thorough joint research work by experts in different disciplines, such as environmental and social sciences. In some developing countries, such as Iran, teamwork is weak due to cultural limitations, and under such circumstances cooperation is not easily achievable. The improvement of teamwork capabilities in the light of these limitations should be reinforced. Finally, with regard to the concept of a "Management Efficiency Index," which can be defined as the ratio of "Utilized Capacity" to "Existing Capacity," recognizing the risk reduction capacities and possibilities is of utmost importance in developing countries.

Notes

1 A.H. Hadjian (1992), "The Spitak Armenia Earthquake: Why So Much Destruction?" in *Proceedings of the Tenth World Conference on Earthquake Engineering*, A. Bernal, ed., pp. 5–10 (Rotterdam, the Netherlands: Balkema).
2 Earthquake Engineering Research Institute (1989), "Armenia Earthquake Reconnaissance Report," Earthquake Spectra Special Supplement (Oakland, CA: Author).
3 "1999 Izmit Earthquake," (n.d.), *Wikipedia*, retrieved June 2013, from http://en.wikipedia.org/wiki/1999_%C4%B0zmit_earthquake
4 "Izmit, Turkey Earthquake of August 17, 1999 (M7.4)" (1999), (Oakland, CA: EQE International).
5 A. Moinfar and A. Naderzadeh (1990), *Technical Report of Manjil Earthquake* (Tehran, Iran: BHRC); M. Ghafory-Ashtiany, M.K. Jafari, and M. Tehranizadeh (2000), "Earthquake Hazard Mitigation Achievement in Iran," paper presented at the 12th World Conference on Earthquake Engineering (12WCEE), January 30–February 4, 2000, Auckland, New Zealand.
6 S. Eshghi and M. Zare (2003), *Preliminary Report of Bam Earthquake* (Tehran, Iran: IIEES).
7 Photos by Izadkhah. Reprinted by permission of the author.
8 D.J. Dowrick (1987), *Earthquake Resistant Design for Engineers and Architects*, 2nd edn. (New York, NY: John Wiley).
9 M. Hosseini and Y.O. Izadkhah (2007), "On the Role of Quality Management in Earthquake Disaster Risk Reduction," in *Proceedings of the 4th International Conference on Excellence Management and Quality Management Systems*, pp. 45–56 (Tehran, Iran: IEEE).
10 Photo by Izadkhah. Reprinted by permission of the author.
11 "Izmit, Turkey Earthquake . . ." (1999).
12 Photos by Hosseini. Reprinted by permission of the author.
13 Thouret and D'Ercole (1996), cited in G.Martine and J.M. Guzman (1999), *Population, Poverty and Vulnerability: Mitigating the Effect of Natural Disasters, Part 1*, retrieved

August 2014, from https://www.wilsoncenter.org/ . . . /Report_8_Martine_%2526_Guzman.pdf

14 Ibid., pp. 48–49.
15 S. Stein, R .J. Geller, and M. Liu (2012), "Why Earthquake Hazard Maps often Fail and What to Do about It," *Tectonophysics*, 562–63, pp. 1–25; M. Wyss, A. Nekrasova, and V. Kossobokov (2012), "Errors in Expected Human Losses Due to Incorrect Seismic Hazard Estimates," *Natural Hazards* 62(3), pp. 927–35.
16 M. Hosseini and Y.O. Izadkhah (2010), "Training Emergency Managers for Earthquake Response: Challenges and Opportunities," *Disaster Prevention and Management Journal* 19(2), pp. 185–98.
17 F. Yaghoobi Vayeghan and M. Hosseini (2009), "Proposing a Quick Seismic Risk Evaluation Method for Inter-City Road Systems," *Journal of Transportation Research* 6(1), pp. 65–86 [Original text in Persian].
18 M. Hosseini (2007), "Reconstruction of Towns and Cities after Destructive Earthquakes: Challenges and Possibilities from the Urban Design Point of View," paper presented at the Ninth Canadian Conference on Earthquake Engineering, June 2007, Ottawa, Canada.
19 Hosseini and Izadkhah (2010).
20 Y.O. Izadkhah and M. Hosseini (2010), "Sustainable Neighborhood Earthquake Emergency Planning in Mega Cities," *Disaster Prevention and Management Journal* 19(3), pp. 345–57.

Bibliography

"1999 Izmit Earthquake." (n.d.) *Wikipedia*. Retrieved June 2013, from http://en.wikipedia.org/wiki/1999_%C4%B0zmit_earthquake

Dowrick, D.J. (1987). *Earthquake Resistant Design for Engineers and Architects*, 2nd edn. New York, NY: John Wiley.

Earthquake Engineering Research Institute. (1989). "Armenia Earthquake Reconnaissance Report." Earthquake Spectra Special Supplement. Oakland, CA: Author.

Eshghi, S., and M. Zare. (2003). *Preliminary Report of Bam Earthquake*. Tehran, Iran: IIEES.

Ghafory-Ashtiany, M., M.K. Jafari, and M. Tehranizadeh. (2000). "Earthquake Hazard Mitigation Achievement in Iran." Paper presented at the 12th World Conference on Earthquake Engineering (12WCEE), January 30–February 4, 2000, Auckland, New Zealand.

Hadjian, A.H. (1992). "The Spitak Armenia Earthquake: Why So Much Destruction?" In *Proceedings of the Tenth World Conference on Earthquake Engineering*. A. Bernal, ed., pp. 5–10. Rotterdam, the Netherlands: Balkema.

Hosseini, M. (2007). "Reconstruction of Towns and Cities after Destructive Earthquakes: Challenges and Possibilities from the Urban Design Point of View." Paper presented at the Ninth Canadian Conference on Earthquake Engineering, June 2007, Ottawa, Canada.

Hosseini, M., and Y.O. Izadkhah. (2007). "On the Role of Quality Management in Earthquake Disaster Risk Reduction." In *Proceedings of the 4th International Conference on Excellence Management and Quality Management Systems*, pp. 45–56. Tehran, Iran: IEEE.

Hosseini, M., and Y.O. Izadkhah. (2010). "Training Emergency Managers for Earthquake Response: Challenges and Opportunities." *Disaster Prevention and Management Journal* 19(2), pp. 185–98.

Izadkhah, Y.O., and M. Hosseini. (2010). "Sustainable Neighborhood Earthquake Emergency Planning in Mega Cities." *Disaster Prevention and Management Journal* 19(3), pp. 345–57.

"Izmit, Turkey Earthquake of August 17, 1999 (M7.4)." (1999). Oakland, CA: EQE International. Retrieved July 2013, from http://www.absconsulting.com/resources/Catastrophe_Reports/izmit-Turkey-1999.pdf

Martine, G., and J.M. Guzman. (1999). *Population, Poverty and Vulnerability: Mitigating the Effect of Natural Disasters, Part 1*. Retrieved August 2014, from https://www.wilsoncenter.org/ . . . /Report_8_Martine_%2526_Guzman.pdf

Moinfar, A., and A. Naderzadeh. (1990). *Technical Report of Manjil Earthquake*. Tehran, Iran: BHRC.

Stein, S., R.J. Geller, and M. Liu. (2012). "Why Earthquake Hazard Maps often Fail and What to Do about It." *Tectonophysics* 562–563, pp. 1–25.

Wyss, M., A. Nekrasova, and V. Kossobokov. (2012). "Errors in Expected Human Losses Due to Incorrect Seismic Hazard Estimates." *Natural Hazards* 62(3), pp. 927–35.

Yaghoobi Vayeghan, F., and M. Hosseini. (2009). "Proposing a Quick Seismic Risk Evaluation Method for Inter-City Road Systems." *Journal of Transportation Research* 6(1), pp. 65–86. [Original text in Persian].

16 Pairing aid systems for disaster management

Case studies of China and Japan

Xuepeng Qian, Weisheng Zhou, and Kenichi Nakagami

Introduction

The history of human society is also the history of human beings fighting various natural disasters. Lots of wisdom has been accumulated from those experiences for disaster mitigation, relief, and recovery. In particular, the technologies and ideas for mitigating disasters such as earthquakes and flood have been developed to a certain level and have worked very well in reducing damage. However, the need for disaster mitigation keeps changing due to the high level of uncertainty in relation to disasters.

In recent years, we have had to face not only natural disasters but also large-scale disease epidemics, chemical pollution, and even terrorist incidents, creating an era of multiple hazards. Some trends can be found in the disasters of the last several decades. Firstly, the frequency of disasters, especially those recognized as low probability/high consequence (LPHC) are increasing. Some seismic zones have entered active periods, as evidenced in several high-magnitude earthquakes. Climate change is causing more extreme weather events, such as hurricanes and floods. Public health has been threatened by new types of viruses and diseases. Terrorist incidents have become an emergent issue for societal security in many countries. The annual numbers of typhoons in or close to Japan have shown an increasing trend, as well as storms with a volume of rainfall over 100mm, which have increased from an average of 2.3 times per year between 1986 and 1995 to 4.7 times per year between 1996 and 2005.

Secondly, disasters or hazards are becoming more and more complex, influencing larger and larger areas. The Great Hanshin Earthquake in 1995 and the Great Sichuan Earthquake in 2008 caused thousands of deaths and extensive damage to both China's infrastructure and society as a whole, and traces of this damage are still evident today. The Flores Earthquake in 1992 and the Sumatra Earthquake in 2004 each induced a tsunami, the latter of which caused more deaths and damage than the earthquake because little attention had been paid to the possibility of a tsunami and preparation had not been made to deal with such an event at that time. In 2011, the Great East Japan Earthquake also produced a deadly tsunami, and attracted worldwide attention and discussion not just because of the high magnitude of the earthquake, but also due to the secondary disaster caused by the damaged nuclear power plants in Fukushima. The damage caused by the earthquake and the tsunami was short-term, but the influence of the nuclear disaster lasted much longer and forced thousands of families to move from their homes. Finally, the accident influenced many countries to change their strategy on using nuclear power.

More technologies and knowledge have been developed and accumulated after experiencing such disasters. Nagamatsu used a disaster cycle model to explain the four important stages related to disasters: from response at the first period, to recovery, to mitigation, to preparedness in the period following a disaster.[1] In recent decades, infrastructure planning and construction technologies have advanced greatly and taken disaster mitigation into consideration. Also, risk management and research have been implemented by administrations, research institutes, and companies at different levels for disaster preparedness. Disaster prevention capabilities have improved significantly, resulting in relatively low losses when a disaster does occur.

However, the inefficiency of the response and recovery activities led by national governments has always been criticized after each disaster, especially following LPHC disasters such as the Great East Japan Earthquake. Some reasons have been put forward for the poor quality of government response, such as complicated political factors, limited fiscal resources, the lack of capable people in management, and the extent of the damage caused by the catastrophe. In this study, we have examined how the pairing aid system worked in disaster relief and recovery after the Wenchuan Earthquake in 2008, when 85,000 people were reported dead or missing. Within two years, about 90 percent of the infrastructure and residential areas were reconstructed, which has proved the effectiveness and efficiency of the pairing aid system. This study starts with an analysis of the establishment, operation, and evaluation of the pairing aid system based on data and a field survey. The ways of improving and institutionalizing the pairing aid system at different levels and forms in each recovery period are then proposed, and the possibility of implementing the system in Japan is discussed. Some lessons on the pairing aid system are drawn from the case studies in order to enhance our understanding of the essence of the pairing aid system.

The pairing aid system for disaster relief and recovery was established immediately after the 2008 Wenchuan Earthquake. Nineteen provinces and (province-level) cities in relatively developed areas were paired with the disaster areas at the city or town level to conduct relief and recovery activities, not only with regard to money and material possessions, but also in terms of knowledge and mental health aid.

After the Great East Japan Earthquake, a counterpart aid program was also proposed and implemented to a limited extent in Japan. "Counterpart aid" was the official term used in related documents and has the same meaning as "the pairing aid system" in Japanese and Chinese. The northeast areas of Japan were severely stricken in the Great East Japan Earthquake on March 11, 2011, and even larger areas were affected by the Fukushima Nuclear Power Plant Accident in the Tohoku region of Japan. Due to the limitation of the capacity in the affected areas and of the central government, the recovery and reconstruction of northeast Japan progressed slowly, in contrast with the Wenchuan Earthquake. The pairing aid system proved an important factor or solution for disaster relief and reconstruction, as introduced and implemented in the counterpart aid program provided by the Union of Kansai Governments. The possibility of a pairing aid system among the local communities and prefectures was also examined through the case study of Iida-Minamisoma.

The pairing system requires a concrete plan for how to conduct human resources, financial, material, and knowledge support at an early stage, which may change some current thinking in disaster prevention and management. This study aims to contribute some ideas and policy implications for creating a sustainable mutual aid system among provinces and cities all over the country for disaster relief and recovery management, especially in the event of a large-scale disaster happening in the future.

Relief and recovery for complex disasters using the pairing system

The pairing aid system in China

The history of the pairing aid system

The idea of a pairing aid system started from the economic development of the border areas mainly inhabited by minorities in 1950s. After decades of practical operation, the pairing aid system that allows economically advanced provinces and cities to provide help for the economic development of the minority-inhabited border areas was officially established by the central government as a national policy in Statement No. 52, 1979. This policy was later expanded to three patterns corresponding to different objectives: for economic development of the minority-inhabited border areas, for grand infrastructure construction projects, and for disaster relief and recovery.[2]

The pairing aid system for the minority-inhabited border areas has been in operation for more than 50 years and has affected all aspects of society in many different ways. As a result, China's central government has accumulated rich experience in applying the idea of pairing aid by mobilizing the whole country. A representative example of pairing aid for grand infrastructure construction is the huge immigration and the areas affected by the construction of the Three Gorges Dam, which involved 22 provinces and began in 1992, when the central government issued the notification for implementing pairing aid in immigration. The 1976 Tangshan Earthquake was the first trial of the pairing aid system for disaster relief and recovery, and showed the advantage of the system in rescue and reconstruction work.

On May 12, 2008, a strong earthquake measuring 8.0 on the Richter scale struck Wenchuan County in Sichuan Province, China. The earthquake caused immense damage. It was confirmed that more than 69,000 people lost their lives, about 18,000 people were missing, and about 375,000 were injured; 7,789,100 houses collapsed, and a further 24,590,000 houses partially collapsed. Some towns, such as those in Beichuan County, were leveled. The main reason for this destruction was that the towns were located on a geological fault. Furthermore, most of those areas are mountainous regions where economic development was quite delayed. Most of the houses were of poor quality and had low resistance to earthquakes.

The establishment of the pairing aid system after the Wenchuan earthquake

The earthquake severely damaged or destroyed the region's infrastructure, which made rescue activities very difficult. Fifteen national and provincial arterial highways and four railways were disconnected. Lifelines such as electric power, telecommunications, and water supply were cut off in large areas. Public facilities, including governmental offices, hospitals, and schools were also damaged. Most of the industrial sectors ceased to operate, including farmland and agriculture facilities.

Besides Sichuan Province, the other nine surrounding provinces and autonomous regions, including Chongqing City, were affected by the earthquake. The total disaster area reached about 500,000 km². The total affected population exceeded 46.25 million. The total amount of direct damage reached CNY ¥845.1 billion, of which

Table 16.1 The pairing aid system[3]

Supporting areas	Coodinators	Supported areas
Shandong	Development and Reform Commission	Beichuan
Guangdong	Department of Transportation	Wenchuan
Zhejiang	Department of Construction	Qingchuan
Jiangsu	Economic Commission	Mianzhu
Beijin	State Assets Administration Committee	Shifang
Shanghai	Department of Education	Dujiangyan
Hebai	Department of Civil Affairs	Pingwu
Liaoning	Department of Finance	Anxian
Henan	Department of Land and Resources	Jiangyou
Fujian	Department of Labor and Social Security	Pengzhou
Shanxi	Department of Water Resources	Maoxian
Hunan	Department of Agriculture	Lixian
Jilin	Department of Forestry	Heishui
Anhui	Department of Commerce	Songpan
Jiangxi	Department of Culture	Xiaojin
Hubei	Department of Health	Hanyuan
Chongqing	Environmental Protection Bureau	Chongzhou
Heilongjiang	Radio and Television Bureau	Jiange

damage to houses accounted for 27.4 percent. The military, the armed police, and domestic and international rescue teams, including a total of 91,300 medical personnel, were involved in emergency rescue at the response stage. Initially, the military and the armed police forces functioned as the main force in the disaster relief operation under the centralized leadership of state.

On May 19, the pairing aid system was established by the Chinese central government, including 18 pairs in Sichuan Province, one in Gansu Province, and one in Shaanxi Province, as shown in Table 16.1. The pairing aid system is also called the "counterpart support system," "partner support system," or "one-to-one assistance and support system." For each pair, one provincial department or commission was assigned to act as the coordinator, helping to initiate the connection between the supporting and the supported areas.

Each pair was decided by the Chinese Central Government, by matching prosperous towns' economic and technological development level with the extent of the need in the affected areas.[4] For example, Beichuan was almost destroyed in the earthquake and had to be reconstructed at a different location, requiring a new design and planning. The Development and Reform Commission was in charge of such matters, and the reconstruction of Beichuan required a large budget, which could be supported by the most economically advanced province, Shandong. Another severely affected city, Dujiangyan, is a famous tourist city with one world heritage site, which was administered by the Department of Education. Shanghai not only had the capability to finance economic development, but also had the most advanced historic architecture and recovery technologies and teams available. Wenchuan needed to recover its secondary industries, an area in which Guangdong is strong. Road reconstruction in Wenchuan was said to be the most difficult part of recovery, so the Department of Transportation was assigned to the pair. The state assets were affected most in Shifang, so the State Assets Administration Committee and Beijin supported that area. Lixian and Heishui have rich agriculture and forest resources, so Hunan and Jilin Province supported these areas, along with the Department of Agriculture and Forestry. Though the basis for deciding on the pairs has never been disclosed in detail, the considerations mentioned above may explain some of the reasons for particular pairings. The other affected areas received pairing aid within Sichuan Province organized by the provincial government.

Contents of the pairing aid system

Supporting provinces and cities were required to provide support for reconstruction of the supported areas for three years, allocating not less than 1 percent of the local fiscal revenue of the previous year for goods and work operations each year. The aid included reconstruction of residential housing; reconstruction of public facilities such as schools and hospitals; reconstruction of infrastructure such as water supply and treatment; reconstruction of agriculture and related infrastructure; and machinery, equipment, and construction materials. Besides the physical reconstruction efforts, social recovery was also included, such as planning, professional consultation, and training of people and education services. In summary, the relief work was implemented to cover all aspects of material and fiscal resources, manpower, and intelligence. For long-term projects, it was also expected that the economic recovery, including agriculture, industries and commercial activities, would be realized by the support of investment for industries, such as collaborative industrial parks.

The onsite headquarters, also called the Offices for Pairing Aid, were formed by members from both the supporting and the supported areas, with assistance from the coordinator of the province department or commission. The headquarters administered the reconstruction and recovery work based on discussions with both sides. The pairing aid system is illustrated in Figure 16.1 (below).

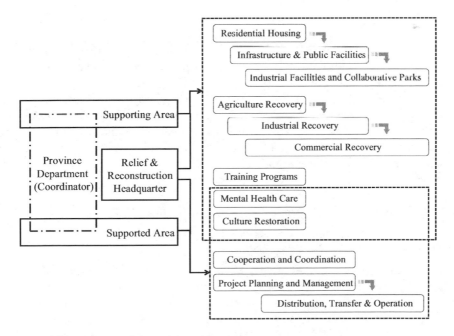

Figure 16.1 The contents of the pairing aid system.

Advantages of the pairing aid system

As reported by China's central government, almost 90 percent of the recovery and reconstruction work had been completed within two years after the earthquake. We visited Dujiangyan City, one of the worst affected areas, in 2008 and 2011 for field surveys. The comparison of the two field surveys also showed that the process of recovery went fairly fast. Dujiangyan received pairing aid from Shanghai. From the interviews and the field surveys, several advantages of the pairing aid system were identified, as explained below.

1   The pairing aid system made the disaster relief and recovery more efficient. Firstly, it established clear accountability for each supporting area or unit, which set up detailed targets and kept to the schedule. Competition resulted in some unnecessary demonstration projects, but also helped to motivate the supporting areas or units. Secondly, the recovery of the area after the devastating earthquake required systematic work covering all the aspects of society, and had to be led by local government. However, most people, including government officials, were unable to cope with routine work after suffering the loss of their families. In fact, it was reported that several officials committed suicide after the catastrophe. The teams sent by the supporting areas were made up of capable officials who could lead reconstruction projects and handle social issues better than the grieving local officials. Thirdly, the pairing aid system ensured that enough goods, machinery, and manpower were available for the relief and recovery work.

2   The pairing aid system helped the transfer of knowledge and technologies. In the Wenchuan case, most supporting areas were more developed areas of China. The local officials we interviewed mentioned that they learned a lot from the supporting officials by working together, such as how to organize and manage projects efficiently. They also brought better technologies and built better facilities. One example explained to us was the wastewater treatment facility constructed by Shanghai, which was much more advanced than Dujiangyan's previous facility. The supporting team also prepared training programs for operation and maintenance. In this way, technological upgrades were realized and the city rebuilt from the ruins was more modern than the former one.

3   The pairing aid system enhanced the connection between the citizens of pairing areas. People in the supported areas are grateful to the supporting areas, and the people in the supporting areas are more concerned with the development of their supported areas. The connection has been enhanced through cooperation and it has attracted more tourists, which also encourages the economy of the supported area. Communication between the paired areas is expected to be long-lasting and to extend to all aspects of society.

Discussion of the pairing aid system in disaster relief and recovery

Fiscal aspect of the pairing aid system

How to obtain enough fiscal resources is always one of the key issues for recovery and reconstruction, and is also closely related to the efficacy of disaster relief. As mentioned before, each supporting area was required to allocate no less than 1 percent of its local fiscal revenue of the previous year for goods and work operations each year. There was some doubt as to whether the requirement would put a burden on supporting areas' economies and whether they would adhere to the requirement. The government fiscal revenues for the years 2007, 2008, and 2009 of each supporting area

Table 16.2 Financial support disbursed through the pairing aid system[5]

Supporting areas	Government revenue 2007 (bn CNY)	Government revenue 2008 (bn CNY)	Government revenue 2009 (bn CNY)	Support (bn CNY)	Supported areas
Shandong	167.5	195.7	219.9	5.8	Beichuan
Guangdong	278.6	331.0	365.0	9.7	Wenchuan
Zhejiang	164.9	193.3	214.3	5.7	Qingchuan
Jiangsu	223.8	273.1	322.9	8.2	Mianzhu
Beijin	149.3	183.7	202.7	5.4	Shifang
Shanghai	207.4	235.9	254.0	7.0	Dujiangyan
Hebai	78.9	94.8	106.7	2.8	Pingwu
Liaoning	108.3	135.6	159.1	4.0	Anxian
Henan	86.2	100.9	112.6	3.0	Jiangyou
Fujian	69.9	83.3	93.2	2.5	Pengzhou
Shanxi	59.8	74.8	80.6	2.2	Maoxian
Hunan	60.7	7203	84.8	2.2	Lixian
Jilin	32.1	42.3	48.7	1.2	Heishui
Anhui	54.4	72.5	86.4	2.1	Songpan
Jiangxi	39.0	48.9	58.1	1.5	Xiaojin
Hubei	59.0	71.1	81.5	2.1	Hanyuan
Chongqing	44.3	57.8	65.5	1.7	Chongzhou
Heilongjiang	44.0	57.8	64.2	1.7	Jiange
		Total Support: No less than 68.7			

are listed in Table 16.2. Supposing that all the supporting areas followed the rule, we calculate that the total support for the three-year recovery and reconstruction plan is CNY ¥68.7bn. In fact, a total of CNY ¥78bn was achieved from 2008 to 2011, even more than had been required. The data shows that the fiscal revenue of each supporting area continued to increase at a fairly high rate for the three years. To understand this mechanism, we need to look into the issue.

Firstly, the monetary flow of the pairing aid system is shown in Figure 16.2. The fiscal resources for pairing aid were not just local government fiscal revenue, but also included a national subsidy, donations, and interest on income. The money was collected and administered by the province's Department of Finance and was transferred to the bank account of field headquarters. It was then used for residential housing projects, public facilities projects, agriculture and related infrastructure projects, and other projects approved by the field headquarters. The national subsidy included a direct subsidy and tax rebates from the central government. In 1994, due to the requirements of market development, China implemented a fundamental fiscal decentralization reform called the tax assignment system reform. From 1994, the central government took more taxes from local governments as national fiscal revenue, especially the consumption taxes (100 percent went to the central government) and value-added taxes (75 percent went to the central government). At the same time, the tax rebate system ensures that the benefit of economic development accrues to the local government. A large proportion of the fiscal resources for pairing aid come from the tax rebate system. Moreover, as most of the goods and materials were produced in the supporting areas and the recovery and reconstruction work was done by teams and

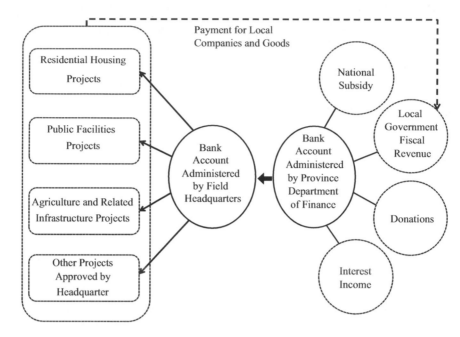

Figure 16.2 The flow of money in the pairing aid system.

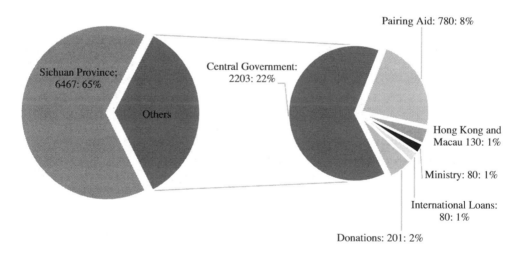

Figure 16.3 The proportion of fiscal support to Sichuan Province (unit: CNY ¥100m).

companies from the supporting areas, parts of the payment finally returned to the supporting areas. In this regard, the mechanism of the pairing aid system is similar to Japan's Official Development Assistance (ODA).

Secondly, the fiscal support of pairing aid comprised a small part of the total fiscal support to Sichuan Province, as shown in Figure 16.3. S. Xiao, the former director of the Policy Research Office of Sichuan Province, explained the situation of fiscal support in May 2011: the fiscal support of pairing aid provided by all the supporting areas made up 8 percent of the total, while the central government contributed 22 percent. Regarding the national subsidy in the pairing aid system, the central government provided most of the direct fiscal support to Sichuan Province. Figure 16.3 shows that Sichuan Province collected CNY ¥646.7 billion, three times the amount provided by the central government, by utilizing policy incentives adjusted by the central government and by loans.

In summary, as in many other countries, China's central government contributed most of the fiscal support directly or indirectly, in addition to devising and supporting the pairing aid system, thus ensuring that the idea of the pairing aid system can be applied if the central government leads and funds recovery activities. An efficient pairing aid system tends to utilize funds more efficiently.

A pair of different scales

Another important factor is the different scales of the supporting area, which is one province or city, and the supported area, which is one city or town, as mentioned earlier. In this study, the difference is studied by comparing the GDPs, per capita GDPs, and populations of each pair, as shown in Table 16.3.

Regarding the GDP for the year 2007, the difference varied from 52 times (Chongqing to Chongzhou) to 1,973 times (Shandong to Beichuan). On average, the scale of the supporting area was 244 times that of the supported area. Regarding the population for the year 2007, the difference varied from 31 times (Shanghai to Dujiangyan) to 1,412 times (Hunan to Lixian). On average, the scale of the supporting area was 146 times that of the supported area. The pairing aid system was between 18 provinces and cities to one province, but the difference in scale was more than 100 times, which made the intensive pairing aid possible in the short term. The pairing aid system is designed to organize the rest of the country in supporting the recovery of affected areas.

Regarding the per capita GDP for the year 2007, the largest difference was 6.1 times (Zhejiang to Qingchuan). On average, the scale of the supporting area was 1.9 times that of the supported area. Another contrast was found among the supporting areas: the financial support from Guangdong was 8 times that of Jilin (Table 16.2). This difference in financial support will probably cause gaps with regard to recovery work in some of the supported areas. Proper communication and coordination among the supporting areas is needed.

Long-term support and evaluation

Pairing aid systems also required the economic cooperation to continue as long-term support to the stricken areas. It was reported that some economically strong provinces sent industry representatives to help with technology and industry transfers, and some new industrial parks were constructed in Sichuan. However, the work lacked a concrete long-term support plan. Consideration was also given to whether long-term support should be included in the evaluation of the work: few positive results could be found in relation to this. Despite the few drawbacks found in the pairing aid system, there is evidence that the system provides a better way to design systems for disaster relief and recovery management, which can be implemented when large-scale disasters happen in the future.

Table 16.3 The contrasts between paired areas[6]

Population (10k)	Capita GDP (CNY)	GDP 2007 (bn CNY)	Supporting areas	Supported areas	GDP 2007 (m CNY)	Scale of GDP	Capita GDP (CNY)	Scale of capita GDP	Population (10k)	Scale of population
9367	27807	2597	Shandong	Beichuan	1316	1973	8598	3.2	16	585
9449	33151	3108	Guangdong	Wenchuan	2877	1080	26204	1.3	10.5	900
5060	37411	1878	Zhejiang	Qingchuan	1378	1363	6107	6.1	24.8	204
7625	33928	2574	Jiangsu	Mianzhu	14252	181	28863	1.2	51.3	149
1633	58204	935	Beijing	Shifang	12728	73	29703	2.0	43.1	38
1858	66367	1219	Shanghai	Dujiangyan	11622	105	18568	3.6	60.9	31
6943	19877	1371	Hebai	Pingwu	1633	839	9366	2.1	18.7	371
4298	25729	1102	Liaoning	Anxian	5073	217	10434	2.5	51	84
9360	16012	1501	Henan	Jiangyou	13844	108	16438	1.0	87.9	106
3581	25908	925	Fujian	Pengzhou	10842	85	14028	1.8	79.5	45
3393	16945	573	Shanxi	Maoxian	1013	566	9512	1.8	10.9	311
6355	14492	920	Hunan	Lixian	633	1453	13245	1.1	4.5	1412
2730	19383	528	Jilin	Heishui	494	1071	8367	2.3	5.9	463
6118	12045	736	Anhui	Songpan	820	898	11596	1.0	7.2	850
4368	12633	550	Jiangxi	Xiaojin	450	1224	5770	2.2	8	546
5699	16206	923	Hubei	Hanyuan	2213	417	6972	2.3	32.1	178
2816	14660	412	Chongqing	Chongzhou	7959	52	12280	1.2	66.5	42
3824	18478	707	Heilongjiang	Jiange	3264	216	5726	3.2	67.5	57
	26069			Average:		244	13432	1.9		146
				Sum:	92411				646.3	
94477		22561		Sum						

Mutual aid and the pairing aid system after the great East Japan earthquake and corresponding policies

Large areas of Northeast Japan were destroyed by the Great East Japan Earthquake and tsunami, and even larger areas were affected by the nuclear power plant accident. Moreover, public works such as roads, bridges, dikes, and railways were completely destroyed. Lifelines such as the water supply system, sewer system, electricity system, and gas lines were also destroyed. The Great East Japan Earthquake caused severe destruction to thousands of householders and to the region. The economic damage to the nation amounted to 3.98 percent of the GDP, slightly less than the cost of the Great Hanshin-Awaji Earthquake in 1995 (4.18 percent). The Great East Japan Earthquake was extreme and unique because of its spatial scale, its complexity, and the resulting widespread social impact. After the earthquake and tsunami, a pairing aid system was proposed by the Science Council of Japan, modeled on the Chinese pairing aid system. The proposal was introduced to and discussed by governments at the prefecture and local levels. Though the nationwide pairing aid system was not adopted in Japan, some good practices relating to mutual aid and pairing aid systems were implemented. In the following section we will discuss three case studies to demonstrate the characteristics of the mutual aid and pairing aid systems adopted in Japan.

Local practices of mutual aid in disaster-affected areas

The administrative response of Tono City (Iwate Prefecture) following the March 2011 earthquake should be remembered as a valuable example of the local practice of mutual aid. Tono City is located in the center inland area of Iwate prefecture (Figure 16.4). In November of 2007, the Sanriku Region Earthquake Disaster Logistic Support Base Provision Implementation Council was established. With the participation and cooperation of 87 prefectural organizations and 8,746 personnel, the advantages and effectiveness of providing logistic support from Tono City was confirmed through training. Furthermore, the strategic concept of the Northeastern Army Disaster Response Training (Michinoku Alert in 2008) for the preparation for an anticipated earthquake and tsunami was formed. In spite of the extensive damage (costing about JPY ¥3.2 bn), Tono City executed the plan under the mayor's strong leadership. At 1:40 am on March 12, 2011, a request for help from Otsuchi Town was received: "There are 500 people seeking refuge at the Otsuchi High School. They have no food or water and need your help." Within a short time, people in Tono procured the necessary materials and the members of the fire department departed for Otsuchi at 4:50 a.m. Otsuchi Town is about 40km away from Tono City; the first group of responders arrived on the morning of March 12, and started rescue activities based on the training they had received on two occasions.

The logistical support project established by relationships between the various providers of assistance were carried out in six projects, including medical, employment, home, community, industry, and mental healthcare networks. These projects overcame several obstacles, such as (1) the large number of appeals for assistance from affected municipalities that are expected to be provided under the terms of the Disaster Relief Act; and (2) the increasing anxiety felt by the affected municipalities in the absence of sufficient information from the national and prefectural governments. Based on the local mutual aid system connected by logistical support, the support was delivered

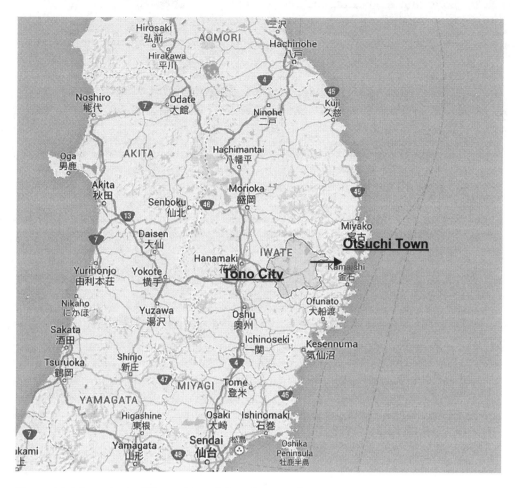

Figure 16.4 Locations of Tono city and Otsuchi town.[7]

immediately to the disaster zone without waiting for appeals. Independent information gathering and other appropriate support activities were conducted.

Pairing aid among local governments

This case study relates to the pairing aid implemented at the local government level. On March 16, 2011, the Iida City Mayor, Mr. Mitsuo Makino, received a telephone call from the Minamisoma City Mayor, Mr. Katsunobu Sakurai, who asked for victims from the stricken city to be accepted in Iida City. Mayor Makino offered informal consent, and after six and a half hours, a formal agreement was issued by Iida City. After another six hours, a transport fleet from Iida left for Minamisoma City to move the victims from the middle to the east part of Japan (as shown in Figure 16.5). One hundred and three victims from Minamisoma City arrived in Iida City at 10.00 pm on March 17. The rescue operation was conducted within two days of the initial request. The long-term friendly relationship between the two cities and

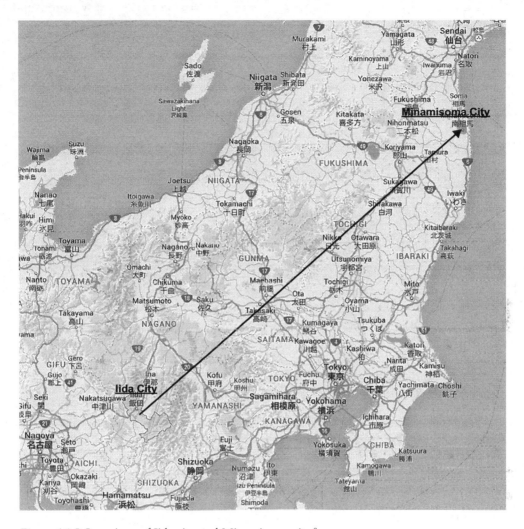

Figure 16.5 Locations of Iida city and Minamisoma city.[8]

the contact between the mayors was regarded as a key factor in making the rescue operation so efficient.

Iida City is not a big city, with a population of about 105,000 people, in 38,000 households. The population ratio of the supporting city to the supported group is 1,019:1, similar to the ratio of the pairing between Guangdong and Wenchuan. The age of the victims rescued ranged from infancy to people in their eighties, so Iida City needed to provide all kinds of public public service resources to support their life in Iida, including education, medical services, social welfare, and even employment. All 103 victims were divided into family groups and were sent to eight different public facilities in five areas of Iida City, with at least one city official taking care of them at each facility. The long-term support did impose a new burden on the administration of Iida City.[9]

According to the Disaster Relief Act, if a group (including administrative group and civic groups) conducts the relief activity at the request of the affected prefecture, it can receive payment from the affected prefecture to cover the expenses incurred, including labor costs, payment for medical services, and so on. The affected prefecture receives a special tax allocation from the national government – that is to say, the financial support can be regarded as being from the national government. Generally speaking, the financial source for disaster relief is guaranteed. This is similar China's pairing aid system: even when the supporting provinces or cities covered the reconstruction costs, they could eventually receive a special tax allocation from the central government to meet the cost. Afterwards, consideration was given to the idea that the Disaster Relief Act should expand its scope to consider formally the direct relief activities at the city level, such as in the Iida-Minamisoma case, which did not follow the prefecture procedure and was contacted directly.[10]

It was reported that Iida City and Minamisoma City signed a Disaster Mutual Aid Agreement on August 7, 2014. The logistic issues and the possibility of another disaster occurring were considered, and the relief activities in 2011 were deemed to be successful. Out of the 103, 18 people from seven families are now living in Iida City.[11] The Disaster Mutual Aid Agreement has become the common method for governments and private enterprises to ensure mutual aid in relief activities. By 2010, 1,571 towns out of 1,750, 20 large cities (including Tokyo), 19 other government ordinance cities, and 26 prefectures had announced such agreements, forming a huge network of mutual aid around the country. In the pairing aid system in China, large-scale provinces or cities supported small-scale cities and reconstruction projects were carried out by different cities from the supporting provinces. This pattern ensured that the aid was from one administrative or operationally unified group, which helped to increase efficiency. Mutual aid in Japan is generally considered for areas of similar scale and at the same administrative level, which is good for ease of communication amongst equals. It will be interesting to seek a suitable pattern for conducting mutual or paired aid.

Paired aid at the prefecture level

Prefecture-level relief activities were also implemented after the Great East Japan Earthquake. We studied the case of counterpart aid from the Union of Kansai Governments. Intended to decentralize power, the Union of Kansai Governments was established by the seven prefectures in the Kansai region, focusing on cooperation in the fields of disaster prevention, promotion of tourism, culture and industries, medical care, preservation of the environment, qualifying examinations and licenses, and training of officials. On March 13, 2011, the Union of Kansai Governments announced a plan for a counterpart aid system to support relief and reconstruction in the Tohoku area. Two or three prefectures were appointed to support one prefecture, as shown in Table 16.4.

The proposal was designed to provide disaster relief measures, supplies, and staff, and to accept victims. Poor communication has always been a problem in relief activities, resulting in lack of clarity about the actual situation in particular localities.[12] The union also required that each supporting prefecture set up a contact office on site to collect the information about the affected area, the logistic situation, and the needs of

Table 16.4 Counterparts of the Union of Kansai Governments[13]

Disaster-stricken area	Union of Kansai governments
Iwate	Osaka, Wakayama
Miyagi	Hyogo, Tokushima, Tottori
Fukushima	Kyoto, Shiga

the stricken areas. Having experienced the Great Hanshin-Awaji Earthquake in 1995 and the slow pace of relief and recovery that followed, the prefectures of the union were capable of providing advice and appropriate measures to the stricken areas. Compared to the pairing aid system in China, the aid intensity and capacity of the counterpart aid system of the Union of Kansai Governments were quite limited with regard to send staff and monetary support. The monetary support, besides the relief expense, was mainly made up of donations from the citizens. For example, Osaka City offered a donation of JPY ¥0.9 bn from its citizens. What is more significant about the relief activities is the long-term plan: it is still an ongoing project and the statistical data are updated every month. In total, 227,600 working days' worth of labor have been provided for the Tohoku area, and 524 households (3,574 people) have been accepted through the counterpart aid system of the Union of Kansai Governments.

Conclusion

This article has explained what was involved in the pairing aid system and its effects in China on disaster relief and recovery after the Wenchuan earthquake. The fiscal aspect and different scales of supporting and supported areas were then discussed to identify the general conditions that may help to promote successful pairing aid systems. China has practiced pairing aid systems for different objectives since the 1950s, and the system has been proved to be effective and efficient on many occasions. There are still many drawbacks to the pairing aid system, such as the difference in the range of support provided and the disparity in scale of pairing partners, the lack of scientific evaluation, and the lack of a concrete long-term support plan. These matters need to be considered in further studies. We also investigated the development of similar pairing aid systems in Japan through the case studies of Tono City supporting Otsuchi Town, Iida City supporting Minamisoma City, and the counterpart aid system implemented by the Union of Kansai Governments after the Great East Japan Earthquake in 2011. From a fiscal perspective, we have found that central and national governments provide substantial support. It may be the case that a huge network of mutual aid has been formed all over the Japan as the result of the acceptance of the Disaster Mutual Aid Agreement; however, the intensity and the scope of mutual or pairing aid in China and Japan are very different. There are many factors that could be used to explain the differences between the Chinese and Japanese experiences, but it is not the purpose of this research to identify the reasons for these differences. A more effective and efficient mechanism should be designed to institutionalize and implement the mutual or pairing aid system. The pairing aid system may be an important solution in disaster relief and recovery management when devastating disasters happen in the future.

Notes

1 S. Nagamatsu (2008), *Introduction of the Policies for Disaster Mitigation* (Tokyo, Japan: Koubundou Press).
2 M. Zhao (2011), "Study on China's Pairing Aid," *Socialism Studies* 2, pp. 56–61.
3 United Nations Centre for Regional Development (2009), *Report on the Great Sichuan Earthquake in China* (Nagoya, Japan: Author), p. 6.
4 J. Xu and C. Hao (2008), "Meta-Synthesis Pattern of Systems Engineering of Corresponding Aided Reconstructions in Post-Wenchuan Earthquake," *Systems Engineering – Theory and Practice* 10, pp. 1–13.
5 *The People's Republic of China Yearbook* (2008), (Beijing, China: National Bureau of Statistics of China); *The People's Republic of China Yearbook* (2009), (Beijing, China: National Bureau of Statistics of China); *The People's Republic of China Yearbook* (2010), (Beijing, China: National Bureau of Statistics of China).
6 Data gathered from the National Bureau of Statistics of the People's Republic of China (n.d.), retrieved November 2014, from http://www.stats.gov.cn/english/
7 Map data ©2015 Google, ZENRIN.
8 Map data © 2015 Google, ZENRIN.
9 T. Mori (2011), "Regional Dispersed Disaster Relief System and Decentralization," *Emergency Response to the Victims of the Great East Japan Earthquake in Minamishinshu: Regional Dispersed Earthquake Relief System* (in Japanese), A. Kamiko, T. Hatori, H. Mori, T. Yoshida, and M. Iwamoto, eds., pp. 27–29 (Kyoto, Japan: College of Policy Science, Ritsumeikan University).
10 A. Kamiko (2011), "Dispersed Disaster Relief System and Financial Security," *Emergency Response to the Victims of the Great East Japan Earthquake in Minamishinshu: Regional Dispersed Earthquake Relief System* (in Japanese), A. Kamiko, T. Hatori, H. Mori, T. Yoshida, and M. Iwamoto, eds., pp. 24–26 (Kyoto, Japan: College of Policy Science, Ritsumeikan University).
11 "Iida City and Minamisoma City Signed a Disaster Mutual Aid Agreement," (2014, August 8), *Minamishinshu News* [Online edition], available at http://minamishinshu.jp/news/
12 N. Funaki, Y. Kawata, and K. Yamori (2006), "Research on Emergency Support System among Prefectural Governments of Large-Scale Disasters: The Case of Niigata Chuetsu Earthquake Disaster," *Journal of Japan Society for Natural Disaster Science* 25(3), pp. 329–49.
13 Union of Kansai Governments Website (n.d.), retrieved January 3, 2016, from https://www.kouiki-kansai.jp/

Bibliography

Bull-Kamanga, L., K. Diagne, A. Lavell, E. Leon, F. Lerise, H. Maskrey, M. Meshack, M. Pelling, H. Reid, D. Satterthwaite, J. Songsore, K. Westgate, and A. Yitambe (2003). "From Everyday Hazards to Disasters: The Accumulation of Risk in Urban Areas." *Environment and Urbanization* 15(1), pp. 193–204.
Database of the Japan Meteorological Agency. (2014). Retrieved December 1, 2014, from http://www.data.jma.go.jp/fcd/yoho/typhoon/statistics
Funaki, N., Y. Kawata, and K. Yamori. (2006). "Research on Emergency Support System among Prefectural Governments of Large-Scale Disasters: The Case of Niigata Chuetsu Earthquake Disaster." *Journal of Japan Society for Natural Disaster Science* 25(3), pp. 329–49.
"Iida City and Minamisoma City Signed a Disaster Mutual Aid Agreement." (2014, August 8). *Minamishinshu News* [Online edition]. Available at http://minamishinshu.jp/news/
Kamiko, A. (2011). "Dispersed Disaster Relief System and Financial Security," In *Emergency Response to the Victims of the Great East Japan Earthquake in Minamishinshu: Regional Dispersed Earthquake Relief System* [in Japanese]. Kyoto, Japan: College of Policy Science, Ritsumeikan University.

Liu, T. (2010). *On the Operation and Legalization of Partner Assistance: An Empirical Study Based on the Post-Disaster Reconstruction of Wenchuan Earthquake.* Beijing, China: Law Press.

Mori, T. (2011). "Regional Dispersed Disaster Relief System and Decentralization," In *Emergency Response to the Victims of the Great East Japan Earthquake in Minamishinshu: Regional Dispersed Earthquake Relief System* [in Japanese]. Kyoto, Japan: College of Policy Science, Ritsumeikan University.

Nagamatsu, S. (2008). *Introduction of the Policies for Disaster Mitigation.* Tokyo, Japan: Koubundou Press.

People's Republic of China Yearbook, The. (2008). Beijing, China: National Bureau of Statistics of China.

———. (2009). Beijing, China: National Bureau of Statistics of China.

———. (2010). Beijing, China: National Bureau of Statistics of China.

Su, M., and Q. Zhao. (2003). *China's Fiscal Decentralization Reform.* Beijing, China: The Research Institute for Fiscal Science, Ministry of Finance, PRC.

Union of Kansai Governments Website. (n.d.). Retrieved January 3, 2016, from https://www.kouiki-kansai.jp/

United Nations Centre for Regional Development. (2009). *Report on the Great Sichuan Earthquake in China.* Nagoya, Japan: Author.

Waller, R., and V.T. Covello, eds. (1984). *Low-Probability High-Consequence Risk Analysis: Issues, Methods, and Case Studies.* New York, NY: Plenum Press.

Xu, J., and C. Hao. (2008). "Meta-Synthesis Pattern of Systems Engineering of Corresponding Aided Reconstructions in Post Wenchuan Earthquake." *Systems Engineering – Theory and Practice* 10, pp. 1–13.

Zhao, M. (2011). "Study on China's Pairing Aid." *Socialism Studies* 2, pp. 56–61.

17 Fukushima dark tourism

Yutaka Sho

Introduction

"Fukushima" has become synonymous with nuclear catastrophe in the aftermath of the 2011 Great East Japan (Tohoku) Earthquake. The disasters in Tohoku included three catastrophic events – the magnitude 9.0 earthquake, the tsunami that reached 130 feet high, and hydrogen explosions at Reactors 1, 3, and 4 at the Fukushima Dai-ichi Nuclear Power Plant (or Fukuichi) – which have altered the region. The effects manifested themselves geologically, economically, socially, politically, and in the built environment, evident in fishing ports, farms, temporary housing, and new construction on higher ground. Nuclear radiation has been the most unpredictable of the three aspects of the disaster, exerting real and imaginary effects of unknown magnitude on the populace. Although Miyagi prefecture is closer to Fukuichi than some parts of Fukushima prefecture, the name "Fukushima" evokes fear of radiation more than any other geographical location. Cultural critics and sociologists have been writing "Fukushima" in *katakana*, one of three syllabaries in the Japanese language reserved for onomatopoeia and foreign words. Reverse-imported after being circulated through the global media, "Fukushima" was stripped of context and content and reduced to a foreign sound. While Tohoku continues its recovery work and battles both the real and imagined contamination, the memory of the earthquake, and the urgency of the debate over nuclear energy are fast fading from the public mind. It is in this context that a group of cultural figures in Japan launched a project to make Fukuichi into a dark tourism destination. Their stated aim is to commemorate the disasters, to challenge the current government reconstruction projects, and to combat public fear of Fukushima.

In hindsight, it was not by chance that radiation threatened Fukushima. I have argued elsewhere that post-WWII neocolonial policies engineered in Tokyo led to the concentration of high-risk nuclear power plants in the Tohoku region.[1] WWII was devastating for Japan: the country lost one-third of its total wealth, up to one-half of its total potential income, and 40 percent of all urban areas.[2] Pressured by immense post-war reconstruction needs, the capital looked to Tohoku and other rural, poor, and aging regions as the suppliers of energy, food, and labor. This policy gained momentum after the failure of the Manchuria colonization, as Japan sought alternative sources of territories and resources.[3] Although secondary compared to their political and economic counterparts, architects and planners played an important role in steering post-war development. Today, five years after the Tohoku earthquake, an interdisciplinary team has been formed to redevelop Fukushima, this time in the form of dark tourism. As the government's post-quake reconstruction plans languish and

174,000 people remain in temporary shelters with little prospect of returning home or relocating to permanent housing, the dark tourism plan to capitalize on the triple disasters is attracting attention.[4] Whether a manifestation of shameless neoliberalism, a pragmatic recovery effort, or conceptual art, the proposal offers an opportunity to reveal and rethink the colonization of rural regions for the benefit of cities.

The discourse around dark tourism was not widely available in Japan prior to the Fukushima "tourization" proposal by the Genron team. Genron (meaning "speech" or "principles" in Japanese, depending on the characters used) is a consortium of cultural critics, artists, and academics. They use Genron Café in Tokyo as their base and to host symposiums, exhibitions, and, more recently, Fukushima and Chernobyl tours. Because of the Genron team's avid use of Twitter and web-based newsletters, in this chapter I have referred to both social media and conventional literature written in Japanese to research Genron's activities, the effects of the disasters, and reconstruction projects. More generally, I have referred to scholarship in museum, heritage, and Jewish Holocaust studies, because dark tourism as a research topic is not widely explored in the design disciplines, despite the fact that many major calamities that have occurred in industrialized countries have memorials and/or museums dedicated to them.

Architecture and planning have become active in post-calamity reconstruction projects in recent years. The Tohoku disasters relief efforts were no exception; hundreds of architects and architecture students contributed.[5] Their projects are often limited to emergency housing and community buildings, however, while typical archival architecture is limited to memorials, exhibitions and museums. Commemorative spaces seldom unite with the everyday.

This existing model will not suffice for Fukushima. Radiation continues to leak into the air, water, and soil today, and the cleanup will continue for years. Safe storage of nuclear waste will be required for far longer. Therefore, any memorial for Fukuichi will not be static, but a living archive that records, retains, and allows access to knowledge about an ongoing crisis. Dark tourism is typically understood as unusual trips to sites of unusual events: tourists "visit sites as innocent outsiders whose actions are believed to have no effect on what they see."[6] Genron's proposal challenges this one-way communication.

The founder of Genron, Hiroki Azuma, is a figure both respected and somewhat disreputable, the latter partly because he is a self-proclaimed *otaku*. Otaku describes social groups and individuals, mostly male, preoccupied with sub-cultures such as anime, video games, and information technology. The otaku stereotype is associated with introverted cliques, collectors of information without a use or value, united in the closed-circuit universe of sub-culture jargons, narratives, and aesthetics.[7] Azuma and his cohorts, however, situate the otaku phenomenon in larger historical and social contexts in their writing, visual, and forum projects, while operating from within the culture. In Fukushima, Genron is criticized for expanding its operation into territories unknown to them, specifically tourism, regional planning, architecture, and infrastructural management. Genron makes it clear, however, that the plan is rooted in a theoretical project. Dark tourism, they assert, was already happening in Fukushima and it was in need of theorization.

Genron's position hovers between that of a reconstruction planner and a theoretical provocateur. At times, they find themselves in trouble for their slippery arguments: when their economic plan is criticized, they take the philosopher's position; when their architectural plan is criticized, they take the economist's position; and so forth. Although their defensive rhetoric is problematic, the dual positioning as a problem solver (and therefore a realist) and a provocateur (and therefore a theorist) is commonplace in architecture. In ideas competitions, design theses, exhibitions, and other speculative architectural projects, designers give shapes to concepts in ways that are intended to invite discourse on their ambitions, critiques, and hypotheses. Take the Metabolist architects' work from 1960s and 1970s – the Genron team's inspiration – for example. Kenzo Tange's "Plan for Tokyo Bay 1960," which suggested that Tokyo's expansion could float over the sea; Arata Isozaki's "Hiroshima Ruined for the Second Time," a drawing depicting the terminal state of all cities as ruins; Kisho Kurokawa's Helix project, a city plan that proposed the cellular organization of space that could grow and expand infinitely as a biological body; and Fumihiko Maki and Masato Otaka's artificial ground and group form – these are some of the theoretical projects that have left a lasting impact on the architectural discipline globally, both in terms of built form and design methodologies. All of them respond to and transcend the historical and immediate context relevant to the built environment of the day. Similarly, Genron presents the Fukushima dark tourism proposal as something to be scrutinized from a multitude of perspectives, including philosophy, economy, and design. Architecture is one of the few media that allow the crossing of these disciplines and the mingling of the real and the fantastical. In this sense the Genron project applies the uniquely architectural interdisciplinary approach to post-disaster reconstruction.

Definitions of dark tourism

The meanings of dark tourism, its history, and implications for society have been well documented, and thus will not be reviewed here.[8] However, the definition published by Foley and Lennon, who originally coined the term, overlaps with the Fukushima case:

> The critical features apparent in the phenomena are, first, that global communication technologies play a major part in creating the initial interest (especially in exploring the territory between the global and the local, thereby introducing a collapse of space and time); second, that the objects of dark tourism themselves appear to introduce anxiety and doubt about the project of modernity (e.g. the use of "rational planning" and technological innovation to undertake the Jewish Holocaust, the industrial scale of death in several wars this century, the failure of "infallible" science and technology at the sinking of the *Titanic*, the impact upon liberal democracy of associations such as those of Kennedy or King, the contradictions apparent in the Cold War tensions and the resolution of these in the progress, rationality and associated so-called meta-narratives); third, the educative elements of sites are accompanied by elements of commodification and a commercial ethic which (whether explicit or implicit) accepts that visitation (whether purposive or incidental) is an opportunity to develop a tourism product.[9]

Global communication technologies, Foley and Lennon's first point, shaped how the Tohoku disasters came to be known, seen, and understood. The Internet carried the pictures of tsunami destruction and hydrogen explosions worldwide, and their effects continue to shape the image of Fukushima and its people. Genron's tourization proposal has been publicly debated among experts and non-experts on social media, attracting much interest. To the researchers' second point, nuclear power has been considered an "infallible" modern technology; hence the absence of emergency protocols in Fukushima. The accident was not supposed to happen.[10] Another example of modernity, neocolonialism used Fukushima to produce electricity for Tokyo and left it with unknown risks while its colonists remains unscathed.[11] Third, visits to Fukushima tourist sites will likely be accompanied by memorabilia that evoke the "consumerism of trauma, fear, and security, and the closely woven relationship of loss to tourism and kitsch,"[12] as are visits to 9/11, the site of the Oklahoma City bombing, and WWII memorials.[13]

Ning Wang, a sociologist from Zhongshan University in China, further elaborates on the types and effects of dark tourism. Wang states that tourists pursue authentic leisure experiences as a reaction to their presumed inauthentic modern life and its routines. Furthermore, Wang analyzes authenticity in tourism in three categories: objective, constructed, and existential.[14] Earlier examples of dark tourism, whether a pilgrimage to a mummified saint, witnessing a public hanging, or visiting a morgue, offered direct exposure to the dead or sites of death. Such tourism offers objective authenticity, "real" objects and sites. Modern dark tourism, on the other hand, offers experiences a degree removed from the actual events, temporally and spatially, in the form of memorials, museum exhibitions, simulations, and reenactments. The numerous memorials and museums for the Jewish Holocaust in the US are such examples. Away from the actual sites of atrocities, they may nonetheless offer symbolically authentic experiences through constructed narratives and aesthetics.

The recent popularity of war tourism to Iraq, Afghanistan, and Ukraine to witness ongoing conflicts provides another perspective.[15] "Adventure tourism" grew by 65 percent from 2009 to 2012 in Europe, the US, and South America, and was valued at $263 billion in 2013.[16] For war and adventure tourists, objective authenticity may be only a backdrop. It certainly helps to heighten the suspense if the conflicts are real, and it helps if the choreography of the experience is well-designed and managed; but the main product of the tour is the experience that inspires feelings that one is alive and true to oneself.[17] The tourists value the real feelings that the thrill affords – what Wang calls existential authenticity.[18] In Japan, the search for existential authenticity appeared as a post-WWII social environment that reemerged in the wake of the Tohoku earthquake, providing the context for Genron's dark tourism proposal.

Life as dark tourism

In a society where 95 percent of its population has smartphones, the disconnect between cyberspace and physical neighborhoods is not science fiction.[19] *Ria-ju*, the abbreviation of *rialu jujitsu* (satisfying reality), has become a popular term in Japan in the last decade. The term enviously describes people who have real (human) friends and live satisfying lives outside of cyberspace. Isolation is a product of modern life in Japan as in the West, with heightened intensity due to the post-war pressure for economic growth. Reading testimonies from post-war and the post-Rapid Economic Growth Era attests that the proliferation of portable Internet technology merely expedited and exacerbated the isolation; it did not cause it. In the 1993 national bestseller *The Complete Manual of Suicide*, Tsurumi described the mood:

> Toward the end of the 80s, there was "the end of the world" boom. *A Dangerous Story* became . . . common sense, the most popular band sang about Chernobyl,

and avant-garde girls began looking for cronies in preparation for the Armageddon. We were ecstatic, thinking "The big one is coming!" "Maybe the world would end tomorrow!"

But the world didn't end. Nuclear power plants wouldn't explode no matter how long we waited and rumours of an all-out nuclear war have disappeared too. Like students who protested against the Japan-U.S. Security Treaty in the 60s, the voyeuristic revolutionaries of the 80s experienced a self-inflicted defeat.[20]

Susan Sontag theorized that the popularity of disaster science fiction is for its "beautification of the unbearable humdrum through destruction on one hand and neutralization of fear for real catastrophe on the other."[21] In 2011, the Tohoku triple disasters provided the hope for an Armageddon in the real world worth shifting attention away from the virtual world: a chance of *ria-ju*. Eiji Otsuka, a social critic, anthropologist, and novelist, criticized individuals who stockpiled food and water and evacuated cities that showed no trace of radiation. Those evacuees, mostly from Tokyo, transferred the regional crisis of Tohoku to the entirety of Japan. The regional crisis was now made available for anyone in the nation to consume, to tap into as a source of existential authenticity, so they could feel alive. People were getting high from the disaster, Otsuka said.[22]

Otsuka showed that the Armageddon highs and lows that occurred in the late 1980s and again following the Tohoku earthquake in 2011 were a mere echo of the same sentiments at the end of WWII.[23] In the 1950s, Yukio Mishima, Ango Sakaguchi, and other influential authors expressed the "war high" brought by the proximity to death, as if life itself was dark tourism, and the subsequent "low" when the war ended. The fear of "the unbearable humdrum" of the everyday, its routines and responsibilities, begins as soon as the existential authenticity – or "high" – subsides. Today we know that post-war Japan's everyday was defined by the Rapid Economic Growth Era, which trampled social and personal fulfillment in the name of national development. Tokyo's amnesia with regard to the Tohoku earthquake and apathy toward the cleanup of radiation in Fukushima is a part of the cyclical "highs" and "lows" that affect Japan, a country prone to every possible disaster. The Fukushima dark tourism proposal by a cohort of otaku, who shepherded the Armageddon boom of the 1980s, is born from the experience of the endless everyday.[24] Genron attempts to bring people to the site of a real catastrophe to combat the apathy and amnesia that the burden of the everyday induces, a condition they know well. Yet if the "low" after a catastrophe feels like déjà vu going back to post-WWII, so does Genron's technique to remedy it – the construction of a mega mall.

Fukushima Gate Village

In the fall of 2012, a team comprising an architect, sociologist, artist, journalist, author, software programmer, and a philosopher (Azuma) established the Fukushima Daiichi Tourization Project Research Committee.[25] Their aim was to combat the fading of the Fukuichi disaster from the public memory, and for that purpose, to make issues surrounding Fukuichi "cool" so that people would want to go to see it.[26] The Fukushima Gate Village is proposed as a multi-use, 42-acre building complex that would turn the site of the nuclear disaster into a dark tourism destination and use it as a means to scrutinize the Fukuichi incident. The Genron team hopes to open the

Village in 2036, 25 years after the disaster. Prior to the 2013 publication of *Tourizing Fukushima: Fukuichi Kanko* [Tourism] *Project*, a manifesto and a research report, the Genron team visited the Chernobyl nuclear museum to learn from Fukushima's predecessor, and added a tourism scholar to the committee.[27] The project schedule is based on the 25-year quarantine period before the Chernobyl nuclear plant was reopened to the public. The Fukushima Gate Village site is the former training ground of the national soccer team, 20 km south of Fukuichi. A direct train line connects Tokyo to the complex in two and a half hours.[28] For the Village's architect, Ryuji Fujimura, mobilization and legibility were the main design directives for this ambitious project.

Mobilization

Fujimura has stated that the true work of architects is to mobilize people.[29] He contends that frequent mobilization inspires better communication and more consumption, which Fukushima needs today. As such, the Village's mass-mobilization program targets people in Tokyo: the high-end "Terminal Mall" mega-shopping center, the hotel complex development called "the Airport to the Future," the eco-tourism farm-stay village for urbanites, and the emphasis on the direct train connectivity to Tokyo all count on visitors bringing their disposable income. The disaster "low" and resultant amnesia are symptoms suffered by Tokyo urbanites, not by people in Tohoku. When Genron maintains that Tokyo benefited from the electricity generated by Fukuichi over the years and is responsible for the disasters, along with TEPCO and the government, they are laying the groundwork for the mega-structures, rail lines, and thematic shopping to allow Tokyo to make amends through consumption that benefits Tohoku.

Genron maintains that because dark memorials and museums do not generate sizable profits, to contribute to the local economy, dark tourism will need to incorporate commercial programs into its development. Another justification for locating shopping on a dark site is that leisure activities give relief to emotionally exhausting experiences.[30] As Genron's own research partner Maki Shoji shows, however, tourists to Tohoku are likely to decrease by 15 percent in 25 years due to Japan's depopulation trend. Tourists must be attracted from abroad, and the amount of spending must increase to make up for the anticipated loss. The Village's estimated initial construction budget is USD$460 million, inclusive of infrastructure and site work.[31] Genron's project budget relies on government investment, with future profits allocated to maintenance costs.[32]

Takashi Kiso, the director of the International Casino Institute in Tokyo, disagrees with architect Fujimura's assertion that mobilization equates with consumption.[33] Kiso studied hotel management in the US and worked for Caesars Entertainment Corporation and Flamingo Las Vegas Hotels and Casino before returning to Japan. He pointed out that tourism based on natural resources and historical heritage, including dark tourism, is fundamentally free of charge to visitors. A native of Hiroshima, Kiso is familiar with dark tourism and endorses the idea of linking commemoration to tourism. The Hiroshima Peace Center stands on a UNESCO registered site, and in 2013 it attracted 1.38 million visitors.[34] However, its direct income is limited to a few yen from each entry fee, Kiso said, similar to what would happen in Fukushima even if Genron could mobilize people to go to visit. Culturally mobilizing people and persuading them to consume are separate challenges.

Instead of investing heavily in infrastructure and mega-structures, Kiso recommends that Genron "tourize" radiated Fukushima as the negative capital that it is, partnering with local hotels and transportation companies and using otaku storytelling skills to mobilize Tokyo.[35] His argument is supported by Shoji's finding that 60 percent of foreign tourists showed an interest in dark tourism in the Tohoku area.[36] Kiso also points out that as the project caters to visitors from Tokyo, and a financing plan relies on federal tax reserves, the project takes resources away from Fukushima's recovery.

The proposed leisure structures in the Village bring to mind the pre-earthquake civic buildings in Tohoku built with the federal government's gift money. During the heyday of nuclear power plant construction, federal policies called the Three Laws of Electric Power Sites subsidized the construction of oversized stadiums and city halls in depopulating and aging towns as indemnity for living with nuclear risks.[37] The Three Laws used expensive modern buildings and infrastructural projects as a political tool to appease host neighborhoods. In not heeding local needs, the Fukushima Gate Village's scale and the program are strikingly reminiscent of the government's post-war approach to development.

Legibility

To create a spatial narrative that is easily legible, Fujimura uses an axial system to organize the Fukushima Gate Village plan.[38] The north-south axis symbolically connects the train station in the center of the Village to Fukuichi and the Asian continent beyond. The axis is meant to communicate that the Village is the entrance gate to the Asian disaster zone.[39] The train station is covered by a roof styled after the Tumulus Period construction, which originated in the third century. Fujimura was inspired by Metabolist Kenzo Tange's use of the same motif for the Hiroshima Peace Center, built in 1954.[40] Other buildings in the Village are modeled after Izumo Taisha, one of the most important Shinto temples in the country. The buildings are for the Fukuichi Disaster Museum, National Museum of Emerging Science and Innovation, and the Tohoku University Graduate Research School of Recovery. Again, Fujimura was inspired by Tange, who referred to the Ise Shrine and the Heian Period's aristocratic villa style in his 1942 and 1943 winning competition entries.[41] The circular shape that repeats throughout the village plan refer to π (3.14), or March 14th, the day of the first Fukuichi hydrogen explosion. The design and planning of the Village use symbols that are easily accessible and appeal to those seeking to reaffirm Japanese identity after the compounded natural, man-made, and economic disasters.

Big, novel, and controversial proposals may help attract Tokyo's attention and make the Village project stand out among other reconstruction projects. Yet consumer-dependent mobilization only implies Fukushima's further dependence on Tokyo. As Kiso argued, mega-structures and mega-infrastructure take away legibility, focus, energy, materials, and space from Fukuichi's main resource – the lessons of the triple disasters. The Village project's schedule, profit turn-around pressure, and scale are synchronized with Tokyo rather than the fishing, agricultural, and housing concerns of rural Tohoku. It is not that people from fishing and farming villages do not go shopping; rather, locals have been organizing dark tours of the affected areas already (which Genron itself has made use of) without relying on large-scale construction projects.[42] The local dark tourism efforts render the Village project obsolete. Genron's proposal does not take advantage of existing expertise and resources or the agencies of dark memories to make Fukushima an instigator for change. The architect Fujimura's design reduces legibility to the lowest common denominator for recycled nostalgia to induce consumption. To fall back on axiality, ancient temples, geometric metaphors, and a mega-mall scheme for a dark tourism plan makes a point only to miss it.

Fujimura's architectural approach contrasts with Genron's philosophical approach, which is nuanced and complex. Genron's interest in dark tourism is not only for its educational and memorialization value but also for its potential to shift the contemporary debate about nuclear power, which is currently divided between simple pros and cons. In both Fukushima and Chernobyl, there are hundreds of workers and thousands of neighborhood residents who rely on the nuclear industry for their livelihoods. Despite the risk, they wait for the day the plants restart so that they can resume their normal lives. Some affected residents resent the anti-nuclear activists, who paid them no attention prior to the disasters but today act as their righteous saviors. They resent those who pity them, glorifying their sacrifices as if to elevate victims to martyrdom.[43] Genron's project has been providing a forum to air many of these issues.

The role of narrative in dark tourism is one such issue that Genron brings to the fore. For an email magazine dedicated to the Fukushima tourization project, a contributor with the username @teramat wrote about his visit to the Yasukuni Temple in Tokyo as a dark tourism case study.[44] Yasukuni is where war criminals – or war heroes, depending on the interpretation – are worshipped. It routinely becomes the target of domestic and international criticism when politicians, especially the heads of state, officially visit. In the last gallery of Yūshūkan, the Yasukuni's exhibition hall, "Gods of Yasukuni" shows walls full of photographs of fallen soldiers and their mementos. @teramat wrote that the patriotic message from the previous galleries becomes internalized in the viewer's mind and makes numerous soldiers appear content to die as heroes for their country. The hall is devoid of images of death, either the soldiers' or their victims'. By eliminating dead bodies and beautifying the war, the exhibition also eliminates the choices that the soldiers may have had.[45] @teramat's account is less a critique of inaccurate narratives or the harmful war ideology than of its singularity. The patriotic narrative deprives the soldiers of the lives "that could have been" even after their deaths, for their deaths have been rigidly labeled by the exhibition as honorable sacrifices.[46]

This desire for simple narratives is not unique to Yasukuni. The controversy surrounding the failed 1995 Smithsonian exhibition on Hiroshima and Nagasaki could serve as an American equivalent. In the exhibition titled "The Last Act: The Atomic Bomb and the End of World War II," the Smithsonian was going to spend ten years and one million dollars to restore Enola Gay, the plane that dropped the A-bomb over Hiroshima, killing 230,000 people.[47] The nature of the exhibition instigated a prolonged battle between the museum, historians, and peace activists, on one side, and organizations including the American Legion, the Air Force Association, and conservative politicians in Congress, on the other.[48] *The New York Times* predicted that the latter would not be satisfied with anything less than "complete vilification of Japanese and uncritical glorification of the American war effort."[49] More than 60 historians signed the petition to realize the exhibition as an open forum, which would offer an opportunity to analyze the justifications for the bombings and their intended outcomes, the effects of radiation on victims, the number of casualties, and numerous other issues over which historians themselves had not reached consensus.[50] The exhibition was never realized, however, because 81 members of the House of Representatives demanded the cancelation of the exhibition, the director's resignation, and the rewriting of history.[51] They did, however, spend ten years and one million dollars to restore the Enola Gay, which today is seen by the American public as having saved many lives by swiftly ending the war.[52]

Similar to the Smithsonian, a federally funded museum, the location of the "Gods of Yasukuni" exhibition leaves little choice but to conform to an official war narrative. The Yasukuni Temple was under the direct administration of the imperial military until the end of WWII and remains an independent religious entity outside the domain of the Association of Shinto Shrines. This arrangement follows from the understanding that the temple will be returned to the federal government someday, despite the post-war "Shinto Directive" that abolished the state Shinto religion.[53] At Hiroshima, Nagasaki, and Okinawa, people who were away from Tokyo's command headquarters may be more likely to support the inclusion of their narratives. For instance, the Himeyuri Peace Museum in Okinawa commemorates the infamous case of the deaths of female high school students and their teachers who were abandoned by the Japanese government in the face of the US invasion. The museum represents various voices of the government officials, military, the emperor, and the survivors, who until recently worked as guides at the museum to share their stories with visitors. Yet at Yasukuni, such autonomy is unattainable.

Conversely, there are many museums and memorials that stand at the site of the event being memorialized, commemorating diverse victims yet offering singular narratives, such as the 9/11 Memorial in New York City, the genocide museums and memorials in Rwanda, and many World War memorials. Others located away from the sites of the event, such as the New England Holocaust Memorial in Boston and the Vietnam War Memorial in Washington, DC, manage to open up the dark event to a multitude of interpretations during and after the design processes.[54] The siting of a memorial, therefore, has little to do with guaranteeing or denying open discourse. The United States Holocaust Memorial Museum in Washington, D.C., uses the opposite curatorial technique from that used at Yasukuni, showing an abundance of dead bodies, yet it achieves the same effect of providing a singular legibility.

In his interview with journalist Philip Gourevitch, the US Holocaust Museum's former project director Michael Berenbaum explained the museum's mission:

> to memorialize the victims of Nazism by providing an exhaustive historical narrative of the Holocaust; and, at the same time, to present visitors with an object lesson in the ethical ideals of American political culture by presenting the negation of those ideals. Berenbaum has coined a phrase to describe the latter part of this mission. He calls it "The Americanization of the Holocaust."[55]

At the US Holocaust Museum, the negation of the American ideal – life, liberty, and pursuit of happiness – translates to the exposition of violence, pain, hunger, shame, hopelessness and the murder of Jews in both still and moving images.[56] The Museum reveals the limit of defining an idea by its negative. In his 1795 secret paper titled "Perpetual Peace," Immanuel Kant asserts that, in order to construct perpetual peace, "the state of peace must be established; for mere cessation of hostilities furnishes no security against their recurrence."[57] From Kant's writing it could be understood that peace, and the desirable quality of life it may guarantee, cannot be defined by the absence of hostilities, violence, and wars. Likewise, Gourevitch stresses that "political and ethical madness, however methodical, teaches nothing about political and ethical sanity. Sanity cannot be asserted by its negative."[58] If this is the case, the essence of dark tourism is called into question, given its stated goal of eternally memorializing violence and insanity. Following Kant, Gourevitch, and other peace scholars, if the goal of dark tourism is to avoid the repetition of the same mistakes by passing on the lessons of history to future generations, its role must be to create an affirmative language that defines peace and to create spaces and experiences where this language can be spoken, heard, and performed. As that language does not yet exist, dark tourism's role may be to gather, display, and make participatory the multitude of narratives and their interpretations. As a physical space that could be shared by strangers, it is conceivable that the sites of dark tourism could offer a crossing of narratives to construct new ones.

The Holocaust Museum in D.C. misses this opportunity of providing and creating new perspectives. Entering the museum, each visitor is issued an identity card of someone who experienced the Holocaust. The ID cards claim to chronicle "the experiences of people who lived in Europe during the Holocaust. These cards are designed to help personalize the historical events of the time."[59] As the visitor proceeds through the exhibitions, one's Holocaust double grows, the information is updated, and the visitor learns whether he or she is incarcerated, tortured, killed, or spared. The narratives of the Holocaust victims, however, are fixed in their victimhood. As dead soldiers at the "Gods of Yasukuni," the ID cards deny the possibility of lives "that could have been" and stabilize the Jews as the prey of an industrialized mass violence, dying their miserable deaths repeatedly as videos shown on loop.[60] The experience is like a primitive video game with a single plot that will be exhausted and discarded. Gourevitch found many ID cards from the Holocaust Museum in trash cans around the National Mall. Some victims of Nazism "had survived the war, only to wind up as part of the litter of a Washington tourist's afternoon."[61] The existential authenticity that this type of Americanized Holocaust may bring will expire as soon as the "violence high" subsides.

Genron's Fukushima dark tourism plan does not rely on such negation. Instead, their proposed Regional Municipal Self-Governance Center included in their Gate Village prompts discussions on alternative futures and tries to find an affirmative language of peace. It is regrettable that the architect Fujimura does not elaborate on the spatial ideas for this program beyond showing a few cartoons and precedents of fashionable office spaces.[62] This Center may be a true dark tourism innovation, due to its critique of the central government's colonialist development decisions for Tohoku and its ability to connect a painful heritage directly to discussions of actionable political programs. Beyond the Google-esque office and conference hall typologies, the Self-Governance Center could inspire design that legibly renders its difficulties, opportunities, and urgent necessities.

Dark tourism's fiction

One of the aims of the Fukushima Gate Village plan is to write a convincing fiction of cool Fukushima, in the hope that this will mobilize consumers to come and shop. The project team trusts that consumption will bring development to Tohoku, despite the fact that a similar consumption-based economic policy brought the nuclear power plants there in post-WWII. The Fukushima Tourization plan uses ancient architectural motifs to make legible the shared past: in the process it reduces architecture to a restorative nostalgia to recreate a past that was never shared or even existed.[63] The project's aim, however, is noble; it attempts to combat the apathy and amnesia that the depression in the aftermath of calamities brings by deliberately focusing on the dark memory to attract tourists from Tokyo, whom Fukuichi historically benefitted. The project challenges those far from Fukushima to grapple with today's disjointed society in order to co-write the shared future, which is always aspirational and fictional. Like all fictions, the project requires a scenario in which distant others and their experiences are made imaginably real. In the case of Fukushima, the distant others are those who fell or continue to struggle in the aftermath of the triple disasters. They are also future citizens who may or may not decide to live with nuclear power.[64] Yet the most difficult "other" to imagine may be the one within oneself in the face of ongoing catastrophes. Risks of global economic crises, radiation poisoning, rising sea levels, and sudden earthquakes occurring within our lifetime are very real, yet the details are unknowable, and therefore our future image is a blur. Dark tourism's task is not to bring the image into focus by predicting the future or by reducing the past event into a stabilized narrative. Instead dark tourism may be able to connect existing players – in this case those in Tokyo and in Fukushima – to imagine new narratives from the past, present, and future, through storytelling via physical space.

Notes

1 Y. Sho (2014), "Thick Food: A Risk-Sharing Network for Post-Fukushima Regional Planning," *Food Studies: An Interdisciplinary Journal* 3(4), pp. 19–32.
2 J.W. Dower (1999), *Embracing Defeat: Japan in the Wake of World War II* (New York, NY: Norton), p. 45.
3 H. Kainuma (2011), *Fukushima Theory: How Nuclear Village Was Born* (Tokyo, Japan: Seidosha).
4 The Japanese Government Reconstruction Agency (2016, February 26), *The Number of Refugees*, retrieved March 1, 2016, from http://www.reconstruction.go.jp/topics/main-cat2/sub-cat2–1/hinanshasuu.html
5 There have been numerous publications documenting reconstruction efforts by architects, planners, and students after the Tohoku disasters. See, for example: T. Igarashi (2011), *3.11/ After: Processes Toward Memory and Resurrection* (Tokyo, Japan: LIXIL); M. Nakamura (2012), *3.11: To Create Is to Live* (Tokyo, Japan: WaWa Project); "The Tohoku Earthquake: 50 Days after 3.11," (2011, April), *Shinkenchiku* 86(5), pp. 30–52; "3.11 Great Earthquake: Memories and Prayers in Photographs," (2011), *X-Knowledge HOME* special edition No. 15; Gakugei Publishing Editorial Department (2011), *East Japan Great Earthquake: Toward Reconstructive City Planning* (Tokyo, Japan: Author); H. Naito and K. Hara, eds. (2012), *3.11: Starting over from Point Zero* (Tokyo, Japan: TOTO Gallery Ma).
6 M. Sturken (2007), *Tourism of History: Memory, Kitsch, and Consumerism from Oklahoma City to Ground Zero* (Durham, SC: Duke University), p. 10.
7 M. Osawa (2008), *The Age of Impossibility* (Tokyo, Japan: Iwanami Bunko), p. 87; H. Azuma (2001), *Otaku: Japan's Database Animal* (Minneapolis: University of Minnesota Press).
8 R. Sharpley and P.R. Stone (2009), *The Darker Side of Travel: The Theory and Practice of Dark Tourism* (Bristol, UK: Channel View); C. Strange and M. Kempa (2003), "Shades of Dark Tourism: Alcatraz and Robben Island," *Annals of Tourism Research* 30(2), pp. 386–405; L.J. Lennon and M. Foley (2000), *Dark Tourism: The Attraction of Death and Disaster* (Ann Arbor, MI: University of Michigan); E. Doss (2010), *Memorial Mania: Public Feeling in America* (Chicago: University of Chicago); M.-R. Trouillot (1997), *Silencing the Past: Power and the Production of History* (Boston, MA: Beacon); K. Walsh

(1992), *The Representation of the Past: Museums and Heritage in the Post-Modern World* (London, UK: Routledge); P. Williams (2007), *Memorial Museums: The Global Rush to Commemorate Atrocities* (Oxford, UK: Berg); J.E. Young (1993), *The Texture of Memory: Holocaust Memorials and Meaning* (New Haven, CT: Yale University Press); K. Savage (2009), *Monument Wars: Washington D.C., the National Mall, and the Transformation of the Memorial Landscape* (Berkeley, CA: University of California); D.D. Meringolo (2012), *Museums, Monuments, and National Parks: Toward a New Genealogy of Public History* (Amherst, MA: University of Massachusetts); L.L. Thomas (2014), *Desire and Disaster in New Orleans: Tourism, Race, and Historical Memory* (Durham, SC: Duke University); S.J. Knell, S. MacLeod, and S. Watson (2007), *Museum Revolutions: How Museums Change and Are Changed* (London, UK: Routledge).

9 Lennon and Foley (2000), p. 11.

10 M. Osawa (2011, May), "Possible Revolution 2: Commune of Fellowship and a Pseudo 'Sophie's Choice'," *AtPlus* 8, p. 15.

11 For the relationship between modernity and risk, see U. Beck (1986), *Risikogesellschaft: Auf Dem Weg in Eine Andere Moderne* (Frankfurt am Main, Germany: Suhrkamp).

12 Sturken (2007), p. 4.

13 J. Brown (2013), "Dark Tourism Shops: Selling 'Dark' and 'Difficult' Products," *International Journal of Culture, Tourism and Hospitality Research* 7(3), pp. 272–80; Sturken (2007).

14 N. Wang (1999), "Rethinking Authenticity in Tourism Experience," *Annals of Tourism Research* 26(2), p. 351.

15 N. Paris (2013, December 17), "Dark Tourism: Why Are We Attracted to Tragedy and Death?" *The Telegraph* [Online edition], available at http://www.telegraph.co.uk/travel/travelnews/10523207/Dark-tourism-why-are-we-attracted-to-tragedy-and-death.html; D. Peisner (2012, September), "Vacations in Dangerous Places," *Departures* [Online edition], retrieved January 3, 2016, from http://www.departures.com/articles/vacations-in-dangerous-places; D. Kamin (2014, July 15), "The Rise of Dark Tourism: When War Zones become Travel Destinations," *The Atlantic* [Online edition], available at http://www.theatlantic.com/international/archive/2014/07/the-rise-of-dark-tourism/374432/; W. Coldwell (2013, October 31), "Dark Tourism: Why Murder Sites and Disaster Zones Are Proving Popular," *The Guardian* [Online edition], available at http://www.theguardian.com/travel/2013/oct/31/dark-tourism-murder-sites-disaster-zones

16 George Washington University International Institute of Tourism Studies and Adventure Travel Trade Association (2013, August 28), *Adventure Tourism Market Study 2013*, available at http://www.gwutourism.org/blog/new-study-reveals-rapid-growth-in-adventure-tourism/; K. Monks (2013, February 12), "The New Tourism: Holidaying in a Warzone," *Metro* [Online edition], available at http://www.metro.us/newyork/lifestyle/travel/2013/02/12/the-new-tourism-holidaying-in-a-warzone/

17 Wang (1999), p. 360.

18 Ibid., p. 358.

19 Japan Ministry of Internal Affairs and Communications (2013), *Annual Survey of Contracts and Penetration Rate of Personal Handy-Phone System*, [2013 data], retrieved September 20, 2014, from http://www.soumu.go.jp/soutsu/tokai/tool/tokeisiryo/idoutai_nenbetu.html; Japan Ministry of Internal Affairs and Communications (2015) Information and Communication Technology White Paper: ICT's Past, Present and Future," Section 2, Chapter 3. Tokyo, Japan: Ministry of Internal Affairs and Communications. Retrieved July 11, 2015, from http://www.soumu.go.jp/johotsusintokei/whitepaper/ja/h27/html/nc372110.html.

20 W. Tsurumi (1993), *The Complete Manual of Suicide* (Tokyo, Japan: Ota Shuppan), p. 3.

21 S. Sontag (1966), "The Imagination of Disaster," in *Against Interpretation and Other Essays*, pp. 209–25 (New York, NY: Farrar, Straus and Giroux).

22 E. Otsuka (2011, May), "A Post-War Literature Theory: High and Low," *AtPlus* 8, p. 98.

23 Ibid., p. 101.

24 The term "endless everyday" was coined by Shinji Miyadai in *Live in the Endless Everyday: Ohm Complete Conquest Manual* (Tokyo, Japan: Chikuma Bunko, 1995).

25 H. Azuma, ed. (2013a), *Tourizing Fukushima: The Fukuichi Kanko Project* (Tokyo, Japan: Genron), p. 12.

26 Ibid., p. 14.
27 H. Azuma, ed. (2013b), *Chernobyl Dark Tourism Guide* (Tokyo, Japan: Genron).
28 Azuma (2013a), p. 95.
29 Ryuji Fujimura, in conversation with Hiroki Azuma and Hajime Yatsuka, Ibid., p. 129.
30 Akira Ide, Ibid., p. 156.
31 Ibid., p. 97.
32 H. Azuma and D. Tsuda (2013, November 26, 12:58), *Twitter Debate with Takashi Kiso, Collected on Togetter* [Website], available at http://togetter.com/li/595181?page=2
33 T. Kiso (2014, December 26, 17:53), "Questions for Fukuichi Tourization Project," *Blogos* [Blog], available at http://blogos.com/article/76714/
34 Hiroshima City (2014, April 16), 2014年4月16日 広島平和記念資料館の入館者等の概況について [Visitors to the Hiroshima Peace Memorial as of April 16, 2014], available at http://www.city.hiroshima.lg.jp/www/contents/0000000000000/1397964653428/index.html
35 Kiso (2014).
36 Maki Shoji, in Azuma (2013a), p. 115.
37 Sho (2014).
38 Fujimura, in Azuma (2013a), p. 96.
39 Ibid., p. 122.
40 H. Yatsuka (2011), *Metabolism Nexus* (Tokyo, Japan: Ohmsha), p. 81. The original design was by Japanese-American artist Isamu Noguchi.
41 Ibid., pp. 44, 48.
42 The Futaba Corporation, for instance, gives tours of former evacuation areas, and Genron has been working with them. Futaba Corporation (n.d.), *Company Website*, retrieved January 5, 2016, from http://311futaba.jimdo.com
43 Kainuma (2011), p. 9; D. Tsuda in Azuma (2013b), pp. 66–79; S. Shimizu (2012), *Looking Back at Nuclear Power Plants: What It Means to Live in Fukushima Today* (Tokyo: Tokyo Shinbun), p. 9.
44 Teramat (2014, August 15), "Weekend Philosophy Research Club Activity Log: Tourism Edition #4: Yasukuni Unofficial Worship," *Genron Tourization Mail-Maga* 19, H. Azuma, ed.
45 Ibid.
46 Ibid.
47 E. Olson (2003, August 19), "Enola Gay Reassembled for Revised Museum Show," *New York Times* [Online edition], available at http://www.nytimes.com/2003/08/19/us/enola-gay-reassembled-for-revised-museum-show.html?pagewanted=print
48 M.J. Hogan (1996), "The Enola Gay Controversy: History, Memory and the Politics of Presentation," in *Hiroshima in History and Memory*, M.J. Hogan, ed., pp. 200–232 (Cambridge, UK: CUP), p. 201.
49 "Smithsonian Alters Plans for Its Exhibit On Hiroshima Bomb," (1994, August 30), *New York Times* [Online edition], available at http://www.nytimes.com/1994/08/30/us/smithsonian-alters-plans-for-its-exhibit-on-hiroshima-bomb.html?pagewanted=print
50 Hogan (1996), pp. 204, 225; K. Bird and M. Sherwin (1995, July 31), *Enola Gay Exhibit: The Historians' Letter to the Smithsonian* [Web page], available at http://www.doug-long.com/letter.htm
51 Hogan (1996), p. 224; P. Blute and S. Johnson (1995, January 24), "Letter from 81 Members of the U.S. House of Representatives," *Lehigh University Digital Library*, available at http://digital.lib.lehigh.edu/trial/enola/files/round3/housemembers.pdf
52 Hogan (1996), pp. 215, 225.
53 W.P. Woodard (1972), *The Allied Occupation of Japan 1945–1952 and Japanese Religions* (Ann Arbor, MI: University of Michigan).
54 Young (2011), pp. 323–35.
55 P. Gourevitch (1993/2014), "Behold Now Behemoth," *Harper's Magazine* 287(1718), pp. 55–62.
56 Gourevitch (1993/2014), p. 60 [5].
57 I. Kant (1795/1891; 2010), *Perpetual Peace: A Philosophical Sketch*, W. Hastie, trans. (Philadelphia, PA: Slought Foundation and Syracuse: Syracuse University Humanities Center), p. 12.
58 Gourevitch (1993/2014), p. 7.

59 United States Holocaust Memorial Museum (n.d.), "Identification Cards," *US Holocaust Memorial Museum Website*, retrieved September 30, 2014, from http://www.ushmm.org/remember/id-cards
60 Gourevitch (1993/2014), p. 8.
61 Ibid., p. 8.
62 Fujimura, in Azuma (2013a), p. 102.
63 S. Boym (2002), "Restorative Nostalgia: Conspiracies and Return to Origins," in *The Future of Nostalgia*, pp. 41–48 (New York: Basic Books).
64 M. Osawa (2013), *Solidarity with Future Is Possible, But in What Sense?* Fukuoka U Booklet No. 4. (Tokyo, Japan: Gen).

Bibliography

2011 Tohoku Earthquake Tsunami Joint Survey Group. (2013). "Research Findings." *Coastal Engineering Committee of Japan Society of Civil Engineers*. Last modified September 10, 2013. Available at http://www.coastal.jp/ttjt/

"3.11 Great Earthquake: Memories and Prayers in Photographs." (2011). *X-Knowledge HOME* Special edition No. 15.

"89,000 Still in Temporary Housing 3.5 Years after the Tohoku Earthquake." (2014, September 11). *Asahi Newspaper*. [Online edition]. Available at http://digital.asahi.com/articles/DA3S11344637.html?iref=comkiji_txt_end_s_kjid_DA3S11344637

Azuma, H. (2001). *Otaku: Japan's Database Animals*. Minneapolis, MN: University of Minnesota.

———— ed. (2013a). *Tourizing Fukushima: The Fukuichi Kanko Project*. Tokyo, Japan: Genron.

———— ed. (2013b). *Chernobyl Dark Tourism Guide*. Tokyo, Japan: Genron.

————, and D. Tsuda. (2013, November 26, 12:58). *Twitter debate with Takashi Kiso*. Collected on *Togetter* [Website]. Available at http://togetter.com/li/595181?page=2

Beck, U. (1986). *Risikogesellschaft: Auf Dem Weg in Eine Andere Moderne*. Frankfurt am Main, Germany: Suhrkamp.

Bird, K., and M. Sherwin. (1995, July 31). *Enola Gay Exhibit: The Historians' Letter to the Smithsonian* [Web page]. Available at http://www.doug-long.com/letter.htm

Blute, P., and S. Johnson. (1995, January 24). "Letter from 81 Members of the U.S. House of Representatives." *Lehigh University Digital Library*. Available at http://digital.lib.lehigh.edu/trial/enola/files/round3/housemembers.pdf

Boym, S. (2002). "Restorative Nostalgia: Conspiracies and Return to Origins." In *The Future of Nostalgia*. pp. 41–48. New York, NY: Basic Books.

Brown, J. (2013). "Dark Tourism Shops: Selling 'Dark' and 'Difficult' Products." *International Journal of Culture, Tourism and Hospitality Research* 7(3), pp. 272–80.

Coldwell, W. (2013, October 31). "Dark Tourism: Why Murder Sites and Disaster Zones Are Proving Popular." *The Guardian*. [Online edition]. Available at http://www.theguardian.com/travel/2013/oct/31/dark-tourism-murder-sites-disaster-zones

Doss, E. (2010). *Memorial Mania: Public Feeling in America*. Chicago, IL: University of Chicago.

Dower, J.W. (1999). *Embracing Defeat: Japan in the Wake of World War II*. New York, NY: Norton.

Futaba Corporation. (n.d.). *Company Website*. Retrieved January 5, 2016, from http://311futaba.jimdo.com

Gakugei Publishing Editorial Department. (2011). *East Japan Great Earthquake: Toward Reconstructive City Planning*. Tokyo, Japan: Author.

George Washington University International Institute of Tourism Studies and Adventure Travel Trade Association. (2013, August 28). *Adventure Tourism Market Study 2013*. Available at http://www.gwutourism.org/blog/new-study-reveals-rapid-growth-in-adventure-tourism/

Gourevitch, P. (1993/2014). "Behold Now Behemoth." *Harper's Magazine* 287(1718), pp. 55–62. ProQuest document link, last updated May 21, 2014. Available at http://search.proquest.com/docview/233485499?accountid=14214, p. 1.

Hamada, T. (2013). *Fishing Industry and Disaster*. Tokyo, Japan: Misuzu Shobo.

Hiroshima City. (2014, April 16). 2014年4月16日 広島平和記念資料館の入館者等の概況について. [Visitors to the Hiroshima Peace Memorial as of April 16, 2014]. Available at http://www.city.hiroshima.lg.jp/www/contents/0000000000000/1397964653428/index.html

Hogan, M.J. (1996). "The Enola Gay Controversy: History, Memory and the Politics of Presentation." In *Hiroshima in History and Memory*. M.J. Hogan, ed., pp. 200–232. Cambridge, UK: CUP.

Igarashi, T. (2011). *3.11/After: Processes Toward Memory and Resurrection*. Tokyo, Japan: LIXIL.

Japan Ministry of Internal Affairs and Communications. (2013). *Annual Survey of Contracts and Penetration Rate of Personal Handy-Phone System* [2013 data]. Retrieved September 20, 2014, from http://www.soumu.go.jp/soutsu/tokai/tool/tokeisiryo/idoutai_nenbetu.html

———. (2015). Information and Communication Technology White Paper: ICT's Past, Present and Future," Section 2, Chapter 3. Tokyo, Japan: Ministry of Internal Affairs and Communications. Retrieved July 11, 2015, from http://www.soumu.go.jp/johotsusintokei/whitepaper/ja/h27/html/nc372110.html

Japanese Government Reconstruction Agency, The. (2016, February 26). *The Number of Refugees*. Retrieved March 1, 2016, from http://www.reconstruction.go.jp/topics/main-cat2/sub-cat2–1/hinanshasuu.html

Kainuma, H. (2011). *Fukushima Theory: How Nuclear Village Was Born*. Tokyo, Japan: Seidosha.

Kamin, D. (2014, July 15). "The Rise of Dark Tourism: When War Zones become Travel Destinations." *The Atlantic*. [Online edition]. Available at http://www.theatlantic.com/international/archive/2014/07/the-rise-of-dark-tourism/374432/

Kant, I. (1795/1891; 2010). *Perpetual Peace: A Philosophical Sketch*. W. Hastie, trans. Philadelphia, PA: Slought Foundation and Syracuse: Syracuse University Humanities Center.

Katayama, L. (2009, July 21). "Love in 2-D." *New York Times*. [Online edition]. Available at http://www.nytimes.com/2009/07/26/magazine/26FOB-2DLove-t.html?pagewanted=all&_r=0

Kiso, T. (2014, December 26, 17:53). "Questions for Fukuichi Tourization Project." *Blogos* [Blog]. Available at http://blogos.com/article/76714/

Klein, N. (2007). *Shock Doctrine: The Rise of Disaster Capitalism*. New York, NY: Picador.

Knell, S.J., S. MacLeod, and S. Watson. (2007). *Museum Revolutions: How Museums Change and Are Changed*. London, UK: Routledge.

Komatsu, R. (2014, August 1). "Hama Douri Journal #7: Starting from Fukushima as a Backyard." *Genron Tourization Mail-Maga* 18.

Lennon, L.J., and M. Foley. (2000). *Dark Tourism: The Attraction of Death and Disaster*. Ann Arbor, MI: University of Michigan.

MacCannell, D. (1999). *The Tourist*. Berkeley, CA: University of California.

Melissinos, C., and P. O'Rourke. (2012). *The Art of Video Games: From Pac-Man to Mass Effect*. New York: Welcome Books/Smithsonian American Art Museum.

Meringolo, D.D. (2012). *Museums, Monuments, and National Parks: Toward a New Genealogy of Public History*. Amherst, MA: University of Massachusetts.

Mishima, Y. (1950). *Confession of a Mask*. Tokyo, Japan: Shinchosha.

Miyadai, S. (1995). *Live in the Endless Everyday: Ohm Complete Conquest Manual*. Tokyo, Japan: Chikuma Bunko.

Monks, K. (2013, February 12). "The New Tourism: Holidaying in a Warzone." *Metro*. [Online edition]. Available at http://www.metro.us/newyork/lifestyle/travel/2013/02/12/the-new-tourism-holidaying-in-a-warzone/

Naito, H., and K. Hara, eds. (2012). *3.11: Starting Over from Point Zero*. Tokyo, Japan: TOTO Gallery Ma.

Nakamura, M. (2012). *3.11: To Create Is to Live*. Tokyo, Japan: WaWa Project.

"Nuclear Power Treaty Will Be Signed: Export of Nuclear Technology to Turkey, UAE." (2014, April 4). *Nikkei Newspaper*. [Online edition]. Available at http://www.nikkei.com/article/DGXNASFS0403H_U4A400C1PP8000/

Olson, E. (2003, August 19). "Enola Gay Reassembled for Revised Museum Show." *New York Times*. [Online edition]. Available at http://www.nytimes.com/2003/08/19/us/enola-gay-reassembled-for-revised-museum-show.html?pagewanted=print

Osawa, M. (2008). *The Age of Impossibility*. Tokyo, Japan: Iwanami Bunko.

———. (2011, May). "Possible Revolution 2: Commune of Fellowship and a Pseudo 'Sophie's Choice'." *AtPlus* 8, pp. 4–17.

———. (2013). *Solidarity with Future Is Possible, but in What Sense?* Fukuoka U Booklet No. 4. Tokyo, Japan: Gen.

Otsuka, E. (2011, May). "A Post-War Literature Theory: High and Low." *AtPlus* 8.

Paris, N. (2013, December 17). "Dark Tourism: Why Are We Attracted to Tragedy and Death?" *The Telegraph*. [Online edition]. Available at http://www.telegraph.co.uk/travel/travelnews/10523207/Dark-tourism-why-are-we-attracted-to-tragedy-and-death.html

Peisner, D. (2012, September). "Vacations in Dangerous Places." *Departures* [Online edition]. Retrieved January 3, 2016, from http://www.departures.com/articles/vacations-in-dangerous-places

Richardson, A. (2009, July 27). "Nemutan's Revenge – Some Fact-Checking and Reaction to the NYT Story on Anime Fetishists." *Mutantfrog Travelogue* [Blog]. Available at http://www.mutantfrog.com/2009/07/27/nemutans-revenge-some-fact-checking-and-reaction-to-the-nyt-story-on-anime-fetishists/

Savage, K. (2009). *Monument Wars: Washington D.C., the National Mall, and the Transformation of the Memorial Landscape*. Berkeley, CA: University of California.

"Sendai Nuclear Power Plant Report Will Be Published Tomorrow, Committee: New Codes Satisfied." (2014, September 9). *Asahi Newspaper*. [Online edition]. Available at http://www.asahi.com/articles/DA3S11340714.html

Sharpley, R., and P.R. Stone. (2009). *The Darker Side of Travel: The Theory and Practice of Dark Tourism*. Bristol, UK: Channel View.

Shimizu, S. (2012). *Looking Back at Nuclear Power Plants: What It Means to Live in Fukushima Today*. Tokyo, Japan: Tokyo Shinbun.

Sho, Y. (2014). "Thick Food: A Risk-Sharing Network for Post-Fukushima Regional Planning." *Food Studies: An Interdisciplinary Journal* 3(4), pp. 19–32.

"Smithsonian Alters Plans for Its Exhibit on Hiroshima Bomb." (1994, August 30). *New York Times*. [Online edition]. Available at http://www.nytimes.com/1994/08/30/us/smithsonian-alters-plans-for-its-exhibit-on-hiroshima-bomb.html?pagewanted=print

Sontag, S. (1966). "The Imagination of Disaster." In *Against Interpretation and Other Essays*. pp. 209–25. New York, NY: Farrar, Straus and Giroux.

Strange, C., and M. Kempa. (2003). "Shades of Dark Tourism: Alcatraz and Robben Island." *Annals of Tourism Research* 30(2), pp. 386–405.

Sturken, M. (2007). *Tourism of History: Memory, Kitsch, and Consumerism from Oklahoma City to Ground Zero*. Durham, SC: Duke University.

Takagi Committee for Fishing Industry Improvement. (2007, July 31). "Recommendation for Expedited Implementation of Fundamental Improvement Strategies for the Fishing Industry to Protect Fish Consumption Culture." *Japan Economic Research Institute*. Available at http://www.nikkeicho.or.jp/result/水産業改革高木委員会　提言発表/

Teramat. (2014, August 15). "Weekend Philosophy Research Club Activity Log: Tourism Edition #4: Yasukuni Unofficial Worship." *Genron Tourization Mail-Maga* 19, H. Azuma, ed.

Thomas, L.L. (2014). *Desire and Disaster in New Orleans: Tourism, Race, and Historical Memory*. Durham, SC: Duke University.

"The Tohoku Earthquake: 50 Days after 3.11." (2011, April). *Shinkenchiku* 86(5), pp. 30–52.

Trouillot, M.-R. (1997). *Silencing the Past: Power and the Production of History*. Boston, MA: Beacon.

Tsurumi, W. (1993). *The Complete Manual of Suicide*. Tokyo, Japan: Ota Shuppan.

United States Holocaust Memorial Museum. (n.d.). "Identification Cards." *US Holocaust Memorial Museum Website*. Retrieved September 30, 2014, from http://www.ushmm.org/remember/id-cards

Walsh, K. (1992). *The Representation of the Past: Museums and Heritage in the Post-Modern World*. London, UK: Routledge.

Wang, N. (1999). "Rethinking Authenticity in Tourism Experience." *Annals of Tourism Research* 26(2), pp. 349–70.

Williams, P. (2007). *Memorial Museums: The Global Rush to Commemorate Atrocities*. Oxford, UK: Berg.

Woodard, W.P. (1972). *The Allied Occupation of Japan 1945–1952 and Japanese Religions*. Ann Arbor, MI: University of Michigan.

Yatsuka, H. (2011). *Metabolism Nexus*. Tokyo, Japan: Ohmsha.

Young, J.E. (1993). *The Texture of Memory: Holocaust Memorials and Meaning*. New Haven, CT: Yale University Press.

18 Still at risk after reconstruction

How does the mode of reconstruction cause new vulnerabilities when rebuilding a vernacular cultural heritage settlement?

Yu Wang and Hans Skotte

Introduction

Natural disasters, along with uncontrolled urbanization, unsustainable tourism, war, and conflicts have caused a significant loss of cultural heritage. The severe damage to ChristChurch Cathedral in New Zealand during the 2011 earthquakes is a recent case in point. The recently published map by RitsDMUCH indicates that a considerable number of world heritage sites are located in active earthquake zones, but the present cultural heritage management has not responded by focusing on disaster risk reduction (DRR) regarding the sites located in the danger zones. For example, UNESCO's World Heritage Committee admits in a survey of 60 world heritage properties located in disaster risk affected areas that only 10 percent have a Risk Preparedness plan.[1],[2] Given the increasing threats to our heritage sites from natural disasters, this paper argues for reducing risk through a systematic risk management program embedded in the current cultural heritage management system.

An integrated disaster risk management is conceptualized as a circle, which starts at "risk prevention and mitigation," goes to "risk preparedness" and "emergency response," and finally "the process of recovery." In the recovery section, post-disaster reconstruction, i.e., repairing and restoring the damage caused by the disaster, is the core practice. The mode of repair and restoration can significantly impact the next round of risk management when similar hazards are encountered. Recovering heritage cannot be done merely through simple physical reconstruction: it requires elaborate professional assistance in the repairing of the damage. Due to the irreplaceability of cultural heritage, post-disaster reconstruction must not only protect the heritage site from present dangers and secure what remains, but also eliminate vulnerabilities and improve the resilience of the site in question in order to mitigate the disaster risks in the future.

Conducting DRR risk assessments of cultural heritage sites is both necessary and urgent. Because Disaster Risk Management (DRM) is a constantly running circle, reconstruction is not the end state. It is merely a phase leading to the next stage. How post-disaster reconstruction is conducted, however, may ultimately determine how it will protect the remaining heritage values from underlying hazards in the future. Our research attempts to tackle how to ensure the sustainable continuity of the "DRM circle" in cultural heritage sites. We will employ evidence from Taoping, a traditional Qiang settlement recently reconstructed after the 2008 Sichuan earthquake, in addressing if and how the reconstruction created new vulnerabilities in Taoping as a vernacular cultural heritage settlement.

In order to answer this question, two modes of analysis have been implemented in this research: the DRR assessment on cultural heritage and the Pressure and Release (PAR) model. The risk assessment is to verify the vulnerabilities of the rebuilt Taoping. Vulnerability was chosen as an investigating factor due to the definition of risk given by Wisner: the "risk of disaster is a compound function of the natural hazard and the number of people, characterized by their varying degrees of vulnerability to that specific hazard, who occupy the space and time of exposure to the hazard event."[3] This is further grounded in the United Nations International Strategy for Disaster Reduction's (UNISDR) claim that risk assessment "is to determine the nature and extent of risk by analyzing potential hazards and evaluating existing conditions of vulnerability that together could potentially harm exposed people, property, services, livelihoods and the environment on which they depend."[4]

The PAR model is "an explanation of disasters [that] requires us to trace the connections that link the impacts of a hazard on people with a series of social factors and processes that generate vulnerability."[5] The PAR model is applied in this paper is in order to identify which factors generate the present vulnerabilities. Also, when applying this model we understand vulnerability not only as a "product" but also is a "process," i.e., we assess how the "vulnerability of the cultural property has increased, decreased or been reinforced over time, especially with respect to disaster situations."[6] This acknowledges the dynamic nature of vulnerability where long-term effects may stem from short-term interventions. In our case, reconstruction-caused alterations have considerable effects on the physical and social environment of Taoping, and these effects may extend to the long-term vulnerability of Taoping as a rural cultural heritage settlement. This is what this paper is about.

The risk assessment of Taoping

The primary measure to prevent a disaster happening is to identify potential disaster risks from relevant information about hazards and vulnerability. Risk assessment is a vital approach to acquiring knowledge of the underlying disaster risks.

A comprehensive risk assessment program comprises: (1) risk identification; (2) disaster risk scenarios simulation; and (3) risk magnitude evaluation.[7] Identifying disaster risk clarifies potential hazards and vulnerabilities. The simulation of scenarios determines possible actions to take when a certain hazard occurs at a specific time based on the results of the risk identification. Evaluating the magnitude of risk involves systematically concluding the level of disaster risk by referring to the results of the scenarios.

Risk identification and risk analysis

Risk identification is about locating potential hazards and investigating the relevant issues that may make Taoping's heritage property vulnerable.

Underlying hazards

Li County, the region where Taoping is situated, is located in the western part of Sichuan Province next to the Tibetan Plateau with a mountain and valley landscape. Zagunao River cuts across the whole county to reach the Minjiang River, one of the main

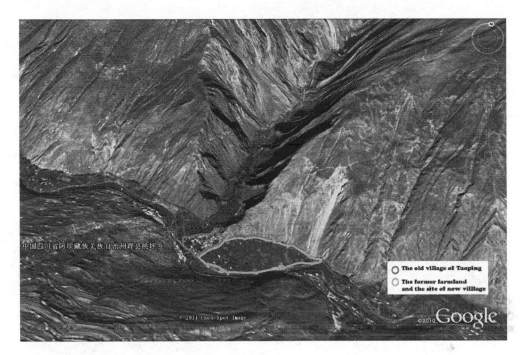

Figure 18.1 Taoping village satellite photo.[8]

rivers in Sichuan Province. The village of Taoping, covering about four hectares, lies on the South side of Dabao Mountain at the bottom of the ridge. Zagunao River runs along the southern edge of Taoping. The annual precipitation in Taoping is between 650–1000 mm a year, most of which occurs in summer and autumn. The area on the eastern side of Taoping village is an alluvial plain created by the Zagunao River. This area used to be the village's agricultural land before the earthquake. Since reconstruction began, this area has been developed as the new residential area of Taoping and the site of Taoping's tourist center (see Figure 18.1).

Pursuing the identification of hazards to Taoping, we also need to investigate the seismic situation and review the history of disasters in the region. Seismic activity is quite frequent in the Taoping region due to its being in the middle of the Longmen Shan Fault Zone, which runs along the Longmen Mountain between the Tibetan Plateau and the Sichuan Basin. The Longmen Shan Fault Zone (see Figure 18.2) is a zone where tectonic plates may collide and release the energy that causes earthquakes. Due to its location in an active seismic zone, Taoping has experienced several catastrophic earthquakes during its history. The magnitude 7.5 earthquake that struck Diexi, Mao County, in August 1933 was the first earthquake to occur following the invention of the scientific means of recording the strength of earthquakes: it killed more than 8,000 people and injured more than 10,000. The epicenter of the Diexi Earthquake was 100 km north of Taoping. This earthquake totally changed the landscape of that region, levelling Diexi and literally converting it into a lake. The 2008 Sichuan Earthquake is the most devastating earthquake recorded in China, having a magnitude of 8.0. It left

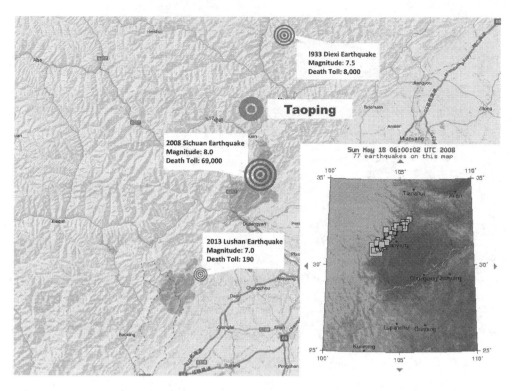

Figure 18.2 Earthquakes around Taoping, along the Longmen Shan Fault line.[9]

more than 69,000 dead and 17,000 missing. The epicenter of the Sichuan quake was 70 km south of Taoping. The most recent earthquake in the region (magnitude 7.0, striking Lushan) happened in 2013, and killed more than 190 people. The epicenter was 100 km south of Taoping. All three of these earthquakes occurred along the Longmen Mountain in the Longmen Shan Fault Zone.

Aside from earthquakes, the Taoping region is also threatened by the hazards of landslides and floods. Landsides happen quite frequently in this area during the rainy season, especially in summer. The latest heavy rain caused several landslides, bringing this hazard into national focus. Four days of heavy rain over July 7–11, 2013, generated landslides and flooding, affecting 13 counties in the Aba Prefecture; over 900 residential buildings were damaged and the national road 317 was blocked, isolating many cities and towns for several days. This disaster killed 16 people and left 20 missing. The economic loss is estimated at approximately 6.8 billion RMB, or about USD$1 billion.[10]

The geographic, topological, hydrological, and seismic information on Taoping shows that this village is located in an active seismic zone, surrounded by gigantic mountains and next to a powerful river, and experiences heavy rains almost every year. The history of disasters in the region reveals that as well as the disasters referred to above, this village has repeatedly experienced several types of natural hazards during the last century, including earthquakes, landslides, and floods. There is no indication that this will not continue to happen in the future.

Relevant issues

After uncovering the potential hazards, it is important to study the relevant issues that may make Taoping vulnerable to underlying hazards, specifically its social and economic situation, heritage factors, the standing of the local community, etc.

The inhabitants of Taoping belong to the Qiang ethnic group, a national minority with its own languages, customs, religions, and lifestyle. The population of Qiang is around 300,000,[11] most of whom are settled in mountainous fortress villages such as Taoping in west Sichuan. This is a transition area between the Han Chinese and Tibet. Because of this, their ethnic character has been influenced by both Han Chinese and Tibetans.[12]

The Qiangs of Taoping have inhabited this fortress village for several centuries. Today, 95 families (more than 500 people) live in the village. The social structure is based on blood relations and clans. This system was previously also the social safety net for Taoping, shoring up the most vulnerable families. However, the clan system gradually lost its power as most households in Taoping overcame dire poverty.

The principal means of livelihood of Taoping until the 2008 earthquake was agriculture. However, the post-quake reconstruction of Taoping fundamentally changed the livelihood of its people. Their agricultural land was converted into a new residential area, making their fortress village a tourist attraction. Without farmland, the Taoping community had no choice but to enter the tourism business.

The pursuit of profits encouraged the townspeople to develop their tourism business quickly, but without considering possible safety issues. Specifically, many families in Taoping have converted their traditional vernacular houses into family inns with enlarged windows, which may make the traditional masonry buildings more vulnerable to earthquakes. The tourism facilities so far lack streetlights, accurate tourist maps, professional tour introductions, credible exhibitions, etc. Furthermore, there are no fire extinguishers in Taoping – a village where people traditionally use open fires in wooden houses. There is no evacuation plan or escape route for tourists in the event of an emergency, in a tourist site famous for its labyrinthine streets.

The landscape of Taoping also changed due to the new uses of the farmland. The Taoping community has gradually moved to the new residential area since reconstruction finished in 2011 (Figure 18.3). Each of these buildings implemented a reinforced concrete structure, designed to be strong enough to withstand magnitude 8.0 earthquakes. Although the new standards are the consequence of lessons learnt from the latest seismic disaster, the very location of the new residential area raises other issues of safety. It is built on agricultural land, a flat area of alluvial plain, very close to the Zagunao River and lower than the original Taoping village. The new residential area is also at the foot of the slope of Dabao Mountain; despite the fact that the new residential area is in a zone prone to floods and landslides, no specific measures or plans have been made for flood and landslide prevention.

Heritage status is also an important issue pertaining to risk identification. Since 2002, Taoping's historical buildings have been identified as a national treasure, which is the highest category of heritage conservation in China.[13] The Sichuan Earthquake damaged all 115 buildings in the Taoping protection zone. After the quake, the State Administration for Cultural Heritage commissioned the Chinese Architecture History Research Institute (CAHRI) to make plans to restore the damaged historical buildings. In June 2008 CAHRI took on the Taoping historical building reconstruction. After three years, in 2011, the Taoping reconstruction project was completed. The work was

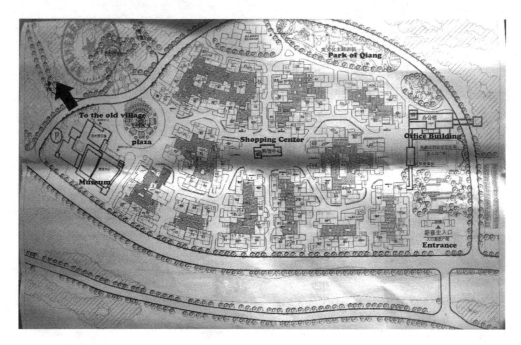

Figure 18.3 The plan of the new village. (For comparison with the old village, please see Figure 18.4).[14]

strictly managed and implemented by professional conservationists, a process constantly disrupted by community members who did not appreciate this professional reconstruction.[15] After the somewhat controversial reconstruction, the management of the heritage of Taoping was left to the local Culture and Sport Office of Li County, which in effect means that there is no management, as the Office's 16 employees are also responsible for all sports and cultural activities in Li County, including libraries, publishing, etc.[16] Hence, there is no professional management or monitoring system for the heritage of Taoping. An emergency plan, a risk preparedness plan, and a risk management plan for the heritage of Taoping are also nonexistent.

Risk analysis

Risk analysis builds upon the identified potential hazards and issues relevant to determining the vulnerabilities of Taoping. It is a study of the impacts of underlying hazards on the heritage of Taoping, based on the reality of those issues. Risk analysis identifies the underlying hazards as earthquake, landslides, and floods.

From our investigation into relevant issues, it is evident that the reconstruction has fundamentally altered Taoping and left it vulnerable to other hazards. During the reconstruction, most of the farmland was taken for the construction of new residential buildings, which may be strong enough to withstand high-magnitude earthquakes but are built on the lower alluvial plain next to the river, leaving the new village open to the risk of flooding. The risk of landslides in the new village is higher than in the old one, because it now directly faces the south slope of Dabao Mountain.

Without farmland, the community was forced to convert its livelihood to tourism. This move partly led to the "second reconstruction" of the old village, which weakened the ability of traditional buildings to resist earthquakes. Meanwhile, due to uncontrolled development and lack of planning, the tourist facilities are not sufficient to support such a large number of tourists (approximately 7,000 at the time of writing).

Based on the investigation of the heritage status, the reconstruction focused only on the rescue and repair of heritage sites. It neglected the opportunity to eliminate vulnerabilities and to introduce DRM into the heritage management. The current heritage management in Taoping is negligible, as there is no specific heritage agency monitoring and preparing emergency interventions or risk assessments.

Building disaster scenarios

Running disaster scenarios is designed to uncover further impacts of the underlying hazards that may threaten the heritage of Taoping by simulating how underlying hazards may strike the village at specific times and/or in certain situations. In these simulations, several sequential events are predicted based on the facts and current situation of Taoping, uncovered through risk identification and risk analysis. These predicted sequential events are addressed as a series of cause-and-effect assumptions. Disaster scenarios are mere predictions, not realities; many uncertainties always remain, despite the assumptions being based on information gathered from risk identification and analysis.

Scenario 1

Scenario 1 is a magnitude 7.5 earthquake occurring in the Longmen Shan Fault line on a mid-summer afternoon. The epicenter of the earthquake is 50 km southwest of Taoping. The energy released violently shakes Taoping, and the mountains of the region are also affected, causing gigantic rocks and debris to roll down the mountainside and setting off subsequent landslides. Because Taoping is located at the bottom of a valley, it is hit by rocks, flows of mud, and debris, which hit the new village (located at the foot of Dabao Mountain) much more severely than they would have the old village. The old town would not be hit by such landslides because it was located on a ridge and in a blind zone for landslides. Earthquake-caused landslides smash the road and block traffic, cutting off National Road 317, the only road connecting Taoping and the other villages in the valley to the outside world. The detached giant stones from landslides also block the river and dam the water; soon after this, the new village is flooded. At the same time fire breaks out in Taoping old town, due to the use of open fires; some of the old buildings collapse due to structural failings caused by inappropriate post-reconstruction restorations implemented by some of the local inhabitants.

The heritage property of Taoping is in danger because of the threats from the earthquake, landslides, and fire. However, no designated people or agency in Taoping are responsible for responding to this emergency situation, because no one has been authorized to manage the heritage property. Furthermore, the local community has no idea how to preserve the collapsed and burning historical buildings because of the lack of training and preparedness for this situation. Even if people wanted to extinguish the fires, there are no fire extinguishers in the old town. The historical buildings are situated close together, allowing the fire spread easily – all the more so because of the new

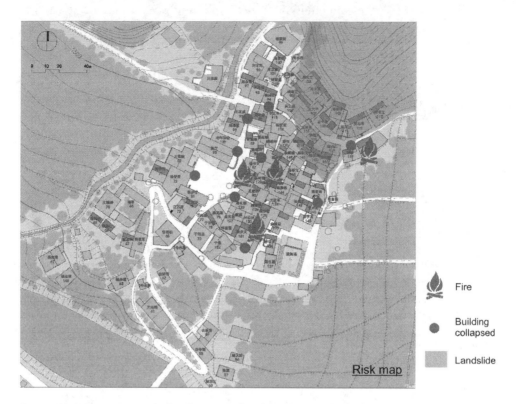

Figure 18.4 Scenario 1 risk simulation in the old village.[17]

and bigger windows and doors. Some streets are tunneled under wooden structures that would collapse if caught in the fire, making escape all the more difficult.

The national road is cut off, delaying the arrival of professionals for several days to advise on how to save the cultural heritage properties.

The local population, including tourists, is also in danger. A few locals are injured and killed when they are caught by the landslide that hit the residential area, burying some of the new houses. Some people are trapped in some of the collapsed buildings; there is no local expertise or equipment to help save the people trapped. On the day of the earthquake, there are around 100 people in the old town of Taoping,[18] mostly tourists visiting Taoping for the first time in groups organized by travel agencies. When the earthquake strikes, the tourists panic and want to get out of the old town, but because of the unfamiliar environment and the labyrinthine streets, they cannot find their way out. There is no map, nor are there signs showing an evacuation route. Most of them are trapped in the old town. Some are injured or killed when they are buried or trapped by collapsing buildings and fire. Those who survive try to help the trapped, but collapsed tunnel streets make certain areas of the old town inaccessible. After a few days, food and other necessities are no longer available because the main road is being blocked. This also prevents emergency medical rescue teams from entering Taoping, thus leaving seriously wounded patients to die. Others are not rescued in time and die trapped under the rubble because the professional rescue teams could not get through to Taoping (see Figure 18.4).

Scenario 2

Scenario 2 envisages heavy rainfall hitting the Taoping area. Intense rainfall (100 mm in three hours) has poured down on the Taoping region for three days. Due to the bad weather, some groups cancel their plans to visit Taoping, so there are fewer tourists than usual. Most of the local population stays indoors in their new houses in the new village. Meanwhile, the Zagunao River has constantly risen, to the level where the local municipality has issued a flood alarm. As the people wait for the rain to stop, a thunderous sound is heard, and seconds later rocks, rubble and mud slide down the north slope of Dabao Mountain. Giant stones and tons of rubble crash into the new residential buildings, some continuing all the way into the Zagunao River. Many new buildings have been hit; most of those closest to the mountain are literally buried. The initial thunderous sound made some of the people believe it was an earthquake, and so they ran out to open spaces such as the square or parking lots, where they were then caught by the sliding mud and debris. Some are injured and some are killed. Those still in their houses are trapped when the houses collapse. Those who are not hurt cannot help others, as there is no efficient equipment in the settlement. The constant rainfall and the debris cause the river to continue to rise, and water soon begins to flood into the new residential area. Survivors from the village inform the local authorities about what is happening, but when they send a professional rescue team they cannot reach Taoping because of a landslide that has blocked the road between the regional head-quarters and Taoping. Days later, the rescue team arrives and is able to save the people still trapped in the collapsed buildings, get the debris and rocks out of the river, and pump water out of the houses that are still flooded.

Evaluating the magnitude of risks

Evaluation of the underlying risks is the third stage of the DRR analysis, where we compare the various underlying risks according to certain criteria such as probability and consequence. The level of those potential risks to Taoping as a rural heritage settlement is then evaluated.[19]

As we showed above, some primary hazards inevitably unleash secondary and even tertiary hazards, either in sequence or simultaneously, as the preceding scenarios show. Although the likelihood of earthquakes is high, the frequency of earthquakes in Taoping is lower than that of other hazards such as landslides and floods caused by extreme weather, which occurs very often in the Taoping area during the rainy season.

Disaster scenarios simulate the severity of the consequences of a likely disaster event. As we saw in Scenario 1, a powerful earthquake hitting the area in the future would damage the historical buildings and significantly deplete the heritage value of the village. The heavy rains, landslides, and flooding in Scenario 2 would not in any significant way damage the old village or the historical buildings, but the local community, who are closely involved in the local heritage, would be severely affected. Taoping's heritage value as an outstanding example of a traditional Qiang settlement will have been indirectly damaged because the disaster will have damaged the local community. Yet all in all, taking the consideration of the local community into account, the second scenario would still have only mild consequences for the heritage property in Taoping.

In Scenario 1, the local community would face even more serious challenges. The likelihood of residents and tourists being killed or injured is much higher in Scenario 1 than in Scenario 2, and in light of the economic and material destruction that would occur, Scenario 1 represents a much more severe disaster for the community.

The hazard of an earthquake occurring in the Taoping area is of low probability, but if it occurs, it could destroy historical buildings and kill and injure numerous people from the local community as well as visitors. It could seriously damage, even annihilate, the tourism-based livelihoods of the local people, and waste the enormous economic investment already made in the area.

Heavy rains causing landslides and floods in Taoping have a high probability. When this happens, it could severely damage the new residential area, killing and injuring local residents and causing heavy economic losses due to the severe destruction in this area. Analyzing all the available information, earthquakes represent a high level of risk to Taoping, whereas disastrous weather causing landslides and floods represent a moderate level of risk.

The pressure and release model of current vulnerability

The factors from which vulnerabilities emanate are closely related to post-quake reconstruction. As mentioned earlier, vulnerability is not a "product" but a process and may increase, decrease, or be reinforced. To investigate further how reconstruction may have increased vulnerabilities in Taoping, the Pressure and Release (PAR) model was applied to analyze this particular issue.

What the PAR model is and how it works requires an introduction. The model was initially introduced in the seminal book, *At Risk: Natural Hazards, People's Vulnerability and Disasters*[20] and is conceptualized and described as follows:

> a disaster is the intersection of two opposing forces: those processes generating vulnerability on one side, and the natural hazard event (or sometimes a slowly unfolding natural process) on the other [. . .] that an explanation of disasters requires us to trace the connections that link the impact of a hazard on people with a series of social factors and processes that generate vulnerability. The explanation of vulnerability has three sets of links that connect the disaster to processes that are located at decreasing levels of specificity from the people impacted upon by a disaster. The most "distant" of these are root causes . . . that give rise to vulnerability (and which reproduce vulnerability over time) are economic, demographic and political processes [. . .] Dynamic pressures are the processes and activities that "translate" the effects of root causes both temporally and spatially into unsafe conditions [. . .] Unsafe conditions are the specific forms in which the vulnerability of a population is expressed in time and space in conjunction with a hazard.[21]

The PAR model overlaps the DRR risk assessment model when "unsafe conditions" in PAR are understood as "relevant issues" in DRR, and post-quake reconstruction is understood as "dynamic pressure" in PAR. This makes the quest for the "root causes" the focus of the PAR analysis.

PAR's "unsafe conditions" for underlying hazards have been shown as relating to the location of a settlement in an active seismic zone in a mountainous landscape. The new residential area is exposed to the hazards of landslides and flood. In the investigation of reconstruction, it was learned that the structure of historical buildings may not be strong enough to resist a powerful earthquake. The new economic livelihood of Taoping is more fragile than it was formerly, due to a lack of training, skills, and experience in tourism. Moreover, the local community primarily focuses on maximizing profits without investing in disaster preparedness measures or equipment for an emergency situation. Finally, no official institution monitors Taoping's heritage properties or prepares the local population for a possible future disaster.

The "dynamic pressure" of the PAR model is, when applied to Taoping, the initial, fast, and top-down managed reconstruction, executed without reference to the DRM cycle. The reconstruction created a series of cause-effect events in Taoping that made the village more vulnerable to underlying risks. The use of established farmland for residential purposes caused the Taoping people to adopt tourism as their new means of earning a living, using the historical buildings and the living environment as a resource to attract tourists, a decision that made the local people hostile towards the official reconstruction, including the repairs of historical buildings. Almost all the families in Taoping therefore conducted a second "reconstruction" on their own in order to transform their buildings into family inns or guesthouses. Those unprofessional restorations have subsequently decreased the "formal" value of the heritage of Taoping.

Reviewing the changes stemming from the reconstruction, most of them have added to the vulnerability of Taoping and its citizens: their new settlement is exposed to the hazards of landslides and floods. Using their agricultural land for new housing deprived the people of their traditional livelihood and made them turn to tourism as their new livelihood, an occupation for which they lack skills and experience. Their new dependence on tourism is driving a second "reconstruction" of the old village, reducing the resilience of the historical buildings in the event of powerful earthquakes in the future.

These dynamic pressures causing unsafe conditions are embedded in PAR's "root causes." The root causes in this case stem from authorities ignoring the crucial role of the local community in reconstructing the cultural heritage of Taoping.

The cultural heritage of Taoping is deeply rooted in the community, which in turn plays a crucial role in upholding Taoping's heritage. Taoping is a rural settlement, categorized as a "vernacular cultural heritage" site. The International Council on Monuments and Sites (ICOMOS) particularly highlights the importance of the community in vernacular cultural heritage: "the appreciation and successful protection of the vernacular heritage depend on the involvement and support of the community, continuing use and maintenance."[22]

The community created the historical and aesthetic values attributed to the rural settlement of Taoping. Agricultural activities generated a sustainable relationship between man and nature, generating stable social practices that continually left traces on the buildings. Over time, these marks have slowly made up what Ruskin called the settlement's "voicefulness"[23] and its value as heritage. Under a livelihood-dominated relationship, the community not only created those values but also maintained them for generations. These social practices protected and accumulated these values long before they were identified and labeled as "heritage."

Local communities play an essential role in conserving the heritage of rural settlements. This follows the unique position the inhabitants hold by actually living *in* their heritage, which places the community in the primary position of "heritage users" and makes them the principal "heritage keepers," along with the heritage administration.

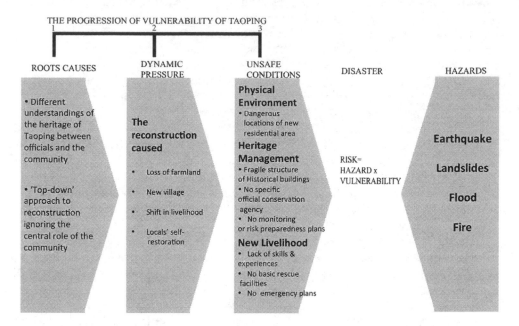

Figure 18.5 The PAR model analysis of Taoping.

Finally, the community "teaches" settlement heritage conservation. Local inhabitants are thoroughly familiar with their living environment and have learnt or inherited the skills and knowledge on how to build and repair their dwelling in ways that experts cannot learn from books.

However, the official understanding of the heritage of Taoping does not recognize the essential role of the community in vernacular heritage sites. The official heritage identification of historical Taoping focuses only on the physical buildings and environment. At the beginning of the reconstruction, the policy makers did not regard the local community as crucial stakeholders. Even the professionals who were responsible for implementing the project to repair heritage sites in Taoping did not recognize the town's heritage as a vernacular cultural heritage, which by definition is very intimately related to the local community. As a result, the reconstruction in Taoping has in effect devalued the heritage, forced a change in locals' livelihoods, and has moved the community to a place exposed to more hazards. As a result, the reconstruction has made Taoping even more vulnerable (see Figure 18.5).

Conclusion

This article has shown that even after the reconstruction following the Sichuan Earthquake, the heritage of Taoping is at serious risk for earthquakes and moderate risk for landslides and floods caused by disastrous weather. The reason lies in the inappropriate post-quake reconstruction program implemented in Taoping, which created a series of impacts on the heritage and community of Taoping that has exacerbated their vulnerability to underlying hazards.

The PAR analysis shows that the different understandings of the concept of heritage on the part of the local community and the policymakers who made decisions on reconstruction is a central root cause of these unsafe conditions. Another root cause is the current "top-down" implementation approach, which largely ignores the crucial role the local community plays in generating and preserving the vernacular cultural heritage of Taoping.

This research addressed the following crucial issues. The Disaster Risk Management (DRM) circle requires post-disaster reconstruction to respond to a variation of vulnerabilities, but neither post-disaster reconstruction nor DRM can work sustainably without recognizing the crucial role of the local community in vernacular cultural heritage settlements. Assessing risk for vernacular heritage settlements is an effective tool for identifying long-term, underlying disaster risks. Combining this with the PAR model places the results of the assessment in a social context, thus revealing the root causes of unsafe conditions.

Notes

1 RitsDMUCH is the acronym for the Institute of Disaster Mitigation for Urban Cultural Heritage, Ritsumeikan University, Tokyo, Japan, available at http://www.rits-dmuch.jp/en/index.html
2 United Nations Office for Disaster Risk Reduction (2013), *Heritage and Resilience Issues and Opportunities for Reducing Disaster Risks* (Geneva, Switzerland: Author).
3 B. Wisner, P. Blaikie, T. Cannon, and I. Davis (2010), *At Risk: Nature Hazards, People's Vulnerability and Disaster*, 2nd edn. (New York: Routledge), p. 49.
4 United Nations Development Program (2010, October), *Disaster Risk Assessment* (New York, NY: Author).
5 Wisner et al. (2010), p. 52.
6 R. Jigyasu (2010), "Risk Assessment of World Cultural Heritage: Tools and Methodology," paper presented at the Risk Assessment in Heritage: A Need or a Luxury? Conference, University of Porto, Portugal, p. 4.
7 Ibid.
8 Image © 2010 Google Inc.
9 Map data © Google Inc.
10 The information about damage caused by landslides and floods is available at the Aba Prefecture Emergency Management Office, available at http://www.abazhou.gov.cn/yjgl/
11 This figure refers to the 2000 Census. The 2008 Sichuan Earthquake, however, is estimated to have killed about one-tenth of the Qiang population.
12 王明珂 [Wang, Mingke] (2008), 羌在藏汉之间，中华书局，北京 [*Qiang, between Tibet and Han*] (Beijing, China: Zhonghua Book Company).
13 中国建筑研究院建筑历史研究所 [The Chinese Architectural History Institute (CAHRI)] (2008), 桃坪羌寨抢险修保护工程——灾后保护工程规划. [*Post-Sichuan Earthquake Reconstruction Planning of Taoping*], unpublished paper on urban planning, 1.
14 Image supplied by the authors, reprinted by permission.
15 Y. Wang and H. Skotte (2014), "Reconstruction after Reconstruction," in *Rebuilding Sustainable Communities after Disasters in China, Japan and Beyond*, Adenrele Awotona, ed., pp. 45–60 (Cambridge, UK: Cambridge Scholars).
16 See the Introduction to the Sport and Culture Office of Li County, available at http://www.abztyj.gov.cn/Article/ShowArticle.asp?ArticleID=77
17 Image produced by Yu Wang, reprinted by permission.
18 This is an average daily number based on annual figures.
19 R. Jigyasu and V. Arora (2013), *Disaster Risk Management of Cultural Heritage in Urban Areas – A Training Guide* (Kyoto, Japan: RitsDMUCH).
20 Wisner et al. (2010).
21 Ibid., pp. 52–53.

22 International Council on Monuments and Sites (1999), *Charter on the Built Vernacular Heritage*, retrieved January 4, 2016, from http://www.icomos.org/fr/chartes-et-normes, p. 1.
23 John Ruskin saw "voicefulness" in buildings as very important, and praised the work of time upon buildings:

> (T)he great glory of a building is not in its stones, nor in its gold. Its glory is in its age, and in that deep sense of voicefulness, of stern watching, of mysterious sympathy, nay, even of approval or condemnation, which we feel in walls that have been washed by the passing wave of humanity.
>
> Source: J. Ruskin (1989), *The Seven Lamps of Architecture*, reprint edition (New York: Dover).

Bibliography

中国建筑研究院建筑历史研究所 [The Chinese Architectural History Institute (CAHRI)]. (2008). 桃坪羌寨抢险修保护工程——灾后保护工程规划. [*Post-Sichuan Earthquake Reconstruction Planning of Taoping*]. Unpublished paper on Urban Planning.

International Council on Monuments and Sites. (1999). *Charter on the Built Vernacular Heritage*. Retrieved January 4, 2016, from http://www.icomos.org/fr/chartes-et-normes

Jigyasu, R. (2010). "Risk Assessment of World Cultural Heritage: Tools & Methodology." Paper presented at the Risk Assessment in Heritage: A Need or a Luxury? Conference, Portugal, University of Porto.

———. (2013). *Proceedings of the Sub-regional Workshop on Disaster Risk Preparedness and Management*, pp. 11-21. (Yogyakarta, Indonesia: UNESCO and Ministry of Education and Cultural Republic of Indonesia).

———, and V. Arora. (2013). *Disaster Risk Management of Cultural Heritage in Urban Areas – A Training Guide*. Kyoto, Japan: RitsDMUCH.

Ruskin, J. (1989). *The Seven Lamps of Architecture*. Reprint edition. New York: Dover.

沈三陵 [Shen, Sanling]. (2001). 桃坪羌寨保护规划，清华大学，北京. [*Taoping Conservation Planning*. Beijing: Tsinghua University.]

孙成民 [Sun, Chengmin]. (2010). 四川地震全记录 公元前26年至公元后2009年, 四川人民出版社，成都. [*Records of Sichuan Earthquake from 26 BC to 2009 AC*. Chengdu: Sichuan People.]

United Nations Development Program. (2010, October). *Disaster Risk Assessment*. New York, NY: Author.

United Nations Office for Disaster Risk Reduction. (2013). *Heritage and Resilience Issues and Opportunities for Reducing Disaster Risks*. Geneva, Switzerland: Author. Available at http://www.unisdr.org/we/inform/publications/33189

王明珂 [Wang, Mingke]. (2008). 羌在藏汉之间，中华书局，北京 [*Qiang, between Tibet and Han*. Beijing, China: Zhonghua Book Company.]

Wang, Y., and H. Skotte. (2014). "Reconstruction after Reconstruction." In *Rebuilding Sustainable Communities after Disasters in China, Japan and Beyond*. Adenrele Awotona, ed. pp. 45–60. Cambridge, UK: Cambridge Scholars.

Wisner, B., P. Blaikie, T. Cannon, and I. Davis. (2010). *At Risk: Nature Hazards, People's Vulnerability and Disaster*. 2nd edn. New York: Routledge.

19 The provision of participatory shelter facilities in adverse environments

A case study of Lahore, Pakistan

Fariha Tariq

Housing – A shelter, a home

Housing is a major place-based infrastructural element and an integral part of the fabric of a community. It has a profound impact on the social, economic, and physical character of a community. The "basic needs" approach introduced by the International Labor Organization in 1976 included shelter as a basic need, along with food and clothing.[1] Provision of appropriate housing is important to the economic and social wellbeing of a society:

> It typically constitutes 15 percent to 20 percent of household expenditures. For all but the wealthy, it is usually the major goal of family saving efforts. Investment in housing represents up to 30 percent of fixed capital formation with vigorous housing programs. For some of the self-employed, housing is the place of work. Apart from physical and economic benefits, housing has also substantial social benefits including welfare impact of shelter from its elements, sanitation facilities and access to health and educational facilities.[2]

Around 900 million urban dwellers worldwide live in settlements that can be characterized as "slums." Their numbers have grown rapidly over the last 20 years, and will continue to do so unless the housing policies of governments and international agencies become far more effective. The urgency of addressing this issue is recognized in the Millennium Development Goals, one of the main targets of which is to significantly improve the lives of at least 100 million slum dwellers by 2020:[3]

> Achieving this will require new ways of addressing such critical issues as access to land for housing, secure tenure, provision of basic services and improvements to housing for the urban poor. It will also mean supporting the incremental processes by which low-income households build, since this is how most dwellings are built or improved. Creative measures and new ways of financing these must be found.[4]

Challenges to sustainable housing development

"The very complicated problem of satisfactorily housing people on a national scale has not been solved in most countries of the world."[5] The situation has not gotten better; as reported in 2005, the housing shortage is still most evident demographically where developing countries have been urbanized considerably since the 1950s.[6] According to the UN Centre for Housing Development, in Third World countries, 25 to 75 percent of urban residents live in absolute poverty, and an active housing policy can be considered a deliberate social and economic investment. This investment would generate multiplier effects in forward and backward linkages and productivity.[7]

In recent years, there has been a growing recognition that providing affordable housing is essential for the physical, social, economic, and environmental futures of nations.[8] In the past, most East and South Asian countries were largely rural; today, they continue to experience rapid urban growth, with many of their urban concentrations reaching a population level of over 1 million.[9] The same is now happening in Pakistan, where the concentration of economic activities in big cities such as Karachi and Lahore is attracting people from rural environments.[10] The process of urbanization, however, is creating a multitude of problems; the shortfall in services, including housing and infrastructure, are just a few among many. During the 1980s, the urban areas of Pakistan experienced rapid growth at a rate of around 77 percent from 1981 to 1993, indicating a high level of "rural-urban transformation."[11]

Translating urban growth into demand for housing

The United Nations projects that developing countries will add approximately 2 billion new urban residents during the next 25 years.[12] This number, along with the existing 1 billion people living in slums, "frames the demand side for the need for housing and infrastructure services in developing countries."[13] By 2030, about 3 billion people, or 40 percent of the world's population, will need housing. The statistical analysis in the 2005 *Global Report on Housing* estimates that 96,150 housing units per day, or 4,000 per hour, will be required.[14] This analysis includes new units and does not include replacements of deteriorated housing stock. As elaborated by Ghaus and Pasha, on the supply side, the slow rate of development of residential plots in the cities and lack of municipal infrastructure had limited the construction of new housing units. On the demand side, inflation in the prices of building materials has reduced levels of affordability for housing, implying on the one hand that residential densities have increased and on the other hand that slums are taking up a progressively higher share of the urban population.[15]

According to United Nations statistics, by the year 2000, half of the population of Asian cities was living in slums and squatter settlements.[16] As Angel said, "the level of informal housing occupancy (houses built without approval, title or adequate urban infrastructure) is normally 20 percent of the total housing stock" in developing countries.[17]

The presence of illegal housing in the form of slum and squatter settlements is a clear indication of the failure of governments to provide adequate housing for the poor.[18] During the last decade, most governments have faced serious fiscal deficits, and local budgets have constrained housing development projects.[19] Hardoy and Satterthwaite[20] examined the housing policies of 17 developing countries and concluded that only two of them had incorporated programs for low-income families in their national housing policy; one of these was Pakistan, where 24 percent of the total population (170 million people) lives below the poverty line. As in other developing countries, poor people in Pakistan are exposed to housing shortages and low incomes. According to the 1980 housing census in Pakistan, the number of housing units increased by only 2.1 percent during a corresponding 3 percent increase in population. According to the International Bank for Reconstruction and Development (IBRD), the rate of residential plot development and provision of infrastructural facilities also failed to keep up with demand. On the demand side, due to the high inflation of building material prices, the ability of poor people to afford to construct a house has decreased significantly.[21] In early 2010, it was estimated that there was a total shortage of 8.8 million

Table 19.1 The main characteristics of the housing stock in Pakistan[22]

Description	1998
Total Housing (units)	19,211,738
Urban (units)	6,031,430
Urban (percentage)	32%
Persons per housing unit (avg)	6.8
Urban	7.1
Persons per room (avg.)	3.1

houses, and this number has continued to grow due to an increased demand for 700,000 houses per year compared to the supply of 330,000 houses per year.[23] The characteristics of the housing stock in Pakistan are shown in Table 19.1.

There has been no census since 1998 upon which one can base precise solutions according to exact figures. For example, in Lahore, according to the Lahore Development Authority, 750,000 people were living in squatter areas in the early 1990s without proper provision of civic amenities, but this is contradicted by much higher unofficial estimates.[24]

Government agencies in Pakistan have experimented with a wide range of low-income housing programs, such as low-cost core units, a sites and services program, and slum-up gradation. However, the government produces a negligible amount of housing as compared to the actual demand. To implement their urban development and housing programs, government agencies in Pakistan turned to land acquisition, which is considered the best means of obtaining land for current projects and reserving it for future use. Attempts by the government to restrain land speculation and to acquire private undeveloped urban land for low-income housing at below-market prices were met with great resistance by private landowners. This caused a huge drop in the supply of low-income housing units by the government. For instance, the Punjab Housing and Town Planning Agency (PHATA) is facing an annual drop of 3,300 housing units because it had to acquire land at market price to comply with the 1985 Land Acquisition Act and offer it on a non-profit basis. As a result, most public agencies have developed low-income housing societies in physically distant locales to avoid high land costs.[25]

On the user end, these housing societies remain unpopular because they are an inconveniently great distance from any major city, as are most low-income housing projects developed by the government. The public sector's failure to anticipate and respond to the housing demand of low-income people continues to result in the creation of irregular settlements and slums, called *Katchi Abadis*, all over the country. For example, in Karachi, which has a population of over 10 million people, more than half of its inhabitants live in squatter or illegally developed informal settlements. Given the basic parameters of the land, land use, land rights, location, and public sector incentives associated with high population pressures and the rapid pace of urbanization,[26] the traditional modes of providing housing on unimproved settlements for the urban poor is unlikely to meet the increased demand. The conventional high level of services has proven to be unaffordable or unsuitable for the poor.

Housing and disasters

Zaman and Ara[27] reported that rapid growth in urban areas caused the proliferation of slums and squatter areas as well as an unhealthy environment. In Pakistan, unplanned housing expansions are characterized as Katchi Abadis, which are often

Figure 19.1 Slum along the river Ravi with contained pond.[28]

overcrowded with unsanitary conditions such as inadequate, inefficient sewage facilities and limited access to water, electricity, and social services.

Qadeer[29] observed that, due to rapid municipal growth, the housing shortage had increased. In the industrial expansion period, increased demand for housing in Lahore outpaced the supply and poor people had no choice but to build Katchi Abadis. Asghar[30] focused his studies on evaluating the socio-economic opportunities and environmental conditions of the slums of Lahore, Pakistan. He found that slums were characterized by a relatively high infant mortality rate, water-borne diseases, poor sanitation, a lack of environmental hygiene concerns, and inadequate basic amenities. Moreover, children and women living in slums were in a more vulnerable condition. Hina[31] revealed the miserable conditions of the inhabitants of the Lahore slums, and she criticized the government for its apparent lack of concern for improving squatter settlements.

Commonly, much of the housing of the poor is vulnerable to serious damage from natural disasters. Housing on steep hillsides and in flood plains may be destroyed by heavy storms, and the flimsy materials available to poor are unlikely to withstand earthquakes, high winds or fire.[32]

Houses in Pakistan typically have walls made of baked or unbaked bricks, with roofs made of reinforced concrete or iron girders. Alvi[33] argued that the differences in construction standards and the level of available services are dependent on household income.

Each year, after the monsoon rains in Pakistan, an emergency situation exists in most areas. The government and armed forces have initiated small-scale housing projects in affected areas with the help of local NGOs and international donations, but every year the destruction is so widespread that, despite these huge international donations and governmental efforts toward relief and reconstruction, the situation is not being handled effectively, and the gap between housing supply and demand widens daily.[34]

The architecture of slums

According to recent statistics, the number of people living in slums in Pakistan reached 26.6 million, while the percentage of urban population in slums is 48 percent.[35] Most of the slums were poorly constructed, using mud as a bonding agent, with baked bricks. Masonry held together with mud cannot withstand a few days of waist-high, stagnant floodwater. Such structures are called "katcha" and "semi-pacca" in the local

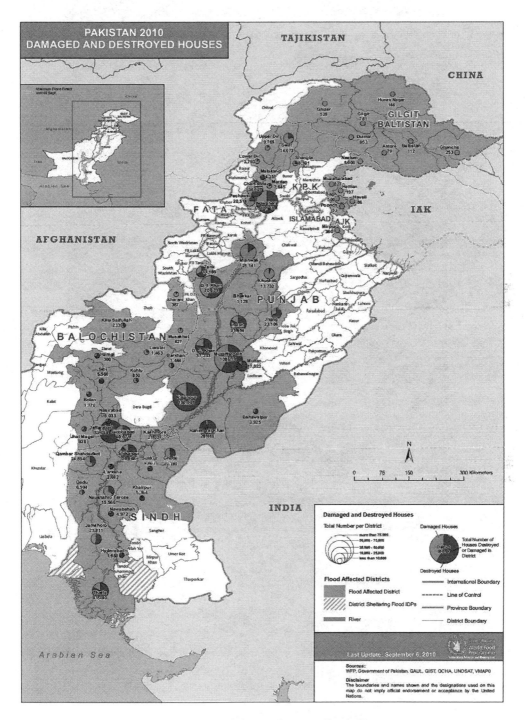

Figure 19.2 Number of houses destroyed during the 2010 floods.[36]

Figure 19.3 A house in a Kamayan slum, constructed of bamboo and cloth.[37]

language. Such houses could not withstand flood conditions, despite the fact that their walls were constructed on foundations. Also, the foundation design and construction did not meet the standards normally practiced in the cities of Punjab.[38]

Roofing materials are typically wooden beams, joists, and mud, or even bamboo. These materials may not securely bond with walls, unless they are properly designed with wooden columns. Such constructions are therefore particularly vulnerable and cannot withstand the impact of a flood. Poorly built houses made of bamboo and those with weak brick and mortar walls with no structural connection to a foundation, as well as the lack of a drainage system, have caused heavy losses in tsunami-hit areas in developing countries.[39]

Designing an inclusive design for slum dwellers

According to Geis,[40] there is an integral relationship between how we design and shape our communities and their capacity to minimize direct and indirect losses from extreme natural hazards. An efficient disaster-resistant design must consider the following:

- The relationship of development to natural (ecological and geological) systems;
- Development and redevelopment patterns;
- The configuration and scale of public infrastructure;
- The design, location, and service capacity of community facilities; and
- Neighborhood and commercial district design.

Before developing and implementing ideas for inexpensive homes in seismically active countries, disaster patterns should be studied. Some architects want to rethink the basic materials used in developing countries. Darcey Donovan advocates replacing concrete walls with load-bearing straw bales. Her nonprofit group, Pakistan Straw Bale and Appropriate Building, erects 7.3 × 7.3 m houses in northwest Pakistan. The bales are stacked and bound together top to bottom with a fishnet, which keeps them from slipping apart during seismic activity; the building is then plastered over.[41]

James Kelly, Professor Emeritus of Civil Engineering at the University of California, Berkeley, expects that people in Haiti will continue to build with concrete because it is cheap and easy to shape into blocks, and because deforestation has left few other materials. Kelly therefore focuses on keeping concrete buildings upright using rubber isolators as shock absorbers. He argues that many buildings in California and Japan sit on hundreds of rubber pads that absorb seismic energy by deforming, as opposed to cracking or shifting. A building on a rubber foundation shakes independently of the ground and less frequently, which helps brittle walls to survive intact.

Van de Lindt is working on a similar design, but with recycled rubber tires instead of strips, and different types of walls and foundations. He says that drilling holes into concrete walls and inserting bamboo buttresses 1.3 m long would keep many modest-sized homes in developing countries standing during quakes.[42]

Nadeem's[43] research revealed that, after the 2010 floods in Pakistan, some of the affected communities in the Muzaffargurh District were rehabilitated by constructing model villages. Some include one-bedroom houses constructed on 1360 ft^2 plots

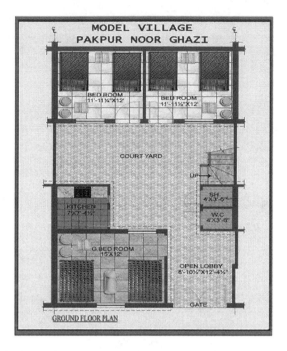

Figure 19.4 A one-bedroom house design with two bedrooms, proposed for future extension.

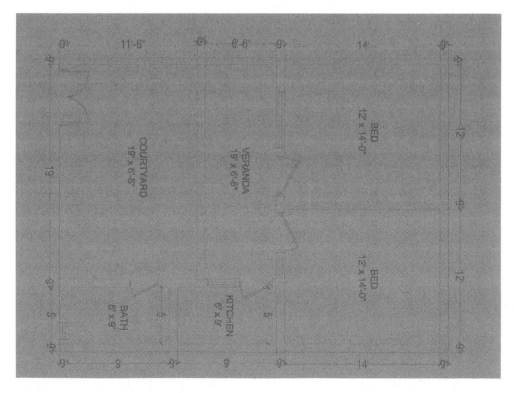

Figure 19.5 A two-bedroom house design.[44]

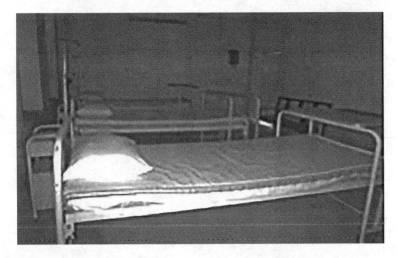

Figure 19.6 Dispensary in miani mallah.

Figure 19.7 Brick-paved street in miani mallah.[45]

(as shown in Figure 19.4), while others have two bedrooms of various sizes on plots ranging from 1360–2720 ft^2 (Figure 19.5). The design of each house provides space for a future addition and includes a kitchen attached to an open courtyard, reflecting the cultural style of the village. Separate washrooms are also provided in every house.

Materials used for home construction include baked bricks and cement. Roofing materials include pre-cast concrete slabs, while doors and windows are made of steel. Utility services such as electricity, water, sewage, and drainage systems, as well as recreational facilities, i.e., parks and open spaces, hospitals, and vocational training centers have also been provided.

Figure 19.8 Shelter designed with minimal requirements.

Figure 19.9 Interior with convertible spaces.[46]

Shelter design and the human body

The architecture students of Beaconhouse National University, Lahore, have undertaken a few exercises in order to design an inclusive slum. The designs are based on the proportions of the human body, as they believe that successfully built environments always complement human body proportions. Le Corbusier also worked out *le modular*, a scale of proportion devised by reference to the human body. He applied this system of proportion in some of his acclaimed buildings, including the Church of Saint Mary De LaTourette, the United Habitation in Marseilles, and the Carpenter Centre for the Visual Arts.

The design is based on the habitation challenge faced by our bodies in spaces in which our dimensions change significantly throughout the day. A number of these spaces are designed specifically for specific minimal requirements, such as a shower or a car. In Figure 19.9, the design has separate and convertible sitting areas, as well as working and resting spaces designed according to minimal body requirements.

The students also explored locally found objects, including recyclable materials such as bamboo, corrugated cardboard, alloy car tire rims, ropes, and parachute material. The first design is called *Firefly* (see Figure 19.10), the base of which is attached to a bamboo raft which can either be stationed on the ground or float on water. The second is called *Ark*, which is made of corrugated cardboard, a material that is readily available. The sheets of cardboard are cut and joined together for stability. The *Ark* can be packed in a box and transferred to any location.

Figure 19.10 Firefly with bamboo base, covered with parachute material.

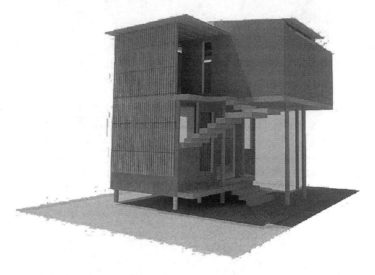

Figure 19.11 Model of a container house.[47]

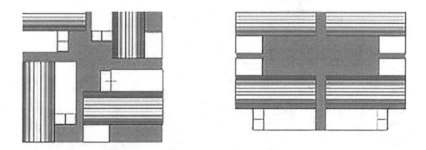

Figure 19.12 Options for container placement.

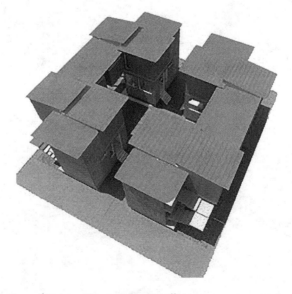

Figure 19.13 Overview of a container community.[48]

Another experiment was done with shipping containers, which seems to solve many of the problems of low-income housing communities. One 20′ × 10′ steel container costs approximately Rs 40,000–80,000 (USD$380–760), and, with modifications, it can accommodate four to seven people. The steel container has a kitchen, laundry room, and commercial store on the first floor, and a washroom, bedroom, and work-shop space on the second floor.

These houses are placed together to form social gathering places. Figures 19.12 and 19.13 illustrate placement options.

House financing for the poor

House financing issues appear to be the most challenging topics in development eco-nomics. A large paradigm shift would be required to improve housing and housing conditions through financing. Housing financial sources in low- and middle-income countries include commercial, private, financial, and banking institutions, which pro-vide credit to upper-income groups at market interest rates upon proof of their income level and the provision of collateral and guarantees. The lower-middle, moderate-, and low-income households, most of which work in the informal economy, have, with few exceptions, been excluded from accessing capital from formal private or public finan-cial institutions. These groups have consistently relied on informal sources, including savings and informal loans from friends and family, remittances from family members working abroad, and the sale of whatever assets they have. In recent years, however, a growing number of non-traditional financial institutions have begun to serve these sectors with innovative financial schemes. These experiences show that the housing needs of the poor can be financed in a way that is economically viable, affordable, and consistent with tested methods of delivering financial services to the poor. Some of these institutions and schemes have been supported by international donor agencies.[49]

The most fundamental financial issue in cities in most developing countries is to obtain a mortgage to finance the purchase of a house. Problems that make housing markets difficult include: (1) lack of formal property rights; (2) poorly functioning land markets; (3) weak regulatory regimes; (4) low incomes and limited access to mortgage funds; (5) lack of housing products; and (6) difficulties in obtaining credit for home improvements. Dowall and Leaf[50] argue that private real estate markets typically produce housing that is affordable to households down to the lowest thirtieth or fortieth percentile of the incomes and housing prices, but private lenders may not wish to enter the market because of risks and the lack of tools to manage them.[51] In this situation, most urban low-income households can only afford to build houses incrementally as and when financial resources become available. Thus, at the inter-national level, micro-finance institutions have started lending for low-income hous-ing, and have become very important in recent years. The Global Report on Human Settlements[52] has also emphasized the fact that finance is the only dimension of secur-ing sustainable solutions that can fill the gap between the two extreme outcomes of current systems and processes: "Affordable shelter that is inadequate and adequate shelter that is unaffordable."[53]

As is the case in other developing countries, owning a decent house seems an unachiev-able dream for almost half of the population of Pakistan. The housing development sectors have failed to provide adequate housing for the poor. Innovative measures, in which the poor can take control of their own housing needs, are required through access to funds for the construction and improvement of their houses. In Pakistan, commercial banks have been reluctant for a variety of reasons – including a lack of faith in the trust-worthiness of the poor and inadequate databases containing pertinent information on potential borrowers – to provide adequate credit for the poor to build houses for them-selves and their families.[54] This accounts for the provision of housing with expected subsequent upgrading, which is to be managed by householders and communities.

Implications for the government

Recognize that poor people build their houses incrementally

In Pakistan, building codes were designed for the construction of complete homes, thus making progressive building illegal despite the fact that this is the most common form of home construction for the poor. These codes also limit the poor's demand for hous-ing finance because of the fear that their "out-of-code structures" will be destroyed. They limit investment in them. Comprehensive housing government policies could better support housing expansion.

Recognize that the poor are creditworthy

As the portfolios of microfinance institutions are growing in Pakistan, it is becoming clear that the poor have sufficient credit to take out and repay loans, and funding policies need to make loans more affordable to the poor. Participants suggested that the government should trust low-income people and should introduce and facilitate more lending organizations.

Provide land, basic infrastructure, and social services to the poor

In the housing sector, government can invest in areas where other institutions cannot, such as providing basic infrastructure such as water, electricity, roads, and buildings, and social infrastructure such as schools and health clinics. According to Akhtar, "an important role of governments is to intervene in land and housing markets to ensure that the lowest income groups in the city have access to secure land and decent housing. Political will within government and civil society is essential to resolve the problems of slum populations."[55]

Provide secure tenure for the poor

Renters hesitate to construct or improve the house in which they live. Banks and lend-ing institutions disburse loans only to those who have legal title to the land and house, so improving tenure rights is necessary to increase improvements in housing standards. When people become property owners, they are more willing to invest larger percent-ages of their income to acquire land or to build or improve their homes.

Create women-specific loan products

Evidence from the literature shows that the wellbeing of families, especially that of children, is affected positively by an increase in women's resources.[56] In *Growing Up in an Urbanizing World*, Moore and Cosco[57] suggested that children are aware of the housing conditions in which they live and the fear of losing their home is a cause of anxiety. Furthermore, the majority of microfinance institutions prefer women as members because of their reputation for being reliable borrowers, thereby contributing to financial viability.[58]

Research in Pakistan indicates that, for a woman, apart from financial and economic functions, a house also serves as a specific cultural place. For example, in the house of an extended Muslim family, female family members require a separate space for practicing purdah. In one case, housing microfinance enabled a widow to construct a boundary wall in order to achieve privacy and security after the death of her husband. In another, it provided a livelihood for a mother to send her children to school.

In some cases, women reported that cultural restrictions prohibit them from participating in outdoor activities. A mindset has developed in which men think that they are responsible for each and every thing and that women need to stay home, even though religion does not restrict women from earning a living. As a result, even if a woman has earning potential and wants to do something to support her family, because of social pressures, she cannot.

Housing microfinance resonates with both Pakistani traditions and the fluctuating incomes of poor women. Houses planned and built according to specific family needs are more productive. To achieve this end, the Pakistani government should encourage microfinance institutions to adopt a targeted, articulated gender focus as part of the national strategy to address women's empowerment. By providing women-only credit services, female staff, and trained personnel adept at serving banks' female clientele, much can be done to foster a female micro-client base.[59]

Improve research opportunities

The government should create opportunities for researchers to review international practices and conduct local surveys relevant to low-income housing, especially after disasters. Housing finance should also be a focus for research, and researchers should then prepare policy recommendations. Apart from producing reports on policy recommendations, researchers could also monitor and evaluate projects and suggest improvements.

Coordinate with building material suppliers

NGOs can act as intermediaries between building material suppliers and households. Allowing materials to be drawn on credit through local commercial suppliers can reduce the burden on low-income families constructing their own houses.

Create community organizations and savings institutions

Localized community organizations may have the capacity to serve the scale of need that exists within their area because each of these entities is tailor-made to suit the particular circumstances of their local constituents, although relatively small, localized operations may be more effective than big centralized ones, allowing for personal knowledge to replace legal requirements and paperwork. Standardized practices often fail to ensure repayment; rather, social pressure within a community guarantees repayment. Community organizations can also channel housing loans through community savings, which can reduce the overheads of lending institutions normally spent on managing loan disbursement and repayment. Community savings also have the power to develop the capacities of the poor to manage and save for future housing and land loans, as well as to establish revolving funds. Women are especially attracted to savings groups because they have more relationships with others in the family and in the community. Savings groups also create a venue for meeting regularly and for learning from others.

Provide technical assistance

If not designed from the start, housing construction, whether it be the addition of a room or the improvement of an existing structure, often involves destruction and can entail a lot of waste of building materials, labor, and all other resources. The lack of involvement of technical advice in incremental construction results in low-quality results and therefore creates environmental and land-use problems (especially after disasters).

Apart from disbursing loans, NGOs should undertake related activities, for example, the hiring of technical staff (architects, civil engineers, etc.) and providing training to household members in the technical skills necessary to maintain the house. Special attention should also be paid to the cost and selection of appropriate construction materials.

Discussion

Severe floods affect not only a country's infrastructure but also the education, health, water and sanitation, transportation, communications, agricultural, trade, and industrial sectors. There is a growing consensus that flood policy formulation must include multidisciplinary, multi-sector, and multi-stakeholder participation, initiatives, and activities to address the flood environment.[60]

In disaster-prone countries such as Pakistan, it is imperative to develop basic disaster risk mitigation knowledge and skills, not only among the policymakers and implementation groups but also among communities. Nonetheless, some weak points also surfaced in the rehabilitation process, including the lack of involvement of the affected communities in construction and preparations for possible flood and mitigation measures. A recent study on rehabilitated settlements revealed several instances of malpractice in the construction and allocation of houses and the maintenance of public facilities.

Communities know their living habits, needs, and priorities best, and how the incremental construction of their houses will work. In his book, *Housing by People*, Turner said:.

> When dwellers control the major decisions and are free to make their own contribution to the design, construction or management of housing, both the process and the environment produced stimulate individual and social wellbeing. And where dwellers are in control, their homes are typically better and cheaper than those built through government programs or large corporations.[61]

Notes

1 J.A. Denton (1990), *Society and the Official World: A Reintroduction to Sociology* (Baltimore, MA: Rowman and Littlefield).
2 The World Bank (1975), *Housing: Sector Policy Paper* (Washington, DC: Author), p. 11.

3 P. Travis, S. Bennett, A. Haines, T. Pang, Z. Bhutta, A.A. Hyder, N.R. Pielemeier, A. Mills, and T. Evans (2004), "Overcoming Health-Systems Constraints to Achieve the Millennium Development Goals," *The Lancet* 364(9437), pp. 900–06.

4 A. Stein and L. Castillo (2005), "Innovative Financing for Low-Income Housing Improvement: Lessons from Programmes in Central America," *Environment and Urbanization* 17(1), p. 34.

5 J.F.C. Turner and R. Fichter (1972), *Freedom to Build: Dweller Control of the Housing Process* (New York, NY: Macmillan), p. 176.

6 United Nations Human Settlements Programme (2005), *Financing Urban Shelter: Global Report on Human Settlements 2005* (London, UK: Earthscan and UN Habitat).

7 K. Jacobsen, S.H. Khan, and A. Alexander (2002), "Building a Foundation Poverty, Development, and Housing in Pakistan," *Harvard International Review* 23(4), pp. 20–25.

8 UN Habitat (2005).

9 Ibid.

10 A. Hasan and M. Raza (2011), *Migration and Small Towns in Pakistan* (Karachi, Pakistan: OUP).

11 C. Pugh, B.C. Aldrich, and R.S. Sandhu (1995), *Housing the Urban Poor: Policy and Practice in Developing Countries* (New Delhi, India: Sage), p. 17.

12 D. Balk, F. Pozzi, G. Yetman, U. Deichmann, and A. Nelson (2005), "The Distribution of People and the Dimension of Place: Methodologies to Improve the Global Estimation of Urban Extents," paper presented at the International Society for Photogrammetry and Remote Sensing, Proceedings of the Urban Remote Sensing Conference, March 14–16, 2005, Tempe, AZ.

13 UN Habitat (2005).

14 Ibid.

15 A. Ghaus and H.A. Pasha (1990), "Magnitude of the Housing Shortage in Pakistan," *The Pakistan Development Review* 29(2), p. 137.

16 D. Murphy and the Asian Coalition for Housing Rights (1990), *A Decent Place to Live: Urban Poor in Asia* (Bangkok, China: Habitat International Coalition-Asia).

17 A. Angel (2000), *Housing Policy Matters: A Global Analysis* (Oxford, UK: OUP), p. 3.

18 Pugh, Aldrich, and Sandhu (1995).

19 UN Habitat (2005).

20 J.E. Hardoy and D. Satterthwaite (1989), *Squatter Citizen: Life in the Urban Third World* (Washington, DC: Earthscan).

21 Ghaus and Pasha (1990).

22 A. Ghaus and H. A. Pasha (1990), "Magnitude of the Housing Shortage in Pakistan," *The Pakistan Development Review* 29(2), pp. 137–53.

23 D. News (2010, March 2), "Housing Statistics of Pakistan," *The Daily Times* [Online edition], available at http://www.dailytimes.com.pk/default.asp?page=2010\02\03\story_3-2-2010_pp.5–8

24 I. Alvi (1997), *The Informal Sector in Urban Economy: Low Income Housing in Lahore* (Karachi, Pakistan: OUP).

25 F. Tariq and A.Q. Butt (2008), "Alternative Approach: The Case of Low-Income Housing," *Science International (Lahore)* 20(1), pp. 67–69.

26 I. Matthäus-Maier and J.D. Pischke (2011), *Financing Housing for the Poor: Connecting Low-Income Groups to Markets* (Berlin, Germany: Springer).

27 A. Zaman and I. Ara (2000), "Rising Urbanization in Pakistan: Some Facts and Suggestions," *The Journal NIPA* 7, pp. 31–46.

28 Photo by Tariq, reprinted by permission of the author.

29 M.A. Qadeer (1983), *Lahore: Urban Development in the Third World* (New York, NY: Vanguard Books).

30 J.H. Asghar (1984), *A Study of Socio-Economic Conditions of Females in Katchi Abadi, Chaudry Colony, Lahore* (Lahore, Pakistan: Department of Social Work, Punjab University).

31 I. Hina (1991), *Role of Government in the Improvement of Katchi Abadis* (Lahore, Pakistan: Department of Economics, Government College).

32 S. Bartlett, R. Hart, D. Satterthwaite, X. de la Barra, and A. Missair (1999), *Cities for Children: Children's Right, Poverty and Urban Management* (London, UK: Earthscan), p. 45.

33 Alvi (1997).

34 F. Tariq (2012), "The Potential of Micro-Financed Housing after Pakistan Disasters," *International Journal of Safety and Security Engineering* 2(3), pp. 265–79.

35 Homeless International (2013), *Organization Website*, retrieved January 5, 2016, from http://www.homeless-international.org/our-work/where-we-work/pakistan
36 WFP (2010).
37 Photo by Tariq, reprinted by permission of the author.
38 Local Government and Community Development Department, Government of the Punjab (2007), *Model Building and Zoning Bylaws for Town Municipal Administrations in Punjab* (Lahore, Pakistan: Author).
39 B. Khazai, G. Franco, J.C. Ingram, C.R. Rio, P. Dias, R. Dissanayake, R. Chandratilake, and J. Kannu (2006), "Post-Tsunami Reconstruction in Sri Lanka and Its Potential Impacts on Future Vulnerability," *Earthquake Spectra* 22(3), pp. 829–44.
40 D.E. Geis (2000), "By Design: The Disaster-Resistant and Quality-of-Life Community," *Natural Hazards Review* 1(3), pp. 151–60.
41 S. Dawang (2010), *Affordable Solution for Earthquake Resistant Building Construction in Haiti* (Calgary, Canada: Southern Alberta Institute of Technology-SAIT).
42 J. Merkel and C. Whitaker (2010), "Rebuilding from Below the Bottom: Haiti," *Architectural Design* 80(5), pp. 128–34.
43 O. Nadeem, A. Jamshed, R. Hameed, G. Abbas Anjum, and M.A. Khan (2013), "Post-Flood Rehabilitation of Affected Communities by Ngos in Punjab, Pakistan-Learning Lessons for Future," *Journal of Faculty of Engineering and Technology* 21(1), pp. 1–20.
44 Image taken from: S. Kean (2010, February 5), "Rebuilding: From the Bottom Up," *Science* 327(5966), p. 639, available at http://www.buffalo.edu/content/dam/www/news/imported/pdf/February10/ScienceHaitiRebuild.pdf
45 Image taken from: O. Nadeem, A. Jamshed, R. Hameed, G. Abbas Anjum, and M.A. Khan (2013), "Post-Flood Rehabilitation of Affected Communities by NGOs in Punjab, Pakistan-Learning Lessons for Future," *Journal of Faculty of Engineering and Technology* 21(1), pp. 1–20.
46 Images taken from Nadeem et al. (2013).
47 Photo by Tariq, reprinted by permission of the author.
48 Ibid.
49 Stein and Castillo (2005).
50 D.E. Dowall and M. Leaf (1991), "The Price of Land for Housing in Jakarta," *Urban Studies* 28(5), p. 707.
51 Matthaus-Maier and Pischke (2011).
52 UN Habitat (2005).
53 Ibid., p. 98.
54 A. Kaleem and S. Ahmed (2010), "The Quran and Poverty Alleviation," *Nonprofit and Voluntary Sector Quarterly* 39(3), pp. 409–28.
55 S. Akhtar (2007), *Expanding Microfinance Outreach in Pakistan* (Karachi, Pakistan: SBP), p. 12.
56 A.M. Goetz and R.S. Gupta (1996), "Who Takes the Credit? Gender, Power, and Control over Loan Use in Rural Credit Programs in Bangladesh," *World Development* 24(1), pp. 45–63.
57 N. Cosco and R. Moore (2011), "Our Neighbourhood Is Like That!" in *Growing up in an Urbanizing World*, L. Chawla, ed., pp. 36–44 (Abingdon, UK: Earthscan).
58 S.M. Hashemi, S.R. Schuler, and A.P. Riley (1996), "Rural Credit Programs and Women's Empowerment in Bangladesh," *World Development* 24(4), pp. 635–53.
59 C. Niethammer, T. Saeed, S.S. Mohamed, and Y. Charafi (2007), "Women Entrepreneurs and Access to Finance in Pakistan," *Women's Policy Journal* 4, pp. 1–12.
60 P. Dorosh, S. Malik, and M. Krausova (2010), *Rehabilitating Agriculture and Promoting Food Security Following the 2010 Pakistan Floods* (Washington, DC: International Food Policy Research Institute), p. 22.
61 J.F.C. Turner (1977), *Housing by People: Towards Autonomy in Building Environments* (New York, NY: Pantheon Books), p. 35.

Bibliography

Akhtar, S. (2007). *Expanding Microfinance Outreach in Pakistan*. Karachi, Pakistan: SBP.
Alvi, I. (1997). *The Informal Sector in Urban Economy: Low-Income Housing in Lahore*. Karachi, Pakistan: OUP.

Angel, S. (2000). *Housing Policy Matters: A Global Analysis*. Oxford, UK: OUP.

Asghar, J.H. (1984). *A Study of Socio-Economic Conditions of Females in Katchi Abadi, Chaudry Colony, Lahore*. Lahore, Pakistan: Department of Social Work, Punjab University.

Balk, D., F. Pozzi, G. Yetman, U. Deichmann, and A. Nelson. (2005). "The Distribution of People and the Dimension of Place: Methodologies to Improve the Global Estimation of Urban Extents." Paper presented at the International Society for Photogrammetry and Remote Sensing, Proceedings of the Urban Remote Sensing Conference. March 14–16, 2005, Tempe, AZ.

Bartlett, S., R. Hart, D. Satterthwaite, X. de la Barra, and A. Missair. (1999). *Cities for Children: Children's Right, Poverty and Urban Management*. London, UK: Earthscan.

Cosco, N., and R. Moore. (2011). "Our Neighbourhood Is Like That!" In *Growing up in an Urbanizing World*. L. Chawla, ed. pp. 36–44. Abingdon, UK: Earthscan.

Dawang, S. (2010). *Affordable Solution for Earthquake Resistant Building Construction in Haiti*. Calgary, Canada: Southern Alberta Institute of Technology-SAIT.

Denton, J.A. (1990). *Society and the Official World: A Reintroduction to Sociology*. Baltimore, MA: Rowman and Littlefield.

Dorosh, P., S. Malik, and M. Krausova. (2010). *Rehabilitating Agriculture and Promoting Food Security Following the 2010 Pakistan Floods*. Washington, DC: International Food Policy Research Institute.

Dowall, D.E., and M. Leaf. (1991). "The Price of Land for Housing in Jakarta." *Urban Studies* 28(5), p. 707.

Geis, D.E. (2000). "By Design: The Disaster-Resistant and Quality-of-Life Community." *Natural Hazards Review* 1(3), pp. 151–60.

Ghaus, A., and H.A. Pasha. (1990). "Magnitude of the Housing Shortage in Pakistan." *The Pakistan Development Review* 29(2), pp. 137–53.

Goetz, A.M., and R.S. Gupta. (1996). "Who Takes the Credit? Gender, Power, and Control over Loan Use in Rural Credit Programs in Bangladesh." *World Development* 24(1), pp. 45–63.

Hardoy, J.E., and D. Satterthwaite. (1989). *Squatter Citizen: Life in the Urban Third World*. Washington, DC: Earthscan.

Hasan, A., and M. Raza. (2011). *Migration and Small Towns in Pakistan*. Karachi, Pakistan: OUP.

Hashemi, S.M., S.R. Schuler, and A.P. Riley. (1996). "Rural Credit Programs and Women's Empowerment in Bangladesh." *World Development* 24(4), pp. 635–53.

Hina, I. (1991). *Role of Government in the Improvement of Katchi Abadis*. Lahore, Pakistan: Department of Economics, Government College.

Homeless International. (2013). *Organization Website*. Retrieved January 5, 2016, from http://www.homeless-international.org/our-work/where-we-work/pakistan

Jacobsen, K., S.H. Khan, and A. Alexander. (2002). "Building a Foundation Poverty, Development, and Housing in Pakistan." *Harvard International Review* 23(4), pp. 20–25.

Kaleem, A., and S. Ahmed. (2010). "The Quran and Poverty Alleviation." *Nonprofit and Voluntary Sector Quarterly* 39(3), pp. 409–28.

Kean, S. (2010, February 5). "Rebuilding: From the Bottom Up." *Science* 327(5966), pp. 638–39. Available at http://www.buffalo.edu/content/dam/www/news/imported/pdf/February10/ScienceHaitiRebuild.pdf

Khazai, B., G. Franco, J.C. Ingram, C.R. Rio, P. Dias, R. Dissanayake, R. Chandratilake, and J. Kannu. (2006). "Post-Tsunami Reconstruction in Sri Lanka and Its Potential Impacts on Future Vulnerability." *Earthquake Spectra* 22(3), pp. 829–44.

Local Government and Community Development Department, Government of the Punjab. (2007). *Model Building and Zoning Bylaws for Town Municipal Administrations in Punjab*. Lahore, Pakistan: Author.

Matthäus-Maier, I., and J.D. Pischke. (2011). *Financing Housing for the Poor: Connecting Low-Income Groups to Markets*. Berlin, Germany: Springer.

Merkel, J., and C. Whitaker. (2010). "Rebuilding from Below the Bottom: Haiti." *Architectural Design* 80(5), pp. 128–34.

Mitlin, D., and D. Satterthwaite. (2014). *Environmental Problems in an Urbanizing World: Finding Solutions in Cities in Africa, Asia and Latin America*. Washington, DC: Earthscan.

Murphy, D., and the Asian Coalition for Housing Rights. (1990). *A Decent Place to Live: Urban Poor in Asia*. Bangkok, China: Habitat International Coalition-Asia.

Nadeem, O., A. Jamshed, R. Hameed, G. Abbas Anjum, and M.A. Khan. (2013). "Post-Flood Rehabilitation of Affected Communities by Ngos in Punjab, Pakistan-Learning Lessons for Future." *Journal of Faculty of Engineering and Technology* 21(1), pp. 1–20.

News, D. (2010, March 2). "Housing Statistics of Pakistan." *The Daily Times* [Online edition]. Available at http://www.dailytimes.com.pk/default.asp?page=2010\02\03\story_3–2–2010_ pp.5–8.

Niethammer, C., T. Saeed, S.S. Mohamed, and Y. Charafi. (2007). "Women Entrepreneurs and Access to Finance in Pakistan." *Women's Policy Journal* 4, pp. 1–12.

Pugh, C., B.C. Aldrich, and R.S. Sandhu. (1995). *Housing the Urban Poor: Policy and Practice in Developing Countries*. New Delhi, India: Sage.

Qadeer, M.A. (1983). *Lahore: Urban Development in the Third World*. New York, NY: Vanguard Books.

Stein, A., and L. Castillo. (2005). "Innovative Financing for Low-Income Housing Improvement: Lessons from Programmes in Central America." *Environment and Urbanization* 17(1), pp. 47–66.

Tariq, F. (2012). "The Potential of Micro-Financed Housing after Pakistan Disasters." *International Journal of Safety and Security Engineering* 2(3), pp. 265–79.

Tariq, F., and A.Q. Butt. (2008). "Alternative Approach: The Case of Low-Income Housing." *Science International (Lahore)* 20(1), pp. 67–69.

Travis, P., S. Bennett, A. Haines, T. Pang, Z. Bhutta, A.A. Hyder, N.R. Pielemeier, A. Mills, and T. Evans. (2004). "Overcoming Health-Systems Constraints to Achieve the Millennium Development Goals." *The Lancet* 364(9437), pp. 900–06.

Turner, J.F.C. (1977). *Housing by People: Towards Autonomy in Building Environments*. New York, NY: Pantheon Books.

———, and R. Fichter. (1972). *Freedom to Build: Dweller Control of the Housing Process*. New York, NY: Macmillan.

United Nations Human Settlements Programme (UN Habitat). (2005). *Financing Urban Shelter: Global Report on Human Settlements 2005*. London, UK: Earthscan and UN Habitat.

World Bank, the. (1975). *Housing: Sector Policy Paper*. Washington, DC: Author.

Zaman, A., and I. Ara. (2000). "Rising Urbanization in Pakistan: Some Facts and Suggestions." *The Journal NIPA* 7, pp. 31–46.

Part 4
Australia

Introduction

Adenrele Awotona

Bushfires, floods, cyclones, severe storms, earthquakes, and landslides are products of the vagaries of the natural environment and human choices whose impacts Australian communities continue to suffer from. The human factors comprise overstocking, vegetation loss, dams, groundwater depletion, and irrigation schemes.[1] For example, "high fuel loads, a change from fire prevention to firefighting measures, and not building adequate buffer zones to protect built assets" have been listed by the Australian government as amongst the human management factors which have contributed to the harshness of bushfires.[2] Table I4.1 shows that the annual economic losses from disasters in Australia between 1980 and 2010 were US$926.5 million. From 2010 to 2013, natural disasters alone caused insurance losses of more than US$6.9 billion.[3]

Eburn and Dovers have noted that, notwithstanding the more than 50 "formal, complex" post-disaster Royal Commissions and other quasi-judicial inquiries in 75 years "to identify how [tragedies] occurred and what can be done to prevent future occurrences," Australian communities continue to suffer from the impact of emergencies and extreme events. For example, they observed that "since 1939, there have been over thirty inquiries into wildfires and wildfire management and at least another 14 into floods, storms, other natural hazards, and emergency management arrangements," all of which were worthy but produced very little "useful learning." One of their key recommendations is that "the community needs to move beyond developing policy by royal commission and instead engage in the realities of life in the Australian context."[4]

Similarly, based on the findings from its case studies of four towns still recovering from flooding, cyclones, and bushfires in Australia, the Regional Australia Institute revealed that "there is inadequate planning for many of the short- and long-term disaster effects."[5]

In Chapter 20, Lisa Gibbs and her colleagues also observe that while there is a dearth of research that has addressed the health impacts of post-disaster relocation after natural disasters, far less attention has been paid to the impact of the decision to stay or relocate on personal wellbeing. They therefore present a case study in this chapter, comparing the experiences of those who stayed and those who moved out of a bushfire-affected community and the impact on personal wellbeing. They conducted interviews with 35 participants and found from their inductive, thematic analysis of the data that the decision to stay in the community was often influenced by a strong commitment to place and people. They further revealed that those who chose to leave did so because their sense of community had been damaged by changes to the physical and social environment. The findings informed the theoretical modeling for analyses

Table 14.1 An overview of natural disasters in Australia from 1980–2010[6]

No. of events	162
No. of people killed	959
Average killed per year	31
No. of people affected	16,051,010
Average affected per year	517,775
Economic damage (US$ × 1,000)	28,719,981
Economic damage per year (US$ × 1,000)	926,451

of survey data from 1,010 participants from bushfire-affected communities (897 stayed in community, 113 relocated). Lisa Gibbs and her colleagues used a structural equation model (SEM) approach to examine associations in the survey data between disaster impact, life stressors, psychological sense of community, and the role that leaving the community may play in moderating these relationships. The study showed that individuals who left the community reported greater disaster exposure and less sense of community in their new location, both of which exerted negative influences on personal wellbeing. This effect was counterbalanced, however, by a moderation effect in which leaving the community lessened the impact of financial and relationship life stressors on personal wellbeing. These findings, according to Lisa Gibbs and her colleagues, have implications in terms of targeted service provision for those who stay and those who leave their communities.

Furthermore, in Chapter 21, John F. Richardson and his colleagues disclose that little is known about the short- and longer-term impacts of the experience of separation from family members during and immediately after a disaster event. While support in reuniting family members has historically been provided internationally by the Red Cross and Red Crescent organizations, the Australian Red Cross has a long involvement in reuniting people post-disaster through the use of the National Registration and Inquiry System (now known as "Register. Find. Reunite."). In this chapter, John F. Richardson and his colleagues present an Australian case study drawing on the Beyond Bushfires research to explore family experiences of separation and not knowing the whereabouts of family members during the Black Saturday bushfires in 2009. Their survey of 1,016 survey participants revealed that 570 said that they had been separated from close family members during the fires. They note that the high levels of reported stress arising from this separation and associated uncertainty, lasting in many cases (45 percent) for more than 12 hours, reflects the anxiety this caused for those involved. Moreover, they also observed that this separation experience was also demonstrated to be associated with symptoms of post-traumatic stress disorder three to four years after the disaster event. John F. Richardson and his colleagues conclude that despite widespread reliance on technology and expectations of constant connection, the lack of contact between family members days after the bushfires shows the importance and relevance of formal reunification services, such as that provided by the Australian Red Cross and other informal avenues such as those provided by word of mouth or local government, police, and emergency services.

Correspondingly, although it is estimated that the percentage of Australians over the age of 65 will increase from 13.5 percent currently to 22.7 percent by 2050,[7] Victoria Cornell discloses in Chapter 22 that little research has been undertaken with older people to explore how they feel about emergency events. Of the existing research, the majority considers the opinion and perspective of aged care service providers or peak bodies, not older people themselves. Cornell also found that most of this research is concerned with the response and recovery phases of emergency events, not the preparedness phase. This chapter discusses research carried out in Australia that explored the meaning of being prepared, for older people, with regard to emergency events. The study found that for the older people who participated, being prepared for an emergency was not a one-off tangible activity; rather, it is an ongoing process and involves achieving a feeling of comfort and security in their world. Cornell notes that the process is something that has been built upon over many years and therefore "being prepared for an emergency event," as a specific activity, is not necessarily something that worried or concerned the older people. Moreover, she observes from her case study that mental strength and the ability to cope was seen by the participants as being of greater importance, helping to build the feeling of comfort and security.

Significantly, the research reported in Chapter 22 by Victoria Cornell highlights the extent to which the emergency management sector has taken for granted the understanding of older people. She concludes that by understanding what influences older people living in the community to prepare and what preparedness means to them, the best methods to assist them in their preparedness planning can be established, rather than making assumptions about what this target group wants or needs.

Notes

1 Australian Government (2014), *Natural Disasters in Australia* [Web page], retrieved March 23, 2015, from http://www.australia.gov.au/about-australia/australian-story/natural-disasters
2 Ibid., para. 15.
3 Insurance Council of Australia (2013), *The Year in Review* (Sydney, Australia: Author), retrieved March 23, 2015, from http://www.insurancecouncil.com.au/assets/report/Year%20 in%20Review%202013%20spreads.pdf, p. 15)
4 M. Eburn and S. Dovers (2015, January 20), "Learning Lessons from Disasters: Alternatives to Royal Commissions and Other Quasi-Judicial Inquiries," *Australian Journal of Public Administration* 74(4), pp. 495–508.
5 Regional Australia Institute (2013), *From Disaster to Renewal: The Centrality of Business Recovery to Community Resilience*, Final Report, retrieved March 19, 2015, from http:// www.regionalaustralia.org.au/wp-content/uploads/2013/08/From-Disaster-to-Renewal.pdf
6 Data gathered from "Australia: Disaster and Risk Profile," *PreventionWeb* [Website], retrieved January 14, 2016, from http://www.preventionweb.net/english/countries/statistics/?cid=9
7 M.T. Sheehan (2012), "The Global Financial Crisis and Natural Disaster: Impact on the Queensland Economy," *Journal of Finance and Management in Public Services* 10(1), p. 30.

Bibliography

Australian Government. (2014). *Natural Disasters in Australia* [Web page]. Retrieved March 23, 2015, from http://www.australia.gov.au/about-australia/australian-story/natural-disasters
Eburn, M., and S. Dovers. (2015, January 20). "Learning Lessons from Disasters: Alternatives to Royal Commissions and Other Quasi-Judicial Inquiries." *Australian Journal of Public Administration* 74(4), pp. 495–508.
Insurance Council of Australia. (2013). *The Year in Review*. Sydney, Australia: Author. Retrieved March 23, 2015, from http://www.insurancecouncil.com.au/assets/report/Year%20 in%20Review%202013%20spreads.pdf

Regional Australia Institute. (2013). *From Disaster to Renewal: The Centrality of Business Recovery to Community Resilience*. Final Report. Retrieved March 19, 2015, from http://www.regionalaustralia.org.au/wp-content/uploads/2013/08/From-Disaster-to-Renewal.pdf
———. (2013, August 28). *Research: Natural Disaster Recovery in Australia*. Available at http://www.regionalaustralia.org.au/archive-blog/research-natural-disaster-recovery-in-australia/
Sheehan, M.T. (2012). "The Global Financial Crisis and Natural Disaster: Impact on the Queensland Economy." *Journal of Finance and Management in Public Services* 10(1), pp. 24–37.

20 Post-bushfire relocation decision-making and personal wellbeing

A case study from Victoria, Australia

*Lisa Gibbs, Hugh Colin Gallagher,
Karen Block, Elyse Baker, Richard Bryant,
Lou Harms, Greg Ireton, Connie Kellett,
Vikki Sinnott, John F. Richardson, Dean Lusher,
David Forbes, Colin MacDougall,
and Elizabeth Waters*

Introduction

Community-based disasters are collectively experienced events that overwhelm local capacity, resulting in the destruction of public, private, and environmental assets, human suffering, and disruption of systems.[1] Affected communities typically face a long period of restoration and recovery. It is often necessary for people to find temporary or permanent alternative accommodation, either within the disaster-affected community or in a different location.[2] This paper explores the impact of post-disaster relocation through a case study of bushfires that occurred in February 2009 in Victoria, Australia. It focuses on decision-making in relation to moving out of a disaster-affected community and the impact of that decision on personal wellbeing. Personal wellbeing is conceptualized in this case study as a combination of life satisfaction (current and anticipated), personal resilience, and subjective overall health rating. The psychological sense of community and its relationship to relocation decisions and experiences will also be explored.

Relocation is a complex phenomenon. The period of displacement may vary considerably; people may have to relocate more than once and experience different types and quality of accommodation. They may also have different levels of desire and control over whether to return and rebuild or relocate, and may experience various financial, environmental, and social stressors.[3]

The decision to relocate is often emotionally charged because of loyalties to neighbors, friends, and the local community[4] and attachments to the natural, social, cultural, and built environments.[5] The disaster experience of disruption and loss may invoke a stronger attachment to the local community amongst those affected.[6] Attachment is also likely to be influenced by length of time spent living in the community.[7] Attachment to place is interconnected with ontological security, social and psychological wellbeing, and sense of identity.[8] Ontological security refers to confidence in the continuity of social and material environments.[9] Dislocation of ontological security resulting from a disaster event can be compounded by relocation.[10]

The term "psychological sense of community" describes the degree of belonging and shared commitment felt by members of a community.[11] It is more than simple attachment to place, as it involves multiple affective, cognitive, and behavioral elements that encompass interactive and dynamic relational aspects between the individual and the community.[12]

Post-disaster recovery efforts tend to have a geographic focus on affected areas. Services are located in a "hub" or recovery center and target local people. For residents to remain in or return to a community, basic services need to be provided, such as health care, schools, banking, and food stores, as well as employment opportunities. This promotes a return to routine and supports rebuilding and re-establishing social connections.[13] Those who move away may not have the same access to services or connection with recovery processes. Their new local services, being outside the affected areas, are unlikely to have the same level of disaster experience or access to information on the range of post-disaster support.

Increased understanding of risk and protective factors for those who stay, for those who are temporarily displaced, and for those who permanently relocate will enable more targeted programming for support services and enable better provision of information to support individual decision-making about whether to relocate. There is some evidence arising from studies of Hurricane Katrina, landslides in Venezuela, and flood events in North Carolina that loss and disruption associated with relocation can be minimized when people are able to see a future in their home and community and understand the steps involved in achieving it.[14]

According to the Internal Displacement Monitoring Center, 143.9 million people were newly displaced by disaster between 2008 and 2012.[15] Investigations of relocation have been relatively sparse, with gaps in the literature in terms of affected populations, disaster types, health and wellbeing outcomes, effective interventions, and even conceptual and definitional issues regarding the phenomenon itself.[16] These evidence gaps are likely to be due to a focus on the short-term effects of disaster, the paucity of pre-event data, considerable variability in the time frame over which relocation may unfold, and the challenges in tracing those involved.[17] There is evidence, however, that relocation entails greater risk for mental health problems.[18] People who are separated from their close social networks can experience reduced access to support and opportunities to provide support to others. Consequences can include loss of self-efficacy and control, a diminished sense of identity, and additional stressors associated with being distant from loved ones.[19] A systematic review of 40 empirical studies highlighted the health effects of dislocation, finding convergent evidence of association with increased psychological morbidity.[20] Various relocation factors that are likely to impact wellbeing outcomes include age,[21] the degree of choice about relocation,[22] distance from the original community,[23] and differential experiences within families over time.[24] Reports of positive relocation experiences are rare in the literature, but studies following Hurricane Katrina have included accounts of the benefits of increased safety and opportunities in new communities, despite cultural differences and the loss of social supports.[25]

"Black Saturday" fires

In February 2009, bushfires raged across the State of Victoria in southern Australia, with the worst occurring on Saturday, February 7, resulting in the "Black Saturday" descriptor. This disaster represents one of Australia's worst, with 173 fatalities, 3,500 buildings damaged or destroyed, significant impact on high-value natural environments, and massive adverse impact on community infrastructure.[26] Two townships were completely destroyed and others sustained significant damage.

Many people in bushfire-affected areas were forced to seek temporary accommodation when their homes were damaged or destroyed. The State Government provided a range of housing assistance options, including temporary living expenses grants; 124 caravans provided for use on people's own blocks of land; movable units (four were utilized); access to public housing (for approximately 118 families); and temporary villages established in the center of bushfire-affected communities (utilized by approximately 314 people). Rental charges for accommodation in the temporary villages (approximately $40 AUD per week for individuals and $100 AUD per week for families) were waived in cases of hardship. The final temporary village closed in June 2012, more than three years after the fires, with remaining families moved into public housing as required.

Others made their own arrangements, staying in privately rented accommodation, with family or friends, in another property they owned, or in caravans or sheds on their bushfire-affected land until homes could be rebuilt or a decision was made to relocate. Residents were free to decide whether to stay in the community or relocate, although, as noted earlier, there were many factors influencing that decision. A non-compulsory buy-back program was introduced by the Victorian government in March 2012, offering money for high-bushfire-risk properties where a house had been destroyed.

The *Beyond Bushfires* study

The *Beyond Bushfires: Community Resilience and Recovery* study is a large-scale, multi-method longitudinal survey of community and individual health, wellbeing, and social connectedness in the wake of the "Black Saturday" fires.[27] It involves academic, community, government, and service provider partners. It recruited over 1,000 participants originating from 25 bushfire-affected communities in ten locations in rural and regional Victoria. A range of communities was selected, including high-impact (many houses lost plus fatalities), medium-impact (a small number or no fatalities, but significant property damage), and low-impact (no evidence of fires being present) areas. Community diversity was also sought in terms of population sociodemographics, size of community, and distance from the capital city of Melbourne.

The *Beyond Bushfires* study team commenced regular community visits in 2010 to ensure an understanding of local issues and contextual influences. Surveys were administered in 2012 and 2014 (only 2012 data were used for this case study), and in-depth interviews were conducted during 2013–14. Tracing those who had relocated was a challenge because of a lack of records of who had moved and where. Privacy regulations prevented agencies, including the police and state government, from releasing relevant contact lists such as those who had been eligible for bushfire relief grants. Following negotiation, the Victorian Electoral Commission (VEC) provided the contact details of both current residents and those who had relocated since the fires (N = 7,467 adults) for an agreed fee. Strict conditions applied, including access to contact details for three weeks only and permitting only one invitation letter to be sent to each potential participant. No reminders could be sent. Recruitment was complemented by further outreach to raise awareness of the study and ways to register, including area-based phone calls, mailbox drops, news media, the study website, emails through community networks, and social media activities.

The emotional intensity of deciding whether to relocate and the need for more information about the impact of doing so first became apparent in community visits. Questions relating to relocation were therefore included in the *Beyond Bushfires* survey. Relocation issues were further highlighted in the in-depth interviews, which helped to direct the analysis of survey data. This flow in the research process will be reflected in the following reporting of findings.

Beyond Bushfires qualitative interviews

The in-depth interviews involved a sub-sample of the survey participants who indicated a willingness to be interviewed. They were purposively selected from high-impact communities, with a variety of perspectives sought in terms of demographics and residential location. Following data collection and concurrent analysis, further participants were sought to explore emerging themes, including the experiences of those who had relocated. In some cases this included community members who had not previously participated in the study survey.

Participant-guided mobile methods were used to explore participants' current sense of place and community. They involved an in-depth interview with the participant, combined with a walking or car-driven tour around their property or local area, guided by the participant, to view and discuss places, things, and events of importance to them. A detailed account of these methods is reported elsewhere.[28] Three researchers (LG, KB, ES) conducted the interviews in pairs. Some people were interviewed with family members, e.g., as a couple or a parent with a child, according to their preference.

Detailed memos were recorded by the interviewers immediately after the interviews to capture initial impressions and insights from the interview.[29] This supported a subsequent thematic coding of the transcripts jointly by the three interviewers, with differing interpretations discussed until consensus was reached. Coding was conducted concurrently with data collection to allow for exploration of emerging themes through subsequent participant recruitment and interview discussions.[30] The data relating specifically to relocation were then further coded and categorized to allow for a more in-depth examination of decision-making and experiences relating to staying or leaving the bushfire-affected community.[31] Categorized data were then compared and contrasted to develop a conceptual understanding of participant perspectives on relocation. Quotes are used in the reporting of results to illuminate the findings. Names have been changed to maintain anonymity as far as possible.

The final interview sample was 35 participants in 25 interviews (18 males and 17 females), ranging in age from 4 to 66 years. All were from high-bushfire-affected communities. Four of the families had relocated to other communities since the bushfires.

Interview findings

The strongest theme emerging from the interviews with regard to relocation was the importance of a sense of community. Choices to stay in the community arose from a strong commitment to place and people, often enhanced by the shared experience of the bushfires. For those who chose to leave, their sense of community had been damaged by physical changes to the built and natural environment; by painful memories associated with the disaster and related events; and by a negative social environment arising from bitter debates and arguments over choices made at the time of the fires and/or decisions about community rebuilding and allocation of recovery funds:

> it'll never be the same town. . .it feels like a really stressful place to be for a lot of different reasons. . . . One of the main things is arguing about where things ought to be and how things ought to happen. . . . We ended up selling our block of land and that was a really hard decision to make. It took a long time.

As previously reported in relation to the experiences of children and young people,[32] some people referred to a need to remain in the community to feel safe; others needed to move away from the risk of future bushfires:

> My thought of going back was needing something familiar and safe, which is probably not at all how it would have been, but in my mind it was "okay let's go and live on our block in a caravan and that will feel safe" for some strange reason.

Those who left recognized the community-wide impact of the decision to relocate and often felt guilty about not supporting the community recovery. Some of those who stayed reported feeling abandoned by the departure of close neighbors and friends, reducing their local community social networks:

> We've had a couple of close friends from here now leave that we didn't know before the fires but we've just got so close to them after the fires in some of the projects I was involved with. When they leave it actually hurts. I know we've not lost them forever and I'll still catch up with them and they come back and we visit each other and all the rest, but . . .
>
> Oh, I feel so guilty for leaving them. Yeah. We try to catch up. It's just hard. It's not the same. And there's this big, windy road to go up and down. . .
>
> There's this whole thing about people that aren't there permanently or people that have moved and aren't living there, have abandoned everybody. Oh there is. There really is. And talking to people that have decided to not go back, there's a lot of negativity about us abandoning our communities and a lot of us have felt that. Some have copped more stick than others. . . .

This also reflects the complexity of community connection, as some of those who stayed because of a commitment to place described a lost sense of community:

> It's like the place that I knew exists only as a legend, as a mythical place and that this place is so different. . . . So you absent yourself because the pain of watching things get done to people and to groups and to the community irrespective of whether it's what's needed or not is too much to bear. . . .

The majority of interview participants had experienced temporary displacement after the fires because of property damage or loss. Participants' memories of this time were often hazy because of the stress and high demands of the period. Although there were some negative experiences with some accommodation options and others did not meet family needs, there were many accounts of people being grateful for what was available until they were able to establish themselves again. Those who planned to return to their community appreciated opportunities to return for community events to reconnect. Temporary displacement commonly involved several changes in accommodation, often because of the competing needs of family members in accessing work, school, and their bushfire-affected property, meaning that some families were separated during the displacement period. Many also reported feeling unsettled and losing connection to place:

> I guess one thing about the fires is I don't plan too much ahead now. The problem is I don't think I'm going to be here. . . . Things change. I don't think we'll ever be in one place for twenty years kind of thing, not now anyway.

Some participants reported feeling torn between the community to which they had relocated and their original, bushfire-affected community:

> Losing a community and not really getting them back, that whole messy feeling of kind of wanting it back, it's never going to be the same. When you go up there, there's a horrible twisted pulling thing of feeling like you're home but not fitting in and it's not home.

Others had a dual sense of connection that was more positive:

> I still talk about it as home but I talk about this as home too. . . . And it is weird to feel at home in more than one place but that's all okay. There's no right or wrong.

All the interview participants who had relocated had carefully weighed their options and made a considered decision to move after the disaster. They may therefore have been more positive about their relocation experience than people who might have experienced less choice in the matter. Those we interviewed experienced relocation as a positive move away from the impact of the disaster and the associated environmental and social negativity, to a new community that matched their needs as a family. Participants described how they had created their own space at their new homes and had developed social networks in their new communities, making the most of new opportunities, activities, and easier access to services and facilities.

The qualitative findings highlighted the opportunity to use the *Beyond Bushfires* survey data to quantify the relationships between relocation, personal wellbeing, psychological sense of community, and other related factors.

The *Beyond Bushfires* survey

The *Beyond Bushfires* survey participants included 1,016 adults (612 females, 404 males) living in the selected communities, representing 16 percent of those eligible. Relative to census data, the sample was disproportionately older, female, and more educated than the general population, which is not unusual in research samples. For further information regarding the characteristics of this sample as a whole, including rates of mental health conditions, see Bryant et al.[33] Six participants were excluded from these analyses due to missing data on exogenous variables within the structural equation modeling (SEM). The final sample for analysis included 1,010 individuals:

897 (88.8 percent) who remained within their community (530 female, 367 male), and 113 (11.2 percent) who left the community (78 female, 35 male). Average age at the time of the bushfires was 53.1 years (56.5 at time of survey), and 350 (34.7 percent) respondents had a tertiary-level education.

Survey questions

Baseline survey data collection took place both online and by telephone, with piloting in late 2011 and data collection for the main study occurring three to four years after the fires from April 2012 to January 2013. The survey included various questions addressing fire exposure, mental health, social networks, resilience, attachment, general health, wellbeing, community hope, and demographics. The items specific to this case study are described below.

Bushfire exposure

Exposure variables included property loss, fear for one's own life, loss of friends and loved ones, and community-level impacts. Participants rated the extent to which they lost personal or business property or possessions, using a scale from 0 (Nothing) to 10 (Everything). Fear for one's own life and the death of a loved one were each measured through dichotomous yes-no questions. Regional impact (high, medium, and low impacts), as described above, was also used in measuring exposure.

Financial and relationship stressors

Participants were presented with a series of potential major life events and asked which they had experienced since February 2009. The current analysis includes negative financial events (changes in income, employment status, and occupation), negative changes in relationship status, and any experience of violence or assault.

Psychological sense of community

Participants' overall cognitive, affective, and behavioral involvement in their current community was measured by means of a six-item scale adapted from Buckner's neighborhood cohesion index (see Appendix 1).[34] Participants indicated their level of agreement using a five-point Likert-type scale (strongly agree/disagree). Internal reliability for the scale was excellent (six items, $\alpha = .81$). Additionally, to assess whether current levels of psychological sense of community were associated with past levels, participants were asked the degree to which they felt they belonged to the community they lived in during January 2009, using a five-point Likert-type scale (strongly agree/disagree).

Personal wellbeing

Wellbeing was defined in terms of life satisfaction (current and anticipated), personal resilience, and subjective overall health rating. The life satisfaction measure was derived from Cummins et al.;[35] overall current life satisfaction was ascertained by asking participants the degree to which they felt satisfied with life as a whole; anticipated life satisfaction (optimism) was measured by asking for participants' expected level of life satisfaction in a year's time. Both questions were scored on a scale from 0 (completely dissatisfied) to 10 (completely satisfied). Resilience was measured through two items taken from the Connor Davidson resilience scale (CD-RISC; see Appendix)[36] using a five-point Likert-type scale ranging from 1 (not at all true) to 5 (true nearly all of the time), and averaged. Finally, participants rated health overall (five-point scale – Poor/Excellent).[37]

Statistical analysis of survey data

This statistical analysis aimed to assess the impact of bushfire exposure on wellbeing. We developed a theoretical model proposing how wellbeing is mediated by subsequent major life stressors and psychological sense of community, comparing those who remained in the community and those who relocated (see Figure 20.1). The theoretical modeling was informed by the interview findings and previous data analyses.[38] Our model proposed that disaster exposure itself not only affects subjective wellbeing directly, but also precipitates subsequent stressful circumstances in the following months and years, which have an additional negative effect on wellbeing. We hypothesized that the psychological sense of community is impacted by a disaster experience. However, as highlighted above, for any single individual, this relationship could reasonably be either positive or negative. Nonetheless, we (weakly) hypothesized that across the sample as a whole, there will be a positive association, with the disaster experience leading to increased sense of community, presumably through the mobilization of social support[39] and a heightened sense of common fate.[40] At the same time, however, the psychological sense of community will itself be diminished by the experience of negative subsequent events that hinder the individual's ability to participate in local community life. In turn, this sense of community is closely tied to wellbeing. Centrally, we consider the role of relocation as a response to the stress of the disaster event and subsequent stressors, which may come with trade-offs in terms of the support and experience of community. In particular, relocation may attenuate the negative influence that disaster

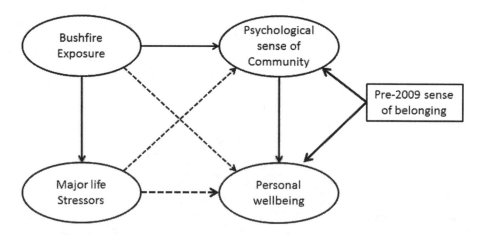

Figure 20.1 Theoretical model.

Solid lines represent positive relationships; dashed lines represent negative relationships.

exposure has on personal wellbeing, but at the same time may lessen a protective influence provided through one's sense of community. The resulting theoretical model to be tested is shown in Figure 20.1. The components of the model include a direct effect of bushfire exposure on personal wellbeing; indirect effects from exposure to wellbeing, as mediated by psychological sense of community and financial/relationship life stressors, respectively; and a direct effect of life stressors on the psychological sense of community. Education level and age were entered as control variables for wellbeing, major life stressors, and the psychological sense of community.[41] Sex was not found to be a significant predictor and so was dropped from the analyses.

The statistical analyses were conducted with SEM using Mplus version 7.3.[42] Through SEM, one can specify a structural model comprising several variables, interlinked through hypothesized/assumed causal paths, with the aim of defining and testing the model which best explains the relationships among these variables. As with all cross-sectional research, SEM cannot determine causality. Instead, causal assumptions are based on an explicit theoretical framework and other strong assumptions made by the researcher and informed by earlier findings. In the current analysis, these variables are bushfire exposure, financial and relationship life stressors, the psychological sense of community, and personal wellbeing. As can be seen in the model (Figure 20.1), a one-way arrow denotes a direct predictive relationship between an independent variable and a dependent variable. Relationships between any two variables can be direct, or indirect as mediated by a third variable.[43]

Given that the data were collected using ordinal scales or dichotomous (yes/no) response formats, Weighted Least Squares Mean Variance (WLSMV) estimation with delta parameterization was used. This method of estimation does not carry assumptions of normality, and is considered the default choice for SEM with ordered categorical data.[44]

Following estimation, the model was assessed to determine whether it was an adequate fit for the data, as ascertained through the calculation of various model fit indices.[45]

Survey findings

Participant characteristics

Chi-square tests revealed that the sample as a whole was disproportionately female, χ^2 (1, 1010) = 42.02, $p < .001$. Comparisons between those who stayed in the community and those who relocated in terms of demographic variables and indicators or bushfire exposure are presented in Table 20.1. These revealed significant differences between the two sub-groups in terms of gender representation. Furthermore, the two sub-groups differed in terms of fear for one's life and property loss, with relocated individuals reporting higher levels of both. Those who relocated also reported lower levels of psychological sense of community (in relation to their current community). Nevertheless, in terms of wellbeing measures, those relocated were only slightly less satisfied with life currently, with no significant differences in terms of optimism, resilience, or subjective self-ratings of health.

The statistical model

The primary aim of the statistical analyses was to see if relocation is *a moderator* of the relationships displayed in the hypothesized model. A moderator is defined as any

Table 20.1 Participant characteristics, by community relocation status

	Community-staying (n = 897)	Community-relocated (n = 113)	
	N (% of sub-group)		χ^2 (1, 1010)
Sex (Female)	530 (59.1%)	78 (68.0%)	4.14*
Tertiary education	302 (33.7%)	48 (39.2%)	3.44
Fear for life	420 (47.8%)	65 (58.6%)	4.58*
Loss of someone close	254 (28.6%)	40 (35.7%)	2.42
	M (SD)		T-test/Spearman's rho
Age	53.3(12.9)	51.4(15.4)	t(132.7) = 1.46ns
Property loss	4.1(3.9)	7.1(3.9)	r_s = .23***
Psychological Sense of Community	3.1(.68)	2.6(.94)	t(127.2) = 5.71***
Financial/relationship stressors	.96(1.1)	1.24(1.41)	r_s = .05
Current life satisfaction	7.0(2.4)	6.2(2.5)	r_s = −.11***
Optimism (anticipated satisfaction)	7.9(2.1)	7.6(1.8)	r_s = −.06
Resilience	4.3(.82)	4.2(.86)	r_s = −.02
Subject health rating	2.2(1.1)	2.2(1.1)	r_s = .002

*** $p < .001$, ** $p <.01$, * $p < .05$; Spearman's rank correlation used instead of t-test for non-continuous data (staying = 0, relocation = 1).

variable that alters the strength or direction of the link between two other variables in the model.[46] In this study, the statistical analyses for detecting moderation effects were three-fold. First, an overall test was necessary to determine whether the hypothesized model held up as an adequate fit for the actual data as a whole. If so, the second step was to test the model independently on the sub-group of relocated individuals and the sub-group of those who stayed in the community, respectively, in order to gain an idea of how the model may differ between sub-groups. A final series of tests was then carried out to determine formally how the model was the same and how it was different, between the two sub-groups. Accordingly, for the first step, the model was first applied to the entire sample, irrespective of sub-group. Various model fit indices suggested an excellent overall fit (Table 20.2). All paths specified in the overall model were significant, with the notable exception of the direct paths from bushfire exposure to wellbeing ($\beta = .05$, $p = .52$) and from bushfire exposure to psychological sense of community ($\beta = .08$, $p = .20$).

Next, the overall model was applied independently to each group. The model was found to be an excellent fit when applied to either the community-staying sub-group or the community-relocating sub-group (Table 20.2). Nevertheless, comparing the two models side by side revealed various apparent differences between the two groups in terms of the strengths of path coefficients, to be tested through multigroup analyses.

Differences between groups: Moderation effects

Given these apparent differences in the model between the two sub-groups, it was necessary to determine how this model was similar and how it differed between those who stayed in the community and those who relocated. A final series of

Table 20.2 Model fit and tests of moderation (invariance)

Model description	χ^2	Df	CFI	NNFI	RMSEA (90%CI)	Model comparison	$\Delta \chi^2$	Δdf	Signif. level
Basic model, complete sample	689.94	387	.967	.968	.039 (.035–.044)				
Basic model, staying sub-group (n = 897)	489.45	156	.964	.957	.049 (.044–.054)				
Basic model, relocated sub-group (n = 113)	196.46	156	.962	.954	.048 (.022–.068)				
Multigroup analysis									
1 Configural model (All parameters free between sub-groups)	661.20	360	.967	.966	.041 (.036–.046)				
2 Equal factor loadings, equal thresholds (Scalar invariance)	663.65	373	.968	.968	.039 (.034–.044)	2–1	11.19	13	.595
3a Equal loadings, Stressors → Wellbeing constrained	667.99	374	.968	.968	.039 (.035–.044)	3a–2	6.05	1	.014
3b Equal loadings, PSC → Pre-2009 Belonging constrained	690.78	374	.965	.965	.041 (.036–.046)	3b–2	24.86	1	<.001
3c Equal loadings, Exposure → Wellbeing constrained	667.92	374	.968	.963	.039 (.035–.044)	3c–2	4.59	1	.032
3d Equal loadings, PSC → Wellbeing constrained	667.48	374	.968	.963	.039 (.035–.044)	3d–2	3.93	1	.048
4 Equal loadings, all paths constrained	689.94	387	.967	.968	.039 (.035–.044)	4–2	36.74	14	<.001
5 Equal loadings, moderated paths unconstrained (3a–3d), all other paths constrained (Figure 20.2)	650.63	383	.971	.971	.037 (.032–.042)	5–2	9.19	10	.514

All χ^2 significant at p <.05 level or more.

steps were conducted to determine whether these differences were statistically significant and which would establish relocation as a moderator of effects between disaster exposure and personal wellbeing.[47] To do so, we followed a general multigroup approach (outlined in detail elsewhere).[48] In brief, the multigroup analysis in Table 20.4 depicts a series of hierarchically ordered steps by which the statistical model was successively merged (made the same) across groups in terms of various statistical parameters (i.e., factor loadings, item thresholds, error terms, etc.). Each model was tested against the preceding model to see if the fit for the model deteriorated significantly. If not, the succeeding model was accepted as a more parsimonious representation of the data.

In particular, Model 2 shows that the data were *scalar invariant*, meaning that the theoretical constructs were adequately equivalent so that it was possible to make the hypothesized statistical comparisons between relocated and non-relocated individuals (i.e., moderation effects). Next, possible moderation effects were tested for four of the paths (as identified by Wald tests): the paths from exposure to wellbeing, from life stressors to wellbeing, from psychological sense of community to wellbeing, and from past sense of belonging to psychological sense of community. In order to test each, four separate models were tested in which the (potentially moderated) path was held constant in order to determine whether the model fit significantly differed (Table 20.2, Models 3a–3d). Models 3a–3d were explicit tests of the moderation effects, showing that the strength of these paths differ significantly between groups. A final model (Model 5) was tested which treats the two sub-groups in exactly the same way, except for the four moderated paths. This final model is depicted in Figure 20.2.

Lasting impact of bushfire exposure: Direct versus indirect effects

As seen above, one apparent difference in the model is that while the effect of disaster exposure on wellbeing was direct among relocated individuals, it was indirect among those who stayed, as mediated by financial and relationship stressors. Accordingly, it was appropriate to conduct an analysis of moderated mediation in order to confirm that this apparent difference was statistically significant.[49] Applied to the equal loading model, these analyses confirmed that among those who stayed in the community, there was a significant negative indirect effect of disaster exposure on personal wellbeing, as mediated by financial/relationships stressors ($\beta = -0.45$, $p = .001$). By contrast, among those who relocated, this indirect effect was not significant ($\beta = -0.07$, $p = .578$). A Wald test indicated that the difference between these (unstandardized) indirect effects was significant ($\theta = -0.424$, $p = .041$).[50] Overall, this analysis confirms that the impact of disaster exposure was direct among those who relocated and indirect among those who stayed (mediated by life stressors), and that the difference between the two was itself statistically significant. In other words, subjective wellbeing among the relocated was a direct outcome of bushfire exposure. This is in contrast to those who stayed in the community, for whom wellbeing was a result of the subsequent major life stressors which exposure had precipitated.

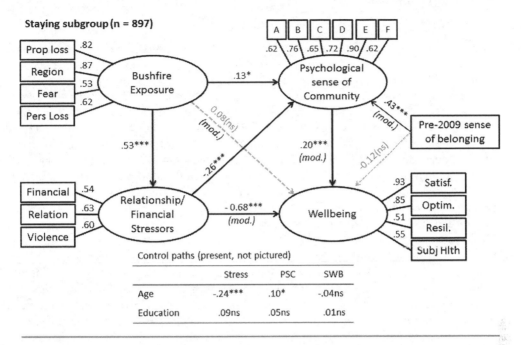

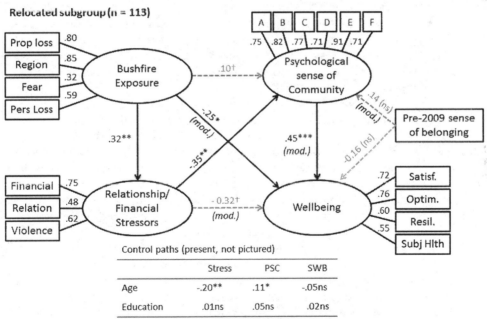

Figure 20.2 Final model: testing similarities and differences between sub-groups.

Parameters are standardized (β). ***p < .001; **p ≤ .01, *P < .05 †p <.10. All factor loadings significant at p <.001. Significant moderation paths are indicated (mod).

Discussion

This case study shows clear relationships between sense of community, relocation, and personal wellbeing. Using mixed methods provided an opportunity to develop a comprehensive understanding of the nature and extent of those relationships.

The findings from the in-depth interviews in the *Beyond Bushfires* study identified the impact of the bushfire experience on the sense of community and how this influenced subsequent decisions regarding whether to remain in the affected community or to relocate. It also provided insight into the individual and community-level impacts of those decisions. These findings informed the development of a theoretical model for statistical analyses of the survey data. The results from the analyses largely supported the theoretical model, albeit with important differences between staying and relocating sub-groups. Overall, personal wellbeing was adversely predicted by negative life events, be that the disaster experience itself, or the negative events that followed. However, statistical analyses showed important differences between those who remained in the community versus those who relocated, revealing two broad effects concerning the ways in which disaster exposure affects current subjective wellbeing.

For those who relocated, current wellbeing was more strongly tied to disaster exposure itself, rather than subsequent life stressors. Conversely, for those who remained within their community, wellbeing was tied more distinctly to subsequent negative life events, over and above initial fire exposure. This is supported by the interview data, which revealed that the post-disaster community environment was often the motivation for leaving and that, for those interviewed, relocation was generally experienced as positive in terms of reducing exposure to repeated negative visual and social bushfire related encounters. It may also have been a means of restoring ontological security by moving to a community with more stability.[51] Thus, while there is no evidence to suggest that those who relocated experienced more or fewer financial and/or relationship stressors (both have been reported in other studies of families who have relocated post-disaster),[52] their relocation may have effectively dealt with or circumvented the impact of these stressors. It may also reflect a difference between the uncertain costs of rebuilding versus known costs of renting/buying an existing home in a new location. Regardless, despite this apparent circumvention, the disaster exposure itself has a lingering effect for those who relocate, independent of subsequent negative events, which is greater than for those who stayed. Interview findings suggest that this may arise due to fewer opportunities for those who relocate for shared processing of the disaster experience and less access to recovery services.[53]

The second pattern concerns the sense of community and its influence on wellbeing. As theorized, a significant relationship from bushfire exposure to sense of community exists among those who stayed in the community, with increased exposure associated with increased sense of community. As initially hypothesized, those who were most impacted may have been primed to think in terms of their need for community and the common fate they share with their neighbors. The interview findings indicated a strong sense of community among those who stayed, enhanced by the shared disaster experience, consistent with studies of wildfires in British Columbia, and Canada, and New Orleans following Hurricane Katrina.[54]

Furthermore, while current psychological sense of community was a significant predictor of wellbeing for both sub-groups, it bore a significantly *stronger* impact on wellbeing among the relocated sub-group. Interview findings suggested that people appreciated the positive physical and social environment of their new communities and were actively seeking opportunities to become involved, consistent with reports of families' experiences of relocation post-Hurricane Katrina.[55] In addition, the interview findings indicated that it is not individual bushfire exposure that influences sense of community but rather the individual experiences of the post-disaster community. For some, sense of community is enhanced by the shared experience of the disaster and the rebuilding processes. For others, it is lost through the damage, disruption, and disharmony.[56]

Altogether, the model indicates a counterbalancing effect of relocation. Much of the model appears to suggest that those who relocate are at greater risk for reduced wellbeing due to a stronger (direct) link with initial disaster exposure. Furthermore, the wellbeing of those who relocate is more strongly intertwined with their current sense of community, shown to be reduced for those who have relocated (see Table 20.1). These heightened risk factors are at least partially offset, however, by the moderating influence of a reduced impact of life stressors. Leaving the community therefore appears to mitigate or interrupt the effect that subsequent stressful life events have on personal wellbeing, to the point where relocated individuals report similar levels of subjective wellbeing to those who stayed (see Table 20.1).

Finally, we see that among those who stayed in their community, current sense of community is associated with recollections of community belonging pre-2009. By contrast, perhaps unsurprisingly, among those who relocated there is no association between current sense of community and sense of belonging to their prior community, indicating that a psychological sense of community is not a constant, behavioral, or personality attribute. Rather, it suggests that one's emotional and behavioral attachments to place are linked keenly to a particular community.[57] Re-establishing oneself elsewhere is likely to be a demanding process that needs careful consideration, but in the interviews it appeared to be embraced by those who decided to leave the challenging environment of a disaster-affected community.

This case study highlights different recovery service needs for those who stayed in their community and those who relocated. Those who stay may benefit from an increased focus on alleviating the subsequent financial and relationship stressors occurring post-disaster. On the other hand, services that support a recovery from the disaster event itself need to be more accessible to those who have relocated. In order to achieve this it would be helpful to establish a register of those who relocate to assist in geographical planning for service delivery. Finally, information about the impacts of staying or relocating on personal wellbeing needs to be made available in disaster-impacted communities to support individuals and families in making informed decisions about whether to stay or to relocate and how to maximise the positives and minimise the potential risks of that decision. There may also be service implications in addressing the disappointment felt by friends and neighbors left behind and the guilt of those who leave.

Limitations

There are a number of limitations to be considered when interpreting the case study findings. First, as in any study, there are limitations in terms of the measures. In particular, the current model lacks an indicator as to when an individual left the community. There may yet be important differences between those who relocated immediately and those who remained for a longer period before relocating. Furthermore, the Buckner scale was originally developed for use at the community level as an aggregate measure of social cohesion. In this study, selected items from the scale were used at an individual level to measure the sense of community. Other measures of individuals' psychological sense of community elucidating the various behavioral, cognitive, and affective dimensions of the sense of community could usefully give a more precise picture of the effect of disaster exposure and relocation.

Second, integrating the survey and interview findings contributes to increased understanding of relocation experiences; however, these should be treated with caution due to the different time periods over which the data were collected.

Third, we could not randomize people to relocating or remaining in the community. As noted, relocating following disaster is influenced by a range of environmental and personal factors, difficult to accommodate into the modeling of psychological outcomes. It is also possible that those who participated in the study had a stronger sense of community than those who did not, thus limiting the generalizability of these findings to the population as a whole and to other contexts.

Conclusion

Decision-making about whether to stay or to relocate from disaster-affected communities is complex and challenging, and, although experienced by individuals and families, has community-wide impacts. This Australian bushfire case study demonstrates the importance of the psychological sense of community in both decision-making and the experience of relocating. For those who feel a strong attachment to their current community, a decision to stay is likely to be most supportive of personal wellbeing. For those whose sense of community has been disrupted by a disaster impact, relocating is likely to reduce the influence of subsequent financial and relationship stressors on personal wellbeing, but will still require them to build social connections with their new community. This has implications in terms of targeted service provision following bushfires to support informed decision-making about relocation; to assist those who stay in their community to deal with financial and relationship stressors; and to increase access to appropriate disaster recovery services for those who have relocated.

Acknowledgments

This chapter reports on findings from the *Beyond Bushfires* study. The authors and investigators wish to thank the community members, local government, and service providers from the participating communities who have supported the development and conduct of the *Beyond Bushfires* study. We acknowledge Professor Philippa Pattison, who, although not an author on this paper, contributed greatly to the *Beyond Bushfires* study as an investigator. We also wish to acknowledge the literature review conducted by Marita Smith as part of her Master of Social Work research, which was helpful background to the work prepared for this paper. We gratefully acknowledge the funding support received from the Australian Research Council for the *Beyond Bushfires* study, and from the Jack Brockhoff Foundation for infrastructure and salary support for Lisa Gibbs and Elizabeth Waters. Finally, we dedicate this chapter to the memory of Professor Elizabeth Waters, whose leadership, vision, and vitality will never be forgotten.

Appendix A

Measures of the psychological sense of community

Selected Items from Buckner's (1988) Neighborhood Cohesion Instrument

1   I plan to remain a resident of this community for a number of years
2   I regularly stop and talk with people in my community
3   I think I agree with most people in my community about what is important in life
4   I would be willing to work together with others on something to improve my community
5   I feel like I belong to this community
6   I am very attached to the local environment and landscape

Connor-Davidson Resilience Scale (selected items)

1   I am able to adapt to change"
2   I tend to bounce back after illness or hardship.

Pre-2009 community belonging

I felt like I belonged to this community (January 2009).

Notes

1   J. Lindy and M. Grace (1986), "The Recovery Environment: Continuing Stressor Versus a Healing Psychosocial Space," in *Disasters and Mental Health: Selected Contemporary Perspectives*, B. Sowder and M. Lystad, eds., pp. 147–60 (Rockville, MD: National Institute of Mental Health); F.H. Norris, M.J. Friedman, P. Watson, C.M. Byrne, E. Diaz, and K. Kaniasty (2002), "60,000 Disaster Victims Speak: Part I. An Empirical Review of the Empirical Literature, 1981–2001," *Psychiatry: Interpersonal and Biological Processes* 65(3), pp. 207–39; D. Guha-Sapir, F. Vos, R. Below, and S. Ponserre (2011), *Annual Disaster Statistical Review 2010: The Numbers and Trends* (Brussels, Belgium: CRED, Université Catholique de Louvain).
2   L. Uscher-Pines (2009), "Health Effects of Relocation Following Disaster: A Systematic Review of the Literature," *Disasters* 33(1), pp. 1–22.
3   E. Chamlee-Wright and V.H. Storr (2009), " 'There's No Place Like New Orleans': Sense of Place and Community Recovery in the Ninth Ward after Hurricane Katrina," *Journal of Urban Affairs* 31(5), pp. 615–34; J.K. Riad and F.H. Norris (1996), "The Influence of Relocation on the Environmental, Social, and Psychological Stress Experienced by Disaster Victims," *Environment and Behavior* 28(2), pp. 163–82.
4   R.L. Hawkins and K. Maurer (2011), " 'You Fix My Community, You Have Fixed My Life': The Disruption and Rebuilding of Ontological Security in New Orleans," *Disasters* 35(1),

pp. 143–59; L. Peek, B. Morrissey, and H. Marlatt (2011), "Disaster Hits Home: A Model of Displaced Family Adjustment after Hurricane Katrina," *Journal of Family Issues* 32(10), pp. 1371–96; M.T.G. Tuason, C.D. Güss, and L. Carroll (2012), "The Disaster Continues: A Qualitative Study on the Experiences of Displaced Hurricane Katrina Survivors," *Professional Psychology: Research and Practice* 43(4), pp. 288–97.

5 M. Lewicka (2011), "Place Attachment: How Far Have We Come in the Last 40 Years?" *Journal of Environmental Psychology* 31, pp. 207–30; J. Buckner (1988), "The Development of an Instrument to Measure Neighborhood Cohesion," *American Journal of Community Psychology* 16(6), pp. 771–91; D.W. McMillan and D.M. Chavis (1986), "Sense of Community: A Definition and Theory," *American Journal of Community Psychology* 14(1), pp. 6–23.

6 R.S. Cox and K.M.E. Perry (2011), "Like a Fish Out of Water: Reconsidering Disaster Recovery and the Role of Place and Social Capital in Community Disaster Resilience," *American Journal of Community Psychology* 48(3–4), pp. 395–411; D.W. Morgan, N.I.M. Morgan, and B. Barrett (2006), "Finding a Place for the Commonplace: Hurricane Katrina, Communities, and Preservation Law," *American Anthropologist* 108(4), pp. 706–18.

7 Chamlee-Wright and Storr (2009); E.L. Kick, J.C. Fraser, G.M. Fulkerson, L.A. McKinney, and D.H. De Vries (2011), "Repetitive Flood Victims and Acceptance of FEMA Mitigation Offers: An Analysis with Community-System Policy Implications," *Disasters* 35(3), pp. 510–39; J.S. Smith and M.R. Cartlidge (2011), "Place Attachment among Retirees in Greensburg, Kansas," *Geographical Review* 101(4), pp. 536–55.

8 Chamlee-Wright and Storr (2009); Morgan, Morgan, and Barrett (2006); Smith and Cartlidge (2011); Hawkins and Maurer (2011); K.A. Roberto, T.L. Henderson, Y. Kamo, and B.R. McCann (2010), "Challenges to Older Women's Sense of Self in the Aftermath of Hurricane Katrina," *Health Care for Women International* 31(11), pp. 981–96.

9 Hawkins and Maurer (2011); Smith and Cartlidge (2011).

10 Hawkins and Maurer (2011); A. Giddens (1990), *The Consequences of Modernity* (Cambridge, UK: Polity).

11 McMillan and Chavis (1986).

12 Lewicka (2011); Buckner (1988); D.W. McMillan (1996), "Sense of Community," *Journal of Community Psychology* 24(4), pp. 315–25.

13 Chamlee-Wright and Storr (2009); Smith and Cartlidge (2011); B. Phillips, P.A. Stukes, and P. Jenkins (2012), "Freedom Hill Is Not for Sale – And Neither Is the Lower Ninth Ward," *Journal of Black Studies* 43(4), pp. 405–26; B. Carrol, H. Morbey, R. Balogh, and G. Araoz (2009), "Flooded Homes, Broken Bonds, the Meaning of Home, Psychological Processes and their Impact on Psychological Health in a Disaster," *Health and Place* 15(2), pp. 540–47; M.T. Fullilove (1996), "Psychiatric Implications of Displacement: Contributions from the Psychology of Place," *American Journal of Psychiatry* 153(12), pp. 1516–23.

14 Chamlee-Wright and Storr (2009); B. Phillips, P.A. Stukes, and P. Jenkins (2012); E. Wiesenfeld and R. Panza (1999), "Environmental Hazards and Home Loss: The Social Construction of Becoming Homeless," *Community, Work and Family* 2(1), pp. 51–65.

15 Internal Displacement Monitoring Centre (2013), *Global Figures* [Web page], retrieved January 7, 2016, from http://www.internal-displacement.org/global-figures#natural

16 Uscher-Pines (2009).

17 Ibid.; C. IJzermans, A. Dirkzwager, and E. Breuning (2005), *Long-Term Health Consequences of Disaster* (Utrecht, the Netherlands: NIVEL).

18 V.J. Carr, T.J. Lewin, R.A. Webster, J.A. Kenardy, P.L. Hazell, and G.L. Carter (1997), "Psychosocial Sequelae of the 1989 Newcastle Earthquake: II. Exposure and Morbidity Profiles during the First 2 Years Postdisaster," *Psychological Medicine* 27, pp. 167–78; R. Acierno, "Risk and Protective Factors for Psychopathology among Older Versus Younger Adults after the 2004 Florida Hurricanes," *American Journal of Geriatric Psychiatry* 14(12), pp. 1051–59; F.H. Norris, A.D. Murphy, C.K. Baker, and J.L. Perilla (2004), "Postdisaster PTSD over Four Waves of a Panel Study of Mexico's 1999 Flood," *Journal of Traumatic Stress* 17(4), pp. 283–92; F. van Griensven, M.L.S. Chakkraband, W. Thienkrua, W. Pengjuntr, B.L. Cardozo, P. Tantipiwatanaskul, P.A. Mock, S. Ekassawin, A. Varangrat, C. Gotway, M. Sabin, and J. W. Tappero (2006), "Mental Health Problems among Adults in Tsunami-Affected Areas in Southern Thailand," *Journal of the American*

Medical Association 296(5), pp. 537–48; Y.L. Chen, C.S. Lai, W.T. Chen, W.Y. Hsu, Y.C. Wu, P.W. Wang, and C.S. Chen (2011), "Risk Factors for PTSD after Typhoon Morakot among Elderly People in Taiwanese Aboriginal Communities," *International Psychogeriatrics* 23(10), pp. 1686–91.

19 Uscher-Pines (2009).

20 Ibid.

21 J.T. Blaze and D.W. Shwalb (2009), "Resource Loss and Relocation: A Follow-Up Study of Adolescents Two Years after Hurricane Katrina," *Psychological Trauma: Theory, Research, Practice and Policy* 1(4), pp. 312–22.

22 Uscher-Pines (2009).

23 Ibid.

24 Peek, Morrissey, and Marlatt (2011).

25 Hawkins and Maurer (2011); Peek, Morrissey, and Marlatt (2011); Tuason, Güss, and Carroll (2012).

26 Victorian Bushfires Royal Commission (2009), *The 2009 Victorian Bushfires Royal Commission Final Report Summary*, retrieved July 8, 2013, from http://www.royalcommission.vic.gov.au/Commission-Reports/Final-Report

27 "Beyond Bushfires," *Melbourne School of Population and Global Health* [Web page], available at www.beyondbushfires.org.au; L. Gibbs, E. Waters, R. Bryant, P. Pattison, D. Lusher, L. Harms, J. Richardson, C. MacDougall, K. Block, E. Snowdon, et al. (2013), "Beyond Bushfires: Community, Resilience and Recovery – A Longitudinal Mixed-Method Study of the Medium to Long-Term Impacts of Bushfires on Mental Health and Social Connectedness," *BMC Public Health* 14(7), pp. 634–43, doi:10.1177/0004867414534476.

28 K. Block, L. Gibbs, E. Snowdon, and C. MacDougall (2014), "Participant-Guided Mobile Methods: Investigating Personal Experiences of Communities," *SAGE Research Methods Cases*, doi:10.4135/978144627305014536373.

29 K. Charmaz (2014), *Constructing Grounded Theory*, 2nd edn. (London, UK: Sage); M.Q. Patton (2002), *Qualitative Research and Evaluation Methods*, 3rd edn. (London, UK: Sage).

30 J. Green, K. Willis, E. Hughes, R. Small, N. Welch, L. Gibbs, and J. Daly (2007), "Generating Best Evidence from Qualitative Research: The Role of Data Analysis," *Australian and New Zealand Journal of Public Health* 31(6), pp. 545–50.

31 Patton (2002); Green, Willis, Hughes, Small, Welch, Gibbs, and Daly (2007).

32 L. Gibbs, K. Block, L. Harms, C. MacDougall, E. Snowdon, G. Ireton, D. Forbes, J. Richardson, and E. Waters (2015), "Children and Young People's Wellbeing Post-Disaster: Safety and Stability Are Critical," *International Journal of Disaster Risk Reduction*, pp. 195–201, doi:10.1016/j.ijdrr.2015.06.006

33 R.A. Bryant, E. Waters, L. Gibbs, H.C. Gallagher, P. Pattison, D. Lusher, C. MacDougall, L. Harms, K. Block, and E. Snowdon (2014), "Psychological Outcomes Following the Victorian Black Saturday Bushfires," *Australian and New Zealand Journal of Psychiatry* 48(7), pp. 634–43, doi:10.1177/0004867414534476

34 J.C. Buckner (1988), "The Development of an Instrument to Measure Neighborhood Cohesion," *American Journal of Community Psychology* 16(6), pp. 771–91.

35 R.A. Cummins, R. Eckersley, J. Pallant, J. Van Vugt, and R. Misajon (2003), "Developing a National Index of Subjective Wellbeing: The Australian Unity Wellbeing Index," *Social Indicators Research* 64(2), pp. 159–90.

36 K.M. Connor and J.R. Davidson (2003), "Development of a New Resilience Scale: The Connor-Davidson Resilience Scale (CD-RISC)," *Depression and Anxiety* 18(2), pp. 76–82.

37 J.E. Ware Jr. and C.D. Sherbourne (1992), "The MOS 36-Item Short-Form Health Survey (SF-36): I. Conceptual Framework and Item Selection," *Medical Care* 30(6), pp. 473–83.

38 D. Forbes, N. Alkemade, E. Waters, L. Gibbs, H.C. Gallagher, P. Pattison, D. Lusher, C. MacDougall, L. Harms, K. Block, et al. (2015), "Anger and Major Life Stressors as Predictors of Psychological Outcomes following the Victorian Black Saturday Bushfires," *Australian and New Zealand Journal of Psychiatry* 49(8), pp. 706–13, doi:10.1177/0004867414565478.

39 K. Kaniasty and F.H. Norris (1995), "In Search of Altruistic Community: Patterns of Social Support Mobilization following Hurricane Hugo," *American Journal of Community Psychology* 23(4), pp. 447–77.

40 K. Lewin (1948), *Resolving Social Conflicts: Selected Papers on Group Dynamics* (Oxford, UK: Harper).

41 M.J. De Silva, K. McKenzie, T. Harpham, and S.R. Huttly (2005), "Social Capital and Mental Illness: A Systematic Review," *Journal of Epidemiology and Community Health* 59(8), pp. 619–27; S. Subramanian and I. Kawachi (2004), "Income Inequality and Health: What Have We Learned So Far?" *Epidemiologic Reviews* 26(1), pp. 78–91; M. Prezza and S. Costantini (1998), "Sense of Community and Life Satisfaction: Investigation in Three Different Territorial Contexts," *Journal of Community and Applied Social Psychology* 8(3), pp. 181–94; P. Butterworth, B. Rodgers, and T.D. Windsor (2009), "Financial Hardship, Socio-economic Position and Depression: Results from the PATH Through Life Survey," *Social Science and Medicine* 69(2), pp. 229–37.

42 L. Muthén and B. Muthén (1998/2013), *Mplus User's Guide*, 7th edn. (Los Angeles, CA: Authors).

43 R.M. Baron and D.A. Kenny (1986), "The Moderator–Mediator Variable Distinction in Social Psychological Research: Conceptual, Strategic, and Statistical Considerations," *Journal of Personal Social Psychology* 51(6), p. 1173.

44 T.A. Brown (2006), *Confirmatory Factor Analysis for Applied Research* (New York, NY: Guilford).

45 Ibid.; R.J. Vandenberg and C.E. Lance (2000), "A Review and Synthesis of the Measurement Invariance Literature: Suggestions, Practices, and Recommendations for Organizational Research," *Organizational Research Methods* 3(1), pp. 4–70. Model fit was assessed using Bentler's comparative fit index (CFI), Bentler and Bonett's non-normed fit index (NNFI), and Steiger and Lind's Root Mean Square Error of Approximation (RMSEA). Given the use of WLSMV estimation, slightly more stringent cutoffs are required; CFI $\geq$.96; NNFI $\geq$ .95; RMSEA $\leq$.05 indicate good fit. C.-Y. Yu (2002), *Evaluating Cutoff Criteria of Model Fit Indices for Latent Variable Models with Binary and Continuous Outcomes* (Los Angeles, CA: University of California); P.M. Bentler (1990), "Comparative Fit Indexes in Structural Models," *Psychological Bulletin* 107(2), p. 238; P.M. Bentler and D.G. Bonett (1980), "Significance Tests and Goodness-of-Fit in the Analysis of Covariance Structures," *Psychological Bulletin* 88(3), p. 588; J.H. Steiger and J.C. Lind (1980), "Statistically Based Tests for the Number of Common Factors," in *Annual Meeting of the Psychometric Society, Iowa City, IA: 1980*, pp. 424–53; L.-T. Hu and P.M. Bentler (1999), "Cutoff Criteria for Fit Indexes in Covariance Structure Analysis: Conventional Criteria versus New Alternatives," *Structural Equation Modeling: A Multidisciplinary Journal* 6(1), pp. 1–55.

46 Baron and Kenny (1986).

47 Ibid.

48 Vandenberg and Lance (2000); X. Koufteros and G.A. Marcoulides (2006), "Product Development Practices and Performance: A Structural Equation Modeling-Based Multi-Group Analysis," *International Journal of Production Economics* 103(1), pp. 286–307; D.A. Sass and T.A. Schmitt (2013), "Testing Measurement and Structural Invariance," in *Handbook of Quantitative Methods for Educational Research*, T. Teo, ed., pp. 315–45 (New York, NY: Springer).

49 Baron and Kenny (1986); K.J. Preacher, D.D. Rucker, and A.F. Hayes (2007), "Addressing Moderated Mediation Hypotheses: Theory, Methods, and Prescriptions," *Multivariate Behavioral Research* 42(1), pp. 185–227.

50 This analysis was conducted using a bootstrap resampling technique with 1,000 samples. D.P. MacKinnon, C.M. Lockwood, and J. Williams (2004), "Confidence Limits for the Indirect Effect: Distribution of the Product and Resampling Methods," *Multivariate Behavioral Research* 39(1), pp. 99–128.

51 Hawkins and Maurer (2011); Uscher-Pines (2009).

52 [7, 9, 65] Hawkins and Maurer (2011); Tuason et al. (2012); L. Peek (2008), "Children and Disasters: Understanding Vulnerability, Developing Capacities and Promoting Resilience – An Introduction," *Children, Youth and Environments* 18(1), pp. 1–29.

53 Chamlee-Wright and Storr (2009); Tuason et al. (2012); Smith and Cartlidge (2011); Phillips et al. (2012); Carrol et al. (2009); Fullilove (1996).

54 Cox and Perry (2011); Morgan et al. (2006).

55 Hawkins and Maurer (2011); Peek et al. (2011).

56 Hawkins and Maurer (2011); Peek et al. (2011); Tuason et al. (2011).
57 Chamlee-Wright and Storr (2009); Morgan et al. (2006); Smith and Cartlidge (2011); Hawkins and Maurer (2011); Roberto et al. (2010)

Bibliography

Acierno, R., K.J. Ruggiero, D.G. Kilpatrick, H.S. Resnick, and S. Galea. (2006). "Risk and Protective Factors for Psychopathology among Older Versus Younger Adults after the 2004 Florida Hurricanes." *American Journal of Geriatric Psychiatry* 14(12), pp. 1051–59.

Baron, R.M., and D.A. Kenny. (1986). "The Moderator–Mediator Variable Distinction in Social Psychological Research: Conceptual, Strategic, and Statistical Considerations." *Journal of Personal Social Psychology* 51(6), p. 1173.

Bentler, P.M. (1990). "Comparative Fit Indexes in Structural Models." *Psychological Bulletin* 107(2), p. 238.

———, and D.G. Bonett. (1980). "Significance Tests and Goodness-of-Fit in the Analysis of Covariance Structures." *Psychological Bulletin* 88(3), p. 588.

Blaze, J.T., and D.W. Shwalb. (2009). "Resource Loss and Relocation: A Follow-Up Study of Adolescents Two Years after Hurricane Katrina." *Psychological Trauma: Theory, Research, Practice and Policy* 1(4), pp. 312–22.

Block, K., L. Gibbs, E. Snowdon, and C. MacDougall. (2014). "Participant-Guided Mobile Methods: Investigating Personal Experiences of Communities." *SAGE Research Methods Cases*, doi:10.4135/978144627305014536373

Brown, T.A. (2006). *Confirmatory Factor Analysis for Applied Research*. New York, NY: Guilford.

Bryant, R.A., E. Waters, L. Gibbs, H.C. Gallagher, P. Pattison, D. Lusher, C. MacDougall, L. Harms, K. Block, and E. Snowdon. (2014). "Psychological Outcomes Following the Victorian Black Saturday Bushfires." *Australian and New Zealand Journal of Psychiatry* 48(7), pp. 634–43, doi:10.1177/0004867414534476

Buckner, J. (1988). "The Development of an Instrument to Measure Neighborhood Cohesion." *American Journal of Community Psychology* 16(6), pp. 771–91.

Butterworth, P., B. Rodgers, and T.D. Windsor. (2009). "Financial Hardship, Socio-economic Position and Depression: Results from the PATH through Life Survey." *Social Science and Medicine* 69(2), pp. 229–37.

Carr, V.J., T.J. Lewin, R.A. Webster, J.A. Kenardy, P.L. Hazell, and G.L. Carter. (1997). "Psychosocial Sequelae of the 1989 Newcastle Earthquake: II. Exposure and Morbidity Profiles during the First 2 Years Postdisaster." *Psychological Medicine* 27, pp. 167–78.

Carrol, B., H. Morbey, R. Balogh, and G. Araoz. (2009). "Flooded Homes, Broken Bonds, the Meaning of Home, Psychological Processes and Their Impact on Psychological Health in a Disaster." *Health and Place* 15(2), pp. 540–47.

Chamlee-Wright, E., and V.H. Storr. (2009). "'There's No Place Like New Orleans': Sense of Place and Community Recovery in the Ninth Ward after Hurricane Katrina." *Journal of Urban Affairs* 31(5), pp. 615–34.

Charmaz, K. (2014). *Constructing Grounded Theory*. 2nd edn. London, UK: Sage.

Chen, Y.L., C.S. Lai, W.T. Chen, W.Y. Hsu, Y.C. Wu, P.W. Wang, and C.S. Chen. (2011). "Risk Factors for PTSD after Typhoon Morakot among Elderly People in Taiwanese Aboriginal Communities." *International Psychogeriatrics* 23(10), pp. 1686–91.

Connor, K.M., and J.R. Davidson. (2003). "Development of a New Resilience Scale: The Connor-Davidson Resilience Scale (CD-RISC)." *Depression and Anxiety* 18(2), pp. 76–82.

Cox, R.S., and K.M.E. Perry. (2011). "Like a Fish Out of Water: Reconsidering Disaster Recovery and the Role of Place and Social Capital in Community Disaster Resilience." *American Journal of Community Psychology* 48(3–4), pp. 395–411.

Cummins, R.A., R. Eckersley, J. Pallant, J. Van Vugt, and R. Misajon. (2003). "Developing a National Index of Subjective Wellbeing: The Australian Unity Wellbeing Index." *Social Indicators Research* 64(2), pp. 159–90.

De Silva, M.J., K. McKenzie, T. Harpham, and S.R. Huttly. (2005). "Social Capital and Mental Illness: A Systematic Review." *Journal of Epidemiology and Community Health* 59(8), pp. 619–27.

Forbes, D., N. Alkemade, E. Waters, L. Gibbs, H.C. Gallagher, P. Pattison, D. Lusher, C. Mac-Dougall, L. Harms, K. Block, et al. (2015). "Anger and Major Life Stressors as Predictors of Psychological Outcomes following the Victorian Black Saturday Bushfires." *Australian and New Zealand Journal of Psychiatry* 49(8), pp. 706–13, doi:10.1177/0004867414565478

Fullilove, M.T. (1996). "Psychiatric Implications of Displacement: Contributions from the Psychology of Place." *American Journal of Psychiatry* 153(12), pp. 1516–23.

Gibbs, L., K. Block, L. Harms, C. MacDougall, E. Snowdon, G. Ireton, D. Forbes, J. Richardson, and E. Waters. (2015). "Children and Young People's Wellbeing Post-Disaster: Safety and Stability Are Critical." *International Journal of Disaster Risk Reduction* 14(2), pp. 195–201, doi:10.1016/j.ijdrr.2015.06.006

———, E. Waters, R. Bryant, P. Pattison, D. Lusher, L. Harms, J. Richardson, C. MacDougall, K. Block, E. Snowdon, et al. (2013). "Beyond Bushfires: Community, Resilience and Recovery – A Longitudinal Mixed-Method Study of the Medium to Long-Term Impacts of Bushfires on Mental Health and Social Connectedness." *BMC Public Health* 14(7), pp. 634–43, doi:10.1177/0004867414534476

Giddens, A. (1990). *The Consequences of Modernity*. Cambridge, UK: Polity.

Green, J., K. Willis, E. Hughes, R. Small, N. Welch, L. Gibbs, and J. Daly. (2007). "Generating Best Evidence from Qualitative Research: The Role of Data Analysis." *Australian and New Zealand Journal of Public Health* 31(6), pp. 545–50.

Guha-Sapir, D., F. Vos, R. Below, and S. Ponserre. (2011). *Annual Disaster Statistical Review 2010: The Numbers and Trends*. Brussels, Belgium: Centre for Research on the Epidemiology of Disasters (CRED), Université Catholique de Louvain.

Hawkins, R.L., and K. Maurer. (2011). "'You Fix My Community, You Have Fixed My Life': The Disruption and Rebuilding of Ontological Security in New Orleans." *Disasters* 35(1), pp. 143–59.

Hu, L.-T., and P.M. Bentler. (1999). "Cutoff Criteria for Fit Indexes in Covariance Structure Analysis: Conventional Criteria Versus New Alternatives." *Structural Equation Modeling: A Multidisciplinary Journal* 6(1), pp. 1–55.

IJzermans, C., A. Dirkzwager, and E. Breuning. (2005). *Long-Term Health Consequences of Disaster*. Utrecht, the Netherlands: NIVEL.

Kaniasty, K., and F.H. Norris. (1995). "In Search of Altruistic Community: Patterns of Social Support Mobilization following Hurricane Hugo." *American Journal of Community Psychology* 23(4), pp. 447–77.

Kick, E.L., J.C. Fraser, G.M. Fulkerson, L.A. McKinney, and D.H. De Vries. (2011). "Repetitive Flood Victims and Acceptance of FEMA Mitigation Offers: An Analysis with Community-System Policy Implications." *Disasters* 35(3), pp. 510–39.

Koufteros, X., and G.A. Marcoulides. (2006). "Product Development Practices and Performance: A Structural Equation Modeling-Based Multi-Group Analysis." *International Journal of Production Economics* 103(1), pp. 286–307.

Lewicka, M. (2011). "Place Attachment: How Far Have We Come in the Last 40 Years?" *Journal of Environmental Psychology* 31, pp. 207–30.

Lewin, K. (1948). *Resolving Social Conflicts: Selected Papers on Group Dynamics*. Oxford, UK: Harper.

Lindy, J., and M. Grace. (1986). "The Recovery Environment: Continuing Stressor Versus a Healing Psychosocial Space." In *Disasters and Mental Health: Selected Contemporary Perspectives*. B. Sowder and M. Lystad, eds. pp. 147–60. Rockville, MD: National Institute of Mental Health.

MacKinnon, D.P., C.M. Lockwood, and J. Williams. (2004). "Confidence Limits for the Indirect Effect: Distribution of the Product and Resampling Methods." *Multivariate Behavioral Research* 39(1), pp. 99–128.

McMillan, D.W. (1996). "Sense of Community." *Journal of Community Psychology* 24(4), pp. 315–25.

———, and D.M. Chavis. (1986). "Sense of Community: A Definition and Theory." *American Journal of Community Psychology* 14(1), pp. 6–23.

Morgan, D.W., N.I.M. Morgan, and B. Barrett. (2006). "Finding a Place for the Commonplace: Hurricane Katrina, Communities, and Preservation Law." *American Anthropologist* 108(4), pp. 706–18.

Muthén, L., and B. Muthén. (1998/2013). *Mplus User's Guide.* 7th ed. Los Angeles, CA: Authors.

Norris, F.H., A.D. Murphy, C.K. Baker, and J.L. Perilla. (2004). "Postdisaster PTSD over Four Waves of a Panel Study of Mexico's 1999 Flood." *Journal of Traumatic Stress* 17(4), pp. 283–92.

———, M.J. Friedman, P. Watson, C.M. Byrne, E. Diaz, and K. Kaniasty. (2002). "60,000 Disaster Victims Speak: Part I. An Empirical Review of the Empirical Literature, 1981–2001." *Psychiatry: Interpersonal and Biological Processes* 65(3), pp. 207–39.

Patton, M.Q. (2002). *Qualitative Research and Evaluation Methods.* 3rd edn. London, UK: Sage.

Peek, L. (2008). "Children and Disasters: Understanding Vulnerability, Developing Capacities and Promoting Resilience – An Introduction." *Children, Youth and Environments* 18(1), pp. 1–29.

———, B. Morrissey, and H. Marlatt. (2011). "Disaster Hits Home: A Model of Displaced Family Adjustment after Hurricane Katrina." *Journal of Family Issues* 32(10), pp. 1371–96.

Phillips, B., P.A. Stukes, and P. Jenkins. (2012). "Freedom Hill Is Not for Sale – And Neither Is the Lower Ninth Ward." *Journal of Black Studies* 43(4), pp. 405–26.

Preacher, K.J., D.D. Rucker, and A.F. Hayes. (2007). "Addressing Moderated Mediation Hypotheses: Theory, Methods, and Prescriptions." *Multivariate Behavioral Research* 42(1), pp. 185–227.

Prezza, M., and S. Costantini. (1998). "Sense of Community and Life Satisfaction: Investigation in Three Different Territorial Contexts." *Journal of Community and Applied Social Psychology* 8(3), pp. 181–94.

Riad, J.K., and F.H. Norris. (1996). "The Influence of Relocation on the Environmental, Social, and Psychological Stress Experienced by Disaster Victims." *Environment and Behavior* 28(2), pp. 163–82.

Roberto, K.A., T.L. Henderson, Y. Kamo, and B.R. McCann. (2010). "Challenges to Older Women's Sense of Self in the Aftermath of Hurricane Katrina." *Health Care for Women International* 31(11), pp. 981–96.

Sass, D.A., and T.A. Schmitt. (2013). "Testing Measurement and Structural Invariance." In *Handbook of Quantitative Methods for Educational Research.* T. Teo, ed. pp. 315–45. New York, NY: Springer.

Smith, J.S., and M.R. Cartlidge. (2011). "Place Attachment among Retirees in Greensburg, Kansas." *Geographical Review* 101(4), pp. 536–55.

Steiger, J.H., and J.C. Lind. (1980). "Statistically Based Tests for the Number of Common Factors." In *Annual Meeting of the Psychometric Society, Iowa City, IA: 1980,* pp. 424–53.

Subramanian, S., and I. Kawachi. (2004). "Income Inequality and Health: What Have We Learned So Far?" *Epidemiologic Reviews* 26(1), pp. 78–91.

Tuason, M.T.G., C.D. Güss, and L. Carroll. (2012). "The Disaster Continues: A Qualitative Study on the Experiences of Displaced Hurricane Katrina Survivors." *Professional Psychology: Research and Practice* 43(4), pp. 288–97.

Uscher-Pines, L. (2009). "Health Effects of Relocation Following Disaster: A Systematic Review of the Literature." *Disasters* 33(1), pp. 1–22.

Vandenberg, R.J., and C.E. Lance. (2000). "A Review and Synthesis of the Measurement Invariance Literature: Suggestions, Practices, and Recommendations for Organizational Research." *Organizational Research Methods* 3(1), pp. 4–70.

van Griensven, F., M.L.S. Chakkraband, W. Thienkrua, W. Pengjuntr, B.L. Cardozo, P. Tantipiwatanaskul, P.A. Mock, S. Ekassawin, A. Varangrat, C. Gotway, et al. (2006). "Mental Health Problems among Adults in Tsunami-Affected Areas in Southern Thailand." *Journal of the American Medical Association* 296(5), pp. 537–48.

Victorian Bushfires Royal Commission. (2009). *The 2009 Victorian Bushfires Royal Commission Final Report Summary*. Retrieved July 8, 2013, from http://www.royalcommission.vic.gov.au/Commission-Reports/Final-Report

Ware Jr., J.E., and C.D. Sherbourne. (1992). "The MOS 36-Item Short-Form Health Survey (SF-36): I. Conceptual Framework and Item Selection." *Medical Care* 30(6), pp. 473–83.

Wiesenfeld, E., and R. Panza. (1999). "Environmental Hazards and Home Loss: The Social Construction of Becoming Homeless." *Community, Work and Family* 2(1), pp. 51–65.

Yu, C.-Y. (2002). *Evaluating Cutoff Criteria of Model Fit Indices for Latent Variable Models with Binary and Continuous Outcomes*. Los Angeles, CA: University of California.

21 Separation and reunification in disasters

The importance of understanding the psychosocial consequences

John F. Richardson,[1] *Elyse Baker,*
Hugh Colin Gallagher, Lisa Gibbs,
Karen Block, Dean Lusher, Connie Kellett,
Colin MacDougall, Louise Harms,
and Marita Smith

Introduction

Disasters, often being sudden-onset and destructive events, produce unpredictable consequences that frequently lead to the separation of family members. People mention that separation from family members is one of the most stressful aspects of such events. Hazards may restrict physical access to an area, both during the threat period and afterwards due to the loss of transport infrastructure (e.g., roads), and communications may be disrupted directly or indirectly through loss of power. As a result of this disruption to people's ability to communicate with each other, many emergency management systems recognize the importance of reuniting families separated by disaster, and therefore provide a means for people to reunite (e.g., Safe and Well operated by the American Red Cross).[2] The term "reunification" is commonly used in the emergency sector. In this chapter, it refers to the point when someone becomes aware of their family member's status and whereabouts (e.g., through phone contact, word of mouth, Internet, face-to-face, or through the Red Cross); it does not necessarily involve being brought together in the same physical location. Despite the rationale for reunification services being supportive for people's wellbeing, there is little empirical evidence relating to the short- and longer-term psychosocial impact of separation from family members during and immediately after a disaster event.

The term "uncertainty" is used in this paper to refer to a lack of knowledge as to the safety and whereabouts of family members. This phenomenon takes place within a broader uncertainty in disasters, occurring not only because of disruption to communication, but contextualized by an "anarchical profusion of information" that occurs when organizational structures are no longer functioning and relaying reliable information to the public.[3] This broader understanding of uncertainty and separation from people other than close family members is recognized as an additional cause of stress and is beyond the scope of this chapter to consider.

Literature review

Evidence has been explored within a previous literature review conducted between 2002 and 2013, focused on stress caused by separation during and immediately after disasters and the uncertainty experienced by family and friends.[4] The review found most papers concerned with separation focused upon longer-term separation caused by displacement, referring to family members living in a new location as a result of the disaster and in some cases living apart from family members. Separation, in these cases caused by displacement, does not necessarily relate to peri-disaster separation – that is, separation that takes place during the threat period – and therefore does not take into account the uncertainty and stress caused by this phenomenon. No studies published during the review time period relating to the short- or long-term psychosocial consequences of separation and reunification were found.[5]

This paper will present a case study from *Beyond Bushfires*, a large study conducted in Victoria, Australia, which explored the specific experiences of people separated from close family members during and after severe bushfires in 2009. The term "bushfires" will be used within this chapter, as it is a commonly used term within Australia to describe forest or grassland fires in rural areas. In other countries they would be commonly known as wildfires.

Disaster planning and behavioral responses to separation

Separation of families can occur in disasters, even with the most prepared plans, due to their unpredictability (e.g., children could be at school when a disaster occurs, parents at work). Disaster planning should consist of multiple scenarios that support keeping the family together, including the safety of children.[6] Practical guidelines suggest that multiple modes of contact should be available within the local community to reunite children and parents post-disaster through family message centers, social media, and hospitals (for injured children).[7] Agencies suggest that families should plan and prepare for disasters that include the possibility of separation.[8]

Encouraging families to stay together is also important to avoid unnecessary risks. One example of an additional risk resulting from separation is convergence, when people become increasingly worried about others they are unable to contact and undertake risky behavior to be reunited, such as attempting to drive to the disaster-affected area to look for their family member(s).[9] This can also lead to road congestion, conflicts, and demands on agencies managing the disaster site. Lives may be put at risk, especially when people attempt to return to a disaster zone, which could lead to emergency personnel undertaking avoidable rescues.

Sorensen and Mileti[10] found that people go through a number of cognitive stages when evacuating in response to a threat:

- Hearing the warning;[11]
- Understanding the contents of the warning;
- Believing the warning to be credible;
- Personalizing the warning to oneself;
- Confirming that the warning is true; and
- Responding by taking protective action.

If people do not go through these cognitive stages, it can result in delayed evacuation. Therefore, if a family is separated during a disaster, evacuation is more likely to be delayed until the status of all family members is known.[12] This delay can lead to a greater risk of death, injury, or exposure to traumatic situations.

Emergency management practitioners report other observed phenomena. Family members unable to contact their loved one(s) will often initially contact the emergency dispatch hotline for police, fire, or ambulance (in Australia this is known as 000). If they are unable to get through, they may try to contact other emergency or public services (e.g., fire services or hospitals). These actions have the effect of overloading the communication lines, resulting in the potential consequence of calls regarding life-threatening situations being blocked.[13]

Whilst it is clear that family separation causes short-term stress and may increase risk behaviors of harm from the hazard,[14] little is known about the long-term impacts on the mental health of family separation during a disaster. Therefore, empirical research is needed to determine whether longer-term effects are played out.

Current practice

Over the last half century, there has been an investment in establishing reunification services for separated families following disasters.[15] This includes practical guidelines supporting family reunification services to reduce possible psychological impacts, stress, and additional injuries resulting from stress during and immediately after disasters.[16]

The international focus on agencies' reunification services in post-disaster settings is on reuniting children with parents.[17] The World Health Organization (WHO) recommends a focus on family reunification post-disaster as part of restoring activities and community participation for positive mental health.[18] The *Sphere Handbook* is a universal guide with contributions from individuals and agencies (including government) for people affected by disasters.[19] One of the indicators discussed in the Sphere Handbook is "[w]hen necessary, a tracing service is established to reunite people and families"; however, reviews of this indicator found limited supportive evidence. This literature mainly focuses on reunification of families after an extended period of time (e.g., refugee families).[20]

Separation and reunification in Australia

In December 1974, a powerful Category 4 cyclone (Cyclone Tracy) destroyed the northern Australian city of Darwin. As a result of the levels of destruction, the subsequent public health issues, and the city's remoteness from the rest of Australia, a decision was taken at the time to evacuate children, women, and the elderly. Over the course of the week, over 30,000 people were evacuated by air and road.[21] The evacuation was rushed and the registration standards of evacuees were variable, resulting in some individuals reporting being separated from their families for significant periods.[22]

The National Registration and Inquiry System (NRIS)[23] was established in 1979 by the then Natural Disaster Organization as a result of the experiences of Cyclone Tracy.[24] The Australian Red Cross was asked to assist state police and other agencies in the operation of the system. This role came about as a result of the Red Cross's wartime experience of assisting with the reunification of families separated by World War II.[25] The system, a paper-based filing and retrieval system, allowed evacuees to register at an evacuation center; copies of the records were then transported to a central point, allowing family and friends to make phone inquiries to determine if people had registered as safe. This manual, paper-based process comprised three main components: registration (of impacted persons); inquiries (by family or friends); and matching (to identify a match and share relevant information on the impacted person).[26]

The Black Saturday bushfires

The southeastern corner of Australia is one of the most bushfire-prone areas in the world. The Wurundjeri people, an indigenous people who traditionally owned the lands around what is now Melbourne, recognize as part of the seasonal calendar that a "big fire" season occurs approximately every seven years.[27] The area has experienced significant bushfires since European settlement, including Red Tuesday in 1898, with the loss of 2,000 buildings; Black Friday in 1939, with the loss of 71 lives and 650 buildings; and Ash Wednesday in 1983, with the loss of 47 lives and 2,000 homes.[28]

In January 2009, following a prolonged decade-long drought, fire conditions became extreme, with fires beginning to ignite in areas east of Melbourne in the state of Victoria and continued to burn for several weeks (see Figure 21.1). Saturday, February 7, saw extreme temperatures, which climbed to 47° Celsius (117° Fahrenheit), while winds gusted at over 100km/h (60mph) and multiple new fires ignited across the state of Victoria. It was described by fire chiefs at the time as the worst fire conditions they had experienced in a generation. As a result of these fires, 173 lives were lost, over 2,200 homes and businesses were destroyed, 109 communities self-identified as being

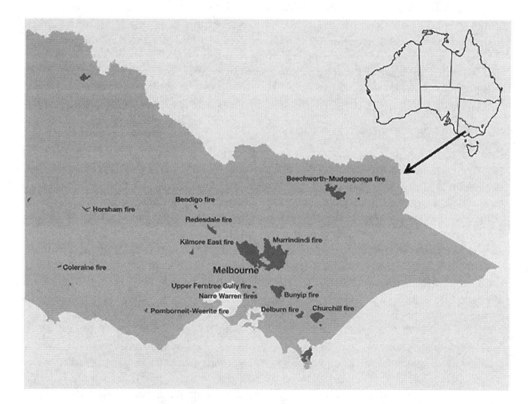

Figure 21.1 Locations of the 2009 victorian bushfires.[29]

impacted by the fires, and 400,000 hectares of landscape were burnt (twice the size of greater Tokyo). Two townships were completely destroyed and others were significantly damaged, resulting in loss of access and loss of public utilities to these areas. Power and communication difficulties in most places were restored within two days; however, it took two and a half weeks to restore power to one of the destroyed townships.[30] More than 4,500 roadblocks were set up, with some roads closed for over a month (Kinglake, Strathewen, Dixons Creek, Steels Creek, and Humevale) in severely impacted areas and nearly two months in one case (Marysville).[31] Access to bushfire areas was restricted for safety reasons and to enable Victoria Police to conduct searches for missing people and dead bodies.

During this period, the Australian Red Cross established the largest NRIS operation in its history. Over a two-week period, 22,000 registrations from affected people were taken and 21,000 inquiries were made.[32]

Importantly for the context of peri-disaster separation, at the time of the Black Saturday fires, fire agencies within Australia had adopted a policy called "prepare, stay and defend, or leave early." This policy advocated that people living in areas at risk of a bushfire should develop a bushfire preparedness plan in which they would assess their capacity to prepare their property for a bushfire so they could stay and actively defend it, or decide to leave early on a day of a high fire danger to remove themselves from the threat.[33] Victoria does not have a mandatory evacuation policy; unless a state of disaster is declared, authorities can only advise that people should relocate in the face of an emergency threat.[34]

The *Beyond Bushfires* study

The Beyond Bushfires: Community, Resilience and Recovery study[35] is a large-scale study tracking the health and social consequences of the 2009 Black Saturday bushfires.[36] Participants for the research were drawn from residents of 25 communities in ten locations in rural and regional Victoria. Communities were selected on the basis of varying levels of disaster impact, including high-impact (many houses lost, plus fatalities), medium-impact (a small number or no fatalities, but significant property damage), and low-impact (no evidence of fires being present). Contact details for both current residents and those who relocated after the fires were obtained through the Victorian Electoral Commission (VEC) (N = 7,693 adults). These individuals received one personalized letter of invitation, in agreement with the VEC.

Various recruitment and engagement efforts were also undertaken to increase awareness and opportunities to participate. Surveys were then conducted between December 2011 and January 2013, approximately three to four years after the disaster.[37] Survey questions gathered data from participants concerning their demographic characteristics; fire exposure; individual and organizational support networks; mental health; general health, wellbeing, and resilience; and attitudes towards their community. Fire exposure questions included four specific questions relating to separation and reunification. These were developed in collaboration between the Australian Red Cross and researchers from a range of universities, agencies, and government departments to explore the impacts during and after the bushfires (including the role of Australian Red Cross), and the link between separation experiences and medium- and longer-term health consequences.

Method

Participants

The original sample consisted of 1,016 individuals; however, 52 participants were excluded, as they reported that neither they, nor anyone they knew, nor their property were at risk, and therefore were not asked about separation. For the purposes of the statistical analyses below, the sub-sample analyzed consisted of 964 adults who lived in one of the *Beyond Bushfires* study communities at the time of the bushfires (585 women and 379 men; average age, 52.8 years).

Measures

Separation, uncertainty, and reunification

Separation was determined by participants' responses to the question "were you separated from close family members during the fire?" Participants who had been separated were then asked to recall how long it was before they had accounted for all their family members. For the purposes of the logistic regression analyses below, an individual was considered "separated" if he or she was separated from a family member and did not remain in contact with the individual from whom they were separated (N = 474).[38] Participants who reported separation were asked to recall the degree to which the separation caused them stress, on a scale of 0 ("not at all") to 10 ("extreme stress"). Finally, participants were asked how they found out about the whereabouts of their close family members (phone contact, face-to-face, word of mouth, Internet/social media, through Red Cross, other).

Disaster exposure

Participants were asked an array of other questions regarding their disaster experiences and subsequent life events. Among these were whether individuals feared for their own life (Yes/No), whether someone close had died as a result of the fires (Yes/No) and the extent of property loss (from 0, "Nothing," to 10, "Everything").

Other life circumstances

Participants were also asked about possible major life events occurring after the fires. For the analyses below, two indices were created, including a count of major life stressors (negative changes in employment status, accommodation and/or relationship status) and a count of subsequent traumatic events (a natural disaster, not including the fires of January and February 2009, a serious accident, or an experience of assault or violence). Also, for secondary analyses, participants were asked whether they had been diagnosed by a doctor as having any of seven specific chronic health conditions.[39]

Mental health

A variety of self-report measures were used to assess probable mental health outcomes. The Patient Health Questions (PHQ-9) were used to assess Major Depressive Episodes (MDE);[40] an abbreviated version of the Post-Traumatic Stress Disorder checklist (PCL) was also used.[41] The K6 scale was used to index non-specific psychological distress.[42] Further information regarding the use of these scales, including information on thresholds used, has been reported elsewhere.[43]

Results and discussion

Patterns of separation

The number of people who reported separation from close family members is noteworthy, at over 50 percent. Without other empirical studies, it is difficult to determine if this is usual for separations during a disaster. A number of significant and important differences between those who reported separation and those who did not were observed (see Table 21.1). On average, those who were separated were older than non-separated individuals. There are also different distributions of education level between the two groups, with the non-separated group having greater numbers of both less- and highly educated individuals. As expected, those who were separated reported higher disaster impacts (fear for life, loss of someone close, property loss). Furthermore, separated individuals also reported more subsequent major life events (major stressors and traumatic events). Finally, when controlling for age, separated individuals reported more chronic health conditions.

Table 21.1 Participant characteristics, by separation

	Non-separated (N = 394) N (%)	Separated (N = 570) N (%)	Difference between groups
Gender			
Female	233 (59.1%)	352 (61.9%)	$z = .81_{ns}$
Male	161 (40.9%)	218 (38.2%)	
Education			
High school or less	151 (38.8%)	190 (33.7%)	
Trade/other postsecondary	88 (22.6%)	184 (32.6%)	$\chi^2(2) - 11.29***$
Tertiary and higher	237 (38.6%)	190 (33.7%)	
Regional affectedness			
High	237 (60.2%)	392 (68.8%)	
Med	66 (16.8%)	111 (19.5%)	$\chi^2(2) = 21.88***$
Low	91 (23.1%)	67 (11.8%)	
Fear for life	171 (44.2%)	317 (56.9%)	$z = 3.85***$
Lost someone close	97 (24.7%)	197 (35.0%)	$z = 3.37***$
	M (SD)	M (SD)	
Age	54.9 (13.6)	51.3 (12.8)	$t(961) = 4.27***$
Extent of property damage	4.2 (4.1)	5.0 (3.9)	Odds Ratio = 1.05**
# Major life stressors	.6 (.7)	.8 (.8)	Odds Ratio = 1.48***
# Traumatic events	.2 (.4)	.4 (.6)	Odds Ratio = 1.68***
# Health conditions[a]	1.5 (1.3)	1.6 (1.4)	Odds Ratio = 1.17**

***$p \leq .001$; **$p \leq .01$; *$p \leq .05$; $_{ns}$ non-significant

[a] controlled for age.

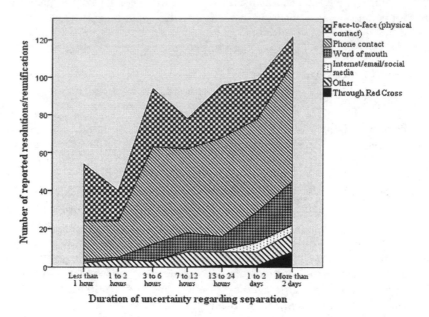

Figure 21.2 Method of resolution of separation uncertainty over time. Note that individuals reported multiple methods which are likely to have occurred at different times.

The results show that most people were able to make contact on their own in the first few hours after being separated, in person or via phone, but as time passed, other methods started gaining importance, including (but not limited to) the reunification service provided by the Australian Red Cross (see Figure 21.2). It is important to note that some people reported multiple methods of reestablishing contact. In situations where greater devastation occurs and there are limited resources available to restore power and communications, phone contact may be greatly reduced. Therefore, having multiple options of reconnection is vital to identify the whereabouts of family members and to support reunification in a timely manner.

Extent of separation and stress

Factors that may have contributed to the high rate of separation (over 50 percent) include the fact that the fires impacted communities in the late afternoon and early evening on a weekend, when people may have been out and about for various work or social reasons. Conditions were hot, dry, and windy, and the fires developed very quickly and moved fast, possibly leaving little time to connect with family members. Additionally, people may have actively decided to separate from their families, as previous research has found family decision-making in regard to the "prepare, stay and defend, or leave early" policy referred to earlier often led men to stay with the property while women left the area with children.[44]

A high number of people (45 percent of those who reported separation) were separated from family members for more than 12 hours, some up to two days or more.[45] The fires had their major impact late in the afternoon and early in the evening;

therefore, being separated for more than 12 hours meant that people probably spent a night not knowing the whereabouts of their family member(s). High levels of stress were reported by those separated from family members as a result of the bushfires (M = 7.90, SD = 2.70), with nearly half of those reporting separation (45.3 percent) reporting the highest level of stress. However, among individuals who did not remain in contact, there was no significant correlation between the *length* of unresolved separation and levels of self-reported stress (r_s = .04, p = .332).

Separation during a disaster could have greater stress impacts because of expectations of constant contact through the use of current technology. Reliance on mobile phones has rapidly increased, changing social impacts, spatial mobility, and planning/scheduling.[46] They have become increasingly common in Australia, especially with young people, and for some their use has resulted in addictive behaviors.[47] Therefore, within a disaster situation where communications and power (to charge mobile devices) are limited or lost, feelings of disconnection and separation may have more powerful impacts because of the usual availability of such connections. This highlights the importance of taking into account the potential for separation, noting the increased dependence on mobile telephony and computing when planning, at both the household and at the emergency management levels. Household emergency plans need to detail how people will act and stay in touch with each other in the event that they are separated by the disaster and unable to rely on technology for communication.

Medium- and longer-term consequences

Earlier findings from the *Beyond Bushfires* study showed that the level of exposure to the bushfires, as well as subsequent negative life events (which were also associated with exposure) were linked to increased rates of depression, post-traumatic stress disorder (PTSD), and general psychological distress.[48] The analyses reported here show that separation with uncertainty was also found to contribute to poorer longer-term mental health outcomes, including PTSD, three to four years after the disaster. This provides empirical evidence for Nager's[49] suggestion that being separated from family members during a disaster can have long-term impacts.

In the current case study, longer periods of uncertainty were not associated with greater reported stress (probably because even short periods of uncertainty were reported to be extremely stressful; see Figure 21.3), but were associated with greater mental health issues after controlling for basic socio-demographic factors and disaster exposure (see Table 21.2). Separation and length of uncertainty are therefore important indicators of risk that can be used by services to identify people who may be in need of mental health support.

Methods of reunification

The most popular method of resolving uncertainty was through phone and face-to-face contact. The Red Cross National Registration and Inquiry Service (NRIS) accounted for 2.5 percent of participants identifying the whereabouts of family members, and the majority of these occurred after one day. The extent of the fires only became fully apparent the next morning after the fires, which is when this service could have been more readily accessed. As shown in Figure 21.2, most people were able to ascertain the whereabouts of their family member(s) quickly through means other than the NRIS.

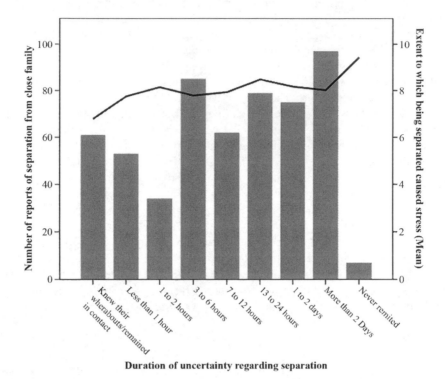

Figure 21.3 Number of cases of reported separation from close family members (N = 560) with extent of reported stress (line).

Table 21.2 Predictors of medium-term psychological outcomes (four years post-disaster). Individuals who reported separation with uncertainty (N = 474)

	PTSD	MDE	Distress (K6)
	Odds ratios (95% confidence interval)		
Sex (female)	1.48 (.89–2.47)[a]	1.10 (.61–1.97)	1.66 (.79–3.47)
Age	.99 (.97–1.01)	.99 (.97–1.02)	.98 (.95–1.00)a
Educational level	.89 (.77–1.02)[a]	.93 (.80–1.09)	.80 (.66–.98)*
Fear for life in bushfires	2.27 (1.34–3.84)**	2.46 (1.32–4.6)**	2.06 (.97–4.40)a
Lost someone in bushfires	1.72 (1.05–2.80)*	1.79 (1.03–3.11)*	2.35 (1.20–4.61)*
Property loss	1.06 (.99–1.14)	1.05 (.97–1.14)	1.06 (.96–1.17)
Major life stressors	2.20 (1.59–3.05)***	1.76 (1.23–2.52)**	2.02 (1.31–3.12)***
Traumatic events	1.46 (.99–2.14)[a]	.83 (.52–1.32)	.89 (.52–1.53)
Length of uncertainty	1.16 (1.02–1.31)*	1.22 (1.05–1.41)**	1.33 (1.10–1.61)**

***$p \leq .001$; **$p \leq .01$; *p

Efforts to find family members were likely to have been hampered by loss of power and communications and by roadblocks that physically prevented contact. It seems that when people were not able to locate their family members through their usual modes of communication, alternative contact services such as NRIS were used. The success rate of inquiries matching registrations was 31 percent over the two weeks that NRIS was operated after the Black Saturday Bushfires.[50] This was the highest rate of reunification that had ever been achieved by the NRIS.

As a result of post-disaster reviews, including the Victorian Bushfires Royal Commission, the Australian Government and Australian Red Cross undertook improvements and updates to the NRIS.[51] The renamed "Register. Find. Reunite." service, informed by the age of mobile telephony, computing, and instantaneous connection, is directly accessible online by the public, which may result in the usage of this service increasing. The online component is mainly in place for people living outside the disaster-affected area to determine the status of friends and family within impacted areas. Services need to be established early to provide certainty and facilitate reunification.[52] As noted earlier, there are changing societal patterns and behaviors relating to immediacy of contact, particularly through social media and mobile telephony.[53] However, this new system can still be undertaken manually to account for loss of power and communications.

Limitations and future directions

The current study contains a number of limitations. First, participants were reporting their recollections of separation and levels of stress several years after the event, with the possibility that these recollections may have been affected by subsequent events, mental health status, or attachment styles. Furthermore, separation was assessed quite broadly in terms of "close family members," and so detail was lacking on potentially important aspects of separation, such as parental and spousal responsibilities. Future research should therefore examine patterns of separation in more detail, including more precise information regarding the roles and relations associated with multiple potential separations (e.g., from spouses, children, friends, etc.).

For future research, it would also be useful to understand individuals' reactions and feelings, how they behaved (did they wait; did they try to seek more information; or did they converge on the emergency site?) and how they felt after reunification, whether their feelings of stress were resolved. Agencies currently advocate that children should receive education about potential separation in disaster preparedness programs; these results suggest that it may also be beneficial to prepare adults for that possible scenario. The current registration service "Register. Find. Reunite." makes use of technology improvements, which could increase its future use and accessibility in a society where technology-based mediums are constantly accessed and utilized. Also, the ability to operate the system manually enables utilization in a catastrophic disaster scenario where there is widespread and prolonged disruption to power communications and transport networks.

Conclusion

Separation from family members is a significant issue, creating high levels of stress in the short term and, for some, serious longer-term mental health consequences, particularly for those who experienced longer periods of separation. Despite the increasing connectivity of the community, separation is experienced when there is a loss of power, transportation, and telecommunications. Expectations of the immediacy of communication can exacerbate the experience of separation for family members.

Therefore, it is important to ensure that reunification services are established early and promoted widely, in order to reduce stress and the potential for the development of post-traumatic stress. This study supports agency policies regarding the importance of timely reunification. Technological advances provide an opportunity to increase the efficiency, accessibility, and immediacy of reunification services, provided they are not also compromised by on-site power and communication failures.

Acknowledgments

This chapter reports on findings from the *Beyond Bushfires* study. The authors and investigators wish to thank the community members, local government, and service providers from the participating communities who have supported the development and conduct of the *Beyond Bushfires* study. We acknowledge the late Professor Elizabeth Waters and Professor Richard Bryant, as well as Professor David Forbes, Greg Ireton, Vikki Sinnott, and Professor Philippa Pattison, who, although they were not involved in writing this paper, contributed greatly to the *Beyond Bushfires* study as investigators. We also wish to acknowledge the literature review conducted by Marita Smith as part of her Master of Social Work research, which was helpful background to the work prepared for this paper. We gratefully acknowledge the funding support received from the Australian Research Council for the *Beyond Bushfires* study and from the Jack Brockhoff Foundation for infrastructure and salary support for Lisa Gibbs and Elizabeth Waters.

Notes

1 Please address any correspondence to John F. Richardson, Australian Red Cross, jfrichardson@redcross.org.au
2 American Red Cross (n.d.), "Safe and Well," [communication platform], retrieved January 10, 2016, from https://safeandwell.communityos.org/cms/index.php
3 C. Gilbert (1995), "Studying Disaster: A Review of the Main Conceptual Tools," *International Journal of Mass Emergencies and Disasters* 13(3), p. 237.
4 M. Smith (2013), "The Psycho-Social Impact of Displacement and Reunification of Families During Disaster," Unpublished thesis, submitted for the degree Master of Social Work (Research), University of Melbourne, Australia.
5 Ibid., p. 44.
6 S. Chung and N. Blake (2014), "Family Reunification after Disasters," *Clinical Pediatric Emergency Medicine* 15(4), p. 336.
7 T.E. Drabek (2013), *The Human Side of Disaster*, 2nd edn. (Boca Raton, FL: CRC), p. 122; Chung and Blake (2014), p. 339.
8 Australian Red Cross (2008), *Emergency Rediplan: Four Steps to Prepare Your Household* (Melbourne, Australia: Author), p. 10; Chung and Blake (2014), p. 334.
9 E. Auf der Heide (2003), "Convergence Behavior in Disasters," *Annals of Emergency Medicine* 41, p. 464; Drabek (2013), p. 122; B. Raphael (1986), *When Disaster Strikes: How Individuals and Communities Cope with Catastrophe* (New York, NY: Basic Books), p. 73.
10 J.H. Sorensen and D.S. Mileti (1988), "Warning and Evacuation: Answering Some Basic Questions," *Organization and Environment* 2(3/4), p. 195.
11 In some circumstances, warnings may not be received due to the suddenness of an event, e.g., an earthquake or technological disaster. People may also not receive a warning if there is no power for radios or warnings are not issued, for example. However, people will respond to triggers to evacuate.
12 Drabek (2013), p. 122; L. Peek (2010), "Age," in *Social Vulnerability to Disasters*, B. Phillips, D. Thomas, A. Fothergill, and L. Blinn-Pike, eds., p. 161 (Boca Raton, FL: CRC Press).

13 Auf der Heide (2003), p. 464.
14 Ibid.; Drabek (2013), p. 234; B. Raphael (1986), *When Disaster Strikes: How Individuals and Communities Cope with Catastrophe* (New York, NY: Basic Books), pp. 72–73.
15 Ibid., p. 259; J.S. Tyhurst (1957), "Psychological and Social Aspects of Civilian Disaster," *Canadian Medical Association Journal* 76(5), p. 389.
16 A.L. Nager (2009), "Family Reunification – Concepts and Challenges," *Clinical Pediatric Emergency Medicine* 10, p. 197.
17 L. Peek and K. Richardson (2010), "In their Own Words: Displaced Children's Educational Recovery Needs after Hurricane Katrina," *Disaster Medicine and Public Health Preparedness* 4, S70.
18 S. Saxena, M. van Ommeren, and B. Saraceno (2006), "Mental Health Assistance to Populations Affected by Disasters: World Health Organization's Role," *International Review of Psychiatry* 18(3), pp. 199–204.
19 Sphere Project (2011), *The Sphere Handbook: Humanitarian Charter and Minimum Standards in Humanitarian Response*, retrieved March 18, 2015, from http://www.sphereproject.org/handbook/, p. 40.
20 R. Batniji, M. Van Ommeren, and B. Saraceno (2006), "Mental and Social Health in Disasters: Relating Qualitative Social Science Research and the Sphere Standard," *Social Science and Medicine* 62, p. 1858.
21 S. Cunningham (2014), *Warning: The Story of Cyclone Tracy* (Melbourne, Australia: Text Publishing), p. 7.
22 Ibid., p. 104.
23 The service is now known as "Register. Find. Reunite." and has been further developed to support web-based registrations and online queries from the public. (See https://register.redcross.org.au/.)
24 L. Avery (1995), "NRIS and Its Use within Victoria," *The Australian Journal of Emergency Management* 10(1), p. 34.
25 A role that continues today.
26 Avery (1995), p. 35.
27 C. Hansen and T. Griffiths (2012), *Living with Fire: People, Nature and History in Steels Creek* (Collingwood, Victoria: CSIRO), p. 120.
28 Department of Environment and Primary Industries (Australia), (2013/2015, March 4), *Bushfire History*, retrieved March 16, 2015, from http://www.depi.vic.gov.au/fire-and-emergencies/managing-risk-and-learning-about-managing-fire/bushfire-history
29 ©State of Victoria, Australia. Copyright for this publication is owned by the Crown in right of the State of Victoria, Australia. This extract is reproduced with the permission of the Crown in right of the State of Victoria, Australia. The State of Victoria accepts no responsibility for the accuracy and completeness of the reproduction of the report. Golbez [pseudonym] (2006), *Blank Map of the States of Australia*, retrieved January 10, 2016, from http://commons.wikimedia.org/wiki/File:Australia_states_blank.png; B. Teague (2010a), *2009 Victorian Bushfires Royal Commission: Final Report Summary* (Melbourne: Government Printer for the State of Victoria), p. 3.
30 Murrindindi Shire Council (2009), "From the SP Ausnet," *Murrindindi Recovery Newsletter*; National Electrical and Communications Association (2014), "Black Saturday . . . 5 Years On," *NECA Victoria Magazine*. March/April, p. 26.
31 Personal Communication from Superintendent Matt Ryan of the Victoria Police (Australia) to John F. Richardson, March 9, 2015; Teague (2009a), p. 9; B. Teague (2010b), *2009 Victorian Bushfires Royal Commission: Volume I. The Fire and the Fire-Related Deaths* (Melbourne, Australia: Government Printer for the State of Victoria), p. 231.
32 B. Teague (2010c), *2009 Victorian Bushfires Royal Commission: Volume II. Fire Preparation, Response, and Recovery* (Melbourne, Australia: Government Printer for the State of Victoria).
33 Hansen and Griffiths (2012), p. 169.
34 Department of Environment and Primary Industries (Victoria, Australia) (2013/2015, March 4), pp. 3–33.
35 "Beyond Bushfires," *Melbourne School of Population and Global Health* [Web page], www.beyondbushfires.org.au

36 Gibbs, L., E. Waters, R. Bryant, P. Pattison, D. Lusher, L. Harms, and D. Forbes (2013), "Beyond Bushfires: Community, Resilience and Recovery – A Longitudinal Mixed-Method Study of the Medium- to Long-Term Impacts of Bushfires on Mental Health and Social Connectedness," *BMC Public Health* 13(1036), doi:10.1186/1471-2458-13-1036
37 Ibid., p. 7.
38 See H.C. Gallagher, J. Richardson, D. Forbes, L. Harms, L. Gibbs, N. Alkemade, C. Mac-Dougall, E. Waters, K. Block, D. Lusher, et al. (2016, January). "Mental Health Following Separation in a Disaster: The Role of Attachment Style," *Journal of Traumatic Stress*, doi:10.1002/jts.22071
39 Heart disease (angina, heart failure, heart attack), asthma, stroke, cancer, diabetes, arthritis, dermatitis, emphysema, back problems, chronic back pain or sciatica, and high cholesterol.
40 K. Kroenke, R.L. Spitzer, and J.B.W. Williams (2001), "The PHQ-9 – Validity of a Brief Depression Severity Measure," *Journal of General Internal Medicine* 16(9), pp. 606–13.
41 P.D. Bliese, K.M. Wright, A.B. Adler, O. Cabrera, C.A. Castro, and C.W. Hoge (2008), "Validating the Primary Care Posttraumatic Stress Disorder Screen and the Posttraumatic Stress Disorder Checklist with Soldiers Returning from Combat," *Journal of Consulting and Clinical Psychology* 76(2), pp. 272–81.
42 R.C. Kessler, P.R. Barker, L.J. Colpe, J.F. Epstein, J.C. Gfroerer, E. Hiripi, and A.M. Zaslavsky (2003), "Screening for Serious Mental Illness in the General Population," *Archives of General Psychiatry* 2, p. 184.
43 R.A. Bryant, E. Waters, L. Gibbs, H.C. Gallagher, P. Pattison, D. Lusher, and D. Forbes (2014), "Psychological Outcomes Following the Victorian Black Saturday Bushfires," *The Australian And New Zealand Journal of Psychiatry* 48(7), pp. 634–43.
44 C. Eriksen (2014), "Gendered Risk Engagement: Challenging the Embedded Vulnerability, Social Norms and Power Relations in Conventional Australian Bushfire Education," *Geographical Research* 52(1), p. 30; B. Teague, R. McLeod, and S. Pascoe (2009), "2009 Victorian Bushfires Royal Commission Interim Report," in *Government Printer for the State of Victoria*, ed., Vol. 225, Session 2006–09, p. 193.
45 In addition, seven people reported never being reunited with those from whom they were separated.
46 E. Thulin and B. Vilhelmson (2007), "Mobiles Everywhere: Youth, the Mobile Phone, and Changes in Everyday Practice," *Young: Nordic Journal of Youth Research* 15(3), p. 236.
47 S.P. Walsh, K.M. White, and R.M. Young (2008), "Over-Connected? A Qualitative Exploration of the Relationship between Australian Youth and Their Mobile Phones," *Journal of Adolescence* 31, p. 78.
48 R.A. Bryant, E. Waters, L. Gibbs, H.C. Gallagher, P. Pattison, D. Lusher, and D. Forbes (2014), "Psychological Outcomes Following the Victorian Black Saturday Bushfires," *The Australian And New Zealand Journal of Psychiatry* 48(7), pp. 634–43.
49 A.L. Nager (2009), "Family Reunification – Concepts and Challenges," *Clinical Pediatric Emergency Medicine* 10, p. 204.
50 Teague et al. (2010).
51 Teague (2010c), p. 332.
52 Avery (1995).
53 Thulin and Vilhelmson (2007); Walsh et al. (2008).

Bibliography

Auf der Heide, E. (2003). "Convergence Behavior in Disasters." *Annals of Emergency Medicine* 41, pp. 463–66, doi:10.1067/mem.2003.126

Australian Red Cross. (2008). *Emergency Rediplan: Four Steps to Prepare Your Household.* Melbourne, Australia: Author.

Avery, L. (1995). "NRIS and Its Use within Victoria." *The Australian Journal of Emergency Management* 10(1), pp. 34–38.

Batniji, R., M. Van Ommeren, and B. Saraceno. (2006). "Mental and Social Health in Disasters: Relating Qualitative Social Science Research and the Sphere Standard." *Social Science and Medicine* 62, pp. 1853–64, doi:10.1016/j.socscimed.2005.08.050

Bliese, P.D., K.M. Wright, A.B. Adler, O. Cabrera, C.A. Castro, and C.W. Hoge. (2008). "Validating the Primary Care Posttraumatic Stress Disorder Screen and the Posttraumatic Stress Disorder Checklist with Soldiers Returning from Combat." *Journal of Consulting and Clinical Psychology* 76(2), pp. 272–81.

Bryant, R.A., E. Waters, L. Gibbs, H.C. Gallagher, P. Pattison, D. Lusher, and D. Forbes. (2014). "Psychological Outcomes Following the Victorian Black Saturday Bushfires." *The Australian And New Zealand Journal of Psychiatry* 48(7), pp. 634–43.

Chung, S., and N. Blake. (2014). "Family Reunification after Disasters." *Clinical Pediatric Emergency Medicine* 15(4), pp. 334–42, doi:10.1016/j.cpem.2014.09.006

Cunningham, S. (2014). *Warning: The Story of Cyclone Tracy*. Melbourne, Australia: Text Publishing.

Department of Environment and Primary Industries (Victoria, Australia). (2013/2015, March 4). *Bushfire History*. Retrieved March 16, 2015, from http://www.depi.vic.gov.au/fire-and-emergencies/managing-risk-and-learning-about-managing-fire/bushfire-history

Drabek, T.E. (2013). *The Human Side of Disaster*. 2nd edn. Boca Raton, FL: CRC.

Eriksen, C. (2014). "Gendered Risk Engagement: Challenging the Embedded Vulnerability, Social Norms and Power Relations in Conventional Australian Bushfire Education." *Geographical Research* 52(1), pp. 23–33.

Gallagher, H.C., J. Richardson, D. Forbes, L. Harms, L. Gibbs, N. Alkemade, C. MacDougall, E. Waters, K. Block, D. Lusher, et al. (2016, January). "Mental Health Following Separation in a Disaster: The Role of Attachment Style." *Journal of Traumatic Stress*, doi:10.1002/jts.22071

Gibbs, L., E. Waters, R. Bryant, P. Pattison, D. Lusher, L. Harms, and D. Forbes. (2013). "Beyond Bushfires: Community, Resilience and Recovery – A Longitudinal Mixed-Method Study of the Medium- to Long-Term Impacts of Bushfires on Mental Health and Social Connectedness." *BMC Public Health* 13(1036), doi:10.1186/1471-2458-13-1036

Gilbert, C. (1995). "Studying Disaster: A Review of the Main Conceptual Tools." *International Journal of Mass Emergencies and Disasters* 13(3), pp. 231–40.

Golbez [pseudonym]. (2006). *Blank Map of the States of Australia*. Retrieved January 10, 2016, from http://commons.wikimedia.org/wiki/File:Australia_states_blank.png

Hansen, C., and T. Griffiths. (2012). *Living with Fire: People, Nature and History in Steels Creek*. Collingwood, Victoria: CSIRO.

Kessler, R.C., P.R. Barker, L.J. Colpe, J.F. Epstein, J.C. Gfroerer, E. Hiripi, and A.M. Zaslavsky. (2003). "Screening for Serious Mental Illness in the General Population." *Archives of General Psychiatry* 2, p. 184.

Kroenke, K., R.L. Spitzer, and J.B.W. Williams. (2001). "The PHQ-9 – Validity of a Brief Depression Severity Measure." *Journal of General Internal Medicine* 16(9), pp. 606–13.

Murrindindi Shire Council. (2009). "From the SP Ausnet." *Murrindindi Recovery Newsletter*, p. 1. Murrindindi, Victoria, Australia: Murrindindi Shire Council.

Nager, A.L. (2009). "Family Reunification – Concepts and Challenges." *Clinical Pediatric Emergency Medicine* 10, pp. 195–207, doi:10.1016/j.cpem.2009.06.003

National Electrical and Communications Association. (2014). "Black Saturday . . . 5 Years On." *NECA Victoria Magazine*. March/April.

Peek, L. (2010). "Age." In *Social Vulnerability to Disasters*. B. Phillips, D. Thomas, A. Fothergill and L. Blinn-Pike, eds., pp. 155–185. Boca Raton, FL: CRC Press.

Peek, L., and K. Richardson. (2010). "In Their Own Words: Displaced Children's Educational Recovery Needs after Hurricane Katrina." *Disaster Medicine and Public Health Preparedness* 4, p. S70.

Raphael, B. (1986). *When Disaster Strikes: How Individuals and Communities Cope with Catastrophe*. New York, NY: Basic Books.

Saxena, S., M. van Ommeren, and B. Saraceno. (2006). "Mental Health Assistance to Populations Affected by Disasters: World Health Organization's Role." *International Review of Psychiatry* 18(3), pp. 199–204, doi:10.1080/09540260600655755

Smith, M. (2013). "The Psycho-Social Impact of Displacement and Reunification of Families During Disaster." Unpublished thesis, submitted for the degree Master of Social Work (Research), University of Melbourne, Australia.

Sorensen, J.H., and D.S. Mileti. (1988). "Warning and Evacuation: Answering Some Basic Questions." *Organization and Environment* 2(3/4), p. 195.

Sphere Project. (2011). *The Sphere Handbook: Humanitarian Charter and Minimum Standards in Humanitarian Response.* Retrieved March 18, 2015, from http://www.sphereproject.org/handbook/

Teague, B. (2010a). *2009 Victorian Bushfires Royal Commission: Final Report Summary.* Melbourne: Government Printer for the State of Victoria.

Teague, B. (2010b). *2009 Victorian Bushfires Royal Commission: Volume I. The Fire and the Fire-Related Deaths.* Melbourne, Australia: Government Printer for the State of Victoria.

Teague, B. (2010c). *2009 Victorian Bushfires Royal Commission: Volume II. Fire Preparation, Response, and Recovery.* Melbourne, Australia: Government Printer for the State of Victoria.

Teague, B., R. McLeod, and S. Pascoe. (2009). "2009 Victorian Bushfires Royal Commission Interim Report." In *Government Printer for the State of Victoria*, ed., Vol. 225, Session 2006–09. Melbourne, Australia: Parliament of Victoria.

Thulin, E., and B. Vilhelmson. (2007). "Mobiles Everywhere: Youth, the Mobile Phone, and Changes in Everyday Practice." *Young: Nordic Journal of Youth Research* 15(3), pp. 235–54.

Tyhurst, J.S. (1957). "Psychological and Social Aspects of Civilian Disaster." *Canadian Medical Association Journal* 76(5), pp. 385–93.

Walsh, S.P., K.M. White, and R.M. Young. (2008). "Over-Connected? A Qualitative Exploration of the Relationship between Australian Youth and Their Mobile Phones." *Journal of Adolescence* 31, pp. 77–92, doi:10.1016/j.adolescence.2007.04.004

22 Disaster preparedness in older people

A case study from Australia

Victoria Cornell

Introduction

Anecdotally, older people are considered particularly vulnerable in emergency events, from the preparation phase through the response phase and into the recovery phase. However, little research has been undertaken in this area.[1] Much of the research that has been undertaken provides conflicting information. When carrying out an extensive literature review to identify the vulnerabilities of older people to disasters, Fernandez et al. found that "[t]he data are contradictory as to whether the elderly groups are more vulnerable than are other age-defined population groups."[2]

In addition, it is not advancing age alone that makes people vulnerable; vulnerabilities are generally due to the issues associated with advancing age, such as "impaired physical mobility, diminished sensory awareness, pre-existing health conditions and social and economic constraints."[3] This may be true, but these issues are not specific to older people and may affect people of all age groups at some stage in their lives.

Recent events and discussions that have taken place during the planning and exercise of disaster management in Australia have triggered consideration of the way older people are engaged in disaster management processes.[4] Most disaster research concerning older people "has focused on displacement and relocation as a consequence of a . . . disaster"[5] – that is, the response and recovery phases. Little research has been undertaken on older people and preparedness for emergency events. Of the preparedness research carried out to date, most has centered on authorities or service providers preparing to respond to an event. For example, there has been a great deal of research on developing social vulnerability indices within communities to assist emergency workers when responding to an event.[6] Similarly, there is literature that considers the preparedness of aged care facilities, such as nursing homes – particularly research and guidance on preparing aged care facilities for emergency events, whether and when to evacuate, and the best way to evacuate facilities.[7] However, there is a lack of research on the preparedness of older people living in their own homes, or even on agencies that provide care to older people in their homes.[8] Discussion thus far has largely considered the development of tools and checklists that are believed to help older people prepare for emergencies, rather than what might influence their decision to prepare. Older people's views are rarely canvassed, with opinion and perspectives sought from aged care service providers or peak bodies.

This chapter discusses research that explored how the lived experience of older people influences their perception of being prepared for emergency events. How has the variety of experiences that older people have had throughout their lives changed them or their understanding of being prepared?

One of the study drivers was the belief that people make decisions with regard to preparing for emergencies based on the context of their own lives and experiences. The study therefore took a qualitative, interpretive approach, using van Manen's[9] hermeneutic phenomenology as a framework. The aim was to explore and understand whether people's life experiences have influenced their perception of preparedness and what it *means* to be prepared for an emergency event.

Method

In-depth interviews were carried out with 11 older people between the ages of 77 and 90 years. The interviews were audio recorded and on average lasted an hour and a half. The in-depth interviews and small sample reflect qualitative methods that allowed for the generation of thick descriptions. The interpretive character of the research focused on why and how something happened, rather than specific behaviors. As noted by Cresswell, "[q]ualitative research is emergent rather than tightly prefigured."[10]

The technique used for the interviews was semi-structured, using open-ended questions that required more than a yes or no answer. This style of interview also allowed questions that arose naturally in the course of the interview to be pursued. The questions developed allowed for responses that provided breadth and richness in data and allowed participants the freedom to respond to questions and probes and to tell their own stories without being tied down to a specific framework. The questions were not necessarily asked in the same order with each participant, and did not use the exact same wording depending on how answers were provided, whether questions were raised by the participant, and how the conversation was progressing in general.

Eight women and three men participated in the study. All were of white Anglo-Australian heritage: nine were born in Australia, two in England. Ten of the 11 participants were or had been married, while one woman had never married. Of the ten who were or had been married, all the men were still married; one of the women was still married, one was divorced, and the five remaining women were widowed.

All participants resided in their own homes, mostly in the greater Adelaide metropolitan area of South Australia, with some living in the more rural Adelaide Hills region to the east of Adelaide. All received low-level in-home care, for example, assistance with shopping or housework. The participants came from a range of professional and socio-economic backgrounds.

The interviews explored the variety of emergency events experienced during the participants' lives, how those events may have changed them, the meaning drawn from the events, and the subsequent influence of their life experiences on the way they prepared (or perhaps chose not to prepare) for emergency events. Participants were assigned pseudonyms during data analysis, and the pseudonyms are used in this paper.

Results

The data analysis followed van Manen's three-stage thematic analysis process, during which the transcripts were read in their entirety several times to identify: (i) keywords and phrases that reflected the experience of the participants and their behaviors with respect to being prepared (holistic phase); and (ii) statements or phrases that helped in representing meaning about being prepared for emergency events (selective phase).[11] Finally, identifying what the keywords and statements revealed about being prepared for an emergency event was uncovered (line-by-line phase). The three following themes were identified.

Understanding my world

The first theme provided shape and context to the participants' lives by firstly highlighting the events that the participants had experienced. As Bonanno et al. stated in their article, which explored the factors that predict psychological resilience after disaster:

> Try as we might, we cannot prevent bad things from happening. During the course of a normal life span, almost everyone is confronted with the painful reality that loved ones die. Most adults are also exposed to at least one potentially traumatic event.[12]

Bonanno et al.'s quote certainly held true for the participants who were interviewed during this research. Indeed, when asked – as an opening question – whether she had lived through a traumatic emergency event, Grace, aged in her mid-eighties, replied: "You don't reach this age without doing so!"

The emergency events that the participants chose to discuss were nominated by *them*; participants were not, for example, asked "have you experienced a bushfire?" or "have you lived through a flood event?" The emergency events that the research participants had encountered during their lives varied greatly; some participants had experienced several events. The traumatic events experienced included family bereavements such as the death of a child; mental and physical illness (either their own or that of a loved one); natural hazard events, such as fires and floods; and perceived socially unacceptable (at least for the time in which they occurred) events, such as teenage pregnancies.

In some cases, one emergency event was enough to change behavior and make the participant feel either completely overwhelmed or much stronger, whereas in other cases it was the combination of events that had an effect. Hattie gained meaning and experience from a number of specific emergency events, but suggested that an accumulation of "everyday" experiences also formed who she was. In her introduction to the question of whether she had lived through any emergency events, she stated:

> Well, ignoring the, you know, obvious things like having four children and. . .the things that happen to children. And especially with my older one who, if she'd been in this era, would have been diagnosed as hyperactive . . . things like catching hold of the tail of a snake that's trying to get up the drainpipe . . . those sort of things. Ignoring those . . .

Eleanor considered that she had been fortunate to not experience one specific very serious emergency event, yet was able to cope with her husband's protracted illness and subsequent death because of "the little bits that have come, dropped their experiences along the road; I've been given enough strength."

Those working in the professional world of emergency management, while espousing the "all hazards" approach, often confine their thinking to a set range of events, traditionally considering natural hazard events and human-induced events such as chemical leaks, terrorism, and major transport accidents. This theme emphasized that older people – who have experienced a variety of such events – have a much broader view. They recognize that experience, strength, and understanding can be gathered from many aspects of their lives, both the big events and the small, and also from an accumulation of events.

The temporal nature of events was highlighted by the participants' stories; an event may not have a clear beginning and certainly may not have an end – it may continue to be a feature of everyday experience, such as living without a loved one after their death, or giving up a child for adoption.

The type of emergency event is also of interest, in terms of whether it was a private family event, such as a teenage pregnancy, a mental health breakdown, or financial hardship; or whether it was a public event that impacted many people, for example, a war or a large-scale bushfire. The participants who had suffered difficult private events admitted that they were particularly trying times. Their life world was one of social stigma, and the event engendered feelings of shame and guilt; it was not accept-able to talk about such problems publicly, and invariably the event was dealt with in secrecy.

Imogen talked about when she fell pregnant as an unmarried 16-year-old. Even though the father was the man who would become her husband on his return from service in World War II, the matter was very distressing to her parents socially. She recalled that "the day after my mother found out, I was in a home . . . I was told that he didn't want anything to do with me, and he was told the same." Imogen remained at the home until after she had the baby. The event was clearly also distressing to her as an individual, especially as she was not involved in any of the decision-making. Following the birth, she was moved to a workhouse that was

> dreadful you know . . . I was scrubbing floors, and . . . really, when you think back, it's true . . . nobody really knows what really goes on in those sorts of places. And then one day, my mother came and we took the baby, and we went to this place and she said to me "oh, you sit there, and I'll take the baby in and see . . . " I don't know who. Anyway, she came out without him and that was it. He was going to be adopted. And because I was only sixteen, I had no say.

In a period of less than a year, Imogen's life world had changed several times. She had started as a carefree teenager in love; moved to being a shamed unmarried preg-nant teenager; to being a young mother, prepared to take on the world with her baby; to being childless without any indication that this was to occur; to finally again being a "normal" teenager back in the community. This had an impact on how Imogen chose to live her life from then on, and particularly on the meaning she gave to preparing for her life in the future. For example, when she and her husband decided to emigrate from England to Australia, she was much more in control and resolute; she did not let either her parents or her parents-in-law intervene.

With these private emergency events, there was no after-care in terms of counseling or advice. In the case of public emergency events, life after the event was very different. With the participants who spoke about wartime experiences – either as service person-nel or citizens – the talk was much more open and free. The event affected not only the participant or his or her family; rather it was a community-wide, or indeed world-wide, issue and therefore not so private. While there was still no post-event care such as counseling and debriefing, as we have become used to today, the very fact that the event could be discussed openly, that there was community camaraderie and that everyone was "in the same boat" was helpful.

Similarly, in the case of events that affected only one family but were "socially acceptable," the event could be talked about and help offered. From early in their marriage, Hattie and her husband owned a farming property in rural South Australia. While they were away from the property on holiday one summer, a young farm laborer started a fire by accident, which destroyed the entire house and most of the contents – only one bed was saved. Her husband returned home immediately while Hattie continued the holiday with their four small children, the youngest not yet two years. The community and community groups such as the local Red Cross rallied to help:

> By the time I went back on to the farm with the children there were about four or five cartons of things – things that you needed for cooking and plates and things like that in the – for ordinary daily living. Clothes . . . and toys for children and, you know, a few other things like towels and linens.

This same level of support and rallying from the community was not provided when, following a blood infection that involved lengthy hospitalization and recuperation, Hattie

> had a nervous collapse. Couldn't eat, couldn't speak – all that sort of thing. And that affected my life because the local doctors didn't really give me anything that would help me in my nervous state.

All of the participants understood that their physical health was not as strong as it once was, and this did not worry them. Rather than dwell on activities they could no longer do, they either adapted previous activities or found new ways to enjoy their world. They were conscious that this reduced physical strength had implications for being prepared for an emergency – in terms of potentially being unable to carry out certain preparedness measures, for example clearing vegetation, and also in terms of reacting to an impending event, due perhaps to no longer having a car. They did not, however, feel vulnerable.

On the whole, the participants were not distressed by the deterioration in physical ability, as it had occurred gradually, allowing them to come to terms with living in a world of changing bodies and altered abilities. There might be a passing frustration that the body does not do what it once could, but there was also acceptance that this is the case. When talking about gardening, for example, Eleanor admitted that she missed being as strong as she was: "Yes, physically. I miss it. I'd just love to be able to put my foot on a fork, or . . . fiddling around in pots is not . . . [quite the same] . . . never mind . . ."

One notable exception in terms of physical health that concerned several of the participants was suffering a health emergency (such as a stroke or heart attack) and subsequently requiring care. None of the participants felt weak or at risk in their day-to-day lives, but when they paused to think about what worried them, in light of their new lived world of reduced ability and altered bodies, a level of vulnerability surfaced. They did not want to become incapacitated and beholden to someone else for care.

This interpretation – of being a burden on someone else – often stemmed from their own experience. Daisy's husband had suffered from Parkinson's disease. He lived at home for many years before suffering a heart attack, after which time he moved into a care home. Towards the end of his time in the family home, Daisy found it a struggle and found she was living on tenterhooks because, in her words, "it was like having a baby in the house." She had to be constantly vigilant as to where he was, what he was doing, and whether he was safe. It was clear from the way that Daisy spoke that she did not wish the same for her family, should anything happen to her. Grace was even clearer when talking of being reliant on someone else for care. When talking about death and dying, she advised that she was not scared of dying, but of being a burden: "I'm scared of how I'm going to *be*; not dying, no, no – not a bit – no, no, I just don't want to be a vegetable for my family, and lay there helpless."

In contrast to being aware of their changing physical health, the participants all felt that their mental health and ability was strong, a benefit that came with age and experience. The participants accepted that physically they were not as able to prepare and cope with an emergency event. However, this was of lesser importance to them; mental strength was their main concern, and feeling sure that they could cope mentally was a comfort to them.

When discussing their mental ability, they were talking less about mental health issues specifically, such as Alzheimer's disease or dementia; rather, they were talking about mental resilience to cope with any emergency event that might occur. They felt prepared mentally, thanks to their age and experience. Daisy, for example, advised: "Yes, you feel you can cope better now."

Mental strength was more important to them in terms of being prepared than physical strength. This feeling of mental strength is a positive finding, and something that can be harnessed by emergency management planners and community groups. These older people were not mentally fragile, and in fact felt stronger due to the events they have lived through. They should therefore be invited to share their experiences, advice, and suggestions as to how they can cope before and during emergencies.

Shrinking my world

The second theme of *shrinking my world* was noticeable in terms of a shrinking social world and shrinking thinking. The participants talked about how they have experienced their social worlds shrinking, with fewer lived human relations and a reduced lived space. From their discussions, two principal reasons for their shrinking social worlds were identified. Firstly, the participants explained that many of their friends and family had died, and, secondly, that their reduced mobility and loss of their driver's licenses had stymied their ability to socialize. The reduction in size of their social world was not something they felt negative about; in fact, in some cases they expressed contentment with their shrinking social world. Art, for example, said:

> Well, most of my associates are dead, I don't really have anybody . . . I don't really have anybody now from . . . all the wine [business associates] boys are dead, naturally, and um . . . no . . . I just have three things that I do. I . . . I'm doing painting, I watch a DVD every night before I go to bed, and I have music, and er . . . that's it, I'm very happy.

Frances was also happy with a smaller social world, saying that her ideal is to have friends close by, but to:

> have your own house. So you could go and visit whenever you wanted to and go on holidays together; go out to dinners together; go to theatres together – that . . . suits me fine but be able to still have time to myself. Because I like company, but I like my own company.

The participants did recognize, however, that a shrinking social world has implications in terms of being prepared for emergencies, both negative and the positive. On the negative side, participants recognized that strong social resources, including good friends and a strong community spirit, are critical supports in times of an emergency. They can provide shelter, food, and psychological support. A shrinking social world reduces access to these resources. On the positive side, the participants said that although their social world was shrinking, the relationships they had with friends and family members were "real," with genuine supportive bonds, rather than casual acquaintances or family who were not close.

The participants displayed shrinking thinking in terms of their considerations of being prepared for emergencies. When considering their lives, the varied events they had experienced, and what those events meant to them, the participants had a narrow view of events that they might consider preparing for. Despite discussing a range of experiences that had occurred during their lives, events that the participants now discussed being prepared for was narrow, sticking to fairly "traditional events" from their lived worlds. At this stage in their lives, their main concern was a health-related emergency, such as a fall in the home, or a more severe issue such as a heart attack or stroke. They did not consider other health issues such as extreme heat or pandemic disease.

With respect to emergency preparedness, there were two particular non-health examples of preparedness that the participants did worry about – the socially acceptable "right" thing to do, and the management and development of land. When discussing what the right thing to do was in terms of being prepared, and activities undertaken to be prepared, several participants raised the issue of insurance. For those living in retirement village settings, the building's insurance was organized by the village management, but contents insurance was the responsibility of the resident. Eleanor believed that having insurance was sensible, and she acknowledged that she was fortunate in being able to pay for it. Imogen, on the other hand, was unusual among the participants, as she did not have any insurance but was not concerned about it.

Two participants were very clear in their opinions about people who choose not to have insurance. Brian said:

> Well, it concerns me, although it's nothing to do with me, some people don't insure themselves. They expect the government to step in and replace their house, or whatever it is. And in a sense, those people might get more than those who've actually prepared. It may not have covered the whole of the house because it's so much more expensive to build today than it was say twenty years ago, but they don't get any help.

Keith was even blunter when discussing insurance policies, suggesting that: "they're things that if you don't take them up, you're being a bit stupid."

These sentiments suggest that, in spite of the fact that the participants' worlds were shrinking, and as they grow older there are some activities and issues that they choose not to engage with or worry about, the comfort and security of insurance was something that they still felt strongly about. Developing and delivering preparedness messaging – indeed, seeking information about being prepared – could therefore be undertaken in conjunction with insurance providers.

The shrinking world provided a state of mind that the participants were content with. They had chosen, for a range of physical and social reasons, to reduce their engagement with the wider world, and focus instead on those issues of interest and concern to them. When looking to the future, they had undertaken a number of activities to prepare for what may come, such as arranging contents insurance and packing a bag for an unexpected hospital visit, but did not feel that they needed to "change the world" – for example, they did not feel the need to consider world affairs in any great detail.

The identification of *shrinking my world* highlights a potential window of opportunity to engage older people in emergency planning activities while they are still happy to share their time, knowledge, and experience, but before they potentially retreat to an increasingly smaller social world and also before their engagement with the wider world – and potentially their thinking – becomes too narrow and focused.

Acceptance of my world

Acceptance of greater dependence was a key issue, and in many cases a release for the participants. In accepting greater dependence, the participants were not relinquishing control; rather, the decision was a sophisticated way of taking control. They understood their need to –and had the confidence to – allow others to "do things for them" and were therefore content to hand over a certain level of control. The sophistication of this control decision was manifest in the fact that it was not a "once and for all" proposition; control could be relinquished and then regained. Imogen talked of a time when she had "Meals on Wheels" delivered to her home, when she was recuperating from surgery. While she was grateful for the support and keen to stress that the service is a good one, she did not enjoy the meals and cancelled them as soon as she was able to cook and prepare food for herself. She has since ensured that she always has frozen meals in the freezer and some extra tins of food in the cupboard. Imogen had both taken control back and learnt from that experience, to be more prepared in the future, as she did not enjoy being dependent on the service.

By accepting greater dependence, the participants were not giving up on life in general and every element within their life. On the contrary, their life experience had taught them that they do not need to control everything; indeed, they had come to learn that you cannot control everything in life. Frances said, during her interview: "I'm a great believer if you can't control the situation there's no point in worrying about it. . . . " Art echoed this with: "You know the old story of putting a marble in a jar . . . with every worry? And at the end of the week, you can't remember what it was for!"

Acceptance of greater dependence was not completely without issue, however, and had both positive and negative connotations for the participants, depending on how that greater dependence had come to be necessary.

Starting with the positive connotations, many of the participants expressed great relief at having moved from the family home in to something smaller and more manageable. This was an example of them taking control and being happy to accept a greater level of dependence – the participants who had downsized from the family home had done so by choice, at a time that suited them. The move was not a rushed affair due to rapidly deteriorating physical ability or the rapid unexpected onset of illness. Participants were happy to be more dependent on others and to have the responsibility of looking after a large home, and in some cases a large garden, removed from them.

The responsibility for upkeep being out of the hands of the participant was a particular relief in terms of emergency preparedness. Several of the participants had lived in the Adelaide Hills region, on semi-rural blocks, with lots of trees and other vegetation that required considerable upkeep with regard to bushfire prevention. The participants were aware that they were no longer able independently to manage their own properties; Brian talked about the work required in his family home, which was a factor in deciding to move to a retirement village setting:

> just to walk up and down was becoming difficult, and I certainly couldn't handle cleaning out the gutters and so on any more, that was . . . and there was lawn to cut, and all that sort of thing, which made it awkward.

Accepting their new world of greater dependence, and accepting a reduction in independence as a result of giving up driving, on the other hand, was seen as negative. When having to stop driving due to macular degeneration, for example, Hattie was clearly particularly upset, and explained that she missed driving terribly. Having spent most of her married life on a farming property, driving all sorts of vehicles, she missed both the physical act of driving, but more importantly the independence that driving brings. She did not feel comfortable having to rely on others:

> And I now suffer a lot from a feeling of so many people doing so many things for me and I have no way of returning it – because I've always been independent. And of course, having lived on a farm, we used to have to take the children places and all that sort of thing as well as ourselves. You didn't go anywhere unless you drove. Now I can't do it.

The mere thought of having to give up driving was also a major concern for some, not only in terms of both their own general independence and a sense of loss, but also in terms of the broader ramifications of not being able to drive, particularly the possible need to reorganize their lives and what being unable to drive might mean to their lives in to the future. Art, for example, when talking about the fact that he might one day have to give up driving, said:

> That would be a catastrophe . . . really. I do think about that . . . and I don't quite know what I'd do. Furthermore, my wife is absolutely dependent upon me. She's housebound – I don't know how she'd get on. No, really . . .

Art was not the only participant to be concerned. Keith was also concerned, as giving up driving would mean that he could no longer take his wife to medical appointments, and said: "Oh, it would make things very difficult, because as it is now [my wife] can't drive, and so whenever she has an appointment or whatever. . . ."

In summary, therefore, while acceptance of a greater dependence could be seen as relinquishing control, or imply a level of helplessness, this was not how the participants felt. For them it was about taking control. The move from the family home to a smaller home, for example, meant that unlike some of their peers, they had made a choice because they could; they had taken control of their lived world and shaped it to fit their requirements. The participants were aware of the changes in their physical health and ability, their environment, and their context. They had made decisions based on their experience and understanding, and they accepted that these choices were protective responses, interacting with their understanding of their capacity, capability, and control.

On the whole, the participants in this research had accepted with great enthusiasm that their time and how they filled it had changed over the course of their lives. As they moved from one stage to the next, the way they had shared this time had changed. For many years, their time may have been taken up with work or family responsibilities. As they got older and perhaps children moved out of the family home, many became keen volunteers in the community, and also looked after friends and family in both formal and informal arrangements.

Some participants had almost had a "second career" after retirement, sharing their time and enthusiasm with others for many years. Quite by accident, Clare found herself teaching dancing on a voluntary basis at a community center after she retired. Dancing was not something she had ever done professionally, just something that she was good at and that she enjoyed. It was clear that people enjoyed her classes, as she ran them for 18 years:

> Oh . . . yes . . . well, I mean there was, at one time, a waiting list a mile long to get in to the two classes at Elizabeth House. It was very . . . very popular, there's no doubt about that. But I tell you what, it took some brain power . . . would have kept my brain active! Cos I'd done Scottish, Irish, Ballroom, you know . . . I had the knack for dancing . . . that was it.

Clare was very committed to her volunteer role and during 18 years of running the classes, she had two hip replacements and recovered from a broken pelvis, and still carried on; it was only a mini-stroke that finally stopped her.

The participants did not consider that "knowledge is power" and were happy to share what they have learned throughout their lives. Keith, who had spent his professional career in the land management/valuation sector, volunteered with the newly formed Uniting Church, assisting them with their property administration, and

> eventually I became Chairman of the Property Committee of the Uniting Church, Chairman of the Property Trust of the Uniting Church. And mind you, with all that there was a lot of practical work on the evaluation side, and so for ten years after I retired I worked voluntarily in the church . . . in the property section. And I was . . . assisted by staff from within the church's property side, but that was quite, a great time I really had in those ten years.

He found the work very rewarding, and after the ten years he "retired again."

The decision to volunteer their time had been a conscious choice by the participants. It was important in allowing them to share their past experience and knowledge, to keep their brains active and to continue to feel part of the wider world, enjoying new encounters and relationships on their own terms.

Acceptance of sharing helped the participants stay connected in their changing lived worlds of downsizing and surrender of control. By sharing their time and knowledge, the participants also felt valuable and vindicated – they knew that they had amassed a great deal of experience, wisdom, and understanding over a range of areas of their lives, and they were keen to pass this on.

When the participants recollected stories from their lives, they were not negative or maudlin about the fact that they were getting older. As touched on in the theme of *understanding my world*, the participants were aware of, and accepted, their deteriorating physical ability. They had become used to their changing bodies, and rather than dwell on what they could no longer do, the participants were keen to emphasize what they still enjoyed and therefore what they were still able to offer. In accepting their advancing age and the restrictions that may come with that, they had adjusted their lives and their activities to suit, rather than stopping their favored activities completely.

Daisy, for example, was less able to leave the house, due to lack of transport. A keen gardener, she used to grow and pot up plants that she would take to the church fête and other community gatherings and then sell for charity. Daisy continued to grow and pot up the plants, but was no longer able to travel to functions to sell them, so she started to sell them from her home. While this meant a much smaller "customer base," her customers were good friends and looked out for her safety and wellbeing – something she found comforting as she lived on a sizeable rural block of land in the Adelaide Hills – even to the point of looking after the plants if she went on holiday:

> they're such good friends now. When I went . . . away for Christmas . . . one of them came and watered the plants. So, she offered to do it, and I offered to pay her, and she said "no way, but I'll take some plants." So she did. I think she said she took about thirty-five . . . and she wanted to pay me for them. I said "no way." You know . . . cos it was so hot.

As well as accepting their advancing age, they also accepted their impending death. The participants felt that they had been lucky to have lived good lives, and in reflecting on their lives, they felt generally positive; the meaning of loss of future time was experienced according to how their entire lives had been lived, and death generally held no fear for them.

Frances, one of the younger participants, had nursed her older sister through the final stages of cancer, and said:

> I think that was a very soul searching and it was a – I was going to say it was a good experience but in some ways it was a good experience because it made you realize that death is not that bad. . . .

Eleanor, a very spritely 90-year-old, still led an active social life. Her husband was a doctor and suffered a lengthy illness before his death. When asked whether she had reached a point in her life where she was accepting of death, she said:

> Well, er, I have written at the end of my will that . . . after I was eighty-six, if at any stage I had, um, something some sort of . . . I worded it much better than this . . . but, that it was a disease or something, or I was in an accident, I, okay, I didn't

want to be resuscitated from that. And I am . . . I mean as much as I'm happy, and I love the children, and I'd love to see the children go on . . . no, I'm . . . it doesn't worry me. I feel that I'm very lucky . . . and, okay, if tonight's the night, it really doesn't . . . worry me.

Daisy used similar words, saying: "I am. If it happens that I go, I'm quite ready . . . I'll be with [my husband]."

To summarize, the participants stressed that they were not concerned about their advancing age or their impending death – they did not feel that these were potential emergency events that they needed to prepare for, particularly in terms of mental preparation. The participants accepted that they might need to make choices and perhaps change the way they lived due to their advancing age and possible deteriorating physical health. While they may have already downsized in preparation, or may have plans in place such as paying into funds to pay for funeral expenses, they were not investing great effort in preparing for or worrying about their advancing age or impending death.

Discussion

This research presents an ontological view of what it means to be prepared for an emergency, having explored this experience within the lives of 11 older people. The study provides new knowledge in understanding that for older people, being prepared for a disaster is not concerned with toolkits and checklists, and tangible items. While it may be important to them to have food in the home and petrol in the car, being mentally prepared and mentally at peace (for example, having insurance) is far more important.

The study also shows that a variety of event types influence behavior and help build a feeling of being prepared. The interview process allowed the participants to define the emergency event and to include the incremental effect of events over an entire lifetime. Several of the participants, for example, said that an accumulation of smaller experiences enabled them to feel prepared and able to cope.

This research also highlights that while older people might not define themselves as "being prepared" in terms of traditional disaster management assessments, they do not feel vulnerable to disasters. They accept their limitations but feel confident they can cope. Given their lack of engagement to date, this is a key finding, as it has implications for how older people may (and should) be engaged in the future. They should not be approached as a "vulnerable" group as such; rather, they comprise a group that has certain specific needs, but also has a wealth of positive attributes in terms of knowledge, experience, and a sense of community. While the older people in this study might not define themselves as being prepared, they certainly consider themselves to be resilient.

In allowing the participants to self-define the emergency events they discussed, events that from a professional "disaster management sector" perspective might not be considered to influence preparedness were raised. This has implications for the development of preparedness messaging and education, and concurs with Graham's suggestion that "[t]he most effective messages are relevant to all hazards and meaningful on a day to day basis, while also effective in an emergency."[13] While Graham made this statement in reference to disaster recovery messaging, there are parallels for preparedness messaging – by being too specific about "being prepared for a flood" or 'being prepared for a bushfire," agencies may miss the opportunity to engage with people about being prepared for emergency events generally.

When considering the development of disaster management plans and policies, in addition to involving older people themselves, the shrinking nature and acceptance of their lived worlds must be taken into account. At this stage in their lives, being prepared for disasters – when understood from the more traditional disaster management sector world of, for example, natural hazards – was not important to the participants. They were confident of their resilience and their ability to cope. Designing preparedness materials for older people on specific hazard types, therefore, may serve no purpose. Ensuring older people are secure, safe, and feel mentally strong is more important.

The findings of this study also suggest that research in being prepared across the life course is important – there is an opportunity to engage those who are on their way to being older people and those who are already aged in order to learn from the understandings gained. In addition, research with those who are in everyday contact with older people, such as family and carers, should be conducted to learn about the nuanced meanings of being prepared for older people.

The tendency of agencies to develop plans for people they consider to be vulnerable – such as older people – and the assumption that everyone must have an emergency plan needs to be challenged. A better understanding on the part of Australian agencies of what it means for people to "feel prepared" is needed. For the individual, most planning and preparation is highly informal, personal, and situational in ways that are not dealt with using the current top-down approaches. Agencies should learn to accept the better judgment of those actually in these situations and develop programs that are supportive and less prescriptive.

More needs to be understood about what constitutes vulnerability in the elderly, and the engagement to build resilience must be genuine. As Cornell, Cusack, and Arbon state, "[e]mergency management planning needs to be less paternalistic and more inclusive if true resilience is to be achieved."[14] In the case of older people, it should be recognized that many older people live in their own homes in their communities and can contribute enormously to community resilience.

Conclusions

Older people are often considered to be a vulnerable group in terms of emergency management planning. They are seen as being under-prepared, and they are considered to be particularly vulnerable during an emergency event (in the response phase) and following an emergency event (in the recovery phase). Little research has been carried out, however, to explore preparedness among older people specifically. Of the research that has been undertaken, much of it considers the aged care service provider perspective, or is concerned with traditional measures of preparedness such as the collection of necessary items, including torches, first aid kits, and extra food and water.

The material gathered in undertaking this research, from the literature review to in-depth semi-structured interviews, shows that for the older people in this study, being prepared is not a one-off tangible activity – it is a process and a feeling of comfort and security in their world. The process is something that has been built upon over many years, and therefore "being prepared for an emergency event," as a specific activity, is not necessarily something that worries or concerns them. They have taken steps through their lives unconsciously – and sometimes almost by serendipity – to be prepared, for example by moving to smaller homes and giving up driving.

In accepting their advancing years and deteriorating physical ability, the participants felt assured in understanding that their life experiences (including the emergency events they had lived through) left them feeling comfortable and resilient in their changing and shrinking world and strong enough mentally to deal with any potential future emergency.

While discussion in the emergency management sector is expanding and considering new approaches, the sector has, as Richardson stated, followed the general approach of having

> someone up the front of a room, telling people what they should do, giving them a booklet and pamphlet, and then going away. Then, later, there is usually some "finger wagging" done, when "they" don't or didn't do what "we" told them to do.[15]

This research, therefore, has implications for emergency management sectors in terms of developing well-informed policy and practice. There is also a need for emergency management sectors to liaise with other sectors that may have knowledge of this group and insights into their world, for example community development teams and volunteering bodies.

By understanding what influences older people living in the community to prepare for emergency events – indeed, understanding that for this group of older participants, being prepared for specific emergency events is less important than feeling resilient and mentally able to cope – the sector can establish how best to assist them in their emergency preparedness planning, rather than making assumptions about what this group wants or needs.

Significantly, this research has shown the extent to which the emergency management sector has a taken for granted understanding of older people that has not previously been exposed like this. The findings of this study indicate that further research could be undertaken to explore the meaning of preparedness by professional groups, such as emergency management planners and policy makers. Other demographic groups, for example different cultural groups, participants from differing socio-economic backgrounds, and even different age groups, could also be explored in similar ways.

Notes

1 E. Ngo (2001), "When Disasters and Age Collide: Reviewing Vulnerability of the Elderly," *Natural Hazards Review* 2(2), pp. 80–89.
2 L. Fernandez, D. Byard, C. Lin, and J. Barbera (2002), "Frail Elderly as Disaster Victims: Emergency Management Strategies," *Prehospital and Disaster Medicine* 17(2), p. 68.
3 Ibid., p. 69.
4 V. Cornell, L. Cusack, and P. Arbon (2012), "Older People and Disaster Preparedness: A Literature Review," *Australian Journal of Emergency Management* 27(3), pp. 49–53.
5 I. Marshall and S. Mathews (2010), "Disaster Preparedness for the Elderly: An Analysis of International Literature Using a Symbolic Interactionist Perspective," *The Journal of Aging in Emerging Economies* 2, p. 80.
6 B.H. Morrow (2007), "Social Vulnerabilities and Hurricane Katrina: An Unnatural Disaster in New Orleans," *Marine Technology Society Journal* 40(4), pp. 16–26; Center on Aging (2005), *Disaster Planning for Older Adults in Palm Beach County* (Miami, FL: Florida

International University); S. Yeletaysi, D. Ozceylan, F. Fiedrich, J. Harrald, and T. Jefferson (2009), "A Framework to Integrate Social Vulnerability into Catastrophic Natural Disaster Preparedness Planning," paper presented at the International Emergency Management Society 2009 Annual Conference, June 9–11, 2009, Istanbul, Turkey; B. Flanagan, E. Gregory, E. Hallisey, J. Heitgerd, and B. Lewis (2011), "A Social Vulnerability Index for Disaster Management," *Journal of Homeland Security and Emergency Management* 8(1), Article 3. DOI: 10.2202/1547-7355.1792.

7 K. Hyer, L. Brown, A. Berman, and L. Polivka-West (2006), "Establishing and Refining Hurricane Response Systems for Long-Term Care Facilities," *Health Affairs* 5, pp. 407–41; Hyer, Polivka-West, and L. Brown (2007), "Nursing Homes and Assisted Living Facilities: Planning and Decision-Making for Sheltering in Place or Evacuation," *Generations* 31(4), pp. 29–33; N. Castle (2008), "Nursing Home Evacuation Plans," *American Journal of Public Health* 98(7), pp. 1235–40.

8 S. Laditka, J. Laditka, C. Cornman, C. Davis, and M. Chandlee (2008), "Disaster Preparedness for Vulnerable Persons Receiving In-Home, Long-Term Care in South Carolina," *Prehospital and Disaster Medicine* 23(2), pp. 133–42.

9 M. van Manen (1990), *Researching Lived Experience: Human Science for an Action-Sensitive Pedagogy* (Albany, NY: SUNY Press).

10 J. Cresswell (2003), *Research Design: Qualitative, Quantitative and Mixed Methods Approaches* (Thousand Oaks, CA: Sage), p. 181.

11 van Manen (1990).

12 G. Bonanno, S. Galea, A. Bucciarelli, and D. Vlahov (2007), "What Predicts Psychological Resilience after Disaster? The Role of Demographics, Resources, and Life Stress," *Journal of Consulting and Clinical Psychology* 75(5), p. 671.

13 W. Graham (2011), *To Study the Effectiveness of Disaster Assistance Programs in Promoting Individual and Community Resilience in Recovery from Disasters* (The Winston Churchill Memorial Trust of Australia), p. 20.

14 Cornell et al. (2012), p. 52.

15 J. Richardson (2013, August 8), "Would You Like Some Steak Knives with That?" *Sastrugi* [Blog], available at http://sastrugi64.wordpress.com/2013/08/08/would-you-like-some-steak-knives-with-that/

Bibliography

Bonanno, G., S. Galea, A. Bucciarelli, and D. Vlahov. (2007). "What Predicts Psychological Resilience after Disaster? The Role of Demographics, Resources, and Life Stress." *Journal of Consulting and Clinical Psychology* 75(5), pp. 671–82.

Castle, N. (2008). "Nursing Home Evacuation Plans." *American Journal of Public Health* 98(7), pp. 1235–40.

Center on Aging. (2005). *Disaster Planning for Older Adults in Palm Beach County*. Miami, FL: Florida International University.

Cornell, V., L. Cusack, and P. Arbon. (2012). "Older People and Disaster Preparedness: A Literature Review." *Australian Journal of Emergency Management* 27(3), pp. 49–53.

Cresswell, J. (2003). *Research Design: Qualitative, Quantitative and Mixed Methods Approaches*. Thousand Oaks, CA: Sage.

Fernandez, L., D. Byard, C. Lin, and J. Barbera. (2002). "Frail Elderly as Disaster Victims: Emergency Management Strategies." *Prehospital and Disaster Medicine* 17(2), pp. 67–74.

Flanagan, B., E. Gregory, E. Hallisey, J. Heitgerd, and B. Lewis. (2011). "A Social Vulnerability Index for Disaster Management." *Journal of Homeland Security and Emergency Management* 8(1), Article 3, doi:10.2202/1547-7355.1792

Graham, W. (2011). *To Study the Effectiveness of Disaster Assistance Programs in Promoting Individual and Community Resilience in Recovery from Disasters*. The Winston Churchill Memorial Trust of Australia. Retrieved October 3, 2013, from http://www.churchilltrust.com.au/fellows/detail/3548/wendy+graham

Hyer, K., L. Brown, A. Berman, and L. Polivka-West. (2006). "Establishing and Refining Hurricane Response Systems for Long-Term Care Facilities." *Health Affairs* 5, pp. 407–41.

———, L. Polivka-West, and L. Brown. (2007). "Nursing Homes and Assisted Living Facilities: Planning and Decision-Making for Sheltering in Place or Evacuation." *Generations* 31(4), pp. 29–33.

Laditka S., J. Laditka, C. Cornman, C. Davis, and M. Chandlee. (2008). "Disaster Preparedness for Vulnerable Persons Receiving In-Home, Long-Term Care in South Carolina." *Prehospital and Disaster Medicine* 23(2), pp. 133–42.

Marshall, I., and S. Mathews. (2010). "Disaster Preparedness for the Elderly: An Analysis of International Literature Using a Symbolic Interactionist Perspective." *The Journal of Aging in Emerging Economies* 2, pp. 79–92.

Morrow, B.H. (2007). "Social Vulnerabilities and Hurricane Katrina: An Unnatural Disaster in New Orleans." *Marine Technology Society Journal* 40(4), pp. 16–26.

Ngo, E. (2001). "When Disasters and Age Collide: Reviewing Vulnerability of the Elderly." *Natural Hazards Review* 2(2), pp. 80–89.

Richardson, J. (2013, August 8). "Would You Like Some Steak Knives with That?" *Sastrugi* [Blog]. Available at: http://sastrugi64.wordpress.com/2013/08/08/would-you-like-some-steak-knives-with-that/

van Manen, M. (1990). *Researching Lived Experience: Human Science for an Action-Sensitive Pedagogy*. Albany, NY: SUNY Press.

Yeletaysi, S., D. Ozceylan, F. Fiedrich, J. Harrald, and T. Jefferson. (2009). "A Framework to Integrate Social Vulnerability into Catastrophic Natural Disaster Preparedness Planning." Paper presented at the International Emergency Management Society 2009 Annual Conference, June 9–11, 2009, Istanbul, Turkey.

Part 5

Europe and multi-continental studies

Introduction

Adenrele Awotona

Studies show that multilateral agencies, the private and public sectors, academia, non-governmental organizations, and local communities are increasingly underscoring the significance of inter-sectoral collaboration and partnerships for the development of sustainable resilient communities worldwide.[1] For example, the 2005 Hyogo Framework for Action (HFA), a comprehensive United Nations ten-year strategy for disaster risk reduction which aimed to reduce human and material losses from disasters by 2015 and to which 168 governments were signatories, recommended and actively promoted the establishment of Public-Private Partnerships as a mechanism for reducing the underlying risk factors that contribute to disasters globally. The HFA also stressed the need to increase international and regional cooperation and support in the field of disaster risk reduction through, *inter alia*:

- The transfer of knowledge, technology, and expertise to enhance capacity-building for disaster risk reduction;
- The sharing of research findings, lessons learned, and best practices;
- The compilation of information on disaster risk and impact for all scales of disasters in a way that can inform sustainable development and disaster risk reduction; and
- Appropriate support in order to enhance governance for disaster risk reduction, awareness-raising initiatives, and capacity-development measures at all levels, in order to improve the disaster resilience of developing countries.[2]

Also, the United Nations Environment Program (UNEP) actively backs partnerships and networks in order to build a global community of decision-makers, advocates, and practitioners who integrate ecosystem management solutions in disaster risk reduction, climate change adaptation, and development strategies. Correspondingly, national governments (such as those of the US, Canada, and Australia) are adopting policy frameworks that stress the significance of partnerships for disaster resilience. In the U.S., the Federal Emergency Management Agency (FEMA) has developed *A Whole Community Approach* to emergency management.[3] The Australian Government's *National Strategy for Disaster Resilience* states that "Disaster resilience is the collective responsibility of all sectors of society, including all levels of government, business, the non-government sector and individuals."[4] In Canada, the Ministers Responsible for Emergency Management state, in their *An Emergency Management Framework for Canada*, that they "support legal and policy frameworks, programs, activities, standards and other measures in order to enable and inspire all emergency management partners in Canada to work in better collaboration to keep Canadians safe."[5]

Similarly, the 2005–2015 Hyogo Framework for Action prioritized disaster risk reduction education in its campaign to develop "disaster-resilient communities" worldwide.[6] It called for, amongst others, the integration and infusion of disaster studies (as well as climate change consciousness and environmental education) into the curricular, co-curricular, and extra-curricular activities of educational institutions at all levels. Also, the *Sendai Framework for Disaster Risk Reduction 2015–2030* calls for the "strengthening of international cooperation and global partnership."[7]

It is in that vein that Petal and Izadkhah explicated the range of formal and informal disaster risk reduction education in schools around the world with a focus on Iran.[8] For instance, earthquake education has been developed and is led by the Ministry of Education and earthquake experts at universities in collaboration with civic organizations, public agency officials, and policy-makers in Iran. There, formal and informal education in disaster preparedness takes place at the "nursery, elementary, secondary, and high school levels on a national scale covering both urban and rural areas" and disaster lessons are incorporated within science, geography, literature, and other curricula.[9]

Thus, these two themes of collaboration and the integration of Disaster Risk Reduction education into the curricula of institutions of higher education are the subjects of the case studies in Chapters 23 and 24. In Chapter 23, Patricia E. Perkins examines climate-related disasters and climate justice in relation to her own work with civil society organizations and community groups in both the Global North (socially disadvantaged areas of Toronto, Ontario) and the Global South (Brazil, Mozambique, South Africa, and Kenya) aimed at developing participatory community-based programs for environmental education and climate-change awareness leading to increased political engagement by socially vulnerable people. Her specific emphases are on watersheds, water management, and gender equity. Perkins proposes that partnerships between university faculty/students and community-based organizations are useful in building climate justice, both locally and globally. She argues that many impacts of global climate change involve weather-related disasters and that socio-economically vulnerable people are simultaneously those most likely to be impacted by such disasters (because of their geographic location) and those least equipped to deal with those impacts (because of their weak economic and political position). She then notes that the field of "climate justice" addresses these problems and is developing indicators of vulnerability, surveys of the extent of climate change-related inequities, and policy proposals to deal with them. Moreover, Perkins observes that just as in development studies more generally, there is a fundamental conceptual and practical difference between top-down mechanisms for income redistribution controlled by international institutions, and grassroots strategies involving fundamental political/economic change and a greater voice for the vulnerable, starting at the local level. Participatory-governance-based ways of addressing inequities are most consistent with climate justice.

In Chapter 24, John "Jack" Lindsay starts with the observation that emergency management is a complex, community-wide activity that intersects in many ways with community planning. However, whilst the research literature, primarily in disaster studies, makes the connection between land-use planning and hazard mitigation, the link is less visible in common planning practice. The opportunity to expand emergency managers' perception of planning beyond just land-use zoning is matched by the need to add comprehensive emergency management to the planning professional's lexicon.

Lindsay's brief review of planning school course offerings shows that, although education will play a key role in improving the planning profession's involvement across the emergency management spectrum, from reducing the creation of risk to ensuring resiliency during disaster recovery, emergency management in general and hazard mitigation in particular is not a frequent topic of discussion. The chapter enables the readers to gain an understanding of how the professional roles of planners and emergency managers should intersect and some of the reasons why this collaboration is underdeveloped. The study draws on research conducted in the US, Canada, and New Zealand, and on Lindsay's own experience as a planner, emergency manager, and university professor.

In Chapter 25, Matías Ezequiel Barberis Rami notes that the possibility to reduce risks in order to ensure resiliency during disaster recovery springs from the people themselves, and it differs in different urban contexts. This is mostly because increasing population density in urban contexts, the growth of mega-cities, environmental degradation, and economic uncertainty, as well as the exponential increase in consumption and climate change are some of the main causes of the proliferation of risks in everyday lives. Some of the other causes of risks include the crisis in models of representative democracy in developed countries, as well as unequal access to wealth in developing countries. Hence, the focus of this chapter is an analysis of some Latin American and European urban contexts, contrasting the strengths and limitations of the possibility of building resilient spaces sustainably. According to Rami, the first results of his analysis serve as a warning that although the existence of international regulatory frameworks around the theme promote integrated risk management, the possibility of building resilient cities should be generated through a process of exchange of experience between different contexts, starting with the similarities between the contexts and the strengths of each city. Moreover, he notes, the different models of urban planning have designed particular relationships among groups so that resilience can be directly linked with urban development.

Still on the topic of the different models of urban planning, in Chapter 26 Giedrius Kaveckis, Benjamin Bechtel, Jürgen Ossenbrügge, and Thomas Pohl examine land-use modeling as a new approach for future hazard-sensitive population mapping in Northern Germany. The authors observe that as the number of people exposed to natural hazards is increasing, there is an urgent need to identify such population groups in order to implement measures to decrease their vulnerability. However, mapping a future population on the local scale still is an unsolved scientific challenge. The chapter explores whether land-use modeling can be used for mapping future hazard-sensitive populations. The basis of the authors' concept is a disaggregation of future population projections using the modeled future land-use data as proxy, elaborated for the Greater Hamburg area in northern Germany. The result of this research, according to the authors, is a new approach, a future population-mapping framework that supports vulnerability assessment and the impacts of climate change.

Notes

1 J. Chen, T.H.Y. Chen, I. Vertinsky, L. Yumagulova, and C. Park (2013), "Public-Private Partnerships for the Development of Disaster Resilient Communities," *Journal of Contingencies and Crisis Management* 21(3), pp. 130–43; J.K. Mitchell (2006), "The Primacy of Partnership: Scoping A New National Disaster Recovery Policy," *The Annals of the American Academy of Political and Social Science* 604(1), p. 237.

2 UNISDR (2005), *Hyogo Framework for Action 2005–2015: International Strategy for Disaster Reduction – Building the Resilience of Nations and Communities to Disasters* (Geneva, Switzerland: Author).
3 Federal Emergency Management Agency (2011), *A Whole Community Approach to Emergency Management: Principles, Themes, and Pathways for Action*, FDOC 104–008–1 / December (Washington, DC: Author).
4 Commonwealth of Australia (2011), *National Strategy for Disaster Resilience: Building the Resilience of Our Nation to Disasters*, Council of Australian Governments, February, retrieved April 9, 2015, from https://www.em.gov.au/Documents/1National%20Strategy%20 for%20Disaster%20Resilience%20-%20pdf.PDF, p. iv.
5 Ministers Responsible for Emergency Management (2011, January), *An Emergency Management Framework for Canada* [Second edition]. Ottawa, Canada: Emergency Management Policy Directorate, Public Safety Canada.
6 UNISDR (2005).
7 United Nations (2015), *Sendai Framework for Disaster Risk Reduction 2015–2030* (New York, NY: Author), p. 5.
8 M.A. Petal and Y.O. Izadkhah (2008), *Concept Note: Formal and Informal Education for Disaster Risk Reduction*, a Contribution from Risk RED to the International Conference on School Safety, May 2008, Islamabad, Pakistan, retrieved April 8, 2015, from http://www. riskred.org/activities/ddredislamabad.pdf
9 Ibid., p. 3.

Bibliography

Chen, J., T.H.Y. Chen, I. Vertinsky, L. Yumagulova, and C. Park. (2013). "Public-Private Partnerships for the Development of Disaster Resilient Communities." *Journal of Contingencies and Crisis Management* 21(3), pp. 130–43.

Commonwealth of Australia. (2011). *National Strategy for Disaster Resilience: Building the Resilience of Our Nation to Disasters*. Council of Australian Governments, February. Retrieved April 9, 2015, from https://www.em.gov.au/Documents/1National%20Strategy%20 for%20Disaster%20Resilience%20-%20pdf.PDF

Federal Emergency Management Agency (FEMA). (2011). *A Whole Community Approach to Emergency Management: Principles, Themes, and Pathways for Action*. FDOC 104–008–1 / December. Washington, DC: Author.

Ministers Responsible for Emergency Management. (2011, January). *An Emergency Management Framework for Canada* [Second edition]. Ottawa, Canada: Emergency Management Policy Directorate, Public Safety Canada. Retrieved April 9, 2015, from https://www.public safety.gc.ca/cnt/rsrcs/pblctns/mrgnc-mngmnt-frmwrk/mrgnc-mngmnt-frmwrk-eng.pdf

Mitchell, J.K. (2006). "The Primacy of Partnership: Scoping a New National Disaster Recovery Policy." *The ANNALS of the American Academy of Political and Social Science* 604(1), pp. 228–55.

Petal, M.A., and Y.O. Izadkhah. (2008). *Concept Note: Formal and Informal Education for Disaster Risk Reduction*. A contribution from Risk RED to the International Conference on School Safety. May 2008. Islamabad, Pakistan. Retrieved April 8, 2015, from http://www. riskred.org/activities/ddredislamabad.pdf

United Nations. (2015). *Sendai Framework for Disaster Risk Reduction 2015–2030*. New York, NY: Author.

UNISDR. (2005). *Hyogo Framework for Action 2005–2015: International Strategy for Disaster Reduction – Building the Resilience of Nations and Communities to Disasters*. Geneva, Switzerland: Author. Retrieved April 10, 2015, from http://www.unisdr.org/2005/wcdr/ intergover/official-doc/L-docs/Hyogo-framework-for-action-english.pdf

23 University-community collaboration for climate justice education and organizing

Partnerships in Canada, Brazil, and Africa[i]

Patricia E. Perkins

Introduction

In the coming decades, countries around the world will face increasingly severe challenges related to global climate change. While the details vary from country to country, the impacts will be especially grave for marginalized people, whose access to food, potable water, and safe shelter may be threatened due to fluctuations in rainfall and temperature and to disasters related to extreme weather events.

International strategies for addressing climate change are in disarray. The complicated financial and carbon-trading mechanisms promoted by the United Nations and other global institutions are far too bureaucratic, weak, internally inconsistent, and scattered to represent meaningful solutions to climate change. Already the housing, health, and livelihoods of marginalized people worldwide are being threatened by the ramifications of climate change. This means that the marginalized in every community, by definition, have expertise in how priorities should be set to address climate change. Their experiences, knowledge, and views must be part of local, regional, national, and international governance – including urban planning and housing, water management, agriculture, health, and finance policies.

Satterthwaite et al.'s recent book on urban climate change adaptation summarizes the major challenges that low- and middle-income countries face as a result of climate change:

> most of the world's urban population live in cities or smaller urban centers ill-equipped for adaptation – with weak and ineffective local governments and with very inadequate provision for the infrastructure and services needed to reduce climate-change-related risks and vulnerabilities. A key part of adaptation concerns infrastructure and buildings – but much of the urban population in Africa, Asia and Latin America have no infrastructure to adapt – no all-weather roads, piped water supplies or drains – and live in poor-quality housing in floodplains or on slopes at risk of landslides. Most international agencies have long refused to support urban programs, especially those that address these problems.[1]

Climate change thus exacerbates already grave sustainable development challenges.

This chapter examines bottom-up strategies for facing these kinds of challenges, especially with regard to how these approaches address social vulnerability. The details of each particular community's situation – ecological, social, political – are crucial to this type of approach. How do communities organize socially and politically to meet the biophysical and weather-related changes that affect their livelihoods? How are the needs of the most vulnerable addressed?

I have been involved with university-community collaborations to address these challenges through two international projects – the *Sister Watersheds* project with Canadian and Brazilian partners (2002–2008) and the *Climate Change Adaptation in Africa* project with partners in Canada, Kenya, Mozambique, and South Africa (2010–2012) – as well as recent green community development initiatives in several marginalized neighborhoods and several additional networking projects. This work has demonstrated the wide applicability of local-level efforts in vulnerable communities to address equity challenges by developing strategies and materials for increasing local residents' knowledge of, interest in, and engagement with water-related and climate-change issues, focusing in particular on women and youth. Collaborative partnerships between university researchers and community activists/organizers can generate fruitful synergies, especially regarding the spread of scientific knowledge, and they can also strengthen educational outreach, build skills, and foster global networking.

After giving some examples of ways in which local community organizations are addressing climate- and water-related challenges through innovative grassroots programs and initiatives, this chapter concludes by noting some similarities in these stories from the Global North and South and some ways in which communication and mutual reinforcement can strengthen and inspire global climate justice work.

Community-based responses to climate change: Green community organizing, North and South

In a warming world, heavier and more frequent precipitation is likely, as are more frequent and severe droughts. All over the world, climate-change-related disasters affect marginalized people first and most severely.

While environmental problems have long been a source of concern in many areas around the world, climate change is raising these issues on local agendas and adding urgency to infrastructure needs. Community groups in many cities are addressing long-standing priorities of job creation, infrastructure development and repair, economic opportunities, and the need for recreational space, while also beginning to respond to the local effects of climate change. These groups' local ecological knowledge and social capital are very important for building disaster resilience strategies, including disaster forecasting, prevention, and early warning; environmental land-use planning; public participation and psychosocial engagement; and progressive approaches to reconstruction.

Attention to the value of vulnerable populations' contributions and ways of engaging them in policy processes should be a global priority. There are many North-South commonalities demonstrating how this can be done. Key to the political and policy impact of all the examples described below is the fact that a more organized and informed citizenry attracts attention from media and government, demands interventions to meet its identified needs and priorities, and is able to articulate its knowledge, perspectives, and rationales in public policy processes, thus contributing to more equitable disaster planning and response.

Sister watersheds: Equity in São Paulo's watershed committees

Brazil has a progressive watershed management system that requires participation on the part of civil society representatives on watershed committees, but low-income people – women in particular – are underrepresented. Watershed committees are formed "so that water users can collectively help to decide issues of allocation, infrastructure and regulation at the watershed level."[2] However, for a variety of reasons, low-income local residents, especially women, are often not motivated to become involved in these processes.

The *Sister Watersheds* project (2002–2008) linked universities and NGOs in Canada and Brazil in developing strategies and materials to increase local residents' knowledge of, interest in, and engagement with water-related issues, focusing on low-income neighborhoods in São Paulo, particularly low-income women. This $1.3 million project, funded by the Canadian International Development Agency (CIDA) through the Association for Universities and Colleges of Canada, combined student exchanges, research, community engagement, and "capacity-building" in local communities and nearby universities. Its novel conceptualization and design were developed by progressive Brazilian environmental educators Dr. Marcos Sorrentino and Larissa da Costa of the Ecoar Institute for Citizenship (ECOAR), a leading environmental education NGO based in São Paulo. The project's design evolved throughout its implementation by organizers at ECOAR and York University in Toronto.

The project developed and tested training programs by conducting workshops led by its local NGO partners with more than 1,450 participants, approximately two-thirds of whom were women, and by partnering with other community organizations to present content on topics related to environmental education and watershed management. For example, staff from ECOAR contacted groups of elementary school teachers, public health extension agents, and other community-based workers, and provided in-service training for them about water and health, basic ecology, and public policy questions related to water in their local communities. The various training programs were shaped and modified to be specifically appropriate for groups of women, children, youth, health agents, school groups, teachers, film/culture/music/arts organizations, and Agenda 21/environmental education groups. The workshops focused on water management, environmental education, community development, and democratic participation, with particular emphasis on gender and socio-economic equity. The methodologies, techniques, and materials developed for these workshops and training programs – made freely available to other organizations through publications and websites – contributed to the capacity of project partner organizations and individual staff members and students to continue related work on watershed policy issues into the future.

The curriculum, materials, and techniques developed by the project were tested and fine-tuned in more than 220 workshops designed and led by project staff, student interns, and university exchange students in three watersheds – two in Brazil and one in Canada – where university campuses are located near low-income residential areas. All of the workshop participants were potential participants in Brazil's watershed committees, as civil society representatives/organizers. The outreach materials developed by the project include an illustrated *Manual on Participatory Methodologies for Community Development*. It contains a set of workshop activities and background materials for participatory community environmental education programs and training sessions focusing on water and gender equity issues; an illustrated guide with practical exercises focusing on urban agroecology; a full-color socio-environmental atlas which brings together ecological, hydrological, and social information about one local watershed in a series of interactive maps; a video about the history and environment of one of the watersheds; a publication outlining Agenda 21 activities in schools; and several blogs and websites with materials and discussion-starters on watershed topics, as well as a book and many journal articles, masters' papers, and other academic publications contributing to the literature on participatory watershed education in Brazil and Canada.

The community environmental perception surveys conducted by the project in each of the Brazilian watersheds established a database of information on public priorities and views on watershed issues. The socio-environmental atlas gathered and made available in one place a wide range of information on ecological, hydrological, social, and political circumstances in the watershed – information which proved very useful to public officials and watershed committee members in understanding the watershed as a whole. The nearly 1,500 participants in workshops conducted by the project gained familiarity and experience with water-related issues and their own ability to influence water management and policy through watershed committee structures, community organizing, community arts, and other means.

One particular contribution of the York/Black Creek "sister watershed" was the evolving arts-based "Black Creek Storytelling Parade," a participatory performance walk held periodically that follows the route of stormwater from the York University campus to the banks of Black Creek, using different storytellers from neighborhood organizations to recount the history of the natural and built environment. Various creative strategies – costumes, sidewalk chalk, and percussion instruments – are employed to engage the audience. The content of the stories includes natural, cultural, and political dimensions: the First Nations' land claim, covering the entire City of Toronto; the Haudenesaunee village buried under electric lines just south of the campus; how the creek acts as a cultural divide between two very distinct neighborhoods; and local ecological restoration efforts, as showcased by young students at a nearby elementary school. The Black Creek Storytelling Parade, which was developed by York graduate students in the area of Community Arts Practice, shakes up conventional understandings of nature by emphasizing social, cultural, and political stories – tales often left untold in the city's official chronicles. This helps build an ecological imagination – the capacity to imagine how local residents could be living in such a way that people, plants, and animals thrive in ecologically sustainable and socially just futures. It does so by sparking dialogue and dreams for a restored creek – restored not just in terms of greenery and cleanliness, but also in terms of social and cultural importance.

This project helped both its university and NGO participants to bridge the gap between academic and community-based methods of environmental education. Graduate exchange students studied and contributed to local training programs; faculty members wrote about the theoretical and practical benefits of public participation in watershed management; NGOs supervised students who received academic credit for their community-organizing work; professors led local watershed governance structures; and innovative methods for environmental education were shared internationally. This collaboration allowed new perspectives on water management to evolve, with benefits for all participants' training/education programs. The University of São

Paulo, York University, and ECOAR developed dozens of new partnerships with other community organizations as a result of this project. Students in both Brazil and Canada played a crucial role in developing the linkages between academic institutions and community-based NGOs. Both locally and internationally, students sought out community organizations for their research and field experiences, and shared the results of their work with both academic and non-academic audiences. The student exchanges of this project thus fueled its interdisciplinary and educational bridging contributions.

The Centro de Saberes: Watershed-level organizing

An inspiring transnational model for intervention to increase public involvement in watershed management has been developed by the Center of Wisdom and Care for the La Plata Basin (*Centro de Saberes*), an organization largely funded through a fraction of the hydroelectric power revenues generated by the huge Iguaçu dam, located on the Paraná River where Brazil meets Paraguay and Argentina.[3] As most climate change impacts are felt through rainfall-related extreme weather events (mainly droughts and floods), water management is crucial for climate justice.

The Paraná watershed, which drains much of central and eastern South America and reaches the Atlantic Ocean via the La Plata River near Montevideo, also includes Bolivia and Uruguay, so the Centro de Saberes works in three languages – Portuguese, Spanish, and Guaraní, an official language of Paraguay. The Centro de Saberes convened a series of "permanent learning circles" attended by media, academic, activist, and government representatives from each of the five countries in the watershed. Each year, like ripples, the "permanent learning circles" expanded, as the participants from the year before invited additional representatives to attend in subsequent years. The circles grew from five participants (one from each country) in 2006 (the first year) to 35 the next year, to hundreds. The agenda and program of the meetings have included social exchanges among participants, discussions of local priorities for environmental and political action, transnational meetings by profession (with the journalists, government officials, teachers, etc., sharing their concerns and strategies), and brainstorming about how to accomplish the goals identified by each group. The "permanent learning circle" meetings thus became an opportunity for participants to take back to their home contexts a watershed perspective on their day-to-day work-related challenges and goals. The Centro de Saberes has also facilitated watershed-based networking and information-sharing for improved public policy and environmental responsibility throughout the Paraná basin.

The Centro de Saberes' five operating principles include: water as the generator theme; the watershed as the operating territory; an ethic of protecting the diversity of life in the watershed and considering the different kinds of knowledge and protection available in the watershed; environmental education as an element capable of engaging society in action; and the collective construction of information, knowledge, and actions.

The political significance of an organization like the Centro de Saberes is that it spans existing political jurisdictions using ecological boundaries – the watershed – as the organizing and motivating principle for its actions. This also makes possible sharing across watersheds (of promising public engagement strategies or policy approaches to specific problems such as how to reduce erosion or limit the use of agricultural chemicals, for example).

Climate justice and water governance in Africa

According to the Intergovernmental Panel on Climate Change (IPCC),

> Africa is one of the most vulnerable continents to climate change and climate variability. This vulnerability is exacerbated by existing developmental challenges such as endemic poverty, limited access to capital, ecosystem degradation and complex disasters and conflicts.[4]

Income inequality in South Africa, Mozambique, and Kenya is among the greatest in the world; in all three countries, equity struggles related to water are growing in social, political, and ecological significance, which is both a symptom and a cause of urban vulnerabilities related to climate change.

In Maputo, Mozambique, climate change is causing coastal erosion and periodic flooding along scenic coastal roadways; saltwater intrusion, wind erosion, and desertification in urban food-producing areas; flooding in coastal slum areas; degradation of water quality in wells and potable water scarcity; and the destruction of mangroves and threats to the locally important shrimp fishery. There are clear signs that the sea level is rising, with concomitant and expensive coastal management problems in the Maputo municipality. On three offshore islands mangroves are disappearing, water quality is declining, and desertification and erosion are increasing.[5] The UN Habitat Cities in Climate Change Initiative, which has begun a pilot project in Maputo, emphasizes local government capacity-building, policy dialogue, climate change awareness, public education, and developing coordination mechanisms between all levels of government as priorities to help address these risks. Mozambique's national water law (1991) considers all water to be state-owned and governed for the benefit of the population, with water access for people, sustainability, and stakeholder participation as priorities. Four water basin committees have been established in Mozambique on the same general model as in Brazil. To make this participatory model more effective, the greatest need is for capacity-building and community-organizing to deepen and strengthen civil society's involvement in water governance.

As in Mozambique, South Africa is implementing watershed committees or "catchment management agencies" (CMAs) to decentralize decision-making and create a framework for integrating the needs of all stakeholders in water governance. Durban's municipal government has already developed a local climate change adaptation strategy; like Maputo, Durban faces coastal inundation and storm surges related to sea level rise, hotter temperatures and heat waves, changed rainfall and storm patterns, slum flooding, and reduced drinking water supplies due to climate change. Local policy initiatives rely on awareness and capacity for effectiveness with regard to climate change risks and adaptive responses in civil society. Environmental education and confidence-building through capacity-raising are recognized as crucial needs in this process; for example, the Inkomati CMA has initiated outreach programs targeting rural poor, emerging farmers, women, and youth. Grounded participatory research leading to accessible public education and responsive community-based programs with civil society organizations are needed to help address these significant water governance challenges. This type of action research is well developed in Durban, partly due to the work of the Centre for Civil Society at the University of KwaZulu-Natal and its partner NGOs.

In Nairobi, severe infrastructure needs are being exacerbated by water supply fluctuations and slum flooding related to climate change. Just as in Maputo and Durban, environmental awareness and education leading to more equitable governance processes are required. As noted by the Kenyan delegation to the 2006 UN Conference on Climate Change in Nairobi, Kenya's adaptation focuses include education, good governance, human resources development and training, institutional capacity building and management change, public finance improvement, and better national resources management. Nairobi, one of the largest and most complex cities in the world, provides a challenging arena for participatory governance research.

The democratic mediation of equity conflicts related to water and sustainable long-term management of water resources in the face of climate change requires public participation, particularly that of low-income marginalized women – the experts.

"Strengthening the role of civil society in water sector governance towards climate change adaptation in African cities – Durban, Maputo, Nairobi" was a three-year project (2010–2012) with African partners in the three cities. Its goal was to improve watershed governance for climate change adaptation and enhance the resilience and adaptive capacity of vulnerable and marginalized groups, especially women. This project was supported by the Climate Change Adaptation in Africa (CCAA) program – a joint initiative of Canada's International Development Research Centre (IDRC) and the UK's Department for International Development (DFID). Like the earlier Sister Watersheds Project, this project's methodology included collaboration between students, NGOs, and academics, as well as community-based research and environmental education. Project partners based in universities and several NGOs in Nairobi, Maputo, and Durban worked together to achieve the following objectives:

- To characterize the institutional framework for urban water governance in the three cities and explain how different actors within this framework cope with climate change and variability;
- To identify and test viable alternatives for enhancing civil society's role towards adaptation to climate change and variability by vulnerable groups (e.g., by developing education, training and awareness programs); and
- To share widely the knowledge generated for potential adoption by other cities in Africa.

The project was implemented by the following community-based NGOs in Africa: the Kilimanjaro Initiative (KI) and Kenya Debt Relief Network (KENDREN) in Nairobi; Women, Gender and Development (MuGeDe) and *Justiça Ambiental* (JA) in Maputo; and *Umphilo waManzi* (Water for Life) and the South Durban Community Environmental Alliance (SDCEA) in Durban. York University (Toronto), the University of Nairobi (Nairobi), Eduardo Mondlane University (Maputo), and the Centre for Civil Society at the University of KwaZulu-Natal (Durban) provided academic research coordination and student supervision for this project.

The project focused on low-income areas of each city, as these tend to be most severely affected by periodic flooding and other climate change impacts. Residents of low-income areas often lack the ability to protect themselves against extreme weather events. The capacity-building aspects of this project included training and research sponsorship for students and faculty in the partner universities; support for community-based research, workshops in low-income communities and secondary schools, curriculum and materials development, and skills development within the partner NGOs and civil society organizations; training of environmental educators and organizers; contributions to the pool of experienced and qualified community workers in each country; strengthening of all the partner institutions' capabilities to carry out international projects; and contributions to the international literature and professional knowledge concerning water issues, environmental education techniques, and community organizing for improved civil society involvement in governance. The networks built extended from local and community-based linkages through regional and national-level policy groupings to international academic and policy networks on civil society, watershed management, and governance.

The political process of policy development and implementation depends on the interchange between civil society groups, researchers generating information on current realities, and government. This project attempted to challenge the conventional notion that only educational institutions "produce" knowledge. Understanding community needs and what helps particular civil society groups to see and act to strengthen their role in democratic governance, for example, is something in which community organizations and NGOs have eminent expertise. This collaborative approach, also known as participatory action research (PAR), is broadly defined as "research by, with and for people affected by a particular problem, which takes place in collaboration with academic researchers. It seeks to democratize knowledge production and foster opportunities for empowerment by those involved."[6]

One objective of this project was to demonstrate how partnerships between academics and non-academics can be very stimulating and effective. This type of partnership encourages and allows the partner NGOs to reflect on and analyze their activities and to document "learning" more systematically than they are often able to do, by bringing student researchers into the NGOs as collaborators/interns. The partnership also encourages universities to be more pragmatic about teaching and research and to "field-test" approaches towards community organization, equity, and capacity building. Students committed to the project's goals of fomenting the participatory engagement of local people in municipal water decision-making are given practical opportunities to develop their skills, as a way of hastening each city's climate-change preparedness. This project aimed to contribute to the integration and meaningful participation of women in formal decision-making processes, as well as to build their adaptive capacity and increase their resilience and ability to cope with climate change.

Specific examples of how climate change responses combine well with gender-aware community organizing, as part of this project, included the following:

- The Kilimanjaro Initiative (KI), a youth-focused NGO, upgraded a sports field in Nairobi's Kibera slum, on the banks of the Nairobi River, which reduced disastrous flooding during extreme weather events because the flood plain was being used for much-needed public recreational space instead of housing. In addition, the KI organized community forums on sustainable water management and environmental education, as well as community and river clean-ups. Young women's leadership was central to this organizing.
- In Durban, women activists from Umphilo waManzi and the South Durban Community Environmental Alliance coordinated "learning journeys" where government officials visited low-income neighborhoods to hear about local

women's experiences with flooding, sanitation, and other types of climate change stresses, helping them to bring these views into policy discourse.

- Maputo University environmental education students worked with intermediate school youth on after-school activities related to climate change, strengthening the school curriculum while developing university students' community development skills and local knowledge. Most participants were women.

Agua Doce: Water and livelihoods near Rio de Janeiro

The potential impacts of both temperature and precipitation extremes for the city of Rio de Janeiro are many. Climate projections show that Rio will experience a 1.5°C increase in temperature by 2050 across all seasons. With warmer temperatures, droughts can be expected in winter, spring, and summer. This is likely to create water scarcity, as well as electricity deficits from shortfalls in hydroelectric production. Increases in ocean surface temperatures, however, could increase the chances of periodic and extreme storms and flooding. With climate change, extremes in precipitation are projected to increase in intensity.

In April 2010, heavy rains partially destroyed several downtown *favelas* (slums) in Rio de Janeiro and its suburb of Niteroi, leaving thousands homeless and forced to relocate with their families and remaining belongings.[7] Also severely hit by heavy rainfall in January 2011 was the *Região Serrana*, a mountainous region in the state of Rio de Janeiro to the north and northeast of Guanabara Bay. Housing infrastructure and social inequities compounded the effects of natural disasters on the marginalized in the Região Serrana; in the town of Nova Friburgo, where several hundred people died due to the floods, 60 percent of the population lived in illegal dwellings.[8]

At the western edge of Guanabara Bay, in an economically depressed area around the Suruí River, the community development organization *Agua Doce* (Fresh Water) has been building social resilience, fostering local cultural pride, and creating green jobs. Started by Vladimir Boff and Maria Regina Maroun in 2001, with seed funding from several Italian labor union and church organizations, Agua Doce's initiatives include community centers and kindergartens; biodigesters and a biomass recycling center; agricultural support offices and fruit-processing centers to process locally grown fruits and create jobs for women; handicrafts training and support for women and girls; public health projects; and a public library and literacy center.[9]

The Agua Doce organization, reinforcing Agenda 21 principles, aims to (re)introduce a more sustainable way of life for the people of the Guanabara region through a range of social and economic community development initiatives. Agua Doce and the local communities are particularly concerned about the imminent effects of climate change in the region, mostly due to their precarious geographical location. Increased rainfall from the mountains and the rising sea level in Guanabara Bay make the area particularly vulnerable to floods and mudslides. Health risks from water-borne illnesses are increasing, and fishing and crabbing are progressively endangered.

Agua Doce nurtures social capital in order to strengthen communities in the region. The organization offers literacy and environmental education workshops and supports schools and children's programs, as well as creating economic opportunities for women. It promotes social transformation and the building up of the capability of local citizens to participate in decision-making processes, especially concerning water, resource management, and local governance.

The organization has mobilized funds from a range of religious, private sector, international development assistance, and other donor organizations for its projects. Their work demonstrates the interlinkages among literacy, economic, and social supports, including job creation, community-building, and the potential for marginalized community members to become involved in collective decision-making. The organization's attention to ecological sustainability, social sustainability, and governance provides a wealth of ideas for ways of integrating community members' environmental priorities with other pressing needs in the community and advancing on many fronts simultaneously through community development.

The green change project: Urban renewal and environmental jobs in Toronto

In Toronto, the effects of climate change are being noted, particularly through high amounts of summer rainfall and sudden storms with intense winds and heavy rain, which are becoming more frequent.[10] Higher amounts of rainfall stress the aging urban water/sewer infrastructure, resulting in overflows of untreated sewage from the combined storm and sanitary sewers in the older parts of downtown Toronto during heavy storms, consequently degrading the water quality in local streams and Lake Ontario.[11] At times, local lakefront beaches are unswimmable for several days following heavy rains due to water pollution. The City of Toronto is undertaking several large infrastructure repairs and has built storage basins to retain sewer overflows until they can be treated, at a projected total cost of $1.047 billion over 25 years.[12] The city has also launched basement flooding studies and programs to help homeowners disconnect their basement drains leading to the municipal sewer system, in order to prevent backups into basements during rainstorms. The increasing numbers of extremely hot days in the summer have led the city to develop a "cooling centers" program where those without air conditioning can come to public libraries, community centers, and other communal spaces, which offer extended hours on very hot days.

From an equity perspective, both the weather extremes and the resulting policies have disproportionately negative effects on Toronto's low-income residents, as they are more likely than higher-income people to occupy basement apartments and to be renters, not house-owners (who can, in contrast, benefit from government infrastructure subsidies). Lower-income people are also more likely to depend on public transit (which is often disrupted during storms) and on public beaches and parks for recreation; they are less likely to have air conditioning in their homes, and they are more likely to have health conditions which are severely exacerbated by heat, such as diabetes, high blood pressure, and heart conditions. These equity implications of disaster risks have been noted in some reports, but have yet to be quantified or estimated in official publications or policy frameworks.

One particular low-income neighborhood in Toronto has borne the brunt of several recent extreme weather events – the Jane-Finch neighborhood, located in northwest Toronto near York University. Due to dense urban growth and university development, the area has become increasingly built-up over the past decade, which has increased the surface runoff into Black Creek, a tributary of the Humber River that empties into Lake Ontario. During an intense storm in August 2005, more than 150 mm of rain fell in the area, and the normally placid Black Creek became a rushing torrent that washed out its culvert under Finch Avenue, a local arterial roadway, leaving a

50-m-wide gaping hole. Construction of a new bridge for Finch Avenue took six months and cost about CND$45 million. During this time, public buses and commuter traffic had to be rerouted through the York University campus, causing delays and problems for the university and local residents alike. This was a graphic example of how extreme weather events – which are increasing in frequency due to climate change – in conjunction with aging infrastructure, urban sprawl (including campus development), and increasing rapid rainfall runoff can have costly and traumatic effects on everyone in the watershed.

Since 2009, the Jane-Finch community has built a vibrant and growing initiative to create jobs while greening the local community. The project began with concerns of residents in local public housing buildings about waste management and energy conservation; about having no control over heating; about not having a place to recycle or compost; and about not being able to see any benefit to switching to more expensive compact fluorescent light bulbs. Residents were concerned about the costs that new "smart meters" would add to their monthly electricity bills. In the larger community, residents were concerned about storm water management and flooding in the area, especially following the 2005 Finch Avenue washout.

The Jane-Finch Community and Family Centre (JFCFC), a community-based social services organization, decided to build on the desire of community residents to be more eco-friendly and seek ways to expand community greening efforts. The JFCFC applied for funding and created partnerships with the Toronto Community Housing Corporation, the Toronto and Region Conservation Authority, and the NGO Zerofootprint to make Green Change happen in the community. Together, they developed a 45-hour training program for Green Change Agents and put out a call for community volunteers to participate in the training. The training program introduced residents to five key areas: energy conservation, waste management and recycling, green active living, social justice, and the green economy. Many residents signed up and participated in different aspects of the training and 60 residents completed the entire program, earning the right to call themselves Green Change Agents.

After finishing their training, the Green Change Agents shared their knowledge with the community by using an online calculator tool developed by Zerofootprint (www.zerofootprint.net) to document the carbon footprint of more than 900 households in the Jane-Finch area. Agents were paid for each household assessment they conducted. The Zerofootprint calculator measured and analyzed the carbon footprints of individuals and households in the Jane-Finch community and also provided tips for Green Change Agents to use to help community members save energy, save money, and be able to reduce their carbon footprints. In 2010, Green Change Agents assisted residents in reducing their carbon footprint by 2,000 tons. Many additional "green changes" have been made in the community, including the planting of a community garden, an eco-friendly Earth Hour event at a local school, tree planting, Earth Day celebrations, a summer farmer's market, and renovation/development of the ground floor of a social housing building into the Centre for Green Change. At the Centre for Green Change, local residents and youth who are concerned about the protection of the environment can become engaged in the process of green change and mobilize others while they increase their own knowledge and skills, initiating individual and collective actions toward building a healthy, safe, prosperous, and environmentally friendly neighborhood. The vision for the Centre includes a Pathways to Green Jobs Program, which will educate and promote environmental stewardship, green jobs, and eco entrepreneurship, modeled on the successful work of the Carpenters Union's CHOICE Apprentice Program and incorporating an environmental training component. The Centre is mobilizing local resources to achieve this vision and working to expand the number of long-term, high-quality green jobs for local residents, especially for youth.

The Green Change Project held successful gala fundraising balls in 2010 and 2011, as part of its celebration of "Twelve Days for Green Change," with the support of a number of community partner organizations, including labor unions, health centers, local politicians, property owners and landlords, charitable organizations, and cultural/media groups. Brooklyn-based activist Majora Carter, who has developed a range of green community development initiatives in low-income neighborhoods across the Global North, was the dynamic keynote speaker at the Green Change 2011 gala.[13]

The Green Change Project continues to win funding, awards, and accolades, including the 2010 Toronto Green Award in the Community category and the 2011 Urban Leader Award for Imagination. York University students work as graduate assistants on Green Change initiatives through a partnership with the Faculty of Environmental Studies, which also facilitates research grant partnerships that benefit the Centre.

Aspects of the project's success include its focus on connecting with the community directly and its emphasis on local skills and needs. The project aims to identify the skills and knowledge already possessed by community members and to empower them to use these assets for green change. The level of skills and knowledge in the Jane-Finch community is significant. Many neighborhood residents are recent immigrants to Canada, and many of them come highly trained and skilled from their home countries, but are unable to break the barrier of employment in Canada. The Green Change project recognizes the skills and abilities of these individuals and aims, through Green Change Agent training, to create jobs. These individuals benefit from being able to support their families financially and also to contribute to the development of their community.

Participants in the Green Change Agent training find themselves being treated with respect in the community for the knowledge and skills they can offer for a change. The Centre space itself creates a focus for green community organizing in the neighborhood and is a model across Toronto.

Watershed education and green community development in the US and Canada

With university-based research input and support, online techniques can be effectively used to disseminate information about watersheds and their management in times of climate change. An example is the Water Atlas, a web portal developed by Dr. Paul Zandbergen and others at the University of South Florida. It brings together hydrological, weather, political and other data related to a number of Florida watersheds, much of which is updated in real time, with the goal of designing a "comprehensive data resource . . . to help citizens and scientists alike make informed decisions concerning our vital water resources." Such online materials may also be translated to be accessible transnationally across bioregional boundaries. By integrating and facilitating exchanges of academic/scientific, government-collected, and participatory local knowledge, such online tools can improve disaster planning and resilience while building social capital and policy relevance.

All of these examples of the interplay among equity-oriented green community development, watershed governance, and policy development, university-community partnerships, and climate justice demonstrate the parallels and similarities in this kind of work in the Global North and South. The following section summarizes and draws out these themes.

Common themes in green community organizing for climate justice

A number of commonalities are evident in the local stories outlined above. Marginalized people everywhere are eager for employment opportunities and public space for recreation, and community-building is almost always in short supply in marginalized neighborhoods. Local governments can play an important role in helping communities meet their needs, especially if people are mobilized and organized politically at the grassroots level – which can be effectively catalyzed by NGOs and civil society organizations and assisted by local universities. Infrastructure funding and disaster-oriented climate change adaptation funding may be available when local requirements are framed in appropriate ways. New, creative techniques for green community organizing are developing rapidly, aided by global networking advances and the hard work of many local and international activists. In combination, all of these trends are helping to strengthen resilience in many communities, allowing them to mobilize socially to meet new ecological and weather-related challenges.

The international commonalities and themes related to green community organizing in the face of climate change include the following.

Green job creation

In marginalized communities people are nearly always interested in increasing their employment opportunities. High levels of unemployment – that is, community members who are eager for training and work opportunities – represent a significant resource for poor communities, because environmental restoration and low-fossil-fuel production require large amounts of people-power. Examples of the kinds of green job creation which are needed in both poor and well-off communities include wetlands reconstruction; rebuilding and strengthening water/sewer systems and other infrastructure; rehabilitation of polluted areas, brownfields, and transportation corridors; construction and maintenance of rooftop, roadside, and other community gardens; production and processing of locally grown food; and development of new recreation options on land which is not appropriate for housing or industry.[14] Funding which is available for infrastructure renewal, disaster prevention, and climate change adaptation can be used to support and catalyze these kinds of green jobs, producing powerful local economic benefits.

Environmental education

Skills needed for evolving green jobs and for effective public involvement in watershed and other environmental policy processes can be developed and shared in marginalized communities via organized processes of continuing education. These can include environmental education programs for youth, seniors, and adults, and in-service training for working professionals such as public health agents/nurses, teachers, and government officials. Local community organizations and NGOs are ideally situated to develop and fine-tune such educational programs, which can lead to increased employment opportunities and/or pay upgrades for local residents. Partnerships with local universities reinforce these initiatives while also improving the local relevance of university programs.

Community-building to develop social resilience and political intervention potential

NGOs with local roots and long-standing community centers and social services organizations have important roles to play in marginalized communities' disaster readiness and response to climate change. Their contributions can include bringing people together to prepare for extreme weather events; collecting knowledge on the local weather, water, and environment, and bringing it to the attention of governments; building social capital and developing social networks for community resilience; and organizing community members politically to express their views and advance their priorities.

Recreation spaces as flood buffers

Low-income communities with many unemployed people, especially youth, need recreation spaces, and these are usually in short supply due to land-use development pressures in urban areas. Increased weather variability and rainfall are expanding the areas that are subject to flooding and thus not appropriate for housing. A good alternative use for these areas is as public spaces, swimming/picnicking areas, and sports fields. If governments can equitably assist people to relocate housing from floodplains to higher land nearby, simultaneously creating jobs and community-controlled recreation spaces in these areas, then climate change adaptation and disaster prevention can accomplish several goals at once.

Mobilizing finance for infrastructure renovation

NGOs and government bodies may be able to use international connections and climate change funding mechanisms, both formal and informal, to carry out projects such as water/sewer infrastructure renewal and development, renewable energy plants, housing reconstruction, and public space provision. The "guilt factor" motivating progressive Northern groups who recognize the responsibility of the wealthy for climate change and who take action to fund remediation projects in the global South, lies behind some of these new global redistribution initiatives. Broader and more radical income redistribution measures to redress the "climate debt" of the rich have also been proposed.[15] Community-academic partnerships may help generate funding proposals and documentation of pilot initiatives that can demonstrate the effectiveness of such redistributive measures.

Creative organizing/workshop techniques/strategies and international sharing of ideas

Ideas, designs, and financing proposals can be shared internationally through green/climate change channels. For example, civil society networking surrounding the UN Framework Convention on Climate Change annual COP meetings involves a wide range of community-based environmental NGOs that have participated in the civil society forums accompanying the governmental negotiations. Academic conferences can bring together university-based researchers and local environmental activists for useful discussions on innovative policy and grassroots solutions to the pressing ecological situations people face on the ground. A conference on the question of "How Will Disenfranchised People Adapt to Climate Change?" at York University in April 2009, organized by the university's Institute for Research and Innovation in Sustainability, focused on local struggles and the importance of traditional ecological knowledge in addressing climate change in Arctic Canada, Brazil, South Africa, and India.[16] Likewise, Majora Carter's advocacy work highlights the common strategies marginalized areas can use in sparking green community development throughout the global North and South. Online networking has vastly broadened the spread of ideas and potential partners.

The role of municipal government and public education

Cities worldwide have begun to develop climate change adaptation plans and to take stock of their new needs – physical, social, economic.[17] When community groups organize to share their expertise and knowledge of challenges, as well as ideas on how to meet them, they may be able to build beneficial partnerships with local governments. Public officials everywhere need training in all aspects of climate change and disaster preparedness, and environmental education for the general public is also crucial in a warming world. Here, too, networking and global communication help policy-makers and local groups to learn from "best practices" and creative ideas that have worked elsewhere.

Rather than top-down, cookie-cutter strategies being imposed from outside, the importance of widespread bottom-up information-sharing lies in the ability of subsequent adapters to use what is most relevant to their situation, building new and locally appropriate policy syntheses. University partners and youth can contribute the computer access, online research, and social media networking skills to make this possible.

Conclusion: Watersheds, equity, and climate justice

Governments and development organizations worldwide are searching for new ideas on how to encourage more public participation in policy decision processes, especially in relation to climate change and disaster readiness. Social and political movements are also expanding their participatory outreach and organizing techniques. In the Great Lakes region of North America, for example, various jurisdictions emphasize public involvement and participation in source water protection.[18] The Council of Canadians, through its Blue Planet project and other organizing initiatives, seeks to focus public opinion globally on the importance of water rights.[19] The European Water Framework Directive is pushing jurisdictions and cross-jurisdictional water basin committees to implement new participatory processes. Researchers, consultants, and activists are generating practices that can be widely discussed and shared. In the end, however, while many insights can come from hearing what has worked and what has not in other places/situations, there is no substitute for locally designed and locally appropriate public participation processes, both within and outside of government.

As the conclusion of the *European Water Framework Directive Guidance Document on Public Participation* states,

> The preamble of the Water Framework Directive includes a very clear statement: active public involvement is most likely the key to success with regard to achieving the desired water quality objectives. This statement reflects several years of accumulated European water management experiences. In simple words: the water users and water polluters need to be turned into part of the solution, not . . . kept outside the considerations as part of the problem. This Guidance has presented a range of recommendations on how to ensure active involvement. It is important, however, to take into account that no blueprint solution can be provided. Each River Basin District has to find its own way to handle this, taking into account the prevailing cultural, socio-economic, democratic and administrative traditions. Careful planning, e.g. stakeholder analysis, is a particular recommendation, but each competent authority has to accept that a dynamic and learning process based on "trial and error" is the challenge to embark on. Experience shows, however, that given sufficient time it will pay off in the long run.[20]

In particular, the roles of class and gender, among other differences, must be acknowledged as determinants of everyone's standpoint and possibilities for participation. For example, are those who are too poor to pay for water "stakeholders"? The substantial consensus among ecological economists, sociologists, political scientists, planners, and radical activists that public participation is essential for good public policy may obscure the vital question of how fairly to elicit, structure, and make use of that participation.[21]

Engaging women is fundamentally important for durable climate change adaptation and disaster readiness. Women possess incomparable knowledge of local, social, and ecological conditions due to long-standing gendered roles and responsibilities, and this knowledge must be shared and utilized in local, national, and international negotiations and decision-making processes – for reasons of both justice and efficiency. Democratic mediation of equity conflicts, adequate disaster preparedness, and sustainable long-term management of resources are only possible through public participation.

While in practice it sometimes seems difficult to organize or elicit public participation (a "bottom-up" problem), and in some cases governments are reluctant to cede control by opening up decision processes for public examination and input (a "top-down" problem), at least in principle, broad public participation is the foundation of sustainable development. It also has a number of quite progressive implications – if it truly gives a political voice to people who have historically been left out of public decision-making.

To address both the "top-down" and the "bottom-up" challenges to broadening public involvement in decision processes, a creative combination of grassroots environmental education and community organizing is needed. Community-based environmental education initiatives which are relevant and interesting for local residents and increase their knowledge of watershed issues, understanding of basic political and ecological principles, and confidence to express their views can serve as the basis of an intervention approach which is progressive, constructive, and democratic. This, in turn, increases the resilience and sustainability of decision-making processes. It also lays the groundwork for community organizing and extension of the environmental education activities to larger constituencies.

Community-based education and organizing are fundamental to creating the conditions for local knowledge to be shared and utilized, through equitable democratic participation. Building inclusive governance structures and strengthening the role of civil society, especially women, are essential for addressing climate change vulnerability and fostering resilience and sustainability in urban centers as well as rural areas. According to the Intergovernmental Panel on Climate Change, "adaptation is shown to be successful and sustainable when linked to effective governance systems, civil and political rights and literacy."[22] NGOs in the Global South have expertise in such initiatives which is potentially transferable to other places, including some in the Global North.

Community-based environmental education initiatives which are relevant and interesting for local residents and increase their job opportunities, knowledge of local environmental issues, understanding of basic political and ecological principles, and confidence to express and act on their views can serve as the basis of a climate change intervention approach which is progressive, constructive, and democratic. This, in turn, increases the resilience and sustainability of watershed, disaster, and climate change decision-making processes. It also lays the groundwork for community organizing and extension of environmental education activities to larger constituencies in local areas affected by climate change. Such grassroots initiatives – and the global sharing of ideas on how to design and implement them, freely available for adaptation in other places – stand in contrast to top-down climate change adaptation mechanisms controlled from the Global North. In this sense, climate justice is a new, more general manifestation of the bottom-up perspective in Development Studies. Furthermore, it is a movement which "best fuses a variety of progressive political-economic and political-ecological currents to combat climate change."[23]

Climate justice – addressing the impacts of climate change on the poorest first – is a powerful imperative at every level, from the local to the global. Civil society groups worldwide, with support from university researchers, are using online and in-person networking tools to share ideas on how to promote climate justice, obtain funding, and press politically for policies addressing the needs of marginalized people. This bottom-up movement builds resilience in the face of the social and political repercussions of extreme weather events – a global priority, as we all inhabit this warming world together.

Notes

i This chapter is a revised version of a paper presented at the International Conference on Renewable Energy, Energy Efficiency and Sustainable Development (REGSA) in Florianópolis, Brazil, May 6–8, 2014.

1 D. Satterthwaite, S. Huq, H. Reid, M. Pelling, and P. Romero Lankao (2007), "Adapting to Climate Change in Urban Areas: The Possibilities and Constraints in Low- and Middle-Income Nations," International Institute for Environment and Development, Human Settlements Discussion Paper Series, Climate Change and Cities – 1, p. vi.

2 F. Hinchcliffe and J. Thompson, eds. (1999), *Fertile Ground: The Impacts of Participatory Watershed Management* (London: Earthscan/IT); P.E. Perkins (2004a), "Participation and Watershed Management: Experiences from Brazil," paper presented at the Conference of the International Society for Ecological Economics (ISEE), July 10–14, 2004, Montreal, Canada, p. 5.

3 Centro de Saberes (2008), *Centro de Saberes e Cuidados Socioambientais da Bacia do Prata* [Web page], retrieved February 5, 2016, from http://www.cultivandoaguaboa.com.br/o-programa/centro-de-saberes-e-cuidados-socioambientais-da-bacia-do-prata

4 Intergovernmental Panel on Climate Change (2007a), *Climate Change 2007: Synthesis Report*, retrieved May 4, 2011, from http://www.ipcc.ch

5 United Nation Human Settlements Program (UN Habitat) (2010), *Climate Change Assessment for Maputo, Mozambique*, Cities and Climate Change Initiative (Nairobi, Kenya: Author), p. 2.

6 S. Kindon, R. Painand, and M. Kesby (2008), "Participatory Action Research," in *International Encyclopedia of Human Geography*, R. Kitchin and N. Thrift, eds., pp. 90–95 (London, UK: Elsevier), p. 90.

7 "Cabral anuncia R$ 1 bilhão em habitação para desabrigados pelas chuva," (2010, April 11), *O Globo* [Online edition], retrieved February 15, 2011, from http://oglobo.globo.com/rio/mat/2010/04/11/cabral-anuncia-1-bilhao-emhabitacao-para-desabrigados-pelas-chuvas-916315287.asp

8 L.E. Magalhães (2011, January 22), "Friburgo tem 60% dos imóveis ilegais; um deles, vizinho a encosta que deslizou, tem 10 andares," *O Globo* [Online edition], retrieved January 22, 2011, from http://oglobo.globo.com/rio/mat/2011/01/21/friburgo-tem-60-dosimoveis-ilegais-um-deles-vizinho-encosta-que-deslizou-tem-10-andares-923586240.asp

9 Centro Clima (2005), *The South-South-North (SSN) Project. Potential Adaptation Projects Report: Brazil* (Rio de Janeiro, Brazil: Author), pp. 57–64.

10 A. Todd (2011), "Climate Change and Water Governance in the Greater Toronto Area," unpublished paper (Toronto, Canada: York University).

11 M. Binstock (2011), *Greening Stormwater Management in Ontario* (Toronto, Canada: CIELAP).

12 City of Toronto (2011a), *WWFMP—Going for the Flow*, retrieved May 8, 2011, from http://www.toronto.ca/water/protecting_quality/wwfmmp/about.htm

13 Majora Carter Group (2011), *Majora Carter Group* [Website], retrieved February 5, 2016, from http://www.majoracartergroup.com/

14 V. Jones (2008), *The Green Collar Economy* (New York, NY: HarperOne).

15 P. Bond and M.K. Dorsey (2010), "Anatomies of Environmental Knowledge and Resistance: Diverse Climate Justice Movements and Waning Eco-Neoliberalism," *Journal of Australian Political Economy* 66(December), pp. 286–316.

16 Institute for Research and Innovation in Sustainability (2009), *Report of the First International Ecojustice Conference: How Will Disenfranchised Peoples Adapt to Climate Change? Strengthening the Ecojustice Movement* (Toronto, Canada: IRIS, York University).

17 O. Lucon and J. Goldemberg (2010), "São Paulo – The 'Other' Brazil: Different Pathways on Climate Change for State and Federal Governments," *The Journal of Environment and Development* 19(3), pp. 335–57.

18 Canadian Institute for Environmental Law and Policy (2004), *Public Participation in Great Lakes Management* (Toronto, Canada: Author).

19 Council of Canadians (2006), *Blue Planet Project* [Website], available at http://www.blueplanetproject.net/Movement/index.html and http://www.canadians.org/

20 European Commission (2002a), *Environment 2010: Our Future, Our Choice. The Sixth Environment Action Programme of the European Community*, retrieved July 9, 2004, from http://europa.eu.int/comm/environment/newprg/, 66.

21 P.E. Perkins (2008), "Public Participation in Watershed Management: International Practices for Inclusiveness," paper presented at the Regional Meeting on Water in the Mediterranean Basin, Near East University, October 9–11, 2008, Lefkosa, Northern Cyprus.

22 M.L. Parry, O.F. Canziani, J.P. Palutikof, P.J. van der Linden, and C.E. Hanson, eds. (2007), *Climate Change 2007: Impacts, Adaptation and Vulnerability* (Cambridge, UK: IPCC), p. 452.

23 Bond and Dorsey (2010).

Bibliography

Abeysuriya, K., C. Mitchell, and J. Willetts. (2008). "Expanding Economic Perspectives for Sustainability in Urban Water and Sanitation." *Development* 51, pp. 23–29.

Anisfeld, S.C. (2010). *Water Resources*. Washington, DC: Island.

Barlow, M. (2009). *Our Water Commons: Toward a New Freshwater Narrative*. Ottawa, Canada: Council of Canadians. Available at http://www.canadians.org/water/publications/water%20commons.html

Barlow, M., and T. Clarke. (2002). *Blue Gold: The Battle against Corporate Theft of the World's Water*. Toronto, Canada: McClelland and Stewart.Bied-Charreton, M. (2008). "Integrating the Combat against Desertification and Land Degradation into Negotiations on Climate Change: A Winning Strategy." *United Nations Convention to Combat Desertification*. Retrieved May 4, 2011, from http://www.unccd.int/science/docs/non_paper_desertif_Climate_eng.pdf

Bielski, Z. (2008, July 12). "Intense Storm Leaves Beaches Flooded, Homes Flooded." *National Post*. Retrieved May 8, 2011, from http://www.nationalpost.com

Binstock, M. (2011). *Greening Stormwater Management in Ontario*. Toronto, Canada: Canadian Institute for Environmental Law and Policy (CIELAP). Retrieved June 23, 2011, from http://cielap.org/pub/pub_greeningstormwaterman.php

Bond, P., and M.K. Dorsey. (2010). "Anatomies of Environmental Knowledge and Resistance: Diverse Climate Justice Movements and Waning Eco-Neoliberalism." *Journal of Australian Political Economy* 66(December), pp. 286–316. Available at http://search.informit.com.au/documentSummary;dn=833845077924067;res=IELBUS> ISSN: 0156–5826

Brooks, N., W.N. Adger, and P.M. Kelly. (2005). "The Determinants of Vulnerability and Adaptive Capacity at the National Level and the Implications for Adaptation." *Global Environmental Change* 15(2), pp. 151–63.

"Cabral anuncia R$ 1 bilhão em habitação para desabrigados pelas chuva." (2010, April 11). *O Globo* [Online edition]. Retrieved February 15, 2011, from http://oglobo.globo.com/rio/mat/2010/04/11/cabral-anuncia-1-bilhao-emhabitacao-para-desabrigados-pelas-chuvas-916315287.asp

Canadian Institute for Environmental Law and Policy. (2004). *Public Participation in Great Lakes Management*. Toronto, Canada: Author.

Centro Clima. (2005). *The South-South-North (SSN) Project. Potential Adaptation Projects Report: Brazil*. Rio de Janeiro, Brazil: Author. Retrieved July 14, 2011, from http://www.southsouthnorth.org/download.asp?name=SSN2_Report_Adaptation%20Context_%20final4.pdf&size=4814863&file=documents/SSN2_Report_Adaptation%20Context_%20final4.pdf

Centro de Saberes. (2008). *Centro de Saberes e Cuidados Socioambientais da Bacia do Prata* [Web page]. Retrieved February 5, 2016, from http://www.cultivandoaguaboa.com.br/o-programa/centro-de-saberes-e-cuidados-socioambientais-da-bacia-do-prata

Chen, S. (2008). "From Community-Based Management to Transboundary Watershed Governance." *Development* 51, pp. 83–88.

Christian Aid. (2005, February). *Unearthing the Truth: Mining in Peru*. London, UK: Author. Retrieved October 26, 2006, from www.christianaid.org.uk

City of Toronto. (2005, September 30). *City Approves Basement Flooding Subsidy for Homeowners*. Available at http://wx.toronto.ca/inter/it/newsrel.nsf/0/3f3a65ce7d283fdd8525708c006aa028?OpenDocument

———. (2011a). *WWFMP – Going for the Flow*. Retrieved May 8, 2011, from http://www.toronto.ca/water/protecting_quality/wwfmmp/about.htm

———. (2011b). *Coatsworth Cut Sewer Upgrades Construction*. Retrieved May 8, 2011, from http://www.toronto.ca/involved/projects/coatsworthcut/index.htm

———. (2011c). *Projects*. Retrieved May 8, 2011, from http://www.toronto.ca/involved/projects.htm

———. (2011d). *Black Creek Boulevard*. Retrieved May 8, 2011, from http://www.toronto.ca/involved/projects/black_creek_blvd/

———. (2011e). *Hogg's Hollow – Road Reconstruction*. Retrieved May 14, 2011, from http://www.toronto.ca/involved/projects/hoggshollow/index.htm#a4

————. (n.d., a). *East Highland Cr EA Study part 1: August 19, 2005 Storm Damage Overview*. Toronto, Canada: Author.

————. (n.d., b). *Basement Flooding Protection Subsidy Program*. Toronto, Canada: Author.

————. (n.d., c). *Basement Flooding*. Toronto, Canada: Author.

City of Toronto Planning Division, York University Development Corporation, and the Planning Partnership. (2008). *York University Background Study: Natural Features and Open Space*. Toronto: Author.

Clean Air Partnership, The. (2006). *A Scan on Climate Change Impacts on Toronto*. Retrieved October 30, 2015, from http://www.cleanairpartnership.org/pdf/climate_change_scan.pdf

Climate Consortium for Research Action Integration. (2011). *Where Does the Water Go? Evapotranspiration in Toronto*. Retrieved May 12, 2011, from http://www.climateconsortium.ca/2011/01/24/where-does-the-water-go-evapotranspiration-in-toronto/

Conca, K. (2006). *Governing Water: Contentious Transnational Politics and Global Institution Building*. Cambridge, MA: MIT.

Conservation Ontario Website. (n.d.). Retrieved January 3, 2016, from http://www.conservation-ontario.on.ca/

Cooke, B., and U. Kothari, eds. (2001). *Participation: The New Tyranny?* London, UK: Zed Books.

Corporate Europe Observatory. (2003). *Alternatives to Privatization: The Power of Participation*. Infobrief no. 4. Available at http://www.corpoprateeurope.org/water/infobrief4.htm

Council of Canadians. (2006). *Blue Planet Project* [Website]. Available at http://www.blueplanet project.net/Movement/index.html and http://www.canadians.org/

de Oliveira, J.A. (2009). "The Implementation of Climate Change Related Policies at the Subnational Level: An Analysis of Three Countries." *Habitat International* 33, pp. 253–59.

de Sherbinin, A., A. Schiller, and A. Pulsipher. (2007). "The Vulnerability of Global Cities to Climate Hazards." *Environment and Urbanization* 19(7), pp. 39–64.

European Commission. (1993). *Towards Sustainability: The European Community Programme of Policy and Action in Relation to the Environment and Sustainable Development (better known as The Fifth EC Environmental Action Programme)*. Retrieved July 9, 2004, from http://europa.eu.int/comm/environment/actionpr.htm

————. (2000). *Directive 2000/80/EC Establishing a Framework of Community Action in the Field of Water Policy*. Retrieved July 9, 2004, from europa.eu.int/eur-lex/pri/en/ oj/dat/2000/ l_327/l_32720001222en00010072.pdf

————. (2002a). *Environment 2010: Our Future, Our Choice. The Sixth Environment Action Programme of the European Community*. Retrieved July 9, 2004, from http://europa.eu.int/ comm/environment/newprg/

————. (2002b). *Guidance on Public Participation in Relation to the Water Framework Directive: Active Involvement, Consultation and Public Access to Information*. Retrieved July 10, 2004, from http://forum.europa.eu.int/Public/irc/env/wfd/library?l=/framework_directive/ guidance_documents/participation_guidance&vm=detailed&sb= Title

Ensor, J., and R. Berger. (2009). *Understanding Climate Change Adaptation: Lessons from Community-Based Approaches*. Bourton on Dunsmore, UK: Practical Action.

Environment Canada. (2008). *All about Watersheds* [Website]. Retrieved January 4, 2016, from http://www.ec.gc.ca/geocache/default.asp?lang=En&n=065E5744–1

European Communities. (2003). *Common Implementation Strategy for the Water Framework Directive (2000/60/EC) Guidance Document No 8: Public Participation in Relation to the Water Framework Directive*. Working Group 2.9: Public Participation. Luxembourg: Author.

Figureido, P., and P.E. Perkins. (2011). "Gender Justice and Climate Justice: Community-Based Strategies to Increase Women's Political Agency in Watershed Management in Times of Climate Change." Paper presented at the Ninth International Conference of the International Development Ethics Association (IDEA), June 9–11, 2011, Bryn Mawr College, Pennsylvania.

Fraser, N. (1993). "Rethinking the Public Sphere: A Contribution to the Critique of Actually Existing Democracy." In *Habermas and the Public Sphere*. Craig Calhoun, ed., pp. 109–142. Cambridge, MA: MIT.

Fuchs, E. Del Carmen. (2004). *Living Dreams: Creating Revolutionary Educational Environments – The Political Education of the Brazilian Landless Rural Workers Movement (MST)*. MES Major Paper. Toronto, Canada: Faculty of Environmental Studies, York University.

Fung, A. (2004). "Deliberation's Darker Side: Six Questions for Iris Marion Young and Jane Mansbridge." *National Civic Review* 93(4), pp. 47–54.

Gedicks, A. (2005). "Mining War in Ecuador." *Z Magazine*. March.

Godden, L. (2005). "Water Law Reform in Australia and South Africa: Sustainability, Efficiency, and Social Justice." *Journal of Environmental Law* 17(2), pp. 181–205.

GreenXChange Project Website. (2011). Retrieved June 23, 2011, from http://www.green xchange.ca/general/green-change-project-notes

Gujit, I., and M. Kaul Shah, eds. (1998). *The Myth of Community: Gender Issues in Participatory Development*. London: IT.

Hinchcliffe, F., and J. Thompson, eds. (1999). *Fertile Ground: The Impacts of Participatory Watershed Management*. London: Earthscan/IT.

Intergovernmental Panel on Climate Change. (2007a). *Climate Change 2007: Synthesis Report*. Retrieved May 4, 2011, from http://www.ipcc.ch

———. (2007b). *Climate Change 2007: The Physical Science Basis*. Contribution of Working Group 1 to the Fourth Assessment Report of the Intergovernmental Panel on Climate Change. S. Solomon, D. Qin, M. Manning, Z. Chen, M. Marquis, K.B. Avery, M. Tignor, and H.L. Miller, eds., Cambridge, UK: CUP.

Institute for Research and Innovation in Sustainability. (2009). *Report of the First International Ecojustice Conference: How Will Disenfranchised Peoples Adapt to Climate Change? Strengthening the Ecojustice Movement*. Toronto, Canada: IRIS, York University. Retrieved December 2, 2011, from www.irisyorku.ca/wp-content/uploads/2010/11/Ecojustice-Conference_final.pdf

International Water Law Project. (2009). *UN Convention on the Law of the Non-Navigational Uses of International Watercourses*. Available at http://internationalwaterlaw.org/documents/intldocs/watercourse_conv.html

Jane-Finch Community and Family Centre Website. (n.d.). Retrieved June 23, 2011, from http://www.janefinchcentre.org/

Jones, V. (2008). *The Green Collar Economy*. New York, NY: HarperOne.

Kapoor, I. (2001). "Towards Participatory Environmental Management?" *Journal of Environmental Management* 63, pp. 269–79.

———. (2002). "Deliberative Democracy or Agonistic Pluralism? The Relevance of the Habermas-Mouffe Debate for Third-World Politics." *Alternatives: Global, Local, Political* 27(4), pp. 459–87.

Kenyon, W. (2003). "Environmental Planning: Scientific-Political Decision Processes, Participatory Decision Processes." Paper presented at the Frontiers 2 Conference of the European Society for Ecological Economics, February 11–15, 2003, Tenerife, Canary Islands.

Kindon, S., R. Painand, and M. Kesby. (2008). "Participatory Action Research." In *International Encyclopedia of Human Geography*. pp. 90–95. London, UK: Elsevier.

Lister, R. (1997). *Citizenship: Feminist Perspectives*. New York, NY: NYU.

Lucon, O., and J. Goldemberg. (2010). "São Paulo – The 'Other' Brazil: Different Pathways on Climate Change for State and Federal Governments." *The Journal of Environment and Development* 19(3), pp. 335–57.

Magalhães, L.E. (2011, January 22). "Friburgo tem 60% dos imóveis ilegais; um deles, vizinho a encosta que deslizou, tem 10 andares." *O Globo* [Online edition]. Retrieved January 22, 2011, from http://oglobo.globo.com/rio/mat/2011/01/21/friburgo-tem-60-dosimoveis-ilegais-um-deles-vizinho-encosta-que-deslizou-tem-10-andares-923586240.asp

Majora Carter Group. (2011). *Majora Carter Group* [Website]. Retrieved February 5, 2016, from http://www.majoracartergroup.com/

Moraes, A., and P.E. Perkins. (2007). "Women and Participatory Water Management." *International Feminist Journal of Politics* 4(Fall), pp. 485–93.

Moraes A., and P.E. Perkins. (2009) "Etica, Genero e Classe Social na Politíca Participativa de Agua." In *Etica, Pesquisa e Políticas Publicas*. G. Aparecida dos Santos and F. Mori Sarti, eds. pp. 107–22. São Paulo, Brazil: Editora Rubio.

Naples, N.A., ed. (1998). *Community Activism and Feminist Politics: Organizing Across Race, Class and Gender*. New York, NY: Routledge.

Neumayer, E., and T. Plümper. (2007). "Catastrophic Events on the Gender Gap in Life Expectancy, 1981–2002." *Annals of the Association of American Geographers* 97(3), pp. 551–66, doi:10.1111/j.1467-8306.2007.00563.x

Ober, J. (2008). *Democracy and Knowledge: Innovation and Learning in Classical Athens*. New Jersey, NJ: Princeton University Press.

Olivera, O., and T. Lewis. (2004). *Cochabamba! Water Rebellion in Bolivia*. Boston, MA: South End Press.

O'Neil, L. (2010, June 27). "Union Subway Station, DVP, Major Downtown Roads Closed Due to Flooding." *Toronto Star* [Online edition]. Retrieved May 8, 2011, from http://www.thestar.com

Paavola, J., W.N. Adger, and S. Huq. (2006). "Multifaceted Justice in Adaptation to Climate Change." In *Fairness in Adaptation to Climate Change*. W.N. Adger, J. Paavola, S. Huq, and J.J. Mace, eds. pp. 263–78. Cambridge, MA: MIT Press.

Parry, M.L., O.F. Canziani, J.P. Palutikof, P.J. van der Linden, and C.E. Hanson, eds. (2007). *Climate Change 2007: Impacts, Adaptation and Vulnerability*. Cambridge, UK: IPCC.

Pereira, E.L. (2007, February). "Brazil and Climate Change: A Country Profile." *SciDev.Net*. Retrieved January 2011, from http://www.scidev.net/en/policybriefs/brazil-climate-change-a-country-profile.html

Perkins, P.E. (2001). "Discourse-Based Valuation and Ecological Economics." Paper presented at the Conference of the Canadian Society for Ecological Economics (CANSEE), McGill University, August 23–25, 2001, Montreal, Canada.

———. (2003). "Public Participation in Ecological Valuation: How Policies Can Help It Happen." Paper presented at the Conference of the Canadian Society for Ecological Economics (CANSEE), October 16–19, 2003, Jasper, Alberta.

———. (2004a). "Participation and Watershed Management: Experiences from Brazil." Paper presented at the Conference of the International Society for Ecological Economics (ISEE), July 10–14, 2004, Montreal, Canada.

———. (2004b). "Public Participation and Ecological Valuation." Paper presented at the Conference of the International Society for Ecological Economics (ISEE), July 10–14, 2004, Montreal, Canada.

———. (2004c). "Women, Public Participation, and Environmental Valuation." Paper presented at the Conference of the International Association for Feminist Economics (IAFFE), St. Hilda's College, Oxford University, August 4–7, 2004, Oxford, UK.

———. (2008). "Public Participation in Watershed Management: International Practices for Inclusiveness." Paper presented at the Regional Meeting on Water in the Mediterranean Basin, Near East University, October 9–11, 2008, Lefkosa, Northern Cyprus.

Pettengell, C. (2010). "Climate Change Adaptation: Enabling People Living in Poverty to Adapt." Oxfam International Research Report. Oxford, UK: Oxfam International. Retrieved May 4, 2011, from http://publications.oxfam.org.uk/results.asp?sf1=contributor&st1=Catherine%20Pettengell&TAG=&CID=

Portal do Governo do Estado de São Paulo. (2010, June 24). *Governo regulamenta lei de mudanças climáticas*. Governo do Estado de São Paulo. Retrieved March 26, 2011, from http://www.saopaulo.sp.gov.br/spnoticias/lenoticia.php?id=210829

Powell, R., E. Lorimer, and E. Perkins. (2011). "The GreenXChange Project: Environmental Development and Jobs in Jane-Finch." Unpublished paper, York University and Jane-Finch Community and Family Centre, Toronto, Canada.

Riversides. (2009). *The Toronto Rain Storm 2005*. Retrieved May 5, 2011, from http://www.riversides.org/rainguide/riversides_hgr.php?cat=1&page=78&subpage=82

Sandink, D. (2007, December). "Diary of Urban Flooding." *Canadian Underwriter*. Retrieved May 5, 2011, from http://www.canadianunderwriter.ca/issues/story.aspx?aid=1000217381

Satterthwaite, D., S. Huq, H. Reid, M. Pelling, and P. Romero Lankao. (2007). "Adapting to Climate Change in Urban Areas: The Possibilities and Constraints in Low- and Middle-Income Nations." International Institute for Environment and Development, Human Settlements Discussion Paper Series, Climate Change and Cities – 1. Retrieved July 14, 2011, from http://www.iied.org/HS/topics/accc.html and www.iied.org/pubs/display.php?o=10549IIED

Sustainable Neighborhood Retrofit Action Plan (SNAP). (2011). "Black Creek SNAP." *SNAP Website*. Retrieved May 7, 2011, from http://sustainableneighbourhoods.ca/wp/black-creek-snap/

Terry, G. ed. (2009). *Climate Change and Gender Justice*. Rugby, UK: Practical Action Publishing/Oxfam GB.

Todd, A. (2011). "Climate Change and Water Governance in the Greater Toronto Area." Unpublished paper. Toronto, Canada: York University.

United Nation Human Settlements Program (UN Habitat). (2010). *Climate Change Assessment for Maputo, Mozambique*. Cities and Climate Change Initiative. Nairobi, Kenya: Author. Retrieved May 5, 2011, from www.unhabitat.org/pmss/getElectronicVersion.aspx?nr=2977&alt=1

United Nations. (1992). *Report of the United Nations Conference on Environment and Development* (Rio de Janeiro, June 3–14). UN document no. A/CONF.151/26 (Vol. I). Retrieved July 8, 2004, from http://www.un.org/documents/ga/conf151/aconf15126–1annex1.htm

United States Environmental Protection Agency. (2009). *Watersheds*. Washington, DC: Author. Available at http://epa.gov/watershed/

van den Hove, S. (2003). "Between Consensus and Compromise: Acknowledging the Negotiation Dimension in Participatory Approaches." Paper presented at the Frontiers 2 Conference of the European Society for Ecological Economics, February 11–15, Tenerife, Canary Islands.

Water Dialogues – South Africa [Website]. (2009). Retrieved July 15, 2011, from http://www.waterdialogues.org/south-africa/

Zapatistas in Cyberspace: A Guide to Analysis and Resources [Web page]. (2003). Retrieved October 13, 2015, from http://la.utexas.edu/users/hcleaver/Chiapas95/zapsincyber.html

24 Preparing planners for prepared communities

Integrating emergency management into planning education

John "Jack" Lindsay

Introduction

Community planning is failing many communities in times of crisis. When an extreme event occurs in a community, whether it is a natural process such as flooding or an accident such as a train derailment, it is the outcome of ongoing environmental interactions amplified by the planning decisions of that community. As Mileti successfully argues, disasters are therefore not random acts but the result of circumstances arising from the design of a community's natural, social, and built environments.[1] "A cornerstone of community planning is that it aims to promote the public interest when coping with the various problems affecting the physical environment,"[2] and yet community planning is not integrated with emergency management's efforts to deal with the most extreme of those problems. Rather, it may undermine community safety due to the potential for risk being generated from planning decisions, and a failure to recognize its potential to improve community resiliency.

Clearly no single profession holds the fate of an entire community in its collective hands. Planning decisions are influenced by a myriad of other factors, and the translation of planning decisions into action involves many different processes and participants. However, this complexity should not be used to disassociate planning decisions from disasters; rather, it should highlight why it is critical for planners to understand how their decisions can contribute to future unintended and undesirable consequences. This is fundamental to the wellbeing of the community and needs to be integrated into the development of planning professionals throughout their education and career.

Consider a simple hypothetical case of how a community might deal with development along its riverside. A proposal is put forward to subdivide a piece of agricultural land in order to develop a residential neighborhood. The land is located on a bend in the river at the urban-rural fringe. The potential costs associated with flooding, if recognized at all, are weighed against the tax revenues and other benefits of the development. Perhaps some structural mitigation measure, such as a levee, is put in place or at least proposed for future consideration. The planning decision is made by the local council, and the community prospers as the development moves ahead.

Many decades later, the setting has changed. The community has grown outwards past the subdivision and is now a truly suburban neighborhood. Erosion and river silting, along with years of deferred maintenance, have lessened the protective capacity of the levee that was built. The homes, now into their third or fourth generation of owners, are slightly rundown, and overall the neighborhood is in economic decline. Single-parent families, new migrants, and low incomes characterize the households. Public transportation is poor but heavily relied on, as many households cannot afford to keep a car on the road. Ongoing planning decisions have supported new developments as the city expands, rather than upgrades to the combined storm-water and sewer system of the existing neighborhoods. However, despite these changes in circumstances, the community still appears safe.

One spring night, a heavy rainfall upstream triggers a high-water event on the river, already swollen with late-season meltwater, overtopping the levees and resulting in a significant breach. The neighborhood is quickly flooded. The residents, unaware of the flood risks and ill prepared, are vulnerable to the initial impact. Many are trapped, unable to self-evacuate and forced to seek safety on their rooftops. Thankfully no lives are lost, but only as the result of expensive and dangerous rescue operations by emergency services.

In the months that follow, residents continue to struggle. Some are uninsured for over-land flooding; others cannot afford to take time off work to coordinate repairs. Homes are abandoned or, worse, occupied despite the growth of mold, as low vacancy rates across the city leave impacted residents with few relocation options. New bylaws to install backwater valves in home sewer connections prove too expensive for many owners, and the building inspection department is too busy to enforce compliance. Several damaged single-family properties are demolished, and higher-density townhouses and apartment buildings are approved to help alleviate the pressure at the lower end of the market. The cycle rolls on: new residents slowly replace those with experience of the flooding; home ownership shifts to rental properties; low rents attract fixed-income seniors; and eventually the stage is reset for the next disaster.

This scenario is not unreasonably fictitious. Many communities face similar challenges and for decades these challenges have been dealt with through various public safety organizations preparing to respond to such impacts. The role of planning is apparent at every step, from the initial decisions as to where and how to develop land, through the ongoing decisions regarding transportation, economic development, social planning, and infrastructure maintenance, to the final decisions guiding the neighborhood's recovery after a disaster. However, the importance of planning decisions in creating the context for an event like this is too often either overlooked or ignored.

Emergency management

The emergency management systems that communities rely on are not infallible either and are often limited in their approach by the organizational perspectives they have inherited. Comprehensive emergency management was introduced in the US in the late 1970s[3] and has since become the dominant model for practice in North America. This approach expanded the traditional civil defense model of planning and preparing for attacks on civilian targets that was first developed in reaction to the aerial bombardments of World War II and institutionalized during the nuclear missile threat of the Cold War. Under comprehensive emergency management, this prior focus on preparedness and response work should be balanced with efforts to mitigate hazard risks and to plan for community recovery following an impact.

This shift also entailed a broadening of the hazards that were considered. While traditional civil defense organizations were aware of the advantage of using natural events as opportunities to improve systems designed for nuclear attack, such as testing urban evacuation plans prior to a hurricane, this is a limiting approach as it considers only hazards with characteristics similar to an attack. The alternatives to simply responding to impacts were recognized once these agencies, found at all levels of government, began to consider other hazards. It also helped to engage the scientific research associated with the various hazard agents, such as the seismology underlying earthquake hazards, to understand the risks better. This created a link between the scientific knowledge of the hazards and decision-making in disaster management, specifically risk reduction, which a comprehensive approach calls for.

Managing risk is fundamental to this new approach. It has become necessary to measure the probability that an event of a specific magnitude may occur and the degree of damage such an impact is likely to cause. Knowing the risk makes it possible to determine what hazards, drawn from an expanding list of threats, should be considered and prioritized. However, calculating the damage component of the equation has in many cases proved more difficult than the probability of occurrence. This is because assessing potential damage involves considering people, the communities they build, their economic and social systems, and other factors that can be difficult to quantify.

Understanding the physical processes behind the hazards – the wind speeds of tornados or the intensity of a wildfire – is important, and this research has contributed greatly to improving our emergency management systems. The past three decades have seen a corresponding, if not equal, effort put into understanding the social processes, such as the determinants of vulnerability[4] or the factors that influence evacuation decisions, that can also lead to better ways of coping before, during, and after an impact.

The social science side of disaster studies has been especially important to making the case for understanding hazards and disasters as societal issues rather than as purely physical phenomena.[5] The range of interactions between the natural, social, and built environments inevitably includes extremes. When these extreme events harm a community's social, physical, economic, and environmental wellbeing, they are considered emergencies and require responses that often go beyond the normal provision of public safety services. In some instances, these impacts, if on a large enough scale and generating sufficient disruption, overwhelm the community's normal coping resources and are recognized as disasters. In these situations, communities must resort to the suspension of civil rights and other extraordinary measures provided for in legislation.

Hazards, then, are the potential interactions of extreme events and vulnerable communities that lead to emergencies and disasters. The US Gulf Coast has a hurricane hazard because there is the potential for extreme meteorological events to disrupt communities significantly and cause more damage than the community can cope with. Hurricane Katrina was a disaster: a specific event where one storm struck one community, causing damage and loss of life that overwhelmed even the emergency management systems. Risk is the term used to encompass the combination of the likelihood of a hazard event occurring and the degree of damage or losses that such an event will create. Keeping these terms (hazard, risk, emergency, and disaster) distinct is crucial to understanding a comprehensive emergency management framework. This, in turn, is the foundation for promoting greater integration between community planning and emergency management.

Hazard mitigation

Accepting that a community's own decisions, taken individually and collectively, underlie the creation of hazards and generate the vulnerability of the population is the first step to seeing how those same decisions can be made to reduce risk and lessen the resulting harm.[6] Hazard mitigation involves taking action to reduce or eliminate the risk of future impacts by either altering the likelihood that an event will occur or that the consequences of an occurrence will exceed the coping capacity of the community and its components. Often these two approaches are combined so that it is the likelihood of a severe event with damaging consequences that is being managed. Mitigation, in contrast to preparedness and response activities, is focused on preventing damage before it can occur, rather than simply limiting the degree of damage once the event has occurred. Mitigation is also the most common link between emergency management and community planning.

Considering how the construction of a floodway, such as the one that protects Winnipeg, Canada, achieves flood mitigation is a useful example. The Winnipeg floodway manages severe flooding by diverting excess water around the city so that the primary stop-banks within the city are not overtopped. However, low-lying areas within Winnipeg are still prone to flooding, even when the floodway is used, and the protection was originally designed to cope only with a 1-in-110-year flood level. Therefore, the risk is lessened, in the sense that it will require a flood of much larger proportions to inundate the city, but not eliminated. Ongoing efforts are required to control this residual risk, such as expanding the floodway to a 1-in-1,000-year level (which is already underway) and land-use planning to manage the development in areas that are not adequately protected.[7]

This example also demonstrates the difference between structural and non-structural mitigation efforts. The floodway is a structural mitigation project designed to physically manage the natural processes that may cause harm to the community. Other structural mitigation examples include construction techniques to withstand earthquake shaking, hurricane winds, or wildfire spread. Non-structural mitigation consists of policy and planning aimed at modifying human activities such as limiting residential development in the floodplain. Planning future development to prevent further construction in liquefaction or storm surge zones, or establishing a fire permit season, are comparable non-structural examples.

The overall goal of mitigation, as part of a wider comprehensive emergency management program, is to increase the community's resiliency. Resiliency is the ability to withstand hazard impacts, so that damage is limited to a level the community can deal with using its own normal resources, while retaining the capacity and capability to recover should overwhelming damage occur. Recognition of the connections between disaster resiliency and sustainable development presents an opportunity for planners to achieve greater integration and improve the outcomes of both.

Mitigation is an integral part of a comprehensive management approach,[8] and yet it has not been widely adopted, despite its proven effectiveness.[9] Over a quarter of a century ago, William Petak recognized that "the lack of sufficient qualified staff personnel in the building and safety and planning departments of local governments has complicated the problem of achieving effective policy adoption and implementation [of natural hazard mitigation]."[10] Unfortunately, this same deficiency continues today. Integrating hazard mitigation into sustainable community decision-making will not happen without the committed support of the planning profession. Researchers have been calling for this in hazard management literature for decades.[11] The challenges to achieving this are not technical questions of how to use land-use zoning and other planning tools to reduce hazards but instead, as Petak[12] also pointed out, social, economic, political, or administrative obstacles. Incorporating more hazard and emergency management information into the planning education curriculum is one step towards overcoming these obstacles.

The emergency management systems in Canada and the US have many similarities and some notable distinctions.[13] While both countries have embraced a comprehensive approach to emergency management that includes mitigation, each has put in place different arrangements to support its application. This is one area of practice that highlights the different relationships between the two countries' federal governments and their respective states and provinces with regard to emergency management.

Canada's National Disaster Mitigation Strategy (NDMS) took over a decade to be developed and adopted.[14] In the end, it was quietly launched in 2008 as a policy statement. The process involved a series of consultations with the provinces and other partners to determine what approach would be acceptable. The progress of the NDMS was also hampered by the reorganizations that the federal ministry went through during the same period.

The Canadian NDMS is mostly concerned with establishing the roles and relationships between the different government partners at a very general level.[15] The NDMS recognizes the importance of integrating mitigation activities with other projects, such as jointly funded infrastructure initiatives, but stops short of requiring specific mitigation actions. It does, however, prompt the higher levels of government to "engage municipalities and other stakeholders to encourage mainstreaming of disaster mitigation considerations into existing programs/activities (e.g., urban planning, public health, [and] community social programming)."[16]

Mitigation has a higher profile in the US, though it still not equal to the response activities of the Federal Emergency Management Agency (FEMA). FEMA manages programs promoting mitigation, including a range of grants that transfer money to states, which then distribute funds to local governments. The importance of mitigation planning goes beyond the direct benefit of reducing risk, as FEMA also places conditions such as having a mitigation plan on other disaster assistance. However, while mitigation plans may be developed at the local level, their influence on local decision-making in the development process may be limited.

Integrating mitigation in planning

The connection between urban planning and hazard mitigation has been growing slowly for several decades. Early writing in the field, especially in the urban planning literature, was limited, and the need for integration was seldom highlighted.[17] By the late 1990s, the hazard and disaster research was clearly identifying hazard mitigation as a key strategy,[18] and both the Canadian and US governments were introducing legislation and policy to promote hazard mitigation. Today, hazard mitigation is an accepted principle and is seen as best practice in emergency management.[19] The challenge is whether or not the practitioner community is prepared to implement these concepts.

The source of this disconnect may lie in the roots of the two professions and the backgrounds of their respective practitioners. Emergency managers, for the most part, come to their positions in lateral career moves based on experience gained in an emergency services or military career. Many of the situations these fields present to practitioners, be it responding to a large fire or reacting to an enemy attack, create skills transferable to disaster response. However, these perspectives are far less related to the policy and planning tasks that mitigation work consists of. This is changing, however, as more university programs in emergency management are developed, and inevitably the next generation of emergency managers will look to their professional partners in fields such as planning to share practices and cooperate on projects.

Planners, on the other hand, tend to come to their positions through a career ladder that has its first rungs in a university degree. This has its own inherent limitations. While planners will be familiar with the kinds of tasks mitigation planning entails, they may not have the knowledge of hazards that is also required. More importantly, higher education also serves to introduce students to their future professions: therefore, if hazard mitigation is not presented at this formative stage as an integral function of urban planning, it may be difficult to convince a mid-career practitioner later that mitigation is even part of a planner's job when the emergency manager calls to discuss it.

Hazard mitigation in planning education

There are 20 university programs in Canada that offer degrees recognized by the Canadian Institute of Planners. The Planning Accreditation Board lists over 70 accredited undergraduate and graduate university planning departments in the US. A brief review of these programs was sufficient to confirm the belief that hazard mitigation is not an integral part of planning education in the US or Canada.

In 2009, each university program's online course calendar was examined by two research assistants, who were instructed to read the program descriptions and look for keywords (hazard, disaster, emergency, mitigation, risk) to identify mitigation-related courses. Any courses identified in this scan were then assessed to determine if they were required or elective, and if they fitted into a specific concentration within a degree.

At the same time, this research also identified required and elective courses in the programs that related to participatory planning (participation, public consultation, public involvement, community decision-making). Participatory planning has become a core principle in community planning over the past several decades,[20] and was selected as a benchmark against which to compare the hazard mitigation offerings.

Not surprisingly, participatory planning commands far more attention in planning education. It is a common topic in most programs, either as a distinct course or as a part of required courses on planning fundamentals. Hazard mitigation, on the other hand, plays a minor role in the programs. A handful of planning programs listed specific natural hazards or similar courses, but these were the exceptions. It is clear from even this cursory scan that understanding hazard mitigation is not seen as a core component of a planner's education.

A few programs require students take a course in a related area, especially around environmental impacts and risk assessment, or perhaps geology or physical geography courses that include natural hazards as a sub-topic. This can be expected, as Miller and Westerlund commented after reviewing land-use planning programs:

> Environmental and natural resources related courses were included because they were an integral part of the land use specialization in many programs and included courses of key importance in land use planning processes such as environmental impact assessment, coastal zone management, and hazards mitigation.[21]

Electives drawn from other fields, such as public health, environmental science, or natural resource management, may also present students with some information regarding hazards.

Options for planning education

While the lack of hazard-related courses is a concern, the imbalance is not necessarily inappropriate: participatory planning is a crucial part of a planner's education and practice. However, a greater understanding of the hazards facing communities and the options for addressing these risks is also important. It deserves greater attention than it now attracts, even if it is never to equal that given to participatory planning or the other common tenets of planning. A heightened level of awareness of hazards and the role of planning in mitigation is a reasonable goal at this time.

Either the requirement of a general course in emergency management or at least the inclusion of an existing elective from another department would be an excellent start. Many universities may have courses available, such as a course in natural hazards in a Geography or Earth Sciences department, or perhaps Sociology of Disasters or a similar social sciences course, which can serve as an introduction. These courses will advance the students' understanding of how hazards are generated by social decisions and in turn how they impact communities.

The development of a specific course in applying planning techniques to hazard mitigation is certainly a more advanced and ultimately preferred option to be explored. Such a course should present both the theoretical and practical aspects of hazard mitigation, just as Burby[22] suggested for developing environmental impact assessment courses. Students of such a course would gain an appreciation of how planning decisions influence risk and how disasters help shape communities. Perhaps most importantly, it will produce a generation of planners who understand their professional relationship to emergency management.

A generic outline for a hazard mitigation course is provided in Appendix A. The course builds from a broad foundation that sets out the fundamental principles of emergency management, and then progresses to a more detailed discussion of mitigation and the role of planning. It is important that planning students gain an appreciation of their explicit role in hazard mitigation projects, as well as how broader decisions around all issues, such as land-use, transportation, economic growth, or social housing, may potentially change a community's risk profile.

Other proposed courses are also available that may be useful to a planning educator intending to develop a Hazard Mitigation course. FEMA's Higher Education Program has supported the development of several course material packages that are available online for instructors to adapt and deliver (http://training.fema.gov/EMIWeb/edu/completeCourses.asp). One of these courses, *Breaking the Disaster Cycle: Future Directions in Natural Hazard Mitigation*, prepared in 2004 by David Godschalk, sets out the US's hazard mitigation policy. Other applicable course packages on the FEMA Higher Education site include *Principles and Practice of Hazard Mitigation (Summary)*,[23] *Hazards Risk Management*[24] and several hazard-specific courses that will help direct students' attention to locally relevant hazards such as flooding or earthquakes.

Reuniting for community safety

Community planning and emergency management share a common goal of ensuring a safe and prosperous future for the community.[25] However, the two professions have fallen out of step in terms of their development and focus. This is making cooperation difficult, and neither is achieving this shared goal alone.

Emergency management, as noted above, is a profession in its early stages of development, not unlike the planning profession in the 1950s and '60s.[26] Its practitioners come from backgrounds different from that of planners and bring different skills to bear on the problems they face. However, emergency managers may see many of the socio-economic aspects of hazard mitigation that appear ordinary to a community planner as unfamiliar.

Planning, in all its forms, influences the risks present in the community. As a profession, planners have helped guide communities towards a desired future, but without sufficient emphasis on guiding those communities around or away from risks. Ideally, planners encourage public participation in decision-making processes, which must include engaging in a conversation about risk with all stakeholders, especially those most likely to be affected and those most vulnerable. Risk reduction depends on this communication regarding the acceptability of risk and mitigation proposals. Planners have the opportunity and the ability to assist in this process.

Furthermore, planners share their neighbor's heartbreak when disasters set back their progress. The Brundtland Commission commented that "our environmental management practices have focused largely upon after-the-fact repair of damage," but

> the ability to anticipate and prevent environmental damage will require that the ecological dimensions of policy be considered at the same time as the economic, trade, energy, agricultural and other dimensions.[27]

So it must be with our community planning decisions. Planners, as professionals and residents, need to embrace aspects of risk management in order to realize their personal goals and to help communities achieve theirs in a sustainable manner.

The opportunity for collaboration is great; however, it requires a catalyst to initiate the fusion. The legislation and policy in Canada and the US can serve this purpose if emergency managers can be encouraged to turn to their community planning colleagues and, in doing so, find a receptive audience willing to see the shared goals and mutual benefits of collaboration. Educating planners about emergency management in general and hazard mitigation in particular is crucial to building this relationship. Supporting the growth of emergency management into a parallel profession will give planners more peers with expertise and resources to contribute to achieving community goals. Emergency managers and planners must bridge the professional gap and cooperate if we are to see our communities become resilient and sustainable.

Appendix A
Hazard mitigation for planners – A generic course outline

Purpose: This course examines existing theories and practices of hazard mitigation as they relate to community planning. These studies explore how human societies deal with long-term risks through planning to reduce existing vulnerabilities, and how communities can leverage all planning decisions to become more resilient to future disasters. To make risk reduction a central objective of community planning, disaster mitigation requires the assessment and analysis of risk and vulnerability to be integral to planning decisions and, in turn, for there to be a recognition of how planning can increase coping capacities and resources. Mitigation, as a central part of planning, is an ongoing process to achieve safe and sustainable communities. Therefore, this course addresses various aspects of human development and consequences of hazards, and discusses the importance of a participatory approach towards hazard mitigation.

Week 1: **Understanding Emergency Management**: overview of hazard, disaster, risk, vulnerability, and the comprehensive emergency management components of mitigation, preparedness, response, and recovery.

Week 2: **Hazard Identification and Assessment**: Models for hazard and environmental assessment.

Week 3: **Social Vulnerability**: the determinants of vulnerability and methods of vulnerability assessment.

Week 4: **Risk Assessment and Management**: Risk management processes and analysis.

Week 5: **Mitigation and Community Planning**: Research supporting integration.

Week 6: **Mitigation Alternatives**: Structural and non-structural mitigation options for different hazards.

Week 7: **Sustainable Mitigation and Disaster-Resistant Communities**: The principles of sustainable mitigation (e.g., from Mileti's (1999) *Disasters by Design* and Geis's (2000) paper on Disaster-Resistant Communities).

Week 8: **Alternative Views**: Critiques of and challenges to sustainable mitigation.

Week 9: **Making the Case for Mitigation**: Support and challenges to mitigation based on benefit cost analysis and other economic and political considerations.

Week 10: **Mitigation Policy and Law**: Relevant mitigation and/or planning legislation and policy supporting or restricting integration.

Week 11: **Case Studies**: Local and national examples of community planning with integrated hazard mitigation.

Week 12: **Path Forward**: Opportunities to increase mitigation through participatory community planning.

Notes

* Acknowledgments: Thank you to Dr. Etsuko Yasui for her insights during the preparation of this paper, and to Sarah Delisle and Kristi Jacques for their research skills and perseverance. This work was prepared with financial assistance from the Brandon University Research Committee.

1 D. Mileti (1999), *Disasters by Design: A Reassessment of Natural Hazards in the United States* (Washington, DC: Joseph Henry).

2 G. Hodge (1986), *Planning Canadian Communities* (Toronto, Canada: Methuen), p. 7.

3 National Governors' Association (1979), *Comprehensive Emergency Management: A Governor's Guide* (Washington, DC: Author); A. Beauchesne (2001), *A Governor's Guide to Emergency Management, Volume One: Natural Hazards* (Washington, DC: National Governors' Association).

4 J. Lindsay (2003), "The Determinants of Disaster Vulnerability: Achieving Sustainable Mitigation through Population Health," *Natural Hazards: Journal of the International Society for the Prevention and Mitigation of Natural Hazards* 28(2–3), pp. 291–301.

5 K. Hewitt, ed. (1983), *Interpretations of Calamity* (Boston, MA: Allen and Unwin); E. Quarantelli (1988), "Planning and Management for the Prevention and Mitigation of Natural Disasters, Especially in a Metropolitan Context: Initial Questions and Issues which Need to be Addressed," *Planning for Crisis Relief: Towards Comprehensive Resource Management and Planning for Natural Disaster Prevention* (Disaster Research Center Preliminary Papers), p. 114.

6 Hewitt (1983).

7 International Joint Commission (1999), *Flood Protection for Winnipeg*, retrieved December 10, 2012, from http://www.floodwayauthority.mb.ca

8 Beauchesne (2001); Public Safety Canada (2011), *An Emergency Management Framework for Canada* (Ottawa, Canada: PSC), retrieved January 12, 2016, from https://www.publicsafety.gc.ca/cnt/rsrcs/pblctns/mrgnc-mngmnt-frmwrk/index-eng.aspx

9 J. Schwab, ed. (2010), *Hazard Mitigation: Integrating Best Practices into Planning* (Planning Advisory Service Report Number 560) (Chicago, IL: American Planning Association).

10 W. Petak (1984), "Natural Hazard Mitigation: Professionalization of the Policy," *International Journal of Mass Emergencies and Disasters* 2(August), p. 300.

11 E.g., H. Foster (1980), *Disaster Planning: The Preservation of Life and Property* (New York, NY: Springer); N. Britton and J. Lindsay (1995a), "Integrating City Planning and Emergency Preparedness: Some of the Reasons Why," *International Journal of Mass Emergencies and Disasters* 13(1), pp. 93–106; R. Burby, ed. (1998), *Cooperating with Nature: Confronting Natural Hazards with Land-Use Planning for Sustainable Communities* (Washington, DC: Joseph Henry); B. Glavovic, W. Saunders, and J. Becker (2010), "Land-Use Planning for Natural Hazards in New Zealand: The Setting, Barriers, 'Burning Issues' and Priority Actions," *Natural Hazards* 54, pp. 679–706.

12 Petak (1984).

13 D. McEntire and J. Lindsay (2012), "One Neighborhood, Two Families: A Comparison of Intergovernmental Emergency Management Relationships," *Journal of Emergency Management* 10(2), pp. 93–107.

14 V. Hwacha (2005), "Canada's Experience in Developing a National Disaster Mitigation Strategy: A Deliberative Dialogue Approach," *Mitigation and Adaptation Strategies for Global Change* 10, pp. 507–23.

15 Public Safety Canada (2008), *National Disaster Mitigation Strategy*, retrieved January 12, 2016, from http://www.publicsafety.gc.ca/prg/em/ndms/strategy-eng.aspx

16 Ibid., p. 2.

17 N. Britton and J. Lindsay (1995a), and (1995b), "Demonstrating the Need to Integrate City Planning and Emergency Preparedness: Two Case Studies," *International Journal of Mass Emergencies and Disasters* 13(2), pp. 161–78.

18 E.g., Burby, ed. (1998); Mileti (1999).

19 D. Godschalk (2003), "Urban Hazard Mitigation: Creating Resilient Cities," *Natural Hazards Review* 4(3), pp. 136–43; M. Lindell, C. Prater, and R. Perry (2007), *Introduction to Emergency Management* (Hoboken, NJ: Wiley); Canadian Standards Association (2008),

CSA Standard Z1600 Emergency Management and Business Continuity Planning (Toronto, Canada: CSA); Godschalk, D. (2004). [0][0]"Breaking the Disaster Cycle: Future Directions in Natural Hazard Mitigation Course." FEMA Emergency Management Higher Education Project. Retrieved December 10, 2012, from http://training.fema.gov/EMIWeb/edu/complete Courses.asp.

20 L. Laurian and M. Shaw (2009), "Evaluation of Public Participation: The Practices of Certi-fied Planners," *Journal of Planning Education and Research* 28, pp. 293–309.

21 D. Miller and F. Westerlund (1990), "Specialized Land Use Curricula in Urban Planning Graduate Programs," *Journal of Planning Education and Research* 9, pp. 203–06.

22 R. Burby (1992), "Comprehensive Impact Assessment in Planning Education and a Course Syllabus," *Journal of Planning Education and Research* 12, pp. 67–75.

23 D. Brower and C. Bohl (2000), *Principles and Practice of Hazard Mitigation Course*, FEMA Emergency Management Higher Education Project, retrieved December 10, 2012, from http://training.fema.gov/EMIWeb/edu/completeCourses.asp

24 G. Shaw, G. Haddow, C. Rubin, and D. Coppola (2000), *Hazards Risk Management Course. FEMA Emergency Management Higher Education Project*, retrieved December 10, 2012, from http://training.fema.gov/EMIWeb/edu/completeCourses.asp

25 J. Lindsay (1998), "Separate Lives or Forgotten Ties: Urban Planning and Emergency Man-agement," *Planning Quarterly* 128(March), pp. 7–12.

26 J. Lindsay and N. Britton. (2010). "Designing Educational Opportunities for the Emergency Management Professional of the 21st Century (Revisited)." In *Integrating Emergency Man-agement Studies into Higher Education: Ideas, Programs and Strategies*. J. Hubbard, ed. Fairfax, VA: Public Entity Risk Institute.

27 World Commission on Environment and Development (1987), *Our Common Future* (Oxford: OUP), p. 39.

Bibliography

Beauchesne, A. (2001). *A Governor's Guide to Emergency Management, Volume One: Natural Hazards*. Washington, DC: National Governors' Association.Britton, N., and J. Lindsay. (1995a). "Integrating City Planning and Emergency Preparedness: Some of the Reasons Why." *International Journal of Mass Emergencies and Disasters* 13(1), pp. 93–106.

———. (1995b). "Demonstrating the Need to Integrate City Planning and Emergency Pre-paredness: Two Case Studies." *International Journal of Mass Emergencies and Disasters* 13(2), pp. 161–78.

Brower, D., and C. Bohl. (2000). *Principles and Practice of Hazard Mitigation Course*. FEMA Emergency Management Higher Education Project. Retrieved December 10, 2012, from http://training.fema.gov/EMIWeb/edu/completeCourses.asp

Burby, R. (1992). "Comprehensive Impact Assessment in Planning Education and a Course Syllabus." *Journal of Planning Education and Research* 12, pp. 67–75.

———, ed. (1998). *Cooperating with Nature: Confronting Natural Hazards with Land-Use Planning for Sustainable Communities*. Washington, DC: Joseph Henry.

Canadian Standards Association. (2008). *CSA Standard Z1600 Emergency Management and Business Continuity Planning*. Toronto, Canada: CSA.

Foster, H. (1980). *Disaster Planning: The Preservation of Life and Property*. New York, NY: Springer.

Glavovic, B., W. Saunders, and J. Becker. (2010). "Land-Use Planning for Natural Hazards in New Zealand: The Setting, Barriers, 'Burning Issues' and Priority Actions." *Natural Hazards* 54, pp. 679–706.

Godschalk, D. (2003). "Urban Hazard Mitigation: Creating Resilient Cities." *Natural Hazards Review* 4(3), pp. 136–43.

———. (2004). "Breaking the Disaster Cycle: Future Directions in Natural Hazard Mitigation Course." *FEMA Emergency Management Higher Education Project*. Retrieved December 10, 2012, from http://training.fema.gov/EMIWeb/edu/completeCourses.asp

Hewitt, K. ed. (1983). *Interpretations of Calamity*. Boston, MA: Allen and Unwin.

Hodge, G. (1986). *Planning Canadian Communities*. Toronto, Canada: Methuen.

Hwacha, V. (2005). "Canada's Experience in Developing a National Disaster Mitigation Strategy: A Deliberative Dialogue Approach." *Mitigation and Adaptation Strategies for Global Change* 10, pp. 507–23.

International Joint Commission. (1999). *Flood Protection for Winnipeg*. Retrieved December 10, 2012, from http://www.floodwayauthority.mb.ca

Laurian, L., and M. Shaw. (2009). "Evaluation of Public Participation: The Practices of Certified Planners." *Journal of Planning Education and Research* 28, pp. 293–309.

Lindell, M., C. Prater, and R. Perry. (2007). *Introduction to Emergency Management*. Hoboken, NJ: Wiley.

Lindsay, J. (1998). "Separate Lives or Forgotten Ties: Urban Planning and Emergency Management." *Planning Quarterly* 128(March).

———. (2003). "The Determinants of Disaster Vulnerability: Achieving Sustainable Mitigation through Population Health." *Natural Hazards: Journal of the International Society for the Prevention and Mitigation of Natural Hazards* 28(2–3).

———, and N. Britton. (2010). "Designing Educational Opportunities for the Emergency Management Professional of the 21st Century (Revisited)." In *Integrating Emergency Management Studies into Higher Education: Ideas, Programs and Strategies*. J. Hubbard, ed. Fairfax, VA: Public Entity Risk Institute.

McEntire, D., and J. Lindsay. (2012). "One Neighborhood, Two Families: A Comparison of Intergovernmental Emergency Management Relationships." *Journal of Emergency Management* 10(2), pp. 93–107.

Milcti, D. (1999). *Disasters by Design: A Reassessment of Natural Hazards in the United States*. Washington, DC: Joseph Henry.

Miller, D., and F. Westerlund. (1990). "Specialized Land Use Curricula in Urban Planning Graduate Programs." *Journal of Planning Education and Research* 9, pp. 203–06.

National Governors' Association. (1979). *Comprehensive Emergency Management: A Governor's Guide*. Washington, DC: Author.

Petak, W. (1984). "Natural Hazard Mitigation: Professionalization of the Policy." *International Journal of Mass Emergencies and Disasters* 2(August), pp. 285–302.

Public Safety Canada. (2008). *National Disaster Mitigation Strategy*. Retrieved January 12, 2016, from http://www.publicsafety.gc.ca/prg/em/ndms/strategy-eng.aspx

———. (2011). *An Emergency Management Framework for Canada*. Ottawa, Canada: PSC. Retrieved January 12, 2016, from https://www.publicsafety.gc.ca/cnt/rsrcs/pblctns/mrgnc-mngmnt-frmwrk/index-eng.aspx

Quarantelli, E. (1988). "Planning and Management for the Prevention and Mitigation of Natural Disasters, Especially in a Metropolitan Context: Initial Questions and Issues which Need to be Addressed." *Planning for Crisis Relief: Towards Comprehensive Resource Management and Planning for Natural Disaster Prevention*. Disaster Research Center Preliminary Papers, 114.

Schwab, J. ed. (2010). *Hazard Mitigation: Integrating Best Practices into Planning* (Planning Advisory Service Report Number 560). Chicago, IL: American Planning Association.

Shaw, G., G. Haddow, C. Rubin, and D. Coppola. (2000). *Hazards Risk Management Course. FEMA Emergency Management Higher Education Project*. Retrieved December 10, 2012, from http://training.fema.gov/EMIWeb/edu/completeCourses.asp

World Commission on Environment and Development. (1987). *Our Common Future*. Oxford: OUP.

25 Resilience in urban contexts

From South America to Europe

Matías Ezequiel Barberis Rami

Introduction

The increasing population density in urban centers, the growth of large cities, environmental degradation, economic uncertainty, the exponential increase in consumption, and climate change are some of the main causes of the proliferation of risks in everyday lives. The notion of risk is post-traditional and post-rational, referring to a period of the history of science characterized by a reflection on the events of modernity.[1] The term risk refers to decisions linked to time, but difficult to know because of the uncertain future; it refers to a contingent phenomenon that offers different perspectives to different observers. In other words, the contingent damage is contingently occasioned and therefore avoidable.[2]

The environmental dimension has been progressively included in the theory and practice of development policies. Protecting the environment is a task that should be addressed within a broad time perspective, as it transcends every human generation. In recent decades, quantitative and qualitative increases in the occurrence of natural and socio-natural disasters have made our societies and government organizations pay special attention not only to the environmental dimension, but also to the social and economic spheres.

Cities around the world are able to withstand great shocks in the short term because of disasters that have weakened their physical, social, and economic structure. However, disasters have always created the possibility of radical changes in people's daily lives.

Urban planning analysis has indicated that any city determines its own risk scenarios.[3] In this sense, a catastrophe that happens in Buenos Aires will not have the same effects that a similar disaster might have in Barcelona or Rome. This idea is directly linked to cultural theories of risk that have explored the ways in which risks are perceived and accepted, showing a tendency to underestimate them. Moreover, the perception of risk is usually based on a historical, cultural, political, and economic path.

Western cities, considered splendid machines that meet the requirements of man, are actually increasingly vulnerable to shocks of nature.[4] When a catastrophic event occurs, its impact does not recognize physical, symbolic, political, or economic boundaries; planning must therefore be approached as a central challenge to the articulation of these areas, which are becoming day by day more vulnerable to disasters.

The aim of this paper is to analyze how urban planning could strengthen resilience in urban contexts, taking into account the case of Buenos Aires and influenced by the exchange of experiences with other European and South American cities. In order to achieve this objective, the paper is divided into five sections. First of all, the perspective of disaster risk management will be presented; second, the link between urban planning and resilience will be analyzed; third, the main trends in terms of urban planning will be presented; fourth, the case of Buenos Aires will be analyzed; and finally, the main conclusions of the work will be explained.

The perspective of disaster risk management

The process of interdependence that exists in the world[5] is an analytical concept that refers to situations characterized by reciprocal effects among countries or among actors in different countries. In this context, the high number of disasters caused by human activity suggests that there is a relationship between different territories and the causes for the occurrence of these events, or the impact they have as a consequence of inadequate disaster risk management.

What does disaster risk management entail? In actual terms, risk may be defined as the result of the interaction between one or multiple threats, vulnerabilities, and response capabilities. A threat refers to the possibility of the occurrence of a physical event that can cause any harm to society. Vulnerability is the relation of a community's level of exposure to its propensity to be damaged and the difficulties it will face in subsequent recuperation.[6] Responsiveness is the combination of strengths, attributes, and resources available within a community, society, or organization to address multiple risks.

Threats can be simple or complex, the latter involving a concatenation or synergy of the former. Simple threats belong to the natural world and are caused by geological factors such as volcanic or oceanic phenomena; socio-natural threats can be hydro-hazards (i.e., floods) or hydrogeological (i.e., landslides), combining natural events with specific social practices and land use and planning; or they may have an anthropogenic origin, being the byproduct of human activity, mainly industrial, such as explosions, pollution, radioactivity, etc.

Similarly, vulnerability is a relative concept, depending on the specific conditions of each area and the type of threat. Still, there are several factors that can be analyzed, including the natural and physical environments, the economic and financial sphere, society, education, ideology, politics, institutions, and technical capabilities. Vulnerabilities depend on:

- Inadequate or insufficient regional planning, urban planning, and land-use regulation;
- The availability, type, and quality of educational programs on the preservation of the environment and the area's cultural and environmental heritage;
- The particular beliefs of a community, the availability or lack of resources, the capacity of governments to secure autonomy and decentralized management, etc.

Moreover, a disaster can and should be understood as a social category, and the degree of a society's vulnerability to the presence of a threat should be constantly updated: disasters occur when a risk is not acknowledged, or when inadequate action is taken against the risks to which that community is exposed.[7]

A disaster is a situation or social process resulting from the manifestation of a phenomenon of natural, technological, or man-made origin that finds favorable conditions of vulnerability in a population and causes intense, serious, and extended changes in that community's normal functioning. Once these concepts are understood, the questions are: what does it mean to manage the risk of disasters, and how is disaster risk management conceived at the local, national, and international levels?

Disaster risk management has begun to feature in government agendas from the consideration of the contribution to development that these processes create: sustainable development regardless of risk control is just an empty slogan.[8] Also, the Principle of Prudence is an integral and crucial part of the Principle of Sustainability, essential in management as it involves the mitigation and reduction of simple or complex risks. Thus, risk management cannot be reduced to a concrete action, but must be linked to a process leading to integrated territorial planning and implementation of policies, strategies, tools, and measures, and should be assumed to be the underlying premise observed by all social actors and by public and private entities.

Local actors primarily carry out local risk management. Social participation, as a part of governance, enables effective risk reduction: in developing countries and in underdeveloped regions of developed countries, urban and regional authorities usually concentrate their action on low-income families and maintain good relations with their citizens, significantly improving the reduction of disaster risk.

In developing countries, the local dimension of risk management is relevant because it generates legitimacy in decision-making processes. Meanwhile, in advanced regions and economies, local, urban, and regional authorities focus their attention both in terms of awareness-raising and in terms of policy issues on efficient risk management.

It is important to know whether local actors manage or co-manage the risk and whether they understand the process as part of sustainable local development. However, to achieve the implementation and expansion of local initiatives, governments need new capacities and skills. The existing political relationship between the international and local levels can be analyzed through the concept of governance. Thus, risk management involves an active relationship between these areas, either for the coordination of joint actions such as obtaining financial support through development programs, or for technical and scientific cooperation to provide guidance to management.

In this framework, capacity-building becomes an instrument with which to approach risk reduction. This category is understood as the process through which individuals, organizations, and societies establish, strengthen, and maintain their ability to set and achieve their own development objectives over time.[9] It is a valid instrument of governance to direct local development.

Why urban planning could be useful for strengthening resilience

The Hyogo Framework introduced into the public agenda the concept of the "resilience" of cities in responding to socio-natural disasters. This refers to the ability of a society exposed to potential hazards to adapt by resisting or changing, in order to achieve or maintain an acceptable level of functioning and structure. It involves components such as social capital, empathy, effectiveness in solving problems, and goals/aspirations balance.

According Godschalk,[10] resilience is a sustainable network between a physical system and human communities. Physical systems comprise infrastructure, communications, and energy facilities, as well as geological and natural systems. During a disaster, physical systems must be able to survive and function under the extreme demands placed on them. A resilient city lacking such systems will be extremely vulnerable to disasters. Human communities are composed of the social and institutional components of the city. During a disaster, community networks should be able to survive and function in extreme and unique conditions; if they fail, decision-making and disaster response could be undermined.

In recent decades, large cities have shown a common trend with regard to the problems of spaces, trivializing contexts, spatial interdependence, increasing social inequality, technological innovation and industrial redevelopment, segmentation and productive specificity, flexibility, etc. The complexity and cultural absorption capacity of these metropolises embody the complexity and variety of the world as a whole,[11] but each metropolis has been planned in response to particular needs and interests in various periods of history. While some European cities were planned in the late nineteenth century based on the idea of progress and power – ideals of modernity – in South American cities, the urban and spatial plans are a hybridization of European ideas and local needs and specificities.

In response to the idea that everyday life is increasingly oriented to consumption, the rationale of planning has shifted towards market logic.[12] This trend of contemporary spatial planning has been implemented in various parts of the world not only in regular planning instances, but also in extraordinary cases.

In emergency periods, some governments tend to plan only on the basis of short-term economic interests. Reconstruction plans in some cases were devised to circumvent the basic rules of planning, disregarding the proper and rational management of territorial transformations. The perspective was of short- and medium-term development, with no regard for sustainable planning.

Why is planning important in terms of building resilience? According to a systemic-ecological perspective, the environment tends to be resilient and will restore its own balance: this resilience is built on the interactions between families, neighborhoods, communities, societies, and culture.[13] Planning answers the need to logically order a society; it also gives rise to regulations that establish an urban plan and determines the behavior patterns of those who inhabit the territory. Planning is therefore a potential tool for managing risk and building resilience, but as a trend it should be linked to a general urban policy that meets territorial specificities.

As explained by Anna Wikström,[14] it is crucial that contemporary urban adaptation planning focuses almost solely on environmental and climate change, but by excluding other vital aspects, such as societal and urban changes discussed above, cities that are planned today will not be able to achieve such effective resilience.

Urban planning: South America and Europe

Urban planning faces a double challenge in responding to the current context: not only should it be aimed at efficiency and reinforcing the distribution of power, but it should also create standards that make resilience-building possible. Cooperation between cities is an effective and widespread trend in the effort to create international policies.[15]

Nowadays, cities are demonstrating certain significantly similar trends in terms of urban planning, aimed mainly at rationalizing, functionalizing, and revitalizing areas.

This contributes to creating different perceptions of risk according to the way spaces are designed and lived in by people all over the world.

The first urban plans in Europe tried to respond to the effects in the century that followed the industrial revolution: urban decay was generated by the settlements that were created around factories, giving rise to overcrowding and hygiene problems. The growth of a working class without resources and the demand of the upper classes to ensure their private space led to demands for urban planning. The Paris of Hauss-mann, the Barcelona of Cerda, and the Rome of Viviani were responses to major labor struggles where workers demanded better living conditions. In these three cases, planning was carried out during periods of authoritarian rule, which allowed a push that sought to respond to popular demands but from the standpoint of the rulers and the technical acuity of city planners.

In recent years, planning in these European cities has focused on new demands associated with the densification of the urban area: the need to improve circulation, administrative decentralization, the growth of the service sector, the accessibility of public administration, and the necessity to develop environmental sustainability. One of the cities that has developed standards of environmental and territorial sustainability is Stockholm, but it has faced social problems such as residential segregation and anomie. While each of these cities has sought to resolve these issues, the effectiveness of their plans has been limited by the recent crisis of representative democracy, new participatory processes from civil society, and the focus of daily life on mass consumption.

Meanwhile, major South American cities such as Buenos Aires, Santiago de Chile, Quito, and São Paolo share the origin of the theoretical model of urbanization that came from the European colonization of the sixteenth and seventeenth centuries: public administration and religious authority were placed around a central square, accompanied by a checkerboard-style outline for residential housing. Due to the transcontinental migrations in the nineteenth and twentieth centuries, as well as migration from the countryside to the city in the late twentieth century associated with the opportunities offered by the city, there has been an increasing densification that has reconfigured the urban model of these metropolises. Some of the main issues present in urban design are related to growing socio-economic inequality, the complexity of vehicular traffic, poverty, and the widespread development of informal settlements in areas that are rapidly changing and are vulnerable to natural disasters.

The case of Buenos Aires

A case of particular relevance is the city of Buenos Aires, characterized by a process of territorial planning that condensed urban elements from different European and South American trends. Throughout its history, the planning of the city, usually sectoral and flexible,[16] has not only responded to social and organizational demands of the everyday life of the population, but has also shaped the evolution of civic life.

The combination of different elements of politics, economics, and planning has made Buenos Aires a city capable of coping with situations of shock in terms of disasters as well as in other areas of daily life. A brief description of the evolution of the planning and urban intervention in the city will allow some understanding of the different elements that make up the variables of vulnerability and the potential of the city in terms of urban risk.

The reflection on the evolution of planning in Buenos Aires and its structure shows that decisions were often made in response to health crises in the old city as well as being prompted by the relationship between the development of the railways and the territory. The proposal of the engineer Luis Huergo and the businessman Eduardo Madero for the construction of a new port gave rise to an in-depth analysis of the need to design and build a logistic matrix structure for the territory and urban space in the 1880s. The expansion of gas lighting, the laying of power lines, and the granting of permission for railway lines completed the set of interventions aimed at improving the fabric of public space.

In 1895, the population census estimated that there were 677,786 inhabitants in Buenos Aires, making it one of the most populous cities in the Southern Hemisphere. By the late nineteenth century, the structuring of the metropolitan area was based on the relationship between environment and society and on building logistical support, as well as on the growth and structure of built space. These developments led to public debates that activated new power relations, giving rise to a new matrix that sustained the growth of the city for over 100 years.

With the centenary celebrations of Argentina's independence, plans were developed for the beautification of the city that incorporated the English and French influence: hygiene and circulation, as the starting point of the first interventions, were complemented by the idea of a beautiful city. The opening of wide boulevards and the design of a system of parks structured a new type of urban fabric.

Meanwhile, the changes following the first World War meant that great numbers of European immigrants moved into the city's tenements. The National Commission on Cheap Housing was created in 1915 (Law 9677) and allowed the city to finance the construction of numerous housing projects. They began to propose interventions throughout the territory of the city that recognized the need to consider "extramural" buildings.

The global crisis of 1930 coincided with a time of political instability and social unrest. Because of the agro-export crisis, domestic industry began to produce products that could no longer be imported. Following the military coup to overthrow the constitutional order, conservative economic policies aimed at reviving the basis of the industry were implemented. Public works were conceived as a way to generate jobs, and the modernization of agriculture accelerated migration to the city. The expansion of the suburbs and the emergence of shantytowns created new social stratifications and conditions in which political struggles developed. Public buildings and housing projects were constructed, and the riverside was expanded. Progress is evident not only in contemporary high-rise buildings but also in buildings that do not comply with the minimum necessary living conditions, reducing air and light inside the block.

Conceptually, the new planning proposals recommended the rational distribution of functions over the territory, incorporating new technologies in construction and systematizing the manufactured housing prototypes produced using industrial processes. Urban development promoted a cultural transformation that was incorporated into daily life, and the new spaces and products brought about greater material comfort and led to increased democratization.

In 1958, the Master Plan for the City of Buenos Aires, which was intended to stimulate capital inflows and the recovery of the domestic industry by multinational companies, was finalized. Processes of social inequality and imbalance between regions had intensified. The state promoted territorial occupation, emphasizing the potential of the river shoreline. The proposal included the restructuring of the port, roads, and the railway, as well as urban balance and renewal of the center, thus creating new centers for industrial development and control of urban sprawl, as well as focusing attention on environmental problems. While recognizing the need for metropolitan coordination, it must be assumed that the imbalance between the city center and its periphery was accentuated during this period.

The transformation of the rural production structure and the development of national industry promoted a migratory movement that marked a turning point in the curve of population growth and greatly expanded the suburbs. The Five Year Plans of 1947 and 1952 were innovative planning tools of state action, defining a policy of nationalization of utility companies, such as railways and gas, phone, and electricity suppliers. It proposed a rational distribution of activities over the territory and promoted the realization of infrastructure to support industrial development in the municipalities of Avellaneda, Lanús, Morón, Vicente López, San Martin, and Lomas de Zamora. The expropriation of large green areas represented intervention on a scale previously unknown. Since 1982, certain failures united the claims of different social sectors for the return of democracy. In December 1983, a cycle of institutional reconstruction started that continues today. Economic difficulties have permanently damaged the recovery of economic prosperity that characterized this society until the mid-1970s.

Since the restoration of democracy in 1985, the territorial planning aimed at creating jobs, improving the habitat, and responding to functional problems, inadequate equipment, and the environmental degradation of the city has been based on participatory processes. The planning process was withdrawn to allow for analysis of the weaknesses of the large-scale plans that were then current. This allowed for the development of intermediate and small-scale plans that involved citizens in the decision-making process. These plans prioritized public spaces over any other issues, supporting the restoration of the public's right to the city, which was fostered by the advent of democracy.

The different territorial interventions and planning ideas for Buenos Aires have increased risk variables, generating a level of flexibility that is directly linked to resilience. Phenomena such as the deterioration of the quality of life in some sectors of society, political change, the lack of long-term policies, unregulated city and land use, the irregularity of growth, and the lack of environmental awareness are some of the factors that have generated a complicated vulnerability map for the city.

However, when survival depends on adapting to innovation in everyday life, an intangible factor develops associated with resilience and the ability to devise different strategies to cope with shock. Factors such as solidarity, the use of social capital, and the capabilities of individuals, have enabled the community to deal with times of instability and crisis. In this sense, although urban interventions have attempted to respond to the needs of society, they have also allowed the social dynamics associated with intangible capital and human resilience to be generated. This does not mean that the city is prepared to face a possible disaster, but it clearly demonstrates a willingness to deal with shocks in the short term through emergency response, which should be addressed with formal intervention and coordination by a governmental agency in order to be better prepared for the future.

Conclusions

Urban interventions have played an essential role in the organization of cities, both in Europe and in South America. The proposed sanitation, the widening of avenues (a model started in Paris and reproduced in several South American cities), the creation of green places, the transit system, and the recent proposals for building sustainable transport and upgrading public spaces are some of the common elements shared by South America and Europe. In terms of resilience, strengthening these elements will not only promote greater social interaction by creating closer social ties even in urban contexts, but will also create an urban space less vulnerable to the possibility of catastrophic events. Those cities that are particularly composed of close social networks have achieved greater strength, adaptability, learning, and resilience.[17]

However, limitations in both contexts have played a significant role in terms of sustainability. The problem of inequality in South America and the crisis in representative democracy in Europe have slowed the effective implementation of urban plans. Proposals are implemented in a piecemeal fashion, or not at all. Recent action plans to deal with disasters – not only in Argentina, but also in Italy, Germany, Brazil, and Ecuador – are characterized by short-term thinking and reactive models. The importance of generational change should be a central element in urban planning: the changes and new interventions in the territory should address the community's needs by focusing on sustainable development. The case analyzed in this paper presents some potential limitations in terms of transferring certain aspects to other socio-cultural contexts. The main strategies for strengthening resilience that could be implemented are related to the articulation of public policies and the creation of a framework in disaster risk management and key aspects of sustainability. Working on disparate areas such as education, health, infrastructure, etc., allows for a sectoral approach that can be adapted to particular issues within each area. However, it is important to develop a flexible framework that permits the articulation of these areas in a matrix capable of standing up to major shocks such as disasters.

Moreover, the lack of resources, ideological divisions, and the striking differences between socio-economical classes limit the development of an integrative policy that strengthens resilience. The concentration of resources and capacities at the national level, the lack of fiscal and human resources, and the limited capacity of local government defined the complex scenario for disaster risk reduction and response. However, social capital and cultural elements represent the key factors that could be exported to other contexts. It is not a particular practice or an existing element; it is the idea of understanding how a society functions and the ability to create plans that take these dynamics into account.

The fact that the State is responsible for carrying out risk management is insufficient and limited, and in states with low budgets, inefficient institutions, and internal political fragmentation, it can be dangerous. Some developing countries have highlighted the importance of community solidarity rather than state intervention as a major value when disasters occur. Changing the vision of management and planning must start from below; the process of governance is empty without the existence of real mechanisms allowing social participation in decision-making and in the implementation of any planned action. Therefore, in the field of policy effects the key issue is the process of horizontal governance: creating partnerships, regulating conflicts of interest, and promoting opportunities for real participation on the part of each actor involved. The difficulty in strengthening real and effective horizontal governance processes in social and political arenas is a cause of the slow progress in achieving the objectives of risk management and mitigation. One of the main objectives is the real and effective capacity of authorities – public and corporate – to deal with planning processes that include disaster risk management as a long-term strategy.

Some variables to consider in terms of planning as an enhancer of resilience in structural terms are:

1   The economic potential for addressing a process of integrated risk management through long-term urban planning;
2   Flexibility and adaptability on the part of governmental authorities when projecting changes in the city;
3   Promoting opportunities for citizen interaction to reverse the trend towards individualism; and
4   Strengthening horizontal cooperation among cities with a strong capacity for action in the territory.

The fact that urban planning is essential to sustainability and risk management is a known issue. However, over time, it has demonstrated that it can work as a full analytical unit. At the moment, we analyze different urban contexts, their problems, and the potential risks involved; we understand the importance of disaggregating the concept of planning. In the transverse axis of planning is where we should discover the potential and the limits of the urban process. Each metropolis, its structure, and particular situation display characteristics that enable or constrain the ability to develop resilience. Each city functions as a laboratory in terms of urban planning,[18] and so provides a particular context for building resilience.

Notes

1 U. Beck (1986), *Risikogesellschaft. Auf dem Weg in eine andere Moderne* (Frankfurt, Germany: Suhrkamp).
2 N. Luhmann (1991), *Soziologie des Risikos* (Berlin, Germany: De Gruyter).
3 E. Salzano (2007), *Fondamenti di urbanistica* (Rome, Italy: Laterza).
4 E. Scandurra (2003), *Città morenti e città viventi* (Rome, Italy: Meltemi Babele Editrice).
5 R. Keohane and J. Nye (1977), *Power and Interdependence* (New York, NY: Longman).
6 R. Basher (2008), "Disaster Impacts: Implications and Policy Responses," *Social Research* 75(3), pp. 937–54.
7 G. Ligi (2009), *Antropologia dei disastri* (Rome, Italy: Editori Laterza); A. Oliver-Smith (1994), "La reconstrucción después del desastre: una visión general de secuelas y problemas," in *Al norte del Río Grande*, Allan Lavell, ed., pp. 3–18 (México: La Red); H. Herzer (2002), *Convivir con el riesgo o la gestión del riesgo*, retrieved August 14, 2013, from www.cesam.org.ar/publicaciones.htm
8 A. Lavell (2002), "Local Level Risk Management: Concepts and Experience in Central America," paper presented at the Disaster Preparedness and Mitigation Summit, American Red Cross, November 21–23, 2002, New Delhi, India.
9 UNDP (2010).
10 D. Godschalk (2003), "Urban Hazard Mitigation: Creating Resilient Cities," *Natural Hazards Review* 4(3), pp. 136–43.
11 Mumford, cited in Scandurra (2003).
12 Sassen, cited in Scandurra (2003).
13 E. Malaguti (2011), "Articolazioni teoriche della resilienza," in *Costruire la resilienza, la riorganizzazione positive della vita e la creazione di legami significativi*, B. Cyrulnik and E. Malaguti, eds., pp. 79–102 (Trento, Italy: Erickson).

14 A. Wikström (2013), "The Challenge of Change: Planning for Social Urban Resilience. An Analysis of Contemporary Planning Aims and Practices," unpublished Master's thesis submitted to Department of Human Geography, University of Stockholm.
15 W. Van Vliet (2002), "Cities in a Globalizing World: From Engines Growth to Agent of Change," *Environment and Urbanization* 14(1), pp. 31–40.
16 N. Clichevsky (2002), "Pobreza y Políticas urbano-ambientales en Argentina," *Serie Medio Ambiente y Desarrollo* 49, pp. 1–32.
17 Godschalk (2003).
18 J. Jacobs (1992), *The Death and Life of Great American Cities* (New York, NY: Vintage).

Bibliography

Basher, R. (2008). "Disaster Impacts: Implications and Policy Responses." *Social Research* 75(3), pp. 937–54.

Beck, U. (1986). *Risikogesellschaft. Auf dem Weg in eine andere Moderne.* Frankfurt, Germany: Suhrkamp.

Caragliano, S. (2007). *Società e Disastri Naturali. La vulnerabilità organizzativa nelle politiche di prevenzione dei rischi.* Bologna, Italy: Pitagora Editrice.

Clichevsky, N. (2002). "Pobreza y Políticas urbano-ambientales en Argentina." *Serie Medio Ambiente y Desarrollo* 49, pp. 1–32.

Douglas, M. (1991). *Come percepiamo il pericolo: antropologia del rischio.* Milan, Italy: Feltrinelli.

Godschalk, D. (2003). "Urban Hazard Mitigation: Creating Resilient Cities." *Natural Hazards Review* 4(3), pp. 136–43.

Herzer, H. (2002). *Convivir con el riesgo o la gestión del riesgo.* Retrieved August 14, 2013, from www.cesam.org.ar/publicaciones.htm

Jacobs, J. (1992). *The Death and Life of Great American Cities.* New York, NY: Vintage.

Keohane, R., and J. Nye. (1977). *Power and Interdependence.* New York, NY: Longman.

Lavell, A. (2002). "Local Level Risk Management: Concepts and Experience in Central America." Paper presented at the Disaster Preparedness and Mitigation Summit, American Red Cross, November 21–23, 2002, New Delhi, India.

Ligi, G. (2009). *Antropologia dei disastri.* Rome, Italy: Editori Laterza.

Luhmann, N. (1991). *Soziologie des Risikos.* Berlin, Germany: De Gruyter.

Malaguti, E. (2011). "Articolazioni teoriche della resilienza." In *Costruire la resilienza, la riorganizzazione positive della vita e la creazione di legami significativi.* B. Cyrulnik and E. Malaguti, eds. pp. 79–102. Trento, Italy: Erickson.

Oliver-Smith, A. (1994). "La reconstrucción después del desastre: una visión general de secuelas y problemas." In *Al norte del Río Grande.* Allan Lavell, ed. pp. 3–18. México: La Red.

Salzano, E. (2007). *Fondamenti di urbanistica.* Rome, Italy: Laterza.

Scandurra, E. (2001). *Gli storni e l'urbanista.* Rome, Italy: Meltemi Editrice.

———. (2003). *Città morenti e città viventi.* Rome, Italy: Meltemi Babele Editrice.Van Vliet, W. (2002). "Cities in a Globalizing World: From Engines Growth to Agent of Change." *Environment and Urbanization* 14(1), pp. 31–40.

Wikström, A. (2013). "The Challenge of Change: Planning for Social Urban Resilience. An Analysis of Contemporary Planning Aims and Practices." Unpublished Master's thesis submitted to Department of Human Geography, University of Stockholm.

26 Land-use modeling as a new approach for future hazard-sensitive population mapping in Northern Germany

Giedrius Kaveckis, Benjamin Bechtel, Jürgen Ossenbrügge, and Thomas Pohl

Introduction

The frequency and severity of natural hazards are well-known reasons for high disaster-related population loss. The urbanization of hazard-prone areas due to the growth of populations and housing also contributes to the increase of exposed populations. Therefore, in order to reduce the impacts of future hazards, it is crucial to know where these sensitive population groups live now and will live in the future.

Demographic data are often aggregated and based on social projections on a rough scale, which is not detailed enough to map a population for risk and emergency applications on the local scale. We aim to overcome this problem by projecting future land(scapes), which are more permanent and "predictable" than populations but are strongly related to them. Hence, landscape projections can serve as ancillary land data for disaggregating population projections. The aim of this research is to discuss and analyze the evidence to establish if land-use modeling can be used for future hazard-sensitive population mapping in the Greater Hamburg area.

In this paper, we present urban climate vulnerability zones by describing hazard-sensitive populations (Section 3) and discussing the term "land use" (Section 4), including the extent and limitations of local climate zones. Afterwards, we present the main disaggregation methods (Section 6), their advantages, disadvantages, and data requirements. This research project is ongoing, and, unfortunately, we do not have a future hazard-sensitive population map yet. However, as an outcome (Section 8), we identify future land-use modeling as a potential solution for future hazard-sensitive population mapping and create a future land-use-based population-mapping framework.

Case study area

Hamburg is a growing city, and like many cities in the world, it was characterized by suburbanization processes at the end of the twentieth century and is now facing shortages in the housing market and increasing rents due to re-urbanization. As a case study area, we selected Greater Hamburg (Figure 26.1), which extends the area of the "Free and Hanseatic City of Hamburg," being a city-state, and includes the neighboring counties of Lower Saxony and Schleswig-Holstein. Greater Hamburg is not the same as the Hamburg Metropolitan area, which includes even more counties.

The reason for our decision was that the Greater Hamburg area represents a daily urban system based on commuting intensities and matches the network of the

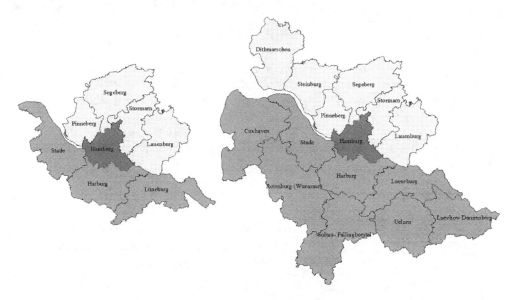

Figure 26.1 Left: Greater Hamburg. Right: the Hamburg Metropolitan area. (The federal state
of Hamburg and the surrounding districts, including parts of the federal states
of Niedersachsen and Schleswig-Holstein).[1]

Hamburg Transport Association (HVV), whereas the bordering districts of Metropolitan Hamburg are greatly influenced by neighboring cities such as Lübeck and Bremen.
According to the national census data for 2011,[2] the Greater Hamburg area accommodates more than 3 million people. Hamburg is a traditional port city, and one of
the biggest load and maritime service centers in Europe. Additional economic clusters
include aerospace, print media, life science, and renewable energies. Though Hamburg
is one of the richest areas in Germany, it is characterized by rather strong socio-
economic segregation within the city and its surroundings. Finally, Hamburg is often
labelled a "green city" because of the wide variety of open spaces within the city and
forests, agricultural land, scrubland, rural villages, and beaches in adjoining counties.
Such variety gives a good opportunity to test land-modeling approaches in order to
map future hazard-sensitive populations.

Hazard-sensitive population

Because all people are sensitive to hazards to one degree or another, it is important to
define the term "hazard-sensitive population" precisely. Several studies[3] identified the
following characteristics which relate population sensitivity to hazards: age, gender,
race, ethnicity, employment and income, education, housing conditions, disease, use
of medication, special needs, and others. However, the dominant and measureable
characteristics are age and gender – older people and children often need physical and
psychological assistance during and after disasters and are more prone to disease. In
France in the summer of 2003, 82 percent of the fatalities caused by the heatwave were
people over 75 years old.[4] Women also tend to be more sensitive, with limited coping

and adaptation capacities. According to the United Nations Development Programme, women and children are 14 times more likely than men to die during a disaster.[5] Furthermore, the sensitivity of a certain population group depends on the hazard: e.g., physically disabled people may be able to withstand heat, but are severely restricted during an evacuation when a flood occurs. Therefore, specific sociodemographic characteristics of a population play major roles during a certain disasters. However, some characteristics, such as age and gender, strongly affect the population's behavior during all types of disasters.

Today, detailed social information, such as diseases, poverty, income, ethnicity, and housing conditions, is available only at very basic and statistically unsophisticated levels. For future projections, the data availability is even poorer, and mainly relies on forecast methods. To give an obvious example, it is hardly possible to model the percentage of sick people on a local scale for 2050 due to the high degree of uncertainty and complexity in attempting to model diseases at the individual level. Hence, for our case study, age and gender were selected as key hazard-sensitive indicators for the future population. Consequently, the question arises as to how these indicators – age and gender – can be related to changes in land use.

Climate zones

The term land use describes the "the total arrangements, activities and inputs that people undertake in a certain land cover type."[6] This definition acknowledges that land use relates people to the land through their activities and the way they use the land. Common land uses are residential, commercial, industrial, agricultural, etc.; however, the term "land use" has to be distinguished from land cover. For instance, forest is often considered as a land use, but is more likely a land cover, not primarily referring to human activity on the land, though it can be used for recreation, logging, or hunting. The semantics of landscape may also vary between regions. What is called a forest in Northern Europe may be called bushland in Australia; a low-density residential area in China would be interpreted as a high-density residential region in Iceland. In general, we can conclude that common land-use schemes encompass a mixture of land use and land cover properties, and have considerable regional variations in the terms used to describe them.

"Land use" and "land cover" are not enough to describe the environment, especially in urban areas; furthermore, some environmental applications take morphology, which characterizes the form and structure of an urban area, into account. In a climatic context, morphology is often used in urban heat-island-related assessments. For instance, Daneke et al.[7] developed a conceptual approach in order to measure the potential of an urban heat island using the morphological characteristics of Hamburg and surrounding counties. Urban morphology describes the forms, their transformations, and spatial structure of a city. Urban areas are filled densely with buildings of irregular height, plenty of impervious surfaces, and many sources of artificial heat. Meanwhile, rural areas contain much more arable land, forests, meadows, and waters, which cause different temperatures in the surroundings.[8] Land use, together with land cover and morphology, can form classes that have distinctive micro-climates. An early climate-based classification was done by Chandler,[9] who separated London into four regions based on climate, built form, and physiography. Auer[10] later classified the city of St. Louis, MO, into 12 land zones according to vegetation and building character-istics. More recently, Oke[11] created a generic city classification scheme of seven homo-geneous regions called "urban climate zones." The zones differ according to their urban structure (dimensions of street canyons and buildings), land cover, fabric (mate-rials), and human activity. However, each of these classifications has some limitations. First of all, not all of them use a full set of surface climate properties. Second, the properties should not be region- or culture-specific. Stewart and Oke[12] tried to over-come these limitations by constructing a new classification system, called local climate zones (LCZs). They defined LCZs as "regions of uniform surface cover, structure, material and human activity."[13] The system consists of 17 classes, each of which is a representation of a thermally homogeneous landscape within the city and is named for a surface property.

Although the LCZ scheme has been successfully applied in many regions of the world, the spectrum of the classes still retains a certain focus on North American architecture. In particular, the high diversity of dense urban structures in Europe, which results from the region's longer development history, is only partly addressed in the scheme. While this might be acceptable for urban climate studies, the sociode-mographic aspects of vulnerability certainly vary too much within these classes (espe-cially LCZ2: compact mid-rise). We therefore complemented Stewart's and Oke's LCZ scheme by introducing subclasses. After comprehensive research on the available local landscape classification systems in Germany, the classification by the German energy scheme[14] was selected as an appropriate source of supplementary information. Accord-ing to this scheme, German urban areas were classified based on their energy factors, such as material, morphology, land cover, heating efficiency, cooling effect, etc. In order to merge it with the LCZ scheme, we reviewed both schemes and analyzed the existing typical urban structures of our case study in Northern Germany. This resulted in a scheme of 24 zones in total, 11 of which were housing-suitable zones; the others were industry, commercial, infrastructure, ports, airports, and natural zones. We will refer to the scheme as "urban climate vulnerability zones" (UCVZs) for Northern Germany.

Urban climate vulnerability zones and population

As the subject of our research is hazard-sensitive populations, we have to relate the urban morphology with socio-demographic patterns. Urban research in the tradition of social ecology suggests that persons with the same social and economic status tended to live closer to each other and to form communities, which is why cities have neighborhoods of rich and poor, young and old, minorities, and religious communi-ties. Numerous studies[15] have emphasized the strong relationship between the social composition of a population, their activities, lifestyle, and environment. Though it is clear that residential land use cannot fully represent the huge variety of different population groups living in a city, we assume that residential land-use classes based on housing type are an appropriate proxy. In our case, Greater Hamburg is represented by our developed UCVZ housing scheme and includes the following seven urban climate vulnerability zones (see Figure 26.2): compact mid-rise, dense compact mid-rise, open high-rise, terraced open mid-rise, perimeter open mid-rise, dense open low-rise and open low-rise.

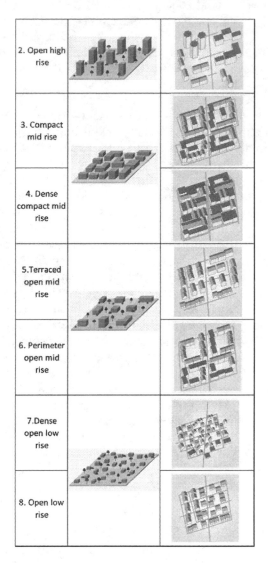

Figure 26.2 Urban Climate Vulnerability Zones (UCVZs) suitable for housing in the Greater Hamburg case study. (Classification based on Stewart and Oke, and Erhorn-Klutting et al.)[16]

Disaggregation

The typical spatial representations of population are based on census data with a rather high aggregation level, which does not allow mapping of the distinct social areas. Therefore, there is a need to research and develop methods that would enable the mapping of the population at the micro-level, and this data would be independent of the administrative areas.

One of the common methods for mapping a population is the spatial disaggregation of the census data. Spatial disaggregation (also called downscaling) is, generally

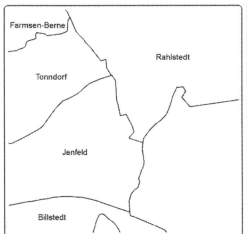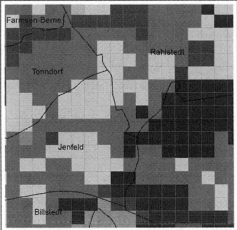

Figure 26.3 Examples of source zones (left) and target zones (right).

speaking, a method that transfers coarse information to a more detailed scale,[17] based on the assumption that the data of an entire case study area can be distributed within the area by means of local parameters.[18] The coarse spatial units with the known data are called source zones, and the finer-scale spatial units the data are assigned to are called target zones (see Figure 26.3).[19]

However, disaggregation can lead to a common problem in geography, known as the modifiable area unit problem (MAUP). MAUP is the phenomenon associated with the use of aggregated data, which will vary depending on how boundaries are drawn; in practice, however, boundaries are not related to the variables of interest.[20] Another issue is that the process of spatial disaggregation is complex due to mismatching and heterogeneity (in terms of density) between the boundaries of the source and target zones. To solve these issues, different disaggregation methods are used.[21]

Simple area weighting (also known as a mass-preserving aerial interpolation) is the simplest spatial disaggregation method. It is based on the assumption that the disaggregated variable is homogeneous within the source zones (which does not occur in the case of population mapping). In order to use a simple area weighting method, a simple overlay operation is done. The main problem with this method is the incorrect assumption that density within the source zone is homogeneous and equally distributed; this method has been demonstrated to be less accurate than others.[22]

The source zone is overlaid with the target zone (see Figure 26.4). Both polygons have the same area, but contain different numbers of persons: Polygon A contains 200 persons and polygon B contains 100 persons. The target zone overlays a quarter of polygon A and a quarter of polygon B. Therefore, assuming that persons are equally distributed within the polygons, there are 50 persons (200/4) in the A target zone and 25 persons (100/4) in the B target zone (75 persons altogether). The simple area weighting method is easy to apply, but the homogeneity assumption is far from reflecting the real distributions of population. Therefore, it does not fit for the more advanced disaggregation applications.

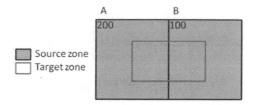

Figure 26.4 The Simple Area Weighting Method.

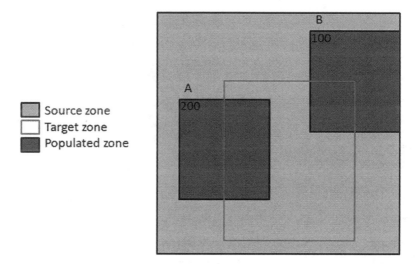

Figure 26.5 A practical example of the binary asymmetric mapping (mask area weighting) method.

Another method is binary asymmetric mapping,[23] also known as a mask area weighting. An improvement over the simple area weighting masks (boundaries), asymmetric mapping is applied within the target zone where the source data should be allocated. The mask or the boundaries can be any spatial ancillary information, such as urban climate vulnerability zones or other spatial units (for example, residential areas), which would be a guideline for the allocation of population.

In binary asymmetric mapping, the source zone is overlaid with populated zones and target zones (Figure 26.5). Compared to simple area weighting (Figure 26.4), where population was displaced within the whole source zone, this time the population is allocated only within the populated zones, Polygons A and B. The principle is the same: instead of the whole source zone, we have to identify the overlay between target and populated zones. One quarter of the populated zone B overlays the target zone, covering 25 persons (100/4). Half of the populated zone A overlays the target zone, covering 100 (200/2). In total, 125 persons (25+100) live in the target zone.

Binary asymmetric mapping masks out unnecessary areas, based on concentrated population with fixed density. However, it cannot be used to represent more complex

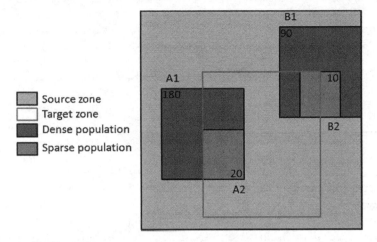

Figure 26.6 A practical example of the classified asymmetric mapping (asymmetric disaggregation) method.

land zones or other units with different densities. Nevertheless, this method is much more advanced than simple area weighting because the disaggregation is focused on area, not the whole source zone.

The binary asymmetric mapping needs ancillary information, such as the land type, populated or other areas, which would mask where the source data should be allocated.

The third method is called classified asymmetric mapping, also known as three-class asymmetric mapping[24] or asymmetric disaggregation. This method uses not only a homogeneous mask but also the different zones with specific weights within the mask. The weighting can represent population density or other population-related variable. Both the weighting and the mask zones are ancillary information, which need to be integrated as well.

In comparison with simple area weighting and binary asymmetric mapping, classified asymmetric mapping uses not only the masked source zone (populated zones), but also allocates weights within these populated zones. This means that different numbers of people can be allocated in parts of populated zones. In Figure 26.6 this is represented as population density (sparse and dense). As in the previous example (Figure 26.5), the source zone is overlaid with populated zones and target zone, but this time the populated zones are split into zones with different population density (the population of dense zones is three times higher than that of sparse density zones). In Figure 26.6, the populated zone A (A1 + A2) contains 200 persons, but only half of it overlays the target zone. The half of this target zone is sparse density (A2, with 20 persons); the other half is dense density, with 180/3 = 60. Therefore 20 + 60 = 80 people can be allocated to the target zone within the populated zone A. Next, the populated zone B (B1 + B2) contains in total 100 people. A quarter of the populated area is within the target zone, and a quarter of it is sparse density (the population is three times smaller). Only half of it falls in the target zone, which means 10/2 = 5 people. The other three quarters of populated zone B is dense population. Only one-sixth of it is within the target zone (90/6 = 15 people). Therefore, in B there are 20 people (5 + 15). There are 100 people in total in the target zone (80 + 20, or A+B).

Classified asymmetric mapping is the most advanced disaggregation method. According to Langford (2006),[25] it outperforms other spatial disaggregation methods and takes advantage of the binary asymmetric mapping method. It uses densities to define weighting and allocate population more precisely, and it uses a loose assumption of homogeneity and is closer to the complexity of the real world. However, the densities (weights) and masked area (as ancillary information) are required.

Due to the heterogeneity of the surface, it is important to obtain ancillary information on the source zone, such as land use, land cover, urban morphology, remote sensing data, etc., in order to represent the disaggregation process more realistically. Such information increases the accuracy of the disaggregation process. Asymmetric mapping requires this ancillary data in order to indicate variations in data distributions of the aggregated source data.[26]

The variations in the data distributions of population-focused applications can be represented as the population density. These densities can be developed using different approaches such as a regression model,[27] which uses ancillary data to improve accuracy. The assumption of regression models is that ancillary data are equally distributed within the source area and the land classes have a uniform area density. This density is related to a specific parameter for each class of ancillary data, based on the population's concentration of the specific class. The common method is to develop regression equations by numerically solving the problem using the combination of values of aggregated source zones and ancillary data with unknown densities.[28] Langford[29] noticed that increasing the complexity of a regression equation improves the accuracy of the interpolation. However, the disadvantage of regression models is that the small errors between the estimated and actual values of source zones emerge.[30] Another issue is that regression analysis is not supported by current GIS and requires additional statistical software. The last (but not the least) obstacle is that the densities of specific classes are spatially stable for the whole study area.[31] The more advanced technique is the global fitted regression model,[32] which was initially created as a control technique to allow the populations within target zones to match the sum within the source zones while also allowing for variability in density for each land class in a specific area. In this way the estimated density of each land class is locally adjusted within each source zone by ratio. Therefore, this simple approach, based on loose homogeneity assumptions of density, is more advanced than the simple regression model. The regression model, which permits a local fit, couples perfectly with the classified asymmetric mapping method and can be used for realistic disaggregation.[33]

According to the disaggregation quality studies carried out by Langford and Li et al.,[34] the classified asymmetric method largely outperforms other techniques. However, this method requires definition of relative densities for each land class. Densities can be defined by the sampling approach or by the regression model.[35] The sampling approach requires small source zone areas, while the regression model can be used for larger source zones.[36]

For our case study in Greater Hamburg, the latter method is most appropriate, because populated areas (residential UCVZ) are available and the population densities can be derived from the statistical social data. However, it is more complicated to acquire future data, for two reasons. First, the future is complex and uncertain; often, future age and gender data exist only at the national level. Fortunately, for our case study in northern Germany, we obtained future age and gender data at the district level. If no future social data exist, the method used by Huang et al.[37] can be applied: they identified the past trend of age and gender changes within the area and assumed that the trend would continue. Second, because there are no future projections of landscape development apart from the limited outlines of urban land planning, information on the target zones (the landscape) are missing. This gap can be filled using future land-modeling techniques.

Land modeling

As discussed earlier, land data, or UCVZ, can be used as ancillary information to improve the accuracy of disaggregated information and to define densities when coupled with historical population data. The future UCVZ can be used as a proxy to disaggregate future population projections. Land (or in our case, the future UCVZ) can be simulated with a cellular automata (CA)-based spatial decision support system called Metronamica.[1]

Metronamica works as a stand-alone application.[38] It is based on the Monitoring Land Use / Cover Dynamics (MOLAND) model and presents the study area as a mosaic of grid cells.

Each of the cells has a state of specific land use, cover, or other entity (Figure 26.7).

As a spatial decision support tool, Metronamica models dynamic land changes, calculating the change of natural area, population growth, relationships with other regions, accessibility (transport), and attractiveness between specific land types, policy, and biophysical factors. It also includes a regional migration and transport model, which simulates future jobs, population, and migration between regions. The Metronamica model is calibrated using historical land changes. The future land change is extrapolated via scenarios and adjusts the historically calibrated land-change trend.

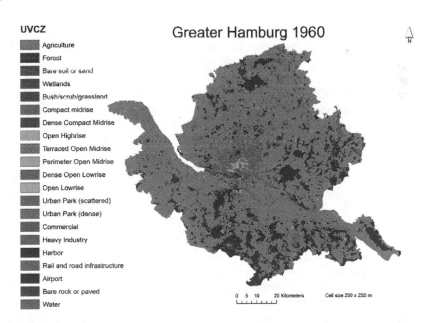

Figure 26.7 Grid raster of Greater Hamburg. Each cell represents a class of UCVZ in 1960.

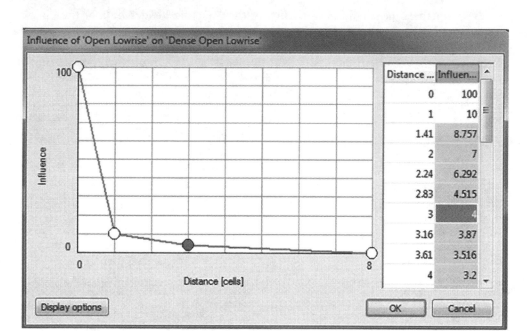

Figure 26.8 A graph of the neighborhood rule between open low-rise and dense open low-rise cells. The attraction of new dense open low-rise cells decreases as distance increases.

The most important factor in Metronamica is the neighborhood potential of the surrounding location. As mentioned before, people of the same or similar social and economic status tend to live close together, and they prefer to build new houses in residential areas where other people live and the infrastructure is developed, whereas businesses prefer logistically better locations, i.e., closer to highways or customers.

The neighborhood potential between the two land types can be represented as a linear function. In Figure 26.8, the function represents decreasing attraction of new open low-rise land type to dense open low-rise. In this case the attraction decreases in the distance. However, the function can represent repulsion as well. The repulsion or attraction of new cells is applied within a distance of eight cells (i.e., people who work in the manufacturing industry prefer to live neither too close to their place of employment due to pollution, nor too far away, so that they can reach it in a short time). The next important factor is physical suitability, which defines the degree to which a certain land type cell can be allocated. Examples of physical suitabilities are slope, water level, soil type, etc. Accessibility or transportation is also considered an important factor in this modeling framework, and is based on the infrastructure network, which consists of lines (roads) and nodes (highway ramps, bus, rail stations, etc.). This factor also influences attractiveness for certain land types – business and industries prefer to be located close to railway stations and highways, while people prefer access to public transport systems. The last factor is zoning (also known as urban development plans), represented as a spatial set of rules for a specific area. If future urban development plans and their restrictions are known, they can be added and made operational for a specific time of the modeling period. All of these factors have to be adjusted in the model between the runs until the outcome is generated.

Results

As a result of this research, we developed a future population mapping framework with a strong focus on land modeling. The framework is presented as a flowchart (Figure 26.9) and contains various activities (shown in ovals) and datasets (shown in rectangles).

All of these elements are connected and should be used sequentially, as the arrows indicate. Although it is not presented in the framework, it is also important to know what future population groups or their properties must be mapped and how they are related to the environment. Afterwards it is essential to review the availability of the historical population and landscape data, which will be used for calibration and population densities. Both of these elements are presented as inputs in the framework (marked with an "I" in the upper left corner of the rectangle). The development of population densities is not strictly described. The studies by Gallego, Li et al., Steinnocher et al., and Gallego et al. could be used as guidelines to obtain the densities.[39]

We called the next important step future land modeling. As discussed earlier, we avoided using the term "land use." For our study, we used the Metronamica landmodeling tool, which requires historical land data at the local grid scale for two periods of time and total population. However, Metronamica can be changed by any other raster- or grid-based future land-modeling tool. By selecting another tool, it is important to understand that its output has to provide a higher resolution than future social projections with the possibility of applying future land development scenarios. Depending on the case study's area and the model's variables, calibration can take a long time, and afterwards the future specific scenarios have to be applied. Despite the future land development, the other future scenarios such as economic and social development, migration, and changes in life expectancy, lifestyle, etc., should be considered, because they can affect both future population and future landscapes. The final (although not the not least important) input element is future population projections,

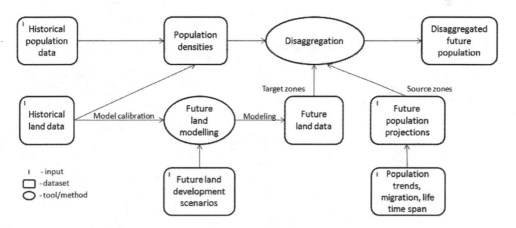

Figure 26.9 A flowchart of the future population mapping framework.

which are typically developed by local or national offices of statistics and based on many indicators observed during past years. In many cases, the indicators are unknown, but they should not differ much from the ones we have already mentioned. If a future land model has a demographic component, then future land and population can be developed using the same model; however, it is still recommended that it should be compared with official population projections. If the projections were developed in the past, it would be useful to examine how closely they match the actual data and assess the similarities and differences between them. When future population data are obtained, it can be used for disaggregation, the methods of which are described in detail in Section 6.

Outlook and conclusion

This research is a part of a project dealing with the assessment of future populations' vulnerability to heatwaves and flash floods in the Greater Hamburg area in Germany. At this time we possess historical population data and are calibrating the Metronamica model using historical land data for the years 1960, 1990, and 2000. The preliminary assessment revealed different trends in urban development during these time periods. After calibration, these trends and future land development scenarios will be integrated in order to project future land data. At the same time, it is important to develop densities for future scenarios and afterwards employ that information to allow the disaggregation of the hazard-sensitive population. During these processes, the factors of migration, life span, population trends, and urban development (redevelopment in urban core/waterfront/urban sprawl, etc.) should be considered, assessed, and applied, if needed.

Notwithstanding the existing issues, our developed framework with analyzed disaggregation methods, input data, and future land modeling contributes to the assessment of future populations' vulnerability to disasters and climate change impacts. Coupled with hazard-sensitive indicators, it is a novel way to map future hazard-sensitive populations. Although the UCVZ scheme has been created for the climate-related case study in Greater Hamburg, it can serve as a basis for other case studies and can also be used to provide ancillary information in disaggregation.

Notes

1 The map was created by the authors using German administrative data from the Institute of Geography, University of Hamburg, Germany (2015).
2 Landesamt für Statistik Niedersachsen Zensus 2011 Bevölkerung Der Kreisfreien Städte Und Landkreise Am 31. (2011, December); Statisches Amt für Hamburg und Schleswig-Holstein (2013), *Übersicht Der Einwohnerzahlen in Hamburg – Im Vergleich Zur Bevölkerungsfortschreibung 30.04.2001 Und Zur Volkszählung 1987.*
3 John Heinz III Center for Science, Economics and the Environment (2000), *The Hidden Costs of Coastal Hazards: Implications for Risk Assessment and Mitigation* (Washington, DC: Island); S.L. Cutter, B.J. Boruff, and W.L. Shirley (2001), "Indicators of Social Vulnerability to Hazards," unpublished paper, and (2003), "Social Vulnerability to Environmental Hazards," *Social Science Quarterly* 84(2), pp. 242–61; B. Wisner, P. Blaikie, and T. Cannon (2004), *At Risk: Natural Hazards, People's Vulnerability and Disasters*, 2nd edn. (London, UK: Routledge); J. Tan (2008), "Commentary: People's Vulnerability to Heat Wave," *International Journal of Epidemiology* 37(2), pp. 318–20.
4 M. Poumadère, C. Mays, S. Le Mer, and R. Blong (2005), "The 2003 Heat Wave in France: Dangerous Climate Change Here and Now," *Risk Analysis* 25(6), pp. 1483–94.

5 FAO (2007), *People-Centred Climate Change Adaptation: Integrating Gender Issues* (Rome, Italy: Author), p. 1; United Nations Development Programme. (2010, October[0] [0]). *Gender and Disasters*. New York, NY: Author; UNDP (2013), "Gender and Adaptation," Report of United Nations Development Programme (New York, NY: Author), p. 2.

6 FAO and the United Nations Environmental Programme (1999), *Terminology for Integrated Resources Planning and Management* (Rome, Italy: Authors).

7 C. Daneke, B. Bechtel, J. Böhner, T. Langkamp, and J. Oßenbrügge (2011), "Conceptual Approach to Measure the Potential of Urban Heat Islands from Landuse Datasets and Landuse Projections," in *Computational Science and Its Applications – ICCSA*, B. Murgante, O. Gervasi, A. Iglesias, D. Taniar, and B.O. Apduhan, eds., pp. 381–93 (Berlin, Germany: Springer).

8 I.D. Stewart, and T.R. Oke (2012), "Local Climate Zones for Urban Temperature Studies," *Bulletin of the American Meteorological Society* 93(12), pp. 1879–900.

9 Oke, T.R. [0](2009), "Chandler, T.J. 1965: *The Climate of London*. London: Hutchinson, 292 pages," [Review], *Progress in Physical Geography* 33(3), pp. 437–42.

10 A.H. Auer (1978), "Correlation of Land Use and Cover with Meteorological Anomalies," *Journal of Applied Meteorology* 17(5), pp. 636–43.

11 T.R. Oke (2004), *Initial Guidance to Obtain Representative Meteorological Observations at Urban Sites* (Geneva, Switzerland: World Meteorological Organization).

12 Stewart, and Oke (2012).

13 Ibid., p. 1884.

14 H. Erhorn-Kluttig, R. Jank, L. Schrempf, A. Dütz, F. Rumpel, J. Schrade, H. Erhorn, C. Beier, C. Sager, and D. Schmidt (2011), *Energetische Quartiersplanung* (Stuttgart, Germany: Fraunhofer IRB).

15 I. Simmons (1987), "Transformation of the Land in Pre-Industrial Time," in *Land Transformation in Agriculture*, M.G. Wolman, and F.G.A. Fournier, eds., pp. 45–75 (New York, NY: Wiley/Scope); G.M. Wolman (1993), "Population, Land Use and Environment: A Long History," in *Population and Land Use in Developing Countries*, National Academy Press, ed., pp. 15–29 (Washington, DC: NAP); B. Entwisle, S.J. Walsh, R.R. Rindfuss, and A. Chamratrithirong (1998), "Land-Use/Land-Cover and Population Dynamics, Nang Rong, Thailand," in *People and Pixels: Linking Remote Sensing and Social Science*, D. Liverman, E.F. Moran, R.R. Rindfuss, and P.C. Stern, eds., pp. 121–44 (Washington, DC: National Academy); J. Fox (2004), *People and the Environment: Approaches for Linking Household and Community Surveys to Remote Sensing and GIS* (Dordrecht, the Netherlands: Kluwer); W.G. Axinn, and D.J. Ghimire (2007), "Social Organization, Population and Land Use," *American Journal of Sociology* 117(1), pp. 209–58.

16 Stewart, and Oke (2012); Erhorn-Klutting et al. (2011).

17 J.J. McCarthy (2001), *Climate Change 2001: Impacts, Adaptation, and Vulnerability: Contribution of Working Group II to the Third Assessment Report of the Intergovernmental Panel on Climate Change* (Cambridge, UK: CUP).

18 K. Steinnocher, I. Kaminger, J. Weichselbaum, and M. Köstl (2010), "Gridded Population – New Datasets for an Improved Disaggregation Approach," paper presented at the European Forum for GeoStatistics, October 5–10, 2010, Tallin, Estonia.

19 T. Li, D. Pullar, J. Corcoran, and R. Stimson (2007), "A Comparison of Spatial Disaggregation Techniques as Applied to Population Estimation for South East Queensland (SEQ), Australia," *Applied GIS* 3(9), pp. 1–16.

20 S. Openshaw (1984), *The Modifiable Areal Unit Problem* (Norwich, UK: Geo Books).

21 Li et al. (2007).

22 J.A. Foley, R. DeFries, G.P. Asner, C. Barford, G. Bonan, S.R. Carpenter, F.S. Chapin, M.T. Coe, G.C. Daily, and H.K. Gibbs (2005), "Global Consequences of Land Use," *Science* 309(5734), pp. 570–74; M. Langford (2006), "Obtaining Population Estimates in Non-Census Reporting Zones: An Evaluation of the 3-Class Dasymetric Method," *Computers, Environment and Urban Systems* 30(2), pp. 161–80.

23 C.L. Eicher, and C.A. Brewer (2001), "Dasymetric Mapping and Areal Interpolation: Implementation and Evaluation," *Cartography and Geographic Information Science* 28(2), pp. 125–38.

24 J. Mennis (2003), "Generating Surface Models of Population Using Dasymetric Mapping," *The Professional Geographer* 55(1), pp. 31–42.

25 Langford (2006).
26 Li et al. (2007).
27 Y. Yuan, R.M. Smith, and W.F. Limp (1997), "Remodeling Census Population with Spatial Information from Landsat TM Imagery," *Computers, Environment and Urban Systems* 21(3), pp. 245–58.
28 Li et al. (2007).
29 Langford (2006).
30 W.R. Tobler (1979), "Cellular Geography," in *Philosophy in Geography*, S. Gale and G. Olsson, eds., pp. 379–86 (Dordrecht, the Netherlands: D. Reidel).
31 Li et al. (2007).
32 Langford (2006).
33 Li et al. (2007).
34 Langford (2006); Li et al. (2007).
35 Mennis (2003); Langford (2006).
36 Li et al. (2007).
37 C. Huang, A.G. Barnett, X. Wang, P. Vaneckova, G. FitzGerald, and S. Tong (2011), "Projecting Future Heat-Related Mortality under Climate Change Scenarios: A Systematic Review," *Environmental Health Perspectives* 119, pp. 1681–90.
38 Research Institute for Knowledge Systems (RIKS BV) (2012), "Metronamica Documentation," *Technical Report of RIKS* (Maastricht, the Netherlands: Author).
39 J. Gallego (2004), "Mapping Rural/Urban Areas from Population Density Grids," in *Institute for Environment and Sustainability* (Ispra, Italy: JRC-EC); Li et al. (2007); Steinnocher et al. (2010); J. Gallego, F. Batista, C. Rocha, and S. Mubareka (2011), "Disaggregating Population Density of the European Union with CORINE Land Cover," *International Journal of Geographical Information Science* 25(12), pp. 2051–69.

Bibliography

Auer, A.H. (1978). "Correlation of Land Use and Cover with Meteorological Anomalies." *Journal of Applied Meteorology* 17(5), pp. 636–43.

Axinn, W.G., and D.J. Ghimire. (2007). "Social Organization, Population and Land Use." *American Journal of Sociology* 117(1), pp. 209–58.

Cutter, S.L., B.J. Boruff, and W.L. Shirley. (2001). "Indicators of Social Vulnerability to Hazards." Unpublished Paper.

———. (2003). "Social Vulnerability to Environmental Hazards." *Social Science Quarterly* 84(2), pp. 242–61.

Daneke, C., B. Bechtel, J. Böhner, T. Langkamp, and J. Oßenbrügge. (2011). "Conceptual Approach to Measure the Potential of Urban Heat Islands from Landuse Datasets and Landuse Projections." In *Computational Science and Its Applications – ICCSA*. B. Murgante, O. Gervasi, A. Iglesias, D. Taniar, and B.O. Apduhan, eds. pp. 381–93. Berlin, Germany: Springer.

Eicher, C.L., and C.A. Brewer. (2001). "Dasymetric Mapping and Areal Interpolation: Implementation and Evaluation." *Cartography and Geographic Information Science* 28(2), pp. 125–38.

Entwisle, B., S.J. Walsh, R.R. Rindfuss, and A. Chamratrithirong. (1998). "Land-Use/Land-Cover and Population Dynamics, Nang Rong, Thailand." In *People and Pixels: Linking Remote Sensing and Social Science*. D. Liverman, E.F. Moran, R.R. Rindfuss, and P.C. Stern, eds. pp. 121–44. Washington, DC: National Academy.

Erhorn-Kluttig, H., R. Jank, L. Schrempf, A. Dütz, F. Rumpel, J. Schrade, H. Erhorn, C. Beier, C. Sager, and D. Schmidt. (2011). *Energetische Quartiersplanung*. Stuttgart, Germany: Fraunhofer IRB.

Foley, J.A., R. DeFries, G.P. Asner, C. Barford, G. Bonan, S.R. Carpenter, F.S. Chapin, M.T. Coe, G.C. Daily, and H.K. Gibbs. (2005). "Global Consequences of Land Use." *Science* 309(5734), pp. 570–74.

Food and Agriculture Organization. (2007). *People-Centred Climate Change Adaptation: Integrating Gender Issues*. Rome, Italy: Author.

————, and United Nations Environmental Programme. (1999). *Terminology for Integrated Resources Planning and Management*. Rome, Italy: Authors.

Fox, J. (2004). *People and the Environment: Approaches for Linking Household and Community Surveys to Remote Sensing and GIS*. Dordrecht, the Netherlands: Kluwer.

Gallego, J. (2004). "Mapping Rural/Urban Areas from Population Density Grids." In *Institute for Environment and Sustainability*. Ispra, Italy: JRC-EC.

Gallego, J., F. Batista, C. Rocha, and S. Mubareka. (2011). "Disaggregating Population Density of the European Union with CORINE Land Cover." *International Journal of Geographical Information Science* 25(12), pp. 2051–69.

Huang, C., A.G. Barnett, X. Wang, P. Vaneckova, G. FitzGerald, and S. Tong. (2011). "Projecting Future Heat-Related Mortality under Climate Change Scenarios: A Systematic Review." *Environmental Health Perspectives* 119, pp. 1681–90.

John Heinz III Center for Science, Economics and the Environment. (2000). *The Hidden Costs of Coastal Hazards: Implications for Risk Assessment and Mitigation*. Washington, DC: Island.

Landesamt für Statistik Niedersachsen Zensus 2011 Bevölkerung Der Kreisfreien Städte Und Landkreise Am 31. (2011, December).

Langford, M. (2006). "Obtaining Population Estimates in Non-Census Reporting Zones: An Evaluation of the 3-Class Dasymetric Method." *Computers, Environment and Urban Systems* 30(2), pp. 161–80.

Li, T., D. Pullar, J. Corcoran, and R. Stimson. (2007). "A Comparison of Spatial Disaggregation Techniques as Applied to Population Estimation for South East Queensland (SEQ), Australia." *Applied GIS* 3(9), pp. 1–16.

McCarthy, J.J. (2001). *Climate Change 2001: Impacts, Adaptation, and Vulnerability: Contribution of Working Group II to the Third Assessment Report of the Intergovernmental Panel on Climate Change*. Cambridge, UK: CUP.

Mennis, J. (2003). "Generating Surface Models of Population Using Dasymetric Mapping." *The Professional Geographer* 55(1), pp. 31–42.

Oke, T.R. (2004). *Initial Guidance to Obtain Representative Meteorological Observations at Urban Sites*. Geneva, Switzerland: World Meteorological Organization.

Oke, T.R. (2009). "Chandler, T.J. 1965: *The Climate of London*. London: Hutchinson, 292 pages." [Review]. *Progress in Physical Geography* 33(3), pp. 437–42.

Openshaw, S. (1984). *The Modifiable Areal Unit Problem*. Norwich, UK: Geo Books.

Poumadère, M., C. Mays, S. Le Mer, and R. Blong. (2005). "The 2003 Heat Wave in France: Dangerous Climate Change Here and Now." *Risk Analysis* 25(6), pp. 1483–94.

Research Institute for Knowledge Systems (RIKS BV). (2012). "Metronamica Documentation." *Technical Report of RIKS*. Maastricht, the Netherlands: Author.

Simmons, I. (1987). "Transformation of the Land in Pre-Industrial Time." In *Land Transformation in Agriculture*. M.G. Wolman, and F.G.A. Fournier, eds. pp. 45–75. New York, NY: Wiley/Scope.

Statisches Amt für Hamburg und Schleswig-Holstein. (2013). *Übersicht Der Einwohnerzahlen in Hamburg – Im Vergleich Zur Bevölkerungsfortschreibung 30.04.2001 Und Zur Volkszählung 1987*. Hamburg, Germany: Statistics Office for Hamburg and Schleswig-Holstein.

Steinnocher, K., I. Kaminger, J. Weichselbaum, and M. Köstl. (2010). "Gridded Population – New Datasets for an Improved Disaggregation Approach." Paper presented at European Forum for GeoStatistics, October 5–10, 2010, Tallin, Estonia.

Stewart, I.D., and T.R. Oke. (2012). "Local Climate Zones for Urban Temperature Studies." *Bulletin of the American Meteorological Society* 93(12), pp. 1879–900.

Tan, J. (2008). "Commentary: People's Vulnerability to Heat Wave." *International Journal of Epidemiology* 37(2), pp. 318–20.

Tobler, W.R. (1979). "Cellular Geography." In *Philosophy in Geography*. pp. 379–86. Dordrecht, the Netherlands: D. Reidel.

United Nations Development Programme. (2010, October). *Gender and Disasters*. New York, NY: Author.

———. (2013). "Gender and Adaptation." Report of United Nations Development Programme. New York, NY: Author.

Wisner, B., P. Blaikie, and T. Cannon. (2004). *At Risk: Natural Hazards, People's Vulnerability and Disasters*. 2nd edn. London, UK: Routledge.

Wolman, G.M. (1993). "Population, Land Use and Environment: A Long History." In *Population and Land Use in Developing Countries*. National Academy Press, ed. pp. 15–29. Washington, DC: NAP.

Yuan, Y., R.M. Smith, and W.F. Limp. (1997). "Remodeling Census Population with Spatial Information from Landsat TM Imagery." *Computers, Environment and Urban Systems* 21(3), pp. 245–58.

27 Summary and conclusion

Planning for a more sustainable and disaster-resilient world

Adenrele Awotona

This book presents diverse perspectives on community-based disaster resilience using case studies from six continents, by authors with practical and academic expertise from a wide range of different sectors in the disaster risk management field. It also demonstrates the intricacies of the interrelationships between humans and nature in preparing for, mitigating, responding to, and rebuilding resilient communities after disasters. The diversity of the contributions is a reflection of the growing commitment of researchers and practitioners to examine and explicate the social, technical, and spatial dimensions of disaster risk management.

The purpose of this concluding chapter is to draw together the common themes discussed throughout the book and propose a way forward on how to plan for and create a more disaster-resilient world.

General themes

Community planning and emergency management need to unite in order to achieve the mutual aim of creating a sustainable disaster-resilient community

Although the two professions of community planning and emergency management are equally concerned with disaster risk management in order to create a safe environment for the community, the potential synergy between them is yet to be fully mobilized and realized. The stumbling blocks to their mutual cooperation include the fact that their practitioners do not share common backgrounds, goals, and skill sets; they perceive and respond to many of the socio-economic aspects of hazard mitigation differently; and they have different development goals and disciplinary emphases. Consequently, neither profession can achieve its expressed shared goal alone.

Thus, as planning in all its forms influences the risks present in a community, professional planners need to embrace aspects of risk management. This is crucial in order to ensure that their plans assist in guiding communities towards achieving the goal of disaster-resiliency planning. However, before emergency managers and planners can bridge the professional gap between them and cooperate to build sustainable and resilient communities, the following two prerequisites should be addressed, amongst others:

- Legislative and policy instruments should be used to encourage the education of community planners about emergency management with a focus on hazard mitigation; and

- Emergency management should be supported to grow into a parallel profession of a comparable status with planning. This would enable its practitioners to harness their expertise and resources to work in tandem towards the common community goals of building community-based, disaster-resilient communities, on an equal professional footing.

A *clear set of urban planning and urban design principles should be developed for improving the built environment's capacities for disaster risk management*

Urban planning is essential to sustainable development and disaster risk management. As a complete and independent analytical unit, it enables us to examine the different urban contexts, their structure, and their problems, as well as the potential and the limits of the urban process. However, in spite of the many benefits, positive contributions, and potential opportunities associated with urban planning, some developing countries continue to invest very few resources in the urbanization process.

Drawing on the case study findings, especially in relation to tsunami evacuation vulnerability, reveals an urgent need to develop a clear set of urban design principles for improving the built environment's capacities for disaster risk management. These principles, which should be rooted in the local political, social, economic, environmental, real estate, and engineering contexts, can guide professionals and decision-makers in the process of developing and arguing for risk reduction changes in the urban realm. Similarly, the principles should also address the factors that lead to the separation of family members from one another during disasters, a significant issue that creates high levels of stress in the short term and for some, serious longer-term mental health consequences, particularly for those who experience longer periods of separation. These factors include those which weaken a community's connections to the rest of the world, such as a loss of power, transport, and telecommunications.

Furthermore, these principles should take into account the need to ensure that reunification services are established early and promoted widely, in order to reduce stress and the potential for the development of post-traumatic stress. Also, technological advances should be employed to increase the efficiency, accessibility, and immediacy of reunification services, provided they are not also compromised by on-site power and communication failures. Moreover, other urban elements need to be strengthened in order to facilitate the timely reunions of community members after disasters, the construction of a disaster-resilient urban space that is less vulnerable to the likelihood of catastrophic events, the promotion of greater social interaction, and the creation of closer social ties. These include, where appropriate, a well-organized system of sanitation, the widening of roads, the design of green urban spaces, the development of affordable and sustainable transport, the upgrading of public spaces, the flexibility and adaptability of governmental authorities when projecting changes in the city, the promotion of opportunities for citizen interaction to reverse the trend towards individualism, and the strengthening of horizontal cooperation among cities with a strong capacity for action in the territory. Indeed, the importance of generational change should be a central element in urban planning: the changes and new interventions to and in the territory should address the region's needs in a sustainable way.

The engagement of the whole community is the foundation of sustainable resilient development

The national government, acting alone, cannot be responsible for carrying out risk management; its efforts are insufficient and limited. Indeed, in countries with low budgets, inefficient institutions and internal political fragmentation can even be dangerous. Community solidarity when a disaster occurs is of the utmost importance for speedy and sustainable recovery. Thus, the change in the vision of management and planning must start from below. The process of governance is empty without the existence of real mechanisms of social participation in decision-making and in the implementation of any planned action. So, in the field of policy effects, the key issue is the process of horizontal governance: creating partnerships, regulating conflicts of interest, and promoting real participation possibilities for each actor involved. The difficulty in strengthening real and effective horizontal governance processes in social and political arenas is a cause of the slow progress on the objectives of risk management and mitigation.

A whole community approach therefore calls for a shift from exclusively "top-down" to fully inclusive approaches. Globally, public and private agencies, social and political movements, researchers, consultants, and activists are all expanding their participatory outreach and organizing techniques, and emphasizing public involvement and participation in relation to climate change and disaster readiness. The case studies demonstrate that there is no substitute for locally designed and locally appropriate all-inclusive community involvement processes; however, each community has to find its own way to handle this, taking into account prevailing cultural, socio-economic, democratic, and administrative traditions.

Grassroots community participation in the recovery process, which creates a sense of ownership, is indispensable for success. The community needs to be involved in the whole planning and decision-making procedure, from policy formulation to implementation. Not only are community members emotionally attached to the area and their property, but they know what their needs are. Grassroots involvement in programs of community participation should be pursued to prevent the dominance of leadership structures.

Older people should be involved in the development of disaster management plans and policies because of their wealth of positive attributes in terms of knowledge, experience, and sense of community. Case studies in this book emphasize that disaster preparedness for older people is "more about being mentally prepared and mentally at peace and less about toolkits, checklists and tangible items." Indeed, while older people might not define themselves as "being prepared" in terms of traditional disaster management assessments, they do not feel vulnerable to disasters. The case studies show that older people are confident of their resilience and their ability to cope.

In sum, broad public participation is the foundation of sustainable development: it gives a political voice to marginalized and vulnerable people who have historically been left out of public decision-making. This bottom-up approach to building resilience, in the face of the social and political repercussions of extreme weather events, should also encourage and support the use of social media and other online and in-person networking tools to share ideas on how to promote disaster risk reduction and environmental justice, to obtain funding and to press politically for policies addressing the needs of people on the margins of society. The quality of participation should also be carefully monitored to ensure the consistent application of the principles of democratic representation, transparency, and accountability. Similarly, the roles played by the different stakeholders in disaster risk management processes should be clearly defined in order to forestall conflict and delays in the implementation of relevant projects. Some of the enduring impediments to, and problems with, community participation, particularly when capacity building and empowerment are objectives, include[1] different value frames between planners and communities, poor literacy levels, the difficulty of identifying community leaders, the difficulty of drawing in the necessary range of interest groups, poor capacity-building with the community through deliberate measures to generate skills and knowledge, and poor traditions of participation. Others include, within the context of global and national processes, the difficulties that communities face in taking responsibility for local disaster risk management issues, in preserving and maintaining their community assets during and after disasters, and in managing their own interests on a sustainable basis.

The enactment and implementation of appropriate laws, policies, and development regulations are essential for disaster risk reduction

The Hyogo Framework for Action has paved the way for legislative developments to take place in different jurisdictions because of the importance of laws, policies, and programs for disaster risk reduction. However, the implementation of the framework requires better integration and collaboration among different levels of the government. It should also take into account the socio-economic circumstances of each country, ranging from the rapid rate of urbanization and urban poverty to existing development-control mechanisms. Legislation should facilitate rather than hinder important developments, such as the improvement of low-income housing and the relocation of poor and low-income people to less risky areas, in order to reduce their disaster risk. In some cases, legislative reform may require the dismantling of traditional and outmoded regulations and the introduction of a streamlined method for regulating development with a view to increasing the scope of regulations and refocusing them on issues of public concern and safety.

There are immense challenges in trying to introduce a culture of compliance and strengthen regulatory systems where none have existed before, such as in several cities in the developing world which have grown organically over centuries with little planning intervention. Where this necessary set of reforms of the development regulation system that is required to build resilient cities and improve planning practice is threatened by vested interests that prefer the status quo and resist any efforts to create systemic changes, it is the task of the reformers to garner political and local support for their objectives. Case studies show that once the public and private sectors, community-based and grassroots organizations, NGOs, and academia come together, positive policy changes can rapidly be realized. With regard to the regulatory environment at the local grassroots level, it is essential to determine which regulations are applicable, how they should be enforced in cooperation with the community, and if there are noticeable undesirable effects of over- or under-regulation.

Integrating disaster risk reduction education into the curricula of colleges and universities is indispensable to building disaster-resilient communities

The levee failures during Hurricane Katrina were one of the largest failures of a civil engineering system in American history. Key failures in the engineering of the system illustrate the importance and relevance of fundamental geotechnical engineering concepts. Undergraduate civil engineering students represent the next generation of engineering professionals and will be responsible for addressing these types of disasters in the future. Consequently, incorporating natural disasters studies into the curricula of colleges and universities should provide student engineers and architects with a keener awareness of the impacts of engineering and architectural decisions and the applicability of technical concepts to real-world events. For undergraduate civil engineering students, for example, case studies such as Hurricane Katrina and the Oso landslide can serve as poignant examples of the importance and relevance of soil mechanics concepts such as soil classification, groundwater flow, settlement, subsurface stress, shear strength, and soil failures, as well as the broader implications of errors in design, maintenance, and public policy. Educating student engineers on the failures of the past is vital to preventing failures of this magnitude from occurring in the future, and in the wake of disasters such as Hurricane Katrina and the Oso landslide, preventing such events should be every engineer's priority.

Similarly, community-based environmental education initiatives which are relevant and interesting for local residents and increase their job opportunities, knowledge of local environmental issues, understanding of basic political and ecological principles, and confidence to express and act on their views can serve as the basis of a climate change intervention approach which is progressive, constructive, and democratic. This, in turn, will increase the resilience and sustainability of watershed, disaster, and climate change decision-making processes. It will also lay the groundwork for community organizing and the extension of environmental education activities to larger constituencies in local areas affected by climate change. Community-based education and organization are fundamental to creating the conditions for local knowledge to be shared and utilized, through equitable democratic participation. Building inclusive governance structures and strengthening the roles of members of civil society, especially women, are essential for addressing climate change vulnerability and fostering resilience and sustainability in urban centers as well as in rural areas.

The intricate and intertwined nature of the social and economic drivers of risk calls for a multidisciplinary and integrative approach to disaster risk management

Finding the appropriate risk mitigation measures for various natural hazards is a multidisciplinary task that needs thorough joint research work by experts in different disciplines, such as the environmental, medical, and social sciences. However, in many developing countries, teamwork is weak due to cultural barriers and limitations, thereby rendering cooperation difficult to achieve; it is therefore necessary to improve and reinforce teamwork capabilities. Similarly, with regard to the concept of the "Management Efficiency Index," defined as the ratio of "utilized capacity" to "existing capacity," recognizing risk reduction capacities and possibilities is of the utmost importance in developing countries.

The need to articulate public policies

The articulation of public policies is an important and necessary approach to reinforce resilience. There has to be a shift from a sectoral strategy of working on disaggregate areas such as education, health, and infrastructure, amongst others, to a flexible framework that enables the cross-sectoral articulation of these areas in a matrix capable of reducing risks and responding comprehensively to major disasters. This integrative policy of building and strengthening resilience calls for a decentralization of resources and capacities from the national to the local community level. It also calls for the development and integration of social capital and cultural elements into creative and dynamic plans for disaster risk management.

The way forward

The common themes, findings, and recommendations from the case studies in this book confirm that disasters are primarily an indicator of unsustainable development. This suggests that there is an urgent need for a shift from disaster management to a development-centered risk management approach which links risk and development. Thus, rather than mainstreaming disaster risk reduction into development, attention should now be primarily focused on the various development-based drivers, forces, and processes associated with the increase of disaster risk.[2] In this vein, Article 3.4 of the United Nations Framework Convention on Climate Change (UNFCCC) states that policies and measures to address climate change "should be integrated with national development programs." This recognizes the fact that focusing exclusively on responding to the effects of climate change (especially in the areas of renewable energy, forest sinks, energy conservation, energy efficiency, adaptation, and low-carbon urbanization) would contribute only little to enabling sustainable development. Rather, the emphasis should be on addressing the fundamental causes of people's powerlessness and vulnerabilities.

Indeed, disasters (including climate-induced ones), unplanned urban development, ecosystem degradation, and weak livelihoods undermine the sustainable development of communities. Hence, some of the drivers of risk in development planning, human development, and investment, as well as in public policy, which should be addressed through multi-level government efforts, cross-sector initiatives, and global actions, include unemployment, underemployment, poverty, and inequalities; weak institutions (poor governance, political instability, an underdeveloped financial market, and lack of a supportive institutional and policy environment); lack of technical support; and unresponsive legal and regulatory frameworks. Others are:

- *Inadequate infrastructural development:* Food and nutrition insecurity; inadequate water supply and squalid sanitary conditions which facilitate the development of a polluted environment; poor sanitation and waste management; unreliable and insufficient power supply; widespread illiteracy and underdeveloped information and communication technologies; lack of healthcare facilities and medical networks; inefficient transport networks; and lack of safeguards of urban areas against erosion, flooding, landslides, and disasters.
- *Inequity and lack of social inclusion:* Unregulated land markets, spatial segregation, extreme income inequalities, and uneven development; absence of fairness and justice; alienation and exclusion; and lack of empowerment and engagement of all social groups, including women, the youth, the elderly, the poor, the differently abled, and minorities of all types.
- *Environmental sustainability and climate change:* Poor information management systems; nonexistence of public education in environmental management, climate change adaptation, and waste management; dearth of capacities at various levels of government and the private sector to ensure environmental sustainability and

resilience; lack of integration of green jobs in the world of work; undeveloped or underdeveloped environmental management tools for sustainable and resilient development planning; unintegrated strategic environmental assessment mechanisms; absence of cross-sectoral collaboration amongst government agencies; and poor building design in the light of the centrality of buildings in climate change mitigation, as they involve disproportionate amounts of natural resources for cooling and heating, waste, and pollution.

I will now briefly examine some of these drivers in a little greater detail.

Development-based drivers of disaster risk accumulation in Africa

Africa is the world's most impoverished continent, with over 50 percent of its population of 1 billion people living in circumstances characterized by urban decay, the lowest levels of infrastructure provision, and substandard housing, as well as food and water insecurity, all of which contribute to the accumulation of disaster risks. Indeed, 20 percent of the world's slum residents are in sub-Saharan Africa. Lack of access to land and the insecurity of land tenure are at the heart of the urban housing problem in Africa, a situation that is made worse by climate change, ineffective policy and institutional structures, and speedy urbanization.[3] In Western Africa, the proliferation of slums and the imbalanced sharing of urban prosperity are some of the expressions of this rapid urbanization. Aluko[4] revealed that poverty results in the poor health of the residents of Lagos, due to exposure to various forms of pollution. It also adversely impacts people's quality of life and housing conditions. Among his recommendations to address these problems are the introduction of operative poverty-alleviation programs, provision of an effective housing loan scheme, partial upgrading of substandard residential areas, implementation of a functioning urban development policy, improvement of sanitary conditions, and the enforcement of housing and building codes. Similarly, the Ebola outbreak in Guinea, Liberia, and Sierra Leone was a consequence of these countries' weak economies and poor health systems. According to the World Bank,[5] more than US$1 billion was pledged at the World Bank Group-IMF Spring Meetings in April 2015 to fight this disease, including $650 million from the World Bank Group. Actually, an Urban Health Updates' publication on Monrovia has reported that barely one-third of the city's 1.5 million residents had access to hygienic toilets in 2009, with 20 to 30 cholera cases being reported weekly.

The United Nations[6] pointed out that the inability of Eastern African governments to supply affordable land to low-income city dwellers is a consequence of four main factors: bureaucratic apathy, costly managerial procedures, allocation inefficiencies, and inappropriate use of public office. Whilst many of the countries in Central Africa are handsomely endowed with natural resources, the United Nations[7] observed that the majority of their citizens continue to experience tremendous poverty, worsening living conditions, extremely unequal sharing of national wealth, and socio-economic and housing segregation.

With 61.7 percent of its population living in urban areas in 2014, Southern Africa remains the most urbanized region on the continent, and this percentage is anticipated to increase to 67 by 2020. The region is, however, characterized by deep-rooted uneven urban development and economic growth, where the consumption of water, electricity, and other urban utilities remains effectively segregated and very unequally distributed between wealthy and poor areas. In South Africa, decades of racial segregation and economic disempowerment mean that strategies aimed at achieving equity for residents of deprived township areas are particularly important. The redevelopment and planning of urban areas in the "new" South Africa offers the country a unique opportunity to evolve more appropriate planning guidelines for the large-scale urban development projects that are planned to take place. The planning guidelines should address the following issues, amongst others: community-based development and sustainability; land tenure; finance; sectoral planning and intersectoral partnerships; transportation; and strategic public sector interventions to facilitate the integration of township environments.

A major threat to Northern Africa's urban populations is climate change, which the United Nations[8] has projected will exacerbate existing desertification and water stress. Indeed, climate change will render most of Northern Africa's cities more vulnerable to disasters associated with extreme weather patterns, especially flooding.

Vulnerable populations and disaster risk

Table 27.1 shows the complex fundamental causes influencing the vulnerability of people and social structures. Policy makers and disaster management officials, therefore, need a thorough understanding of these distinctions and what it means for vulnerable populations to "feel prepared." With regard to vulnerable populations everywhere, Donner and Rodríguez[9] note that some of the major disaster risk factors are population growth and distribution, especially increased population density and urbanization. These result in increasing population concentration in coastal communities, the consequences of which are increased human exposure to coastal flooding, hurricanes, and tsunamis; urban earthquakes; congestion; limited escape routes in times of catastrophes; dense infrastructure; and poverty.

Communities in the Caribbean, one of the most disaster-prone areas in the world, are highly exposed to floods, hurricanes, tsunamis, and earthquakes. In 2015, the European Union, which has always advocated the integration of disaster risk reduction efforts into Caribbean development policy, allocated US$10 million for the implementation of a total of 14 risk-reduction projects in the region between 2015 and 2016, in order to reduce vulnerability to natural hazards. The goal is to reduce urban risks, strengthen the

Table 27.1 The fundamental causes of vulnerability affecting people and social structures[10]

Vulnerability	Causes
Material/economic vulnerability	Lack of access to resources
Social vulnerability	Disintegration of social patterns
Ecological vulnerability	Degradation of the environment and inability to protect it
Organizational vulnerability	Lack of strong national and local institutional structures
Educational vulnerability	Lack of access to information and knowledge
Attitudinal and motivational vulnerability	Lack of public awareness
Political vulnerability	Limited access to political power and representation
Cultural vulnerability	Certain beliefs and customs
Physical vulnerability	Weak buildings or weak individuals

resilience of the populations, and enable the communities themselves to carry out simple and inexpensive preparatory measures, such as risk mapping, emergency plans, early warning systems, education campaigns, and small infrastructure projects, all of which are aimed at avoiding the loss of lives, property and livelihoods.[11]

The major risk factors that exacerbate vulnerability for the elderly comprise poverty; limited resources; having a sensory, physiological, or cognitive disability; and a lack of social infrastructure, which results in social isolation.[12] Others, as identified by the Centers for Disease Control and Prevention's (CDC) *Disaster Planning Goal* include multiple chronic conditions, risk of trauma, loss of pets, transportation problems (inability to drive, no longer owning a car, lack of access to public transportation), major disruptions to essential services, reluctance to seek assistance, health risks from inadequate nutrition, fraud by fraudulent contractors and "con men" who exploit victims financially and physically, and mental abuse.[13] Hence, more needs to be understood about what constitutes vulnerability. Also, further research is needed to determine the disaster risk factors in other demographic groups, especially in socially diverse settings.

Gender and disaster risk management

Amongst the poorest members of the community are women and girls, because of their social, economic, and political status. They are predominantly unprotected from climate-related risk and are most likely to suffer higher rates of mortality, morbidity, and threats to life and livelihood. However, "they are not helpless victims".[14] Their unique experiences and informal community involvement in disaster reduction and response, which are still routinely largely excluded from formal planning and decision-making, include the organization of the provision of food, shelter, child and family care, aid-delivery systems, and entitlement procedures, as well as the distribution of supplies. They also commonly organize themselves to pool labor and resources to create community support services to meet basic family and community needs. The engagement of women is, therefore, fundamentally imperative for durable institutional arrangements, long-term management of resources, legal frameworks, climate change adaptation, and the linking of disaster risks with development.

Finally

The problem of disaster risk is universal, regardless of the level of human and economic development. Consequently, in the final analysis, people should be the central focus of disaster resiliency planning and sustainable development. Plans for holistic disaster risk management must therefore be a fusion of equitable planning for sustainable development and equitable planning of resources for disaster risk reduction. Such formation must have, as its central objective, the need to mobilize all sectors of the community at all levels to actively participate in the whole process. The poor cannot be left out: they are also significant stakeholders.

Notes

1 A. Awotona, D. Japha, M. Huchzermeyer, and O. Uduku (1995, October), "Community-Based Strategies for Local Development: Lessons from South Africa," Working Paper No. 8, The Integration and Urbanization of Existing Townships in the Republic of South Africa (ODA Research Scheme Number R6266), Center for Architectural Research

and Development Overseas, University of Newcastle, UK, and the School of Architecture and Planning, University of Cape Town, South Africa.

2 United Nations (2015), *Global Assessment Report on Disaster Risk Reduction – Making Development Sustainable: The Future of Disaster Risk Management*, retrieved May 10, 2015, from http://www.preventionweb.net/english/hyogo/gar/2015/en/gar-pdf/GAR2015_EN.pdf; A. Lavell and A. Maskrey (2013), *The Future of Disaster Risk Management: An Ongoing Discussion*, draft synthesis document, meeting notes, background papers and additional materials from a Scoping Meeting for the 2015 UN Global Assessment Report on Disaster Risk Reduction (GAR 2015), April 18–19, 2013, San Jose, Costa Rica, retrieved May 9, 2015, from http://www. unisdr.org/files/35715_thefutureofdisasterriskmanagement.pdf

3 A. Awotona (2012), "Housing Abroad: Africa," in *The Encyclopedia of Housing*, Vol. 1, 2nd edn., Andrew T. Carswell, ed., pp. 309–12 (Newbury Park, CA: Sage).

4 O. Aluko (2012, April), "Impact of Poverty on Housing Conditions in Nigeria: A Case Study of Mushin Local Government Area of Lagos State," *Journal of African Studies and Development* 4(3), pp. 81–89.

5 D. Barne (2015, April 17), "Ebola: $1 Billion So Far for a Recovery Plan for Guinea, Liberia, and Sierra Leone," *Voices: Perspectives on Development* [Blog], The World Bank, retrieved May 15, 2015, from http://blogs.worldbank.org/voices/ebola-1-billion-so-far-recovery-plan-for-guinea-liberia-and-sierra-leone

6 UN Habitat (2014), *The State of African Cities 2014: Re-Imagining Sustainable Urban Transitions*, Nairobi, Kenya, retrieved May 15, 2015, from http://www.citiesalliance.org/sites/citiesalliance.org/files/SoAC2014.pdf

7 Ibid.

8 Ibid.

9 W. Donner and H. Rodríguez (2011, January), *Disaster Risk and Vulnerability: The Role and Impact of Population and Society*, Population Reference Bureau, available at http://www.prb.org/Publications/Articles/2011/disaster-risk.aspx

10 V. Gokhale (2008), "Role of Women in Disaster Management: An Analytical Study with Reference to Indian Society," paper presented at the 14th World Conference on Earthquake Engineering, October 12–17, 2008, Beijing, China, retrieved May 15, 2015, from http://www.iitk.ac.in/nicee/wcee/article/14_10–0049.PDF, adapted with the permission of the author.

11 "EU Allocates US$10m for Disaster Risk Reduction in Caribbean" (2015, May 19), *Jamaica Observer*, available at http://www.jamaicaobserver.com/news/EU-allocates-US-10m-for-disaster-risk-reduction-in-C-bean

12 S.A. Harvey (2015, March 20), "Natural Disasters Are Especially Hard on Seniors," *Cornell Chronicle*, available at http://news.cornell.edu/stories/2013/03/natural-disasters-are-especially-hard-seniors

13 W.F. Benson (n.d.), *Centers for Disease Control and Prevention's (CDC) Disaster Planning Goal: Protect Vulnerable Older Adults*, retrieved May 21, 2015, from http://www.cdc.gov/aging/pdf/disaster_planning_goal.pdf

14 Ibid.

Bibliography

Aluko, O. (2012, April). "Impact of Poverty on Housing Conditions in Nigeria: A Case Study of Mushin Local Government Area of Lagos State." *Journal of African Studies and Development* 4(3), pp. 81–89. Available at http://www.academicjournals.org/JASD

Awotona, A. (2012). "Housing Abroad: Africa." In *The Encyclopedia of Housing*. Vol. 1. 2nd edn. Andrew T. Carswell, ed. pp. 309–312. Newbury Park, CA: Sage.

———, D. Japha, M. Huchzermeyer, and O. Uduku. (1995, October). "Community-Based Strategies for Local Development: Lessons from South Africa." Working Paper No. 8, The Integration and Urbanization of Existing Townships in the Republic of South Africa (ODA Research Scheme Number R6266). Center for Architectural Research and Development Overseas, University of Newcastle, UK, and the School of Architecture and Planning, University of Cape Town, South Africa.

Barne, D. (2015, April 17). "Ebola: $1 Billion So Far for a Recovery Plan for Guinea, Liberia, and Sierra Leone." *Voices: Perspectives on Development* [Blog]. The World Bank. Retrieved May 15, 2015, from http://blogs.worldbank.org/voices/ebola-1-billion-so-far-recovery-plan-for-guinea-liberia-and-sierra-leone

Benson, W.F. (n.d.). *Centers for Disease Control and Prevention's (CDC) Disaster Planning Goal: Protect Vulnerable Older Adults.* Retrieved May 21, 2015, from http://www.cdc.gov/aging/pdf/disaster_planning_goal.pdf

Donner, W., and H. Rodríguez. (2011). *Disaster Risk and Vulnerability: The Role and Impact of Population and Society.* Population Reference Bureau, January. Available at http://www.prb.org/Publications/Articles/2011/disaster-risk.aspx

"EU Allocates US$10m for Disaster Risk Reduction in Caribbean." (2015, May 19). *Jamaica Observer.* Available at http://www.jamaicaobserver.com/news/EU-allocates-US-10m-for-disaster-risk-reduction-in-C-bean

Gokhale, V. (2008). "Role of Women in Disaster Management: An Analytical Study with Reference to Indian Society." Paper presented at the 14th World Conference on Earthquake Engineering, October 12–17, 2008, Beijing, China. Retrieved May 15, 2015, from http://www.iitk.ac.in/nicee/wcee/article/14_10–0049.PDF

Harvey, S.A. (2015, March 20). "Natural Disasters Are Especially Hard on Seniors." *Cornell Chronicle.* Available at http://news.cornell.edu/stories/2013/03/natural-disasters-are-especially-hard-seniors

Lavell, A., and A. Maskrey. (2013). *The Future of Disaster Risk Management: An Ongoing Discussion.* Draft synthesis document, meeting notes, background papers and additional materials from a Scoping Meeting for the 2015 UN Global Assessment Report on Disaster Risk Reduction (GAR 2015), April 18–19, 2013, San Jose, Costa Rica. Retrieved May 9, 2015, from http://www.unisdr.org/files/35715_thefutureofdisasterriskmanagement.pdf

United Nations. (2015). *Global Assessment Report on Disaster Risk Reduction – Making Development Sustainable: The Future of Disaster Risk Management.* Retrieved May 10, 2015, from http://www.preventionweb.net/english/hyogo/gar/2015/en/gar-pdf/GAR2015_EN.pdf

UN Human Settlements Program (UN-Habitat). (2012). *State of the World's Cities 2012/2013: Prosperity of Cities*, Nairobi, Kenya. Retrieved May 18, 2015, from https://sustainabledevelopment.un.org/content/documents/745habitat.pdf

———. (2014). *The State of African Cities 2014: Re-Imagining Sustainable Urban Transitions.* Nairobi, Kenya. Retrieved May 15, 2015, from http://www.citiesalliance.org/sites/cities alliance.org/files/SoAC2014.pdf

Index